The Skull of *Australopithecus afarensis*

HUMAN EVOLUTION SERIES

THE SKULL OF *Australopithecus afarensis*

William H. Kimbel

Yoel Rak

Donald C. Johanson

With a contribution on the brain endocast by

Ralph L. Holloway and Michael S. Yuan

2004

OXFORD
UNIVERSITY PRESS

Oxford New York
Auckland Bangkok Buenos Aires Cape Town Chennai
Dar es Salaam Delhi Hong Kong Istanbul Karachi Kolkata
Kuala Lumpur Madrid Melbourne Mexico City Mumbai Nairobi
São Paulo Shanghai Taipei Tokyo Toronto

Published by Oxford University Press, Inc.,
198 Madison Avenue, New York, New York 10016

www.oup.com

Library of Congress Cataloging-in-Publication Data
Kimbel, William H.
The skull of Australopithecus afarensis / by William H. Kimbel,
Yoel Rak, Donald C. Johanson.
p. cm. — (Human evolution series)
Includes bibliographical references and index.
ISBN 0-19-515706-0
1. Australopithecus afarensis. 2. Skull. 3. Craniology. I. Rak, Yoel.
II. Johanson, Donald C. III. Title. IV. Series.
GN283.25.K56 2003
599.9'48—dc22 2003060872

9 8 7 6 5 4 3 2 1

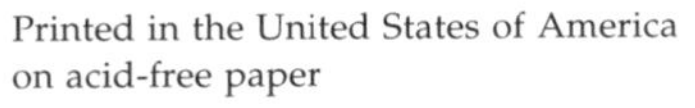

Printed in the United States of America
on acid-free paper

We dedicate this work to the memory of four dear friends:

Meles Kassa

Dato Ahmedu

Wubishet Fantu

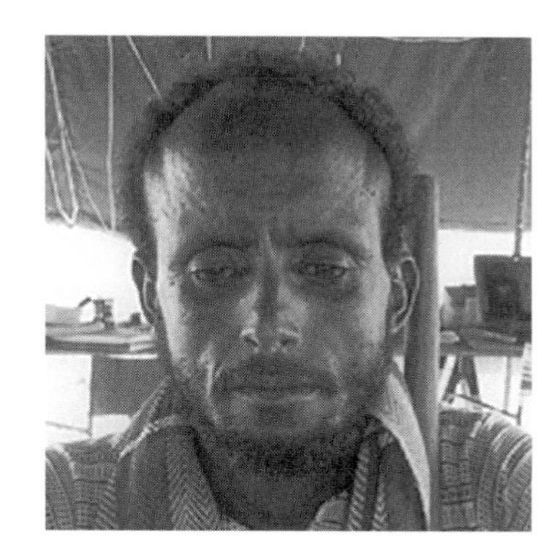

Dato Adan

Acknowledgments

We have incurred numerous debts of gratitude during the field and laboratory research that have culminated in this book. First and foremost, we are grateful to the Authority for Research and Conservation of Cultural Heritage (of the Ethiopian Ministry of Youth, Sports and Culture), and its head, Mr. Jara HaileMariam, for permission to carry out fieldwork at the Hadar site; to the Culture and Tourism Bureau of the Afar Regional State government for local permissions and assistance; and to the General Manager and staff of the National Museum of Ethiopia for permitting and facilitating our logistical preparations and laboratory research in Addis Ababa.

Fieldwork at Hadar and lab research in Addis Ababa during the past decade would not have been possible without the generous and consistent financial support of the National Science Foundation (grants BNS-9113066, SBR-9222604, EAR-92106515, SBR-9511172/9996020, and BCS-0080378), the National Geographic Society, and the Institute of Human Origins (IHO) at Arizona State University.

The IHO board of directors has been an invaluable source of support and enthusiasm during all phases of our research at Hadar. Members of the IHO board and other close friends have made timely special gifts that substantially enhanced our ability to carry out the Hadar fieldwork, which led to several significant discoveries, including the skull that forms the centerpiece of this book. We are particularly grateful for the special support of Mr. David Koch, Mr. Thomas P. Jones III, Mr. Thomas F. Hill, Mr. David Deniger and Ms. Mara Batlin, and Mrs. Ann Lurie.

Our access to fossil and extant comparative collections has been facilitated through the kind cooperation of the following institutions and individuals: in Kenya, the National Museums of Kenya, Dr. Meave Leakey and Mr. Christopher Kiarie; in South Africa, the University of the Witwatersrand Medical School, Professor Phillip Tobias, Dr. Lee Berger, and the late Mr. Alun Hughes; the Transvaal Museum, former Director Dr. C. K. (Bob) Brain, Dr. Francis Thackeray, and Mr. David Panagos; in Switzerland, the Adolf Schultz Collection at the University of Zurich, Dr. Robert Martin. Special thanks are due to Drs. Christoph Zollikofer and Marcia S. Ponce de Leon of the Anthropological Institute and Multimedia Laboratory (Department of Computer Science), University of Zurich, who were instrumental in providing us with the stereolithographic elements incorporated into the final reconstruction of *A. afarensis* skulls A.L. 444-2 and A.L. 417-1—a project generously funded by the National Geographic Society.

As the Igor Orenstein Chair at Tel Aviv University, Y. Rak is grateful for the continuous support that this fund has provided. He also thanks the teaching and administrative staff of the Department of Anatomy and Anthropology of the Sackler Faculty of Medicine, Tel Aviv University, who, by enthusiastically supporting this endeavor and patiently accommodating the extensive changes to the teaching schedule that it has necessitated, have enabled him to participate in the fieldwork for this project.

We have benefited from discussions with numerous colleagues during the preparation of this book. First, we thank Drs. Ralph Holloway and Michael Yuan for their dedicated efforts to decipher the A.L. 444-2 endocast and for providing Chapter 4 of this monograph. We are grateful to Drs. Zeresenay Alemseged, Fred Grine, Meave Leakey, Charles Lockwood, Fred Spoor, Alan Walker, Carol Ward, and Bernard Wood for sharing valuable information and insight. We thank Fred Spoor for allowing us to generate craniograms from his CT scans of fossil hominin specimens.

Illustrations are at the heart of this book. We are extremely grateful for the talent, persistence, and patience of the following illustrators who contributed their graphical expertise: Anna Behar (all figures except as noted below), Gerald Eck (Figure 1.2), Yehudit Sherman (Figures 3.1, 3.5, 3.6, 3.31, 3.36, 3.47, 3.50, and 5.27), and Daniel Arsen (Figures 3.55 and 5.15). All photographs were taken by the authors.

At the Institute of Human Origins, Arizona State University, Ms. Elizabeth Harmon, Mr. Scott Burnett, Ms. Kim Stout, and Ms. Ann Silvers provided invaluable technical, logistical, and research support. At the Sackler Faculty of Medicine, Tel Aviv University, Ms. Avishag Ginzburg, Mr. Mendel Schatz, and Ms. Valentina Litinski were of great technical assistance. In the National Museum of Ethiopia's paleoanthropology laboratory, casting technician Mr. Alemu Ademasu provided us with superb casts of the fossils. The entire manuscript benefited from the expert editorial skills of Ricka Rak. We thank Amy Rector for her heroic work on the index.

Since we recommenced working at Hadar in 1990, many American, Israeli, and Ethiopian scientists and students, as well as Ethiopian government representatives, have participated in the research. We are grateful to them all for helping to make the Hadar project successful:

Geology: Drs. James Aronson, Kay Behrensmeyer, Craig Feibel, Mulugeta Feseha, Million HaileMichael, Paul Renne, Carl Vondra, Robert Walter, John Kappelman, and Tesfaye Yemane; Mr. Christopher Campisano, Mr. Tony Troutman, the late Mr. Solomon Teshome

Paleontology: Drs. Gerald Eck, Kaye Reed, Rene Bobe, Charles Lockwood, Zeresenay Alemseged, Ray Bernor, and Michelle Drapeau; Ms. Elizabeth Harmon, Mr. Mohamed Ahamadin, Mr. Tadiwos Asebework, Mr. Alemayehu Asfaw, Mr. Michael Black, Mr. Kebede Geleta, Mr. Tekele Hagos, Mr. Tesfaye Hailu, Mr. Ambachew Kebede, Mr. Caley Orr, Ms. Amy Rector, and Mr. Tamrat Wodajo

Archeology: Drs. Erella Hovers and Zelalem Assefa; Ms. Karen Schollmeyer, Ms. Talia Goldman, and Mr. Essayas (Bruk) GebreMariam

Our Addis Ababa–based camp crew seems to make all things possible, for which we are eternally grateful: Messrs. Alayu Kassa, Mesfin Mekonen, Getachew Senbato, Assaye Zerihun, Achamyeleh Teklu, and Abebe Dileligne. The late Wubishet Fantu was a dear friend and member of our camp crew from 1990 to 1999.

In Addis Ababa, Kebede Worke, Esq., has been of incalculable assistance in the administrative and bureaucratic aspects of our project since 1995. We are grateful for his friendship and hard work above and beyond the call of duty.

Last, but certainly not least, we acknowledge the Afar people of Eloaha village and the surrounding countryside, which includes Hadar, without whose close cooperation and warm friendship our project could not possibly succeed. We are especially grateful to the following participants in the Hadar fieldwork: Abdulla Mohamed, Abdu Mohamed, the late Ahmed Bidaru, Ali Mohamed Ware, Ali Welleli, Ali Yussef, Dato Adan, the late Dato Ahmedu, Dawid Ebrahim, Ebrahim Digra, Ebrahim Habib, Ebrahim Nore, Edris Ahmed, Ese Hamadu, Hamadu Mohamed, Hamadu Humed, Hamadu Meter, Humed Michael, Humed Waleno, Kaloyta Ese, Maumin Alehandu, Meter Dato, Michael Dato, Mohamed Ahmed Bidaru, Mohamed Ese, Mohamed Gofre, Mohamed Omar, Nore Ali, and Omar Abdulla. Mr. Mohamed Ahamadin of the Culture and Tourism Bureau of the Afar Regional State government, in addition to being a dedicated field team member, has been a close friend and advisor in many matters concerning our relationship with the Afar people.

During the Hadar field work, our families have put up with our long absences, unpredictable communications, and unusual illnesses. To Patricia and Arren; Ricka, Ariel, Benjamin, and Carmi; Lenora and Tesfaye: Our work places great demands on you, and we are incredibly lucky to have your support and love when it counts.

W. H. K.
Tempe
Y. R.
Tel Aviv
D. C. J.
Tempe

Contents

The Skull of *Australopithecus afarensis*

1

Background

Australopithecus afarensis is a fossil hominin species known from at least four East African Rift Valley sites ranging from northern Ethiopia in the north to northern Tanzania in the south and bridging the time period between approximately 3.6 and 3.0 million years ago (Ma) (see Figure 1.1).[1] First identified in the late 1970s as the bipedal but craniodentally apelike rootstock from which later *Australopithecus* and *Homo* evolved (Johanson et al., 1978; Johanson and White, 1979), *A. afarensis* constituted the first substantial record of unequivocal human ancestors older than 3.0 million years (Myr). An array of more recently made discoveries have placed *A. afarensis* in a pivotal position in early hominin phylogeny, bracketed in time between, on the one hand, two temporally successive species, *A. anamensis* and *Ardipithecus ramidus*, that jointly extend the hominin record back to 4.4 Ma (M. Leakey et al., 1995, 1998; White et al., 1994, 1995), and, on the other hand, the earliest strong (stratigraphic) evidence for hominin lineage diversification, with the first known records of *A. africanus* (ca. 2.7 Ma) in southern Africa, and of *A. aethiopicus* (ca. 2.7 Ma) and *A. garhi* (2.5 Ma) in eastern Africa (Walker et al., 1986; Asfaw et al., 1999).[2] The task of sorting out the relationships among all of these species hinges on the interpretation of *A. afarensis* itself, from its alpha taxonomy and phylogenetic role to its pattern of evolution over time. A prerequisite to achieving this goal is a more complete knowledge of the *A. afarensis* fossil record, narrowing gaps in our knowledge of anatomy and variation, as well as of distributions in space and time.

On sample size alone, *A. afarensis* is the best-known hominin species in the eastern African fossil record. The vast majority of fossils in the *A. afarensis* hypodigm, some 360 specimens, or approximately 90% of the total, have been recovered at the Hadar site, from the 200+ meter sequence of silts, sands, and clays that comprise the Hadar Formation, which is exposed along the drainages of the Awash River in the Afar Depression of northern Ethiopia (Johanson et al., 1982a; Kimbel et al., 1994) (Figure 1.2). The Hadar sample of *A. afarensis* spans 3.4 to 3.0 Ma and includes the iconic partial skeleton known as

Distribution of *Australopithecus afarensis*

Ma	Hadar	Middle Awash (Maka)	Koobi Fora	Laetoli
3.0	Kada Hadar			
3.2	Denen Dora			
			Tulu Bor	
3.4	Sidi Hakoma	Matabaietu		
3.6				Upper Laetolil
3.8				
4.0				

Figure 1.1. Temporal and geographic distribution in Africa of *Australopithecus afarensis*. Ma = millions of years ago. Geographic sites run from north to south across the top; the entries for temporal distribution are the *A. afarensis*-bearing stratigraphic units represented at each site.

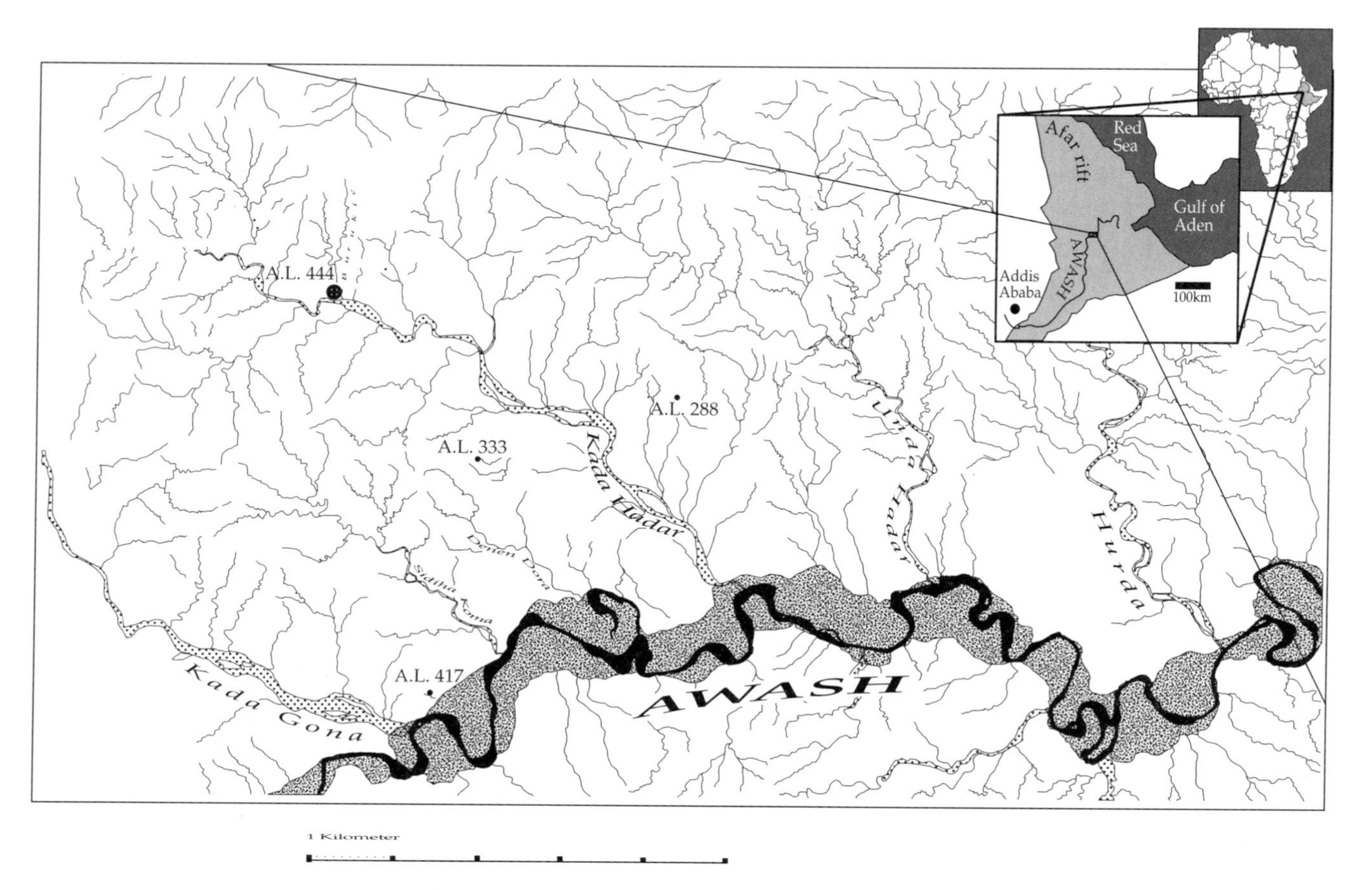

Figure 1.2. Map of the Hadar site showing geographic position of Afar Locality (A.L.) 444 and other key hominin-bearing localities. Base map courtesy of Dr. Gerald Eck.

"Lucy," as well as a collection of some 250 cranial, dental, and postcranial specimens—possibly representing a single biological population—from a hillside outcrop called Afar Locality (A.L.) 333. Yet despite (or perhaps because of) the unusual richness of the Hadar hominin sample, the taxonomy of *A. afarensis* has been debated ever since the species was first described.[3]

After comparative studies conducted in 1977–1978 first established the taxonomic hypothesis that the Hadar specimens were conspecific with diagnostic jaws and teeth from 3.5 to 3.7 Myr-old sediments at Laetoli, Tanzania (Johanson et al., 1978; Johanson and White, 1979), paleoanthropological reaction featured an alternative hypothesis that the *A. afarensis* hypodigm included more than one hominin species (R. Leakey and Walker, 1980; Olson, 1981, 1985; Coppens, 1983; Schmid, 1983; Senut, 1983; Tardieu, 1983; Zihlman, 1985; Shipman, 1986; McKee, 1989). Some critics focused on the high level of size or morphological variation (or both) in the Hadar sample, while others saw taxonomic differences between the temporally and geographically disjunct Hadar and Laetoli site samples. Still others decried the absence of a complete skull in the otherwise comprehensive hominin sample. Although the latter void was filled temporarily by a composite reconstruction that incorporated a dozen adult specimens from Hadar, mostly from the A.L. 333 sample (Kimbel et al., 1984; Kimbel and White, 1988a), large gaps in knowledge of *A. afarensis* adult skull morphology persisted (partly remedied by the 1981 discovery in the Middle Awash of the 3.8 Myr-old Belohdelie frontal fragment [Clark et al., 1984] attributed by Asfaw [1987] to *A.* aff. *afarensis*).

The 1980s witnessed intensive testing of the taxonomic hypothesis of Johanson and White (1979), which featured detailed investigations of quantitative and morphological variation in the *A. afarensis* hypodigm (Blumenberg and Lloyd, 1983; Blumenberg, 1985; White, 1985; McHenry, 1986, 1991; Cole and Smith, 1987; Kimbel and White, 1988b; Lovejoy et al., 1989). While all of these studies found *A. afarensis* to be characterized by moderate to very high levels of size variation, none of them found a strong statistical basis for dividing the sample taxonomically. Thus, out of this body of work developed the paleoanthropological consensus, which persists to the present day, that

Table 1.1 *Australopithecus afarensis* **Specimens from the 1990–1994 and 1999–2001 Hadar Field Seasons**

Locality/Specimen	Year	Discoverer	Identification	Stratigraphic Position
A.L. 58-22b	1993	Alemayehu Asfaw	Right maxilla frag. I^1–I^2 roots	Shu-DD-2
A.L. 125-11	1990	Alemayehu Asfaw	Left maxilla frag. P^3–M^1 frags.	SH, 2 m > SHT
A.L. 137-50	1990	Tim White	Right humerus	SH, 5 m > SHT
A.L. 152-2	1993	Alemayehu Asfaw	Right proximal femur	SH, 10 m < KMT
A.L. 176-35	1992	Abdu Mohamed	Left lower P4	DD-1
A.L. 198-22	1993	Dato Adan	Left mandible frag. M1–M2 f.	SH, 5.5 m < SH-2s
A.L. 207-17	1999	Mohamed Hussein	Right lower M3	SH, 1 m < TT-4
A.L. 224-9	1993	Alemayehu Asfaw	Occipital frag.	SH, 13 m < SH-2s
A.L. 225-8	1990	Dato Ahmedu	Left mandible frag. M1–M3	SH, 10 m > SHT
A.L. 228-2	1990	Mohamed Gofre	Left mandible frag. M1, isol. P4	SH-3s
A.L. 237-3	1993	Zelalem Assefa	Left mandible frag.	SH, 1.5 > SHT
A.L. 293-3	1992	Dato Adan	Upper central I	DD, 16 m > TT-5
A.L. 309-8	1993	Meles Kassa	Upper molar frag.	DD-3s
A.L. 315-22	1990	Alemayehu Asfaw	Right mandible frag., P3, M1	DD, 7.5 m < DD-3s
A.L. 330-5	1993	Alemayehu Asfaw	Mandible LC f., RP4–M3	SH, 4 m < TT-4
A.L. 330-6	1993	Alemayehu Asfaw	Right proximal tibia	DD-3s
A.L. 330-7	1999	Hamadu Meter	Right mandible frag. P4, M1	SH, 4 m < TT-4
A.L. 333-140	1993	Bill Kimbel	Subadult right distal femur	DD-2
A.L. 333-141	1993	Meles Kassa	Right metacarpal V	DD-2
A.L. 333-142	1994	Team	Subadult right proximal femur	DD-2
A.L. 333-144	1994	Team	Proximal frag. metacarpal III	DD-2
A.L. 333-145	1994	Team	Pedal proximal phalanx II	DD-2
A.L. 333-147	1994	Team	Right talus	DD-2
A.L. 333-148	1994	Team	Head intermed. hand phalanx	DD-2
A.L. 333-149	1994	Team	Intermed. hand phalanx	DD-2
A.L. 333-150	1994	Team	Intermed. hand phalanx	DD-2
A.L. 333-152	1994	Team	Thoracic vertebra	DD-2
A.L. 333-153	2000	Team	Proximal frag. left metacarpal III	DD-2
A.L. 333-154	2000	Team	Proximal foot phalanx	DD-2
A.L. 333-155	2000	Team	Rib fragment	DD-2
A.L. 333-156	2000	Team	Rib fragment	DD-2
A.L. 333-157	2000	Team	Proximal frag. left metatarsal III	DD-2
A.L. 333-158	2000	Team	Prox. frag. prox. foot phalanx I	DD-2
A.L. 333-159	2000	Team	Terminal phalanx	DD-2
A.L. 333-160	2000	Team	Left metatarsal IV	DD-2
A.L. 333-161	2000	Team	Rib fragment	DD-2
A.L. 333-162	1994	Team	Left proximal femur shaft	DD-2
A.L. 333-163	2001	Team	Proximal right MT II	DD-2
A.L. 333-164	2001	Team	Rib fragment	DD-2
A.L. 333-165	2001	Team	Tooth crown fragment	DD-2
A.L. 333-166	2001	Team	Tooth crown fragment	DD-2
A.L. 333n-1	1999	Edris Ahmed	Right juvenile mandible, dm2	DD-2
A.L. 333n-2	2000	Abrahim Nore	Distal frag. hand phalanx	DD-2
A.L. 413-1	1990	Alemayehu Asfaw	Right maxilla frag., C–M^3 roots	DD, 13 m > TT-5
A.L. 417-1a–d	1990–1993, 1999	Dato Adan, team	a: Left mandible frag. C–M3	SH, 33 m > SHT
			b: Right mandible frag. M2–M3	
			c: Basioccipital, basisphenoid, right alisphenoid	
			d: Maxilla, RI^2–M^3, LC–M^3	
A.L. 418-1	1990	Dato Ahmedu	Left mandible frag. M2	SH, 31 m > SHT
A.L. 423-1	1990	Dato Adan	Right maxilla frag., P^4–M^1	SH, 4 m < KMT
A.L. 427-1a–c	1990	Ray Bernor, team	a: Maxilla LM^3 f.	DD/KH, 0.5 m > DD-3s
			b: Occipital frag.	
			c: Molar frag.	
A.L. 432-1	1990	Alemayehu Asfaw	Right mandible frag. M2–M3 ff.	DD-3s

(continued)

Table 1.1 ***(continued)***

Locality/Specimen	Year	Discoverer	Identification	Stratigraphic Position
A.L. 433-1a–c	1990	Dato Adan, A. Asfaw	a: Right mandible frag. P4 f.	DD/KH, 1 m > DD-3s
			b: Left mandible frag.	
			c: Molar root frag.	
A.L. 436-1	1990	Dato Ahmedu	Right mandible frag., M2–M3 roots	SH, SH-2s
A.L. 437-1	1992	Dato Ahmedu	Left mandible frag., P4–M3	KH, 17 m < BKT-2
A.L. 437-2a–c	1992	Zelalem Assefa	a: Mandible, LI1–C frags., RI1–I2 frags., RM2–3 frags.	KH, 17 m < BKT-2
			b, c: Isol. LP3, M2 frags.	
A.L. 438-1a–v	1992	Don Johanson, team	a: Left ulna	KH, 10–12 m < BKT-2
			b: Frontal frag.	
			c: Right proximal humerus shaft	
			d: Left metacarpal III	
			e: Left metacarpal II	
			f: Right metacarpal II	
			g: Right mandible w/ramus	
			h: Right lower M1	
			i: Right lower M3	
			j: Right lower I1	
			k: Right P^4 frag.	
			l: Proximal right radius frag.	
			m: Shaft frag. right ulna	
			n: Humeral shaft frag.	
			o: Humeral shaft frag.	
			q: Upper molar root frag.	
			s: Maxilla frag.	
			u: Left lower molar root	
			v: Clavicle frag.	
A.L. 438-2	1992	Team	Right lower P3	
A.L. 438-3	1992	Team	Left lower P3 frag.	
A.L. 438-4	1992	Team	Proximal hand phalanx	
A.L. 439-1	1992	Zelalem Assefa	Occipital	KH, 17.5 m < BKT-2
A.L. 440-1	1992	Dato Ahmedu, team	Right mandible frag. C, P3; isol. LI1, C, P4–M2	KH, 7.5 m < BKT-2
A.L. 441-1	1992	Abdu Mohamed	Molar frag.	DD, 5 m > TT-5
A.L. 442-1	1992	Dato Ahmedu	Right maxilla, M2	DD, 5 m > TT-5
A.L. 443-1	1993	Hamadu Meter	Left mandible frag., P4, M2	SH lower
A.L. 444-1a, b	1992	Yoel Rak	a: Right occipital frag.	KH, 12 m < BKT-2
			b: Left occipital frag.	
A.L. 444-2a–h	1992	Yoel Rak, team	a: Maxilla, RI^1, C, P^4–M^3 LI^1, C–M^3	KH, 10.5 m < BKT-2
			b: Mandible, I ff., RC, P4–M1	
			c: Right zygomatic	
			d: Frontal w/right parietal frag.	
			e: Left parietal	
			f: Occipital + right and left temporals	
			g: Right parietal fragment	
			h: Nasal bones	
A.L. 444-3	1992	Team	Right lunate	
A.L. 444-4	1992	Team	Manual proximal phalanx	
A.L. 444-5	1992	Team	Phalanx frag.	
A.L. 444-6	1992	Team	Right lower dm2 frag.	
A.L. 444-7	1992	Team	Last lumbar vertebra body	
A.L. 444-8	1992	Team	Thoracic vertebra spine	

(continued)

Table 1.1 ***(continued)***

Locality/Specimen	Year	Discoverer	Identification	Stratigraphic Position
A.L. 444-9	1992	Team	Cervical vertebra frag.	
A.L. 444-10	1992	Team	Thoracic vertebra frag.	
A.L. 444-11	1992	Team	Thoracic vertebra frag.	
A.L. 444-12	1992	Team	Thoracic or cervical vertebra frag.	
A.L. 444-13	1992	Team	Left humerus shaft frag.	
A.L. 444-14	1992	Team	Left humeral epiphysis frag.	
A.L. 444-15	1992	Team	Right distal humerus shaft frag.	
A.L. 444-16	2000	Elizabeth Harmon	Frag. $LM_{1/2}$	
A.L. 444-29	1992	Team	Lower deciduous central incisor	
A.L. 444-30	1992	Team	Lower molar frag.	
A.L. 452-18	1999	Zeresenay Alemseged	Molar frag.	KH, 9 m < BKT-2
A.L. 457-2	1994	Abdu Mohamed	Right parietal fragment	KH, 17 m < BKT-2
A.L. 462-7	1999	Hamadu Mohamed	Left lower M3	KH, 16.1 M < BKT-2
A.L. 465-5	1993	Hamadu Meter	Lower molar frag.	SH, 1.5 m > SHT
A.L. 466-1	1993	Hamadu Meter	Maxillary M^2–M^3 ff.	SH, 5 m < SH-3s
A.L. 486-1	1993	Meles Kassa	Left maxilla, I^1–I^2, P^3–M^3	DD-3s
A.L. 487-1a–g	1993	Abrahim Nore, team	a: Right mandible frag., M3 f. b: Two left mandible frags., P3–M3 roots c: Left maxilla frag., C d: Right maxilla frag., P^3–P^4 ff. e: Left palate frag. f: Lower LC g: Lower RC	DD, 2.4 m < KHT
A.L. 545-3	1993	Don Johanson	Right distal tibia	DD, 12.5 m < DD-3s
A.L. 557-1	1993	Dato Adan	Maxillary molar	KH "upper"
A.L. 582-1	1993	Meles Kassa	Mandible frag., I1, LP4–M1, RP3–M1 ff.	DD, 4.6 m < CC
A.L. 604-1	1993	Zelalem Assefa	Left mandible ramus frag.	Shu–DD-2
A.L. 620-1	1994	Abrahim Nore	Left mandible, M3	DD-2
A.L. 651-1	1990	Dato Ahmedu	Left maxillary frag., P^3–M^3	SH, 4 m < SH-2s
A.L. 655-1	1994	Million H/Michael	Left lower P3 w/roots	SH, 5.4 m < TT-4
A.L. 660-1	1994	Ali Samla	Lower LM2 frag.	SH-1
A.L. 697-1	1994	Dato Adan	Molar fragment	SH-1
A.L. 699-1	1994	Abrahim Nore	Right lower P4 frag.	DD-3s
A.L. 701-1	1994	Dato Adan	Left frontoparietal frag.	DD-3u/KH-1
A.L. 724-1	1999	Hamadu Meter	Proximal phalanx	KH, 16.7 m < BKT-2
A.L. 724-3	1999	Hamadu Meter	Proximal phalanx	KH, ~ 15 m < BKT-2
A.L. 729-1	1999	Charles Lockwood	Mandible	KH, 6.2 m < BKT-2
A.L. 762-1	1999	Maumin Alahandu	Right lower M3 frag.	DD, 2.75 m > TT-4
A.L. 763-1	1999	Hamadu Meter	Left upper canine with root	DD, 4.75 m < KHT
A.L. 766-1	1999	Abrahim Nore	Mandible symphysis frag.	DD, 4 m < KHT
A.L. 769-1	1999	Dato Adan	Right distal humerus frag.	SH, 5.7 m < SH-2s
A.L. 770-1a, b	1999	Nore Ali, Mohamed Ahamedin	a: Left maxilla frag. b: Right maxilla frag.	SH, 0.75–2 m > SHT
A.L. 772-1	1999	Charles Lockwood	Right lower dm2 frag.	DD, 1.5 m < KHT
A.L. 777-1	1999	Hamadu Meter	Right lower dm1 frag.	SH, 10.25 m < SH-2s
A.L. 822-1	2000	Dato Adan	Partial skull	KH, 5 m < BKT-1
A.L. 827-1	2000	Dato Adan	Femur	KH, 2 m > BKT-1
A.L. 922-1	2000	Dato Adan	Maxilla, frag. LI^1–M^1, M^2–M^3; frag. RI^1–M^3	KH, ~11m > BKT-1
A.L. 996-1	2001	Hamadu Meter	Left mandible frag., P4–M1	KH, ~11m > BKT-1
A.L. 1017-1	2001	Team	Premolar fragments	KH, ~11m > BKT-1

SH, Sidi Hakoma member; DD, Denen Dora member; KH, Kada Hadar member; >, above; <, below.

Table 1.2. Distribution of 1990–2001 Hadar Hominins by Skeletal Part (n = 112)

	Postcranial							
Skeleton[a]	Upper	Lower	Indet.	Axial	Skull	Cranium	Mandible	Teeth[b]
438-1	137-50	152-2	333-148	333-152	417-1	58-22b	198-22	176-35
	333-141	330-6	333-159	333-155	444-2	125-11	225-8	207-17
	333-144	333-140	444-5	333-156	487-1	224-9	228-2	293-3
	333-149	333-142	724-1	333-161	822-1	413-1	237-3	309-8
	333-150	333-145	724-3	333-164		423-1	315-22	333-165
	333-153	333-147		444-7		427-1	330-5	333-166
	333n-2	333-154		444-8		439-1	330-7	438-2
	438-4	333-157		444-9		442-1	333n-1	438-3
	444-3	333-158		444-10		444-1	418-1	441-1
	444-4	333-160		444-11		457-2	432-1	444-6
	444-13	333-162		444-12		486-1	433-1	444-16
	444-14	333-163				651-1	436-1	444-29
	444-15	545-3				701-1	437-1	444-30
		827-1				770-1	437-2	452-18
						922-1	440-1	462-7
							443-1	465-5
							582-1	466-1
							604-1	557-1
							620-1	655-1
							729-1	660-1
							766-1	697-1
							996-1	699-1
								762-1
								763-1
								772-1
								777-1
								1017-1
Totals:								
1	13	14	5	11	4	15	22	27

[a]Includes associated postcranial, mandible, and cranial elements.
[b]Specimens listed comprise isolated teeth only. Specimens listed under Cranium and Mandible may also include teeth.

A. afarensis is, indeed, both biologically and statistically speaking, a "good" species.

At the same time, the first round of numerical cladistic analyses in paleoanthropology produced ambiguous results about the relationships of *A. afarensis* to subsequent hominin species (Skelton et al., 1986; Wood and Chamberlain, 1986, 1987; Chamberlain and Wood, 1987). Johanson and White (1979) had proposed that the southern African species *A. africanus* was phylogenetically linked exclusively to *A. robustus* and *A. boisei* by a series of shared-derived craniodental characters related to heavy mastication (see also White et al., 1981; Rak, 1983; Kimbel et al., 1984). While almost all of these early numerical cladistic analyses supported the basal position of *A. afarensis* relative to *A. africanus*, opinion divided on several mutually exclusive phylogenetic hypotheses:

1. *A. africanus* is the plesiomorphic sister taxon to an *A. robustus* + *A. boisei* clade (indicated chiefly by the teeth, jaws and face—this is the Johanson and White [1979] hypothesis; see Chamberlain and Wood, 1987).
2. *A. afarensis* is the plesiomorphic sister taxon to an *A. robustus* + *A. boisei* clade, with a separate *A. africanus* + *Homo* clade (as suggested by some aspects of calvarial morphology and cranial venous outflow systems; originally proposed by Olson, 1981, 1985, and Falk and Conroy, 1983; see Wood and Chamberlain, 1986).
3. *A. africanus* is the plesiomorphic sister taxon to a clade comprising *A. robustus* + *A. boisei* and *Homo habilis* (based on derived character states shared by these taxa, such as fixation of the large P_3 metaconid, but for which *A. afarensis* retained the primitive state [see list in Kimbel et al., 1984: 375]; Skelton et al., 1986).

Combining plesiomorphic characters shared with *A. afarensis* but *not* with *A. africanus*, and derived characters shared uniquely with *A. robustus* and *A. boisei*, the 2.5

Myr-old KNM-WT 17000 cranium of *A. aethiopicus*, described by Walker et al. in 1986, showed that neither the Johanson and White (1979) hypothesis nor the Skelton et al. (1986) hypothesis was likely to be true (e.g., Kimbel et al., 1988). This specimen also underscored the likelihood that the early evolution of the hominin skull was characterized by significant homoplasy, the nature and degree of which depended on the phylogenetic role accorded *A. africanus* (which, in turn, was affected by growing doubts concerning its taxonomic unity; see, e.g., Clarke, 1988; Kimbel and White, 1988b) and on whether southern and eastern "robust" *Australopithecus* species constituted a monophyletic or polyphyletic assemblage (see contributions in Grine, 1988; Skelton and McHenry, 1992). Thus, at the close of the 1980s, the state of early hominin taxonomy and phylogeny was unsettled.

Beginning in 1990, renewed fieldwork in Ethiopia, both at Hadar and in the Middle Awash, sought to address these unresolved issues with new fossil data recovered in refined chronostratigraphic and paleoenvironmental contexts (White et al., 1993; Kimbel et al., 1994). A major goal of Hadar fieldwork was to augment the *A. afarensis* sample, in part by following its record up-section through scrutiny of large tracts of previously unsurveyed outcrops sampling the upper strata of the Kada Hadar Member, the youngest of the four currently recognized members of the Hadar Formation (Figure 1.3). These strata are now dated to <3.18 Ma, the $^{40}Ar/^{39}Ar$ age obtained for the Kada Hadar Tuff (KHT) that marks the base of the Kada Hadar Member (Walter, 1994). The goal of exploring these particular sediments was closing the chronological gap between *A. afarensis* and the earliest species of the post-3.0-Myr lineage diversification, *A. africanus* and *A. aethiopicus*, which would potentially clarify the above-mentioned uncertainties about hominin phylogeny.

The upper sediments of the Kada Hadar Member were essentially a blank at the cessation of fieldwork in 1976–1977, but they postdated the youngest of the *A. afarensis* specimens then known ("Lucy"—A.L. 288-1; see Figure 1.3). Before 1992, when systematic paleontological surveying of these young sediments began, only 2 of 28 (ca. 7%) hominin localities were known from sediments stratigraphically above KHT; in contrast, 16 of 47 (ca. 34%) new hominin-bearing localities logged during the 1990s lie within the Kada Hadar Member. Discoveries made during a 10-day period in 1992 at five localities in the upper reaches of the main Kada Hadar tributary (A.L. 437, 438, 439, 440, and 444) nearly doubled the temporal range of *A. afarensis* in the Hadar Formation from ca. 0.2 to ca. 0.4 Myr by extending the species' last known appearance datum to approximately 3.0 Ma. All of the hominin fossils from this geographically concentrated cluster of localities lie between 7.5 and 17 m stratigraphically below the BKT-2 tephra, which has an $^{40}Ar/^{39}Ar$ age of 2.94

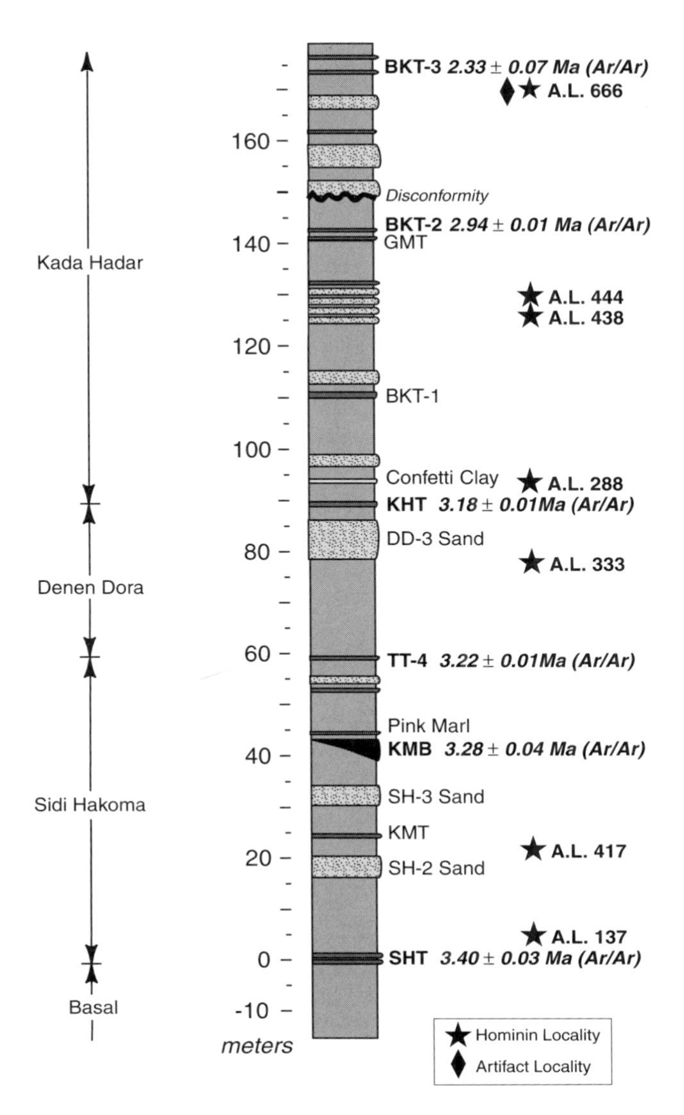

Figure 1.3. Composite stratigraphic section of the Hadar Formation showing position of A.L. 444 and other key hominin-bearing localities. Tephra dated by $^{40}Ar/^{39}Ar$ are indicated (see Renne et al., 1993; Walter and Aronson, 1993; Walter, 1994; Kimbel et al., 1996; Semaw et al., 1997).

Myr (Kimbel et al., 1994; Semaw et al., 1997). This important series of discoveries includes the first, much anticipated, fairly complete adult skull of *A. afarensis* (A.L. 444-2), as well as other jaws, teeth, and cranial fragments and a partial upper limb skeleton associated with cranial and mandibular elements (Kimbel et al., 1994; Drapeau, 2001).

The A.L. 444-2 skull is the center of the work presented in this volume. It is important for three reasons, beyond the mere fact of its relative completeness:

1. It includes cranial regions that were previously poorly represented in adult *A. afarensis* from Hadar, such as the frontal bone (only a small

supraorbital fragment and a posterior segment of the squama remain of the "Lucy" frontal bone).

2. It is the first *A. afarensis* adult specimen to preserve the upper face and mandible of a single individual; this combination was repeated with the discovery of the A.L. 417-1 maxilla and mandible during fieldwork at Hadar in the 1990s (Kimbel et al., 1994).
3. It is among the geologically youngest known specimens of *A. afarensis*, which, as noted, ranges in time at Hadar from 3.4 to 3.0 Ma, about double the temporal extent of the species based on the stratigraphic spread of the 1970s sample.

Thus, A.L. 444-2 presents an unprecedentedly full portrait of the *A. afarensis* skull at one end of the species' temporal range. At the other end there remain open questions (see Lockwood et al., 2000a). The Laetoli sample of *A. afarensis*, centered in time around 3.6 Ma, includes no adult cranial remains except for the original Garusi I maxillary fragment of 1939, while the partial frontal bone from the Middle Awash site of Belohdelie, isolated at 3.9 Ma, may represent the oldest fossil attributable to the species (Asfaw, 1987; Kimbel et al., 1994).

This volume presents the results of our comparative morphological study of A.L. 444-2. The monograph is organized in chapters, as follows:

Chapter 2: Recovery and Reconstruction of A.L. 444-2. The skull was recovered in a fragmentary and partially damaged state. After summarizing the recovery procedure in the field, in this chapter we describe the process of restoring the specimen to its final state, which featured the use of three-dimensional computerized tomography and stereolithography (Zollikofer et al., 1995).

Chapter 3: A.L. 444-2: The Skull as a Whole. In this chapter we describe and compare A.L. 444-2 from the perspective of observations that transcend the morphology of the individual skull bones, focusing on the size and shape relationships among the different anatomical regions of the skull. Here the specimen is examined in standard anatomical views, which is followed by a brief comparison of the skull with the 1984/1988 White-Kimbel skull reconstruction, and then by analyses of the cranial cresting pattern and the endocranial venous drainage pattern.

Chapter 4: Endocranial Morphology of A.L. 444-2. In this chapter Ralph Holloway and Michael Yuan join us to detail the results of their efforts to estimate the endocranial volume of the new skull and to record their observations on the form of the endocast.

Chapter 5: A.L. 444-2: Elements of the Disarticulated Skull. In this chapter we describe in detail the anatomy of the individual cranial bones, the mandible, and the dentition, following Sherlock Holmes's axiom that "the little things are infinitely the most important." Each element is first described and then submitted to comparative analysis. One objective of this chapter is the assessment of the effect of A.L. 444-2 and other new additions to the Hadar sample on previous characterizations of *A. afarensis* skull and dental variation.

Chapter 6: Implications for the Taxonomic and Phylogenetic Status of *Australopithecus afarensis*. After summarizing the main conclusions of the preceding chapters, we focus discussion on their implications for the taxonomy of *A. afarensis* and its role in hominin phylogeny. The ≥30% increase in the size of the Hadar hominin sample over the 1970s collection affords a fresh look at old arguments about the "acceptable" degree of size variation for a single hominin species. Finally, we reexamine the phylogenetic position of *A. afarensis* vis-à-vis other *Australopithecus* species, especially in light of the recently described *A. anamensis* from the early Pliocene of eastern Africa.

Our method is classical comparative anatomy. Although we are currently engaged in analytical projects that apply recently developed quantitative methods to the description and comparison of cranial form (e.g., Lockwood et al., 2002), here our primary concern is to create a document of record, including description, measurement, and comparison of the first complete adult *A. afarensis* skull.

2

Recovery and Reconstruction of A.L. 444-2

Recovery

The A.L. 444-2 skull was found on 26 February 1992, during a strategic paleontological survey of Kada Hadar Member sediments that are stratigraphically situated between BKT-1 and BKT-2 tephras, on the eastern edge of the Awash River's Kada Hadar tributary (Figures 1.2, 2.1, and 2.2). Yoel Rak discovered two fragments of hominin occipital bone (A.L. 444-1) at the base of a steep hill composed of Kada Hadar Member silts and clays capped by a weathered sandstone remnant (Figure 2.2). Subsequent examination of the upslope surface revealed additional hominin skull fragments (the temporal bones and maxillae) clustered together and partially exposed in a narrow gully that dissected the face of the hill. During the next seven days, probing and dry sieving of the gully infill and hillside colluvium over a 77 m^2 area led to the recovery of fragments representing about 75%–80% of a single hominin skull. It was immediately apparent that the upslope finds duplicated the anatomical parts represented by the two A.L. 444-1 occipital fragments and therefore constituted a second hominin individual, cataloged as A.L. 444-2. In addition, the lambdoidal suture of the A.L. 444-1 occipital is completely unfused, suggest-

Figure 2.1 View to the north of the A.L. 444 hill showing the location of the probable in situ horizon (arrow) relative to the 2.94 Myr-old BKT-2 tephra, which lies stratigraphically ca. 10.5 meters above it. The fossil-bearing sediments at A.L. 444 are estimated to be 3.0±0.02 Myr old.

Figure 2.2 View to the southeast of A.L. 444, on whose surface the remains of skull A.L. 444–2 were discovered by Yoel Rak on 26 February 1992. The dry Kada Hadar tributary of the Awash River is in the background.

ing subadult status, whereas fused cranial sutures and extreme dental occlusal wear indicate an advanced ontogenetic age for A.L. 444-2.

Stratigraphic Provenance and Geological Age

In February–March 1993 the A.L. 444 hillside was excavated in an effort to locate missing parts of the A.L. 444-2 skull and to determine its precise stratigraphic provenance (Figure 2.2). No further remains of the hominin skull were encountered in situ, but a complete viverrid cranium and indeterminate fragments of large mammal bone with preservation and patina (mottled dark gray, white, and yellowish gray) identical to those of the hominin were excavated in an unstratified, cemented carbonate silt that exactly matches the matrix adhering to A.L. 444-2. We are confident that the hominin skull is from this sedimentary horizon. It is approximately 10.5 m stratigraphically below the BKT-2 tephra, which outcrops in the immediate vicinity of A.L. 444 (Figure 2.1).

Single-crystal laser fusion (SCLF) $^{40}Ar/^{39}Ar$ ages for BKT-2 and Kada Hadar Tuff (KHT) bracket the geological age of A.L. 444-2 between 2.94 and 3.18 Myr (Kimbel et al., 1994; Walter, 1994; Semaw et al., 1997) (see Figure 1.3). Further refinement of the age of the skull can be achieved based on the stratigraphic position of the top of the Kaena paleomagnetic subchron, dated to 3.04 Myr (Cande and Kent, 1995), ca. 22.5 m below BKT-2 (J. L. Aronson, personal communication). Assuming constant sedimentation, interpolation yields an age estimate of 2.99 Myr for the inferred in situ horizon. Using the stratigraphic position of KHT (62.5 m below BKT-2) as the reference point for the calculation yields an age estimate for the skull horizon of 2.98 Myr. We consider 3.0 ± 0.02 Myr to be the best estimate of the geological age of A.L. 444-2, making it approximately 160 kiloyears (Kyr) younger than "Lucy," 180 Kyr younger than the A.L. 333 hominin accumulation, and 380 Kyr younger than the oldest hominin fossils from the Hadar Formation (Walter and Aronson, 1993; Walter, 1994) (Figure 1.3).

The 444 locality is situated in a dense cluster of faunal localities that sample a fluviatile depositional cycle of the Kada Hadar Member between BKT-1 and BKT-2 volcanic marker beds. The BKT-2 tephra, dated to 2.94 myr (Kimbel et al., 1994; Semaw et al. 1997), is widely exposed in this area, and the fossil-bearing units lie stratigraphically between 7.5 and 17 m below it. Outcropping in this region are fossiliferous sands and silts related to a major stream channel and its overbank and floodplain. At A.L. 444 the fossil hominin skull derives from a floodplain silt proximal to the stream channel (possibly its levee), denoted by a sand/pebble conglomerate unit that outcrops at equivalent stratigraphic position some 50 m to the west of the locality (Christopher Campisano, personal communication).

Hominin fossils were found at eight localities in this region (A.L. 437, A.L. 438, A.L. 439, A.L. 440, A.L. 444, A.L. 452, A.L. 457, A.L. 462) spanning a maximum linear distance of ca. 0.9 km. From three of these (A.L. 437, A.L. 438, A.L. 444) the remains of more than one hominin individual were recovered. In addition to the adult skull, the sample from A.L. 444 contains at least one, and perhaps as many as three subadult individuals (A.L. 444-1, occipital fragments; A.L. 444-6 and A.L. 444-29, deciduous teeth).

Besides hominins, the faunal sample from these deposits includes a wide variety of micro- and macromammalian taxa. Both cercopithecines and colobines have been recovered, including the partial skeleton of a juvenile individual at A.L. 437. Bovid taxa include the tribes Alcelaphini, Antilopini, Reduncini, Bovini and Tragelaphini. Suids, giraffids, equids, deinotherids, murid rodents, and viverrids were also recovered. Although these localities are associated with stream channel sedimentation—as suggested by the lithology as well as fossils indicative of riverine forest habitats (monkeys, *Tragelaphus* cf. *T. pricei*, and frequently encountered fossil wood)—the relatively high abundance of antilopine and alcelphine bovids argue for an overall drier habitat with fewer trees than in deposits stratigraphically beneath the BKT-1 marker bed (Reed, in prep).

The local ecological setting around A.L. 444 evidently was attractive, but potentially risky for hominins, as testified to by the frequency with which their remains are found here. At least one specimen bears clear carnivore puncture marks (A.L. 437-1, an adult mandible corpus). However, lengthy postmortem exposure on the surface does not appear to have been typical. Most of the hominin as well as nonhominin bones in the area appear relatively fresh, retain good to excellent surface detail and sharp edges where broken, and lack obvious signs of prolonged surface weathering or significant transport in water. In one or two cases, hominin individuals were probably interred as partial carcasses, which implies minimal transport after death. At A.L. 438 parts of both arms and hands were associated on the surface and in situ with an adult's mandible, partial frontal bone, and maxillary fragments (A.L. 438-1), while at A.L. 444 a large adult's vertebral, humeral, wrist, and hand elements were recovered from the surface deposits with the A.L. 444-2 adult skull (although there is no direct way to associate these A.L. 444 specimens with one another).

Taphonomic Aspects and Reconstruction of the Skull

The A.L. 444-2 skull was recovered in approximately 50 fragments, not including isolated teeth, tooth crown and root fragments, and indeterminate bone scraps. These fragments have been joined to form eight major parts (Figure 2.3 a–c):

1. Frontal bone with attached anterosuperior fragment of right parietal
2. Left parietal with adhering superior fragment of squamous temporal
3. Small posterior fragment of right parietal, located approximately midway along bregma–lambda arc and in contact with the left parietal along the sagittal suture
4. Posterior calvaria, composed of the occipital squama and both temporal bones
5. Right zygomatic bone
6. Maxilla, with RI^1, R^C, RP^4–M^3, LI^1, L^C, and LP^3–M^3
7. Partial nasal bones
8. Right mandible corpus and symphyseal region, with left and right incisors, partial R_C, damaged RP_4–M_1

Details of Preservation

1. The frontal bone, composed of eight fragments, is missing the ends of both zygomatic processes, the glabellar mass and adjacent medial orbital walls, small patches of the squama in the supraglabellar region, the inferior part of the right temporal surface (pterionic region), and, internally, almost all of the floor of the anterior cranial fossa. The anterosuperior piece of the right parietal bone, consisting of four fragments, is attached to the frontal along the coronal suture, running from bregma laterally for about 65 mm.

The left side of the frontoparietal fragment is mostly intact and minimally deformed. Plastic deformation has affected the right side of the specimen through moderate superomedial rotation of the supraorbital region and adjacent temporal surface, artificially accentuating the arch of the supraorbital torus and elevating the squama on this side. The line along which this deformation occurs is clearly visible as a crack through the internal and external surfaces; it begins anteriorly at the supraorbital torus about 16 mm to the right of the midline, passes posteriorly to the temporal line, and then drops posteroinferiorly on the medial wall of the temporal fossa. The attached parietal is also flattened and plastically deformed upward just posterior to the (completely fused) coronal suture, and much of its endocranial surface is destroyed. Significant damage occurs in the glabellar region, which is lost except for three very thin fragments of the supraglabellar platform (recovered from a crushed block), consisting of the fragile walls of the frontal sinus and superior midfacial bones (including the nasals), which had been pushed down onto the left maxilla. These fragments were inserted into the supraglabellar region. Cortical bone has been abraided from most surfaces, both internally and externally, reducing thickness by 0.5–1.0 mm in some areas and obliterating some anatomical detail.

2. The left parietal consists of five pieces that articulate perfectly to form most of the central portion of the bone. A superior fragment of the temporal squama is pushed up and away from its articulation along the squamosal suture; because it is very delicate, it has been left

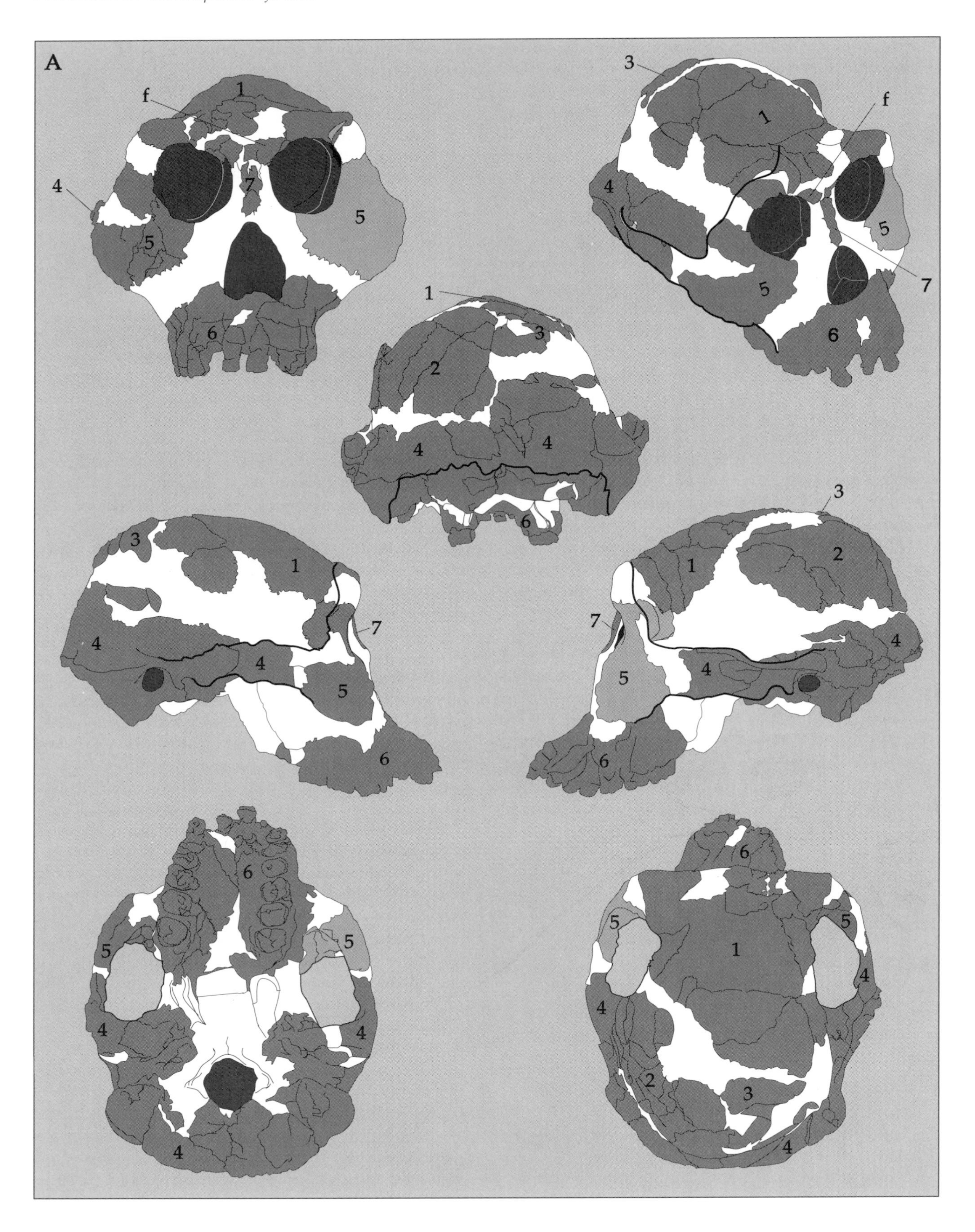
A
1
f
4
7
5
5
6
3
f
1
4
5
7
5
6
1
3
2
4
4
6
3
3
1
1
2
7
7
4
4
5
4
5
6
6
6
6
5
5
5
5
1
4
4
4
4
4
2
3
4
4

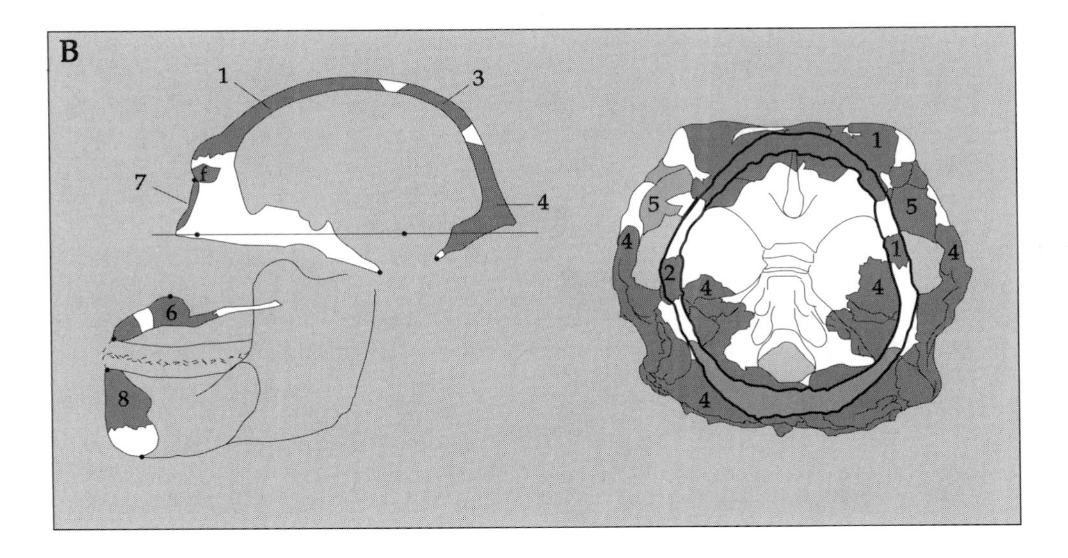

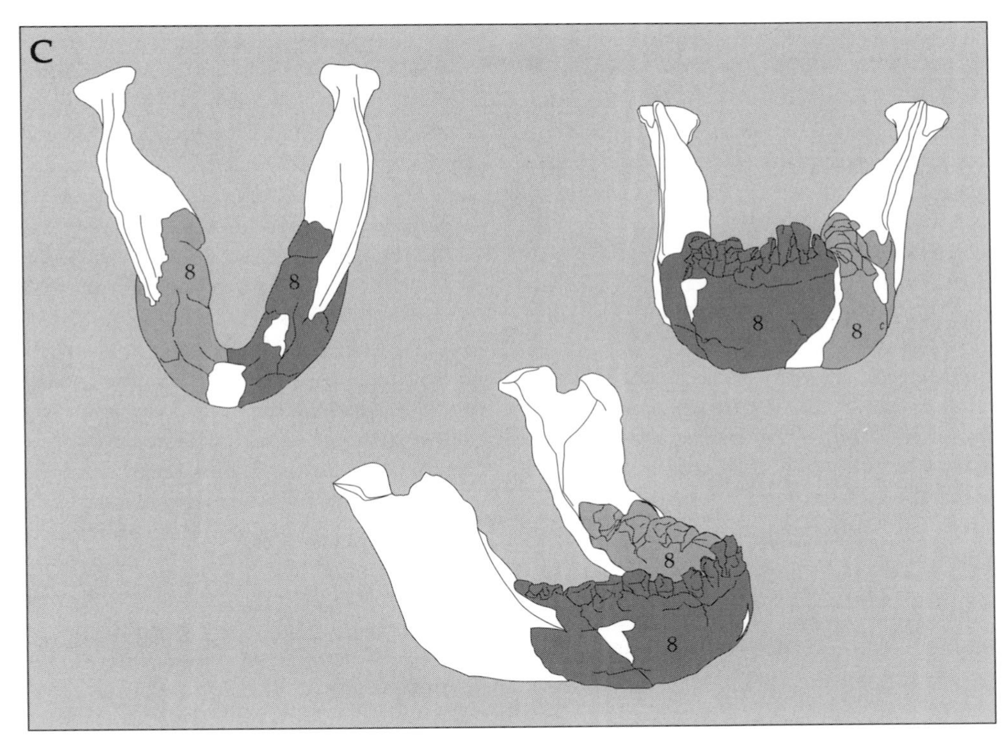

Figure 2.3 Schematic views of the reconstructed A.L. 444-2 skull. Dark gray areas represent original bone, light gray areas represent stereolithographic reconstructions, and the white areas represent wax reconstructions of missing bone. The skull is preserved in eight major segments, not including the teeth (see numbered segments in figure, which correspond to numbered fragments in the description in the text). The letter "f" refers to parts of the frontal bone based on specimen A.L. 438–1b (see text). (A) Seven views of the cranium. (B) *Left*, median cross section of the skull; *right*, endocranial aspect of the cranial base. (C) Three views of the mandible.

in this incorrect anatomical position. The coronal margin and most of the temporal margin are missing. The posterior half of the sagittal margin is intact, as is the medial part of the lambdoidal margin, to which is fused a small section of the occipital squama. The mastoid and sphenoid angles are broken. Endocranial surface and diploë have been lost along the sagittal and broken coronal edges. The preserved detail is fair to good on most surfaces.

3. The small, roughly triangular portion of the posterior right parietal bone, consisting of four fragments, has been flattened somewhat by plastic deformation. It bears a short section of sagittal suture and temporal line, and it contacts the left parietal approximately midway along bregma–lambda arc.

4. The posterior calvaria originally consisted of three major pieces: the occipital squama with an attached fragment of the posterior right parietal bone and the two temporal bones. The right temporal was separated at the occipitomastoid suture, whereas the left temporal had broken just medial to the suture; both temporals articulate with the occipital. The occipital bone preserves most of the squamous part and the right pars lateralis. A short segment of the foramen magnum margin just anterior to opisthion is present bilaterally (6.0 mm long on the left and 10.5 mm on the right), but opisthion itself is missing.

The two halves of the occipital were originally compressed toward the midline, resulting in the slight overlapping of the nuchal plane along a sagittally oriented "fault" that approximates the midline. This break was cleaned, and the two halves were separated and reattached along the break in their proper position. At present, the left and right halves of the nuchal plane are offset vertically by 2 to 3 mm; this separation increases posteriorly toward the external occipital protuberance, which is crushed. The right posterior cranial fossa is "folded" along a crack that runs diagonally from the occipitomastoid suture, across the cerebellar fossa, to the superior limb of the cruciate eminence. The transition across the endocranial bone table from the cerebellar fossa to the cerebral fossa on this side is artificially sharp along the fold.

An undistorted posterior section of right parietal bone is intact along the (externally fused) lambdoidal suture between lambda and asterion. This segment includes a narrow strip, built from three separate fragments, that runs anteriorly above the supramastoid region to form the posterior part of the vault's lateral contour.

Both temporal bones are fairly complete and well preserved. The right temporal is in better condition than the left, which has survived moderate plastic deformation and bone plate displacement. Bone loss due to breakage on both sides chiefly affects the squamosal portions of the calvarial walls. On the left side this part of the squama is completely destroyed down to the level of the temporal shelf, whereas on the right it is preserved for about 25–35 mm above the shelf. Both zygomatic processes, broken at their roots, were recovered separately and have been reattached. The left zygomatic process, although intact, is plastically deformed and rotated laterally such that its medial surface faces superomedially. Although it is much less deformed than the left, the anterior half of the right process is artificially flared laterally and pushed inferiorly to about 5 mm below its correct articulation with the posterior half.

The left petrous pyramid sits in a more superior position than the right, with the left porion approximately 8 mm lower than the right porion. This discrepancy can be explained by noting that, relative to the undeformed right temporal, the left has been artificially rotated inferolaterally on the sagittal plane and pivoted about the midline break through the nuchal plane of the occipital, lowering porion while elevating the petrous.

Basal anatomy is fairly complete and very well preserved, especially on the right side, where only a small triangular section in the center of the mandibular fossa's roof and the apical half of the petrous pyramid are missing. The entire left petrous is present but is badly crushed anterior to the carotid foramen. Also lost on the left side is the preglenoid plane, the tip of the entoglenoid process, the middle third of the articular eminence, part of the roof of the mandibular fossa, and the medial two-thirds of the postglenoid process. Internally the right temporal is in good condition, except for the aforementioned loss of the petrous apex and moderate loss of surface bone from the petrous's anterior surface. The left petrous is in poor condition; its anterior and posterior surfaces are mostly broken medial to the internal auditory meatus. The matrix that secures crushed and displaced pieces of the petrous has not been disturbed.

5. The right zygomatic bone was recovered in two pieces: the frontal process and the body of the bone. Both frontal and temporal processes are missing their tips. The frontal process, although separated along its base from the body, is in perfect contact with the body at the inferior margin of the orbit. The body of the zygomatic suffers from deformation at three sites: (1) along the inferior margin of the orbit, where the maxillary process has been displaced about 5 mm laterally; (2) along the posterior wall of the maxillary sinus, which has been bent posterolaterally to form a right angle with the rest of the temporal surface; and (3) along the zygomaticomaxillary suture, where there has been modest superficial bone displacement.

6. Excluding teeth and dental fragments (see Chapter 5 for details of tooth preservation), the maxilla was recovered in three primary pieces: the entire right half, the left half anterior to the M^2 position, and the alveolar

bone and maxillary sinus walls above left M^2 and M^3, which were part of a separate block that also contained the supraglabellar plates and upper midfacial bones, which had been crushed downward into the maxillary sinus. On both sides the lateral ends of the horizontal plates of the palatine bones are in place; on the right the inferior tip of the pterygoid process of the palatine is also present. Most of the zygomatic and frontal processes are broken; on the left they are preserved to a more superior level than on the right. Although the left maxilla is somewhat more complete, it is less well preserved than the right. It is both plastically deformed and broken along a major crack that runs inferiorly from the root of the zygomatic process to the alveolar margin at P^3/P^4; along this crack there is a considerable overlapping of the bone table. Another significant crack runs from the alveolar margin at right I^1 into the nasal cavity at the corner of the nasal aperture, where it separates the lateral margin of the aperture from its inferior margin. The relatively undeformed right maxilla joins perfectly with the left along a break that approximates the midline. Missing bone leaves gaps between the two sides on the nasoalveolar clivus and the posterior half of the palatal roof. It is at the latter point that there is a discrepancy in the elevation of the two sides of the palatal roof along an artificial step in the midline, the left side being higher than the right. This discrepancy gradually decreases anteriorly and is negligible anterior to the P^3 level. When all the pieces are assembled, it is evident that plastic deformation compressed the maxilla along the midline step and skewed it backward and to the left of the midsagittal plane, rendering it asymmetric.

7. The nasal bridge fragment, composed of the fused left and right nasal bones, was recovered from within the block of upper midfacial fragments that was crushed down onto the left maxilla. Superior, inferior, and lateral margins are broken. The preserved fragment is 28 mm tall and 10 mm wide.

8. The right side and anterior corpus of the mandible were recovered in three pieces, not including teeth and dental fragments. The left side is represented by alveolar bone below the I_1 and I_2 positions. The corpus is broken along a line running from the left I_2/C interdental septum diagonally to the base below the right canine. The posterior part of the right corpus (containing M_2–M_3 and the root of the ascending ramus) is displaced laterally, disrupting the natural path of the postcanine tooth row and creating a gap and slight step on the lateral surface that measures 3.5 mm wide just below the alveolar margin at the M_1/M_2 level and running anteroinferiorly to the base below P_4/M_1, where it narrows to about 1 mm. The medial surface of the corpus is crushed inward slightly between P_4 and M_2 just below midcorpus level.

Reconstructing the Skull

Initial joining of skull fragments that showed fresh, clean breaks occurred in the field during the 1992 season. This was followed by three months of preparation, reconstruction, molding, and study at the National Museum of Ethiopia, Addis Ababa, in 1993–1995.

Almost all of the skull fragments were coated to one degree or another with a hard, calcareous silt matrix. Large-scale reduction of this matrix, which was sometimes loosened through the parsimonious application of acetic acid, was achieved through the use of electric scribes mounted with sharpened carbide points. Removal of the matrix close to the bone surface was accomplished under a binocular microscope with a sharpened steel sewing needle mounted in a pin vise.

The nature and pattern of bone breakage and displacement in the original state of discovery reveal the trajectory of the geological forces that deformed the cranium while it was still intact. It is clear that the force was directed diagonally downward from above the right orbit toward the left posterior region of the cranial base (see Figures 2.4, 2.5, 2.6).

The major portions of the skull were assembled in the following stages:

1. The left parietal bears the sagittal suture with both sets of temporal lines, establishing both the

Figure 2.4 The effect of postmortem deformation on A.L. 444-2. The arrow represents the direction of the postmortem geological pressure on the skull.

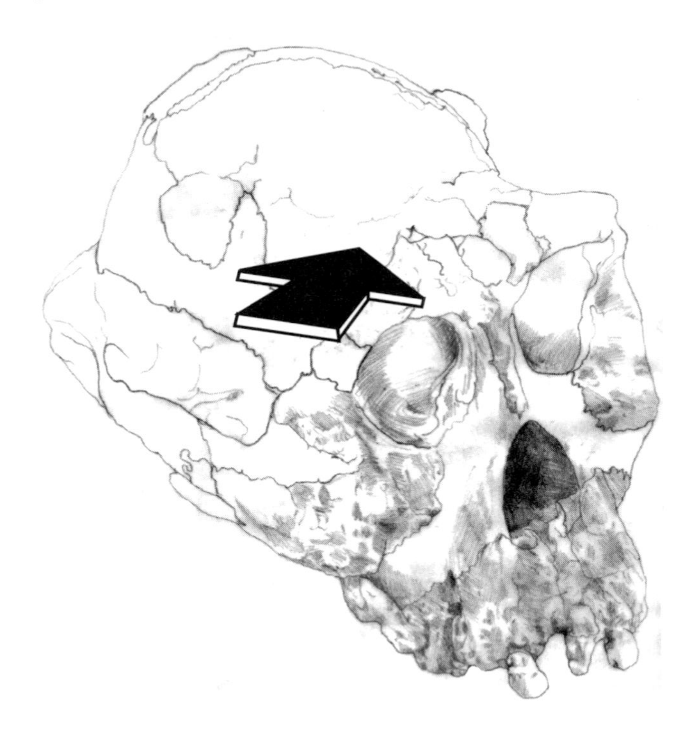

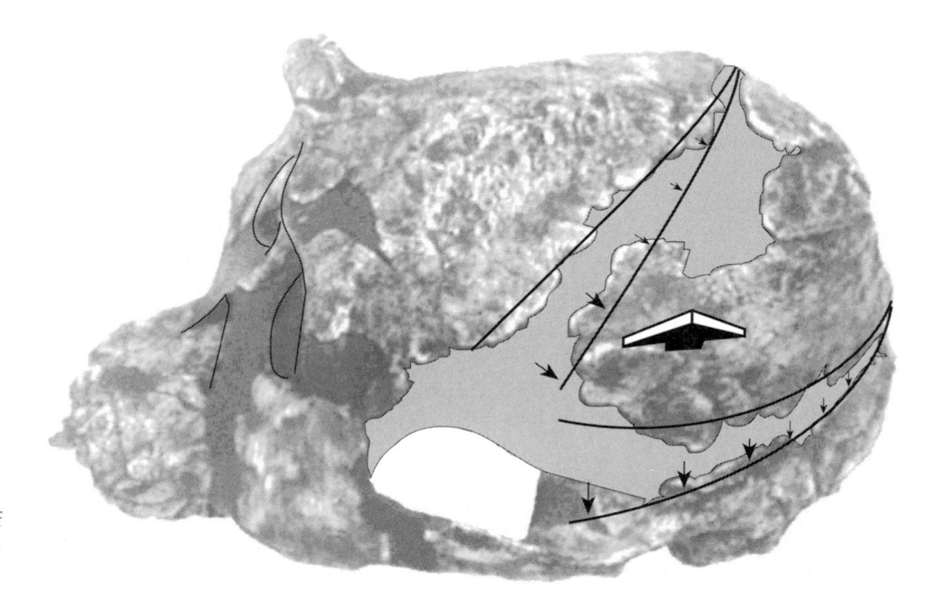

Figure 2.5 Postmortem displacement of cranial segments due to deformation. See text for details.

midline and the continuity of the laterally diverging temporal lines on each side of the occipital squama, and securing the contact between the parietal and the occipital bones. This positioning of the parietal is confirmed by the proper alignment of its squamosal sutural surface with the suture's continuation in the asterionic region of the left temporal bone. From a secure contact near lambda, an anteriorly widening gap (about 9 mm wide above porion) intervenes between the inferior margin of the parietal and the broken base of the temporal squama. This gap should be bridged by rotating the temporal medially about an anteroposterior axis, elevating the base of the squama to reach the parietal (rather than lowering the parietal). This correction will also have the positive effect of elevating the left porion exactly the amount needed to bring it to the correct level of porion on the undeformed right side (Figure 2.6).

2. As recovered, the isolated right parietal fragment was rotated 90° vertically and pushed forward against and fused with the frontoparietal segment. Our placement of the piece is reinforced by the following observations: (a) when cleaned and pivoted posteriorly to conform to the natural curvature of the calvaria, the prominent temporal crest and the sagittal sutural surface that it bears are aligned with the midsagittal plane; (b) the relative strength of the temporal crest in this position matches the temporal line's appearance at the same position on the opposite side; (c) the artificially deformed, flattened external arc of the fragment matches that of the right side of the frontoparietal segment, to which the fragment was

Figure 2.6 Coronal cross section of A.L. 444-2 calvaria on the biporion line. *Upper*: Due to deformation, the left porion was dropped by about 11 mm, opening up a crack in the lower part of the temporal squama (marked by the dotted line on left). *Lower*: Coronal outline of the calvaria after the effect of the deformation has been eliminated.

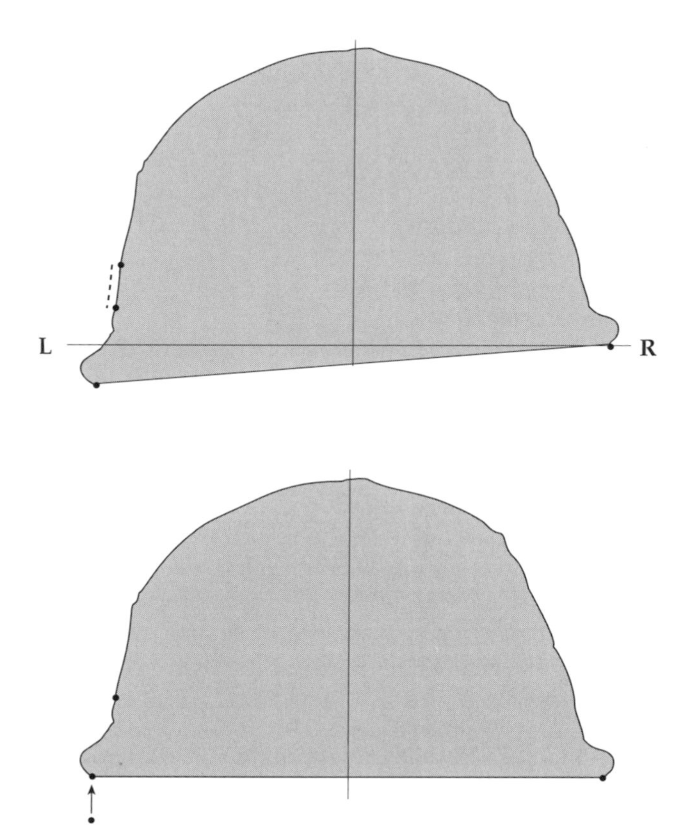

originally fused, whereas more posterior portions of the calvaria are unaffected by such deformation. The fragment is now separated ectocranially from the frontoparietal segment by a gap of about 5 mm, owing to the loss of intervening bone at the time of deformation. However, the process of "unhinging" the fragment from its vertically rotated position leaves no gap between its *endocranial* surface and that of the frontoparietal segment. The fragment has a 34-mm long articulation with the left parietal across the sagittal suture. This is of major importance as it permits the restoration of the entire sagittal arc of the calvaria.

3. The frontoparietal segment was aligned in the midline and oriented to approximate the natural contours of the calvaria provided by the parietal portions.
4. The zygomatic bone was aligned according to the positions of the zygomatic processes of the frontal and temporal bones. Although the zygomatic bone does not contact these processes, our placement of the bone is secured by the continuity of its orbital and temporal surfaces with those of the frontal. The position and orientation of the zygomatic bone are also constrained by two relationships that typify hominoid crania: the superior and inferior orbital margins are aligned on the same coronal plane when the cranium is oriented on the Frankfurt Horizontal (hereafter, FH), and the anterior origin of the masseter muscle (masseteric tubercle) and the lateral margin of the orbit lie in the same sagittal plane.
5. Several factors regulate the anteroposterior position of the maxilla, which lacks contacts with the zygomatic and frontal bones in the reconstruction. First, the coronal plane of the zygomatic process of the maxilla is aligned with that of the maxillary process of the zygomatic bone. Second, a vertical line from orbitale intersects the tooth row within the P^4 to M^1 interval, a relation that generally holds among the hominoids. The vertical position of the maxilla is controlled primarily by the need for continuity in the contour of the zygomaticoalveolar crests (established by the preserved roots of the maxillary zygomatic processes), which would be disrupted by movement of the maxilla more than about 5 mm inferiorly or superiorly. The maxilla cannot be in a significantly more inferior position, as an unnaturally tall nasal aperture and an exaggerated vertical distance between the back of the palate and the cranial base will result.
6. The nasal bone block was inserted into the nasal bridge of the reconstruction, although, as neither nasion nor rhinion is preserved, its vertical position is constrained only to within ±5 mm. For the width of the nasal bridge (i.e., the interorbital distance) we used the measurement (21 mm) provided by A.L. 438-1b, the frontal fragment associated with a large mandible and partial upper limb skeleton from the Kada Hadar Member (Kimbel et al., 1994). The contour and position of the medial parts of each orbital roof on this fragment match the preserved morphology of the A.L. 444-2 frontal bone.
7. The gap between the posterior two fragments of the right mandible corpus was eliminated on a cast by detaching the displaced posterior piece and properly realigning it with the more anterior segment along the fracture line. This eliminated the artificial gap between the M_1 and M_2 positions.

After all of the major fragments were assembled, plastic deformation was corrected on a cast. This was accomplished using the following method. First, parts of a mold corresponding to the undeformed areas of the cranium were coated with a rigid urethane casting resin (Smooth-On Co., C-1506) and permitted to set for 24 hours. Then, the parts of the mold corresponding to the deformed areas were coated and the mold was closed, allowing the newly coated areas to set for five hours. The mold was opened before the newly coated areas could cool; at this stage, taking advantage of the still malleable resin, we bent the deformed parts into place. The hardened, undeformed parts of the cast formed the rigid framework governing the corrections. The cast was then allowed to cure for 24 hours.

Plastic deformation was corrected by this method at three sites. The nuchal plane was slightly twisted around an axis approximating the biporion line (with no effect on the topography of the nuchal plane itself) to bring the anterior part of the left temporal into vertical alignment with the right. The exaggerated lateral flare and elevation of the anterior parts of the temporal bones' zygomatic processes were corrected: on the right by pushing the process medially to align it with the curve of the undeformed root of the process, and on the left by pushing it medially and twisting it to bring the superomedially directed inner surface to face the calvarial wall. Finally, the artificially elevated and flattened right frontoparietal segment (including the supraorbital torus) and the attached right parietal fragments were bent downward to conform to the contours of the left side.

In spite of our efforts, some original deformation can still be detected on the A.L. 444-2 reconstruction (Figure 2.7):

1. It was impossible to correct the skewedness of the maxilla, although if its left side is ignored, there is little apparent deformation. Owing to uncorrected

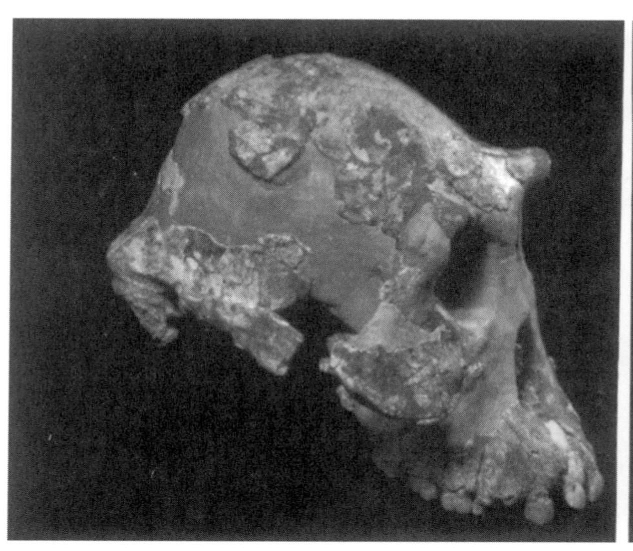
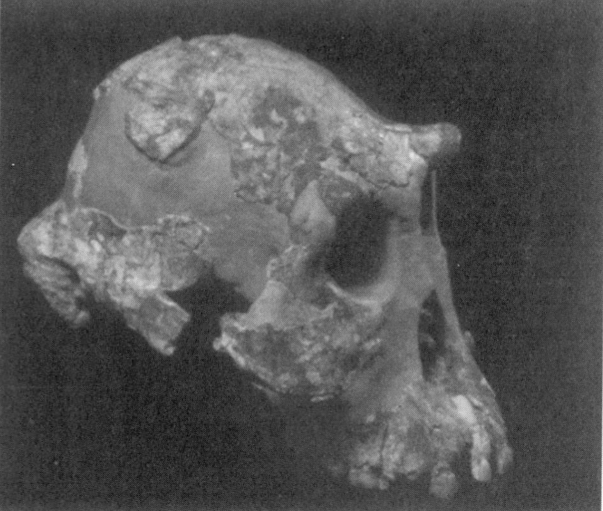

Figure 2.7 Stereoscopic photograph depicting final reconstruction of A.L. 444-2 cranium, in right oblique view (approximately one-third natural size). The deformation remaining on the right side of the cranium is evident in the superiorly displaced supraorbital element, the laterally deviating zygomatic arch, and the upwardly bent parietal segment.

compression in the midline, the measurements of palatal width are artificially narrow posterior to P^3: by 2 mm at the P^3/P^4 position to nearly 10 mm at M^2/M^3.
2. When the cranium is viewed from the superior aspect, a slightly more bulging profile of the calvarial wall is noted on the left side.
3. The right parietal fragments, even though restored in the reconstruction to their original positions, still show a bit of the original flattening of ectocranial curvature.
4. The left porion still resides approximately 8 mm inferior to the right porion; thus, based on our understanding of the vectors of the deforming forces, measurements that depend on the position of the FH should come from the undeformed right side of the cranium.

The final stage of the reconstruction process entailed the mirror-image flipping of the right zygomatic bone and right mandible corpus to the left side through computerized tomography and stereolithography (see Zollikofer et al., 1995). The left and right mandibular tooth rows were occluded with those of the deformed maxilla; in this orientation the mandible segments were joined. Although this yielded in the mandibular arch a mirror image of the distortion in the maxillary dental arcade, occlusion of the upper and lower teeth would have otherwise been impossible. With the now bilaterally complete and occluded dental arcades as a guide, ascending rami were carved of plaster following the contours of *A. afarensis* specimen A.L. 333-108 (White and Johanson, 1982; Kimbel et al., 1984). To accommodate the deformed alignment of the maxilla relative to the cranial base, the reconstructed rami (especially the left) deviate to the left of the midsagittal plane.

Ontogenetic Age and Sex of A.L. 444-2

Heavy wear on the occlusal surfaces of all preserved teeth, as well as the marked rugosity of masticatory muscle origin and insertion sites, leaves little doubt that the A.L. 444-2 skull belonged to an old adult individual. Size comparisons within the Hadar hominin sample show that A.L. 444-2 consistently occupies a position at the high end of the range of variation for skull and dental metrics. We take this as evidence of male status. Our finding of a statistically significant increase in hominin mandible size (and, by inference, overall skull size) in the time horizon with which the A.L. 444-2 skull is associated at Hadar (Lockwood et al., 2000) does not undermine our confidence in assigning the new skull to a male individual. See the descriptive and analytical sections in subsequent chapters for further details.

3

A.L. 444-2: The Skull as a Whole

The Cranium with the Occluded Mandible

Among the largest Plio-Pleistocene hominin skulls found to date, A.L. 444-2 is bigger, though not by much, than an average female gorilla's skull. At first glance, A.L. 444-2 assumes a somewhat simian appearance (Figure 3.1), the outcome of a relatively small braincase combined with an inclined frontal squama and prognathic jaws. However, this apelike appearance is offset by several distinctive hominin features: a very tall face that is much less prognathic than would be expected from the skull's general simian-like appearance; a deep, vertical mandibulosymphyseal profile; delicate supraorbital elements; and the absence of a supratoral sulcus intervening between the frontal squama and the forward-jutting supraorbital element. Nevertheless, the characteristics that account for the skull's hominin appearance demonstrate a certain uniqueness, which is manifested in the disproportion between the considerable total height of the face and the great size of its constituent elements (primarily the zygomatic and maxillary bones), on the one hand, and the delicateness of the supraorbital element and the almost negligible degree of its anterior projection, on the other. An apparent unevenness emerges along the vertical axis of the face between its upper portion—the orbits, including the elements above and between them—and its lower portion, that is, the elements below the level of orbitale down to gnathion. Undoubtedly, part of this appearance stems from the heavy, somewhat vertical, deep, and anteriorly bulbous symphyseal region of the mandible. The corresponding region in the African apes, in contrast, is transversely pinched, as its two sides converge downward toward the midline. Furthermore, the region slopes inferoposteriorly; in anterior view, it is tucked under the alveolar element and hence is less exposed than in A.L. 444-2.

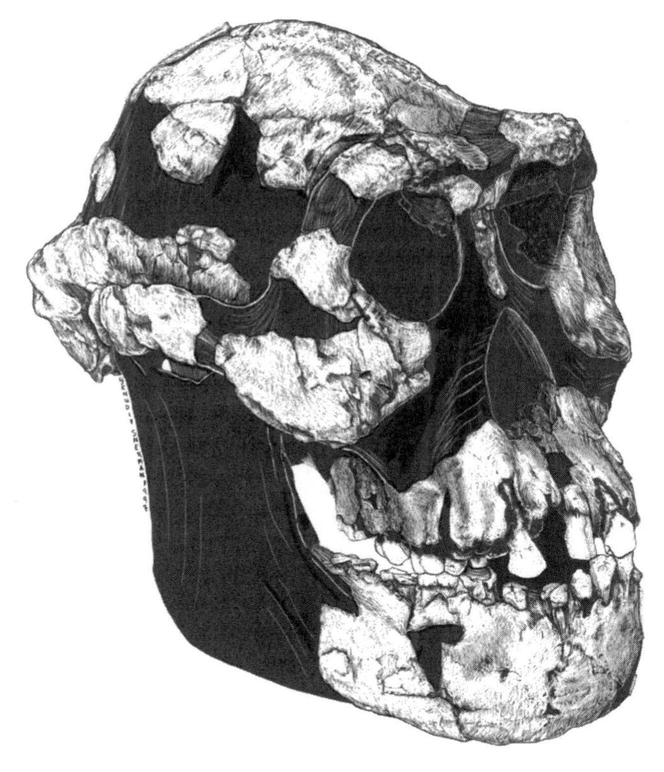

Figure 3.1 Artist's rendering of the reconstructed skull of A.L. 444-2, anterolateral view. See the picture gallery following page 122 for an enlarged view (75% natural size).

The preservation of the mandible of A.L. 444-2 and its occlusion with the upper dental arcade afford a unique opportunity to evaluate some of the characteristics of an entire *A. afarensis* skull. Two standard measurements can be recorded: the distance between gnathion and the estimated site of nasion—a measure of the total height of the face—which is 150 mm, and the distance between gnathion and basion, estimated at 157 mm (Figure 3.2 and Table 3.1). The nasion to basion distance is 105 mm. The gnathion to nasion and the gnathion to basion dimensions are substantially greater than in our sample of modern humans, where facial height measures 121 mm and the gnathion to basion distance is 107 mm. However, nasion to basion distance in modern humans, at 101 mm, is not much different from that of A.L. 444-2. The low values obtained from modern humans reflect both orthognathism and the proportionately small size of the human face. The differences between A.L. 444-2 and modern humans are even greater when the measurements are corrected for overall skull size (as represented by the biorbital breath; see Table 3.1). In other words, the area of nasion–gnathion–basion triangle is very small in modern humans, especially in relation to the total size of the skull.

Specimen A.L. 444-2 exhibits smaller facial height dimensions than male gorillas, which yield measurements of 179 mm from nasion to gnathion, 188 mm from gnathion to basion, and 142 mm from nasion to basion. The fossil is much more similar to our sample of female gorillas, which yield means of 140 mm, 144 mm, and 125 mm, respectively (Table 3.1.) (Nevertheless, note the discrepancy between nasion to basion measurements of A.L. 444-2 and the female gorillas. Indeed, the former exhibits a small nasion to basion distance. This disproportion influences the proportions of nasion–gnathion–basion triangle [see below].) Correction for skull size does not greatly affect these comparisons between the Hadar specimen and gorillas. Total facial height and the gnathion to basion distance in A.L. 444-2 are larger, both absolutely and relatively, than in our chimpanzee sample.

The considerable depth of the mandibular corpus in A.L. 444-2 combines with the steepness of its symphysis to place gnathion anteroinferior to its position in the apes. This position of gnathion in the Hadar specimen, along with the relatively short nasion to basion distance that apparently results from basion's anterior position, pro-

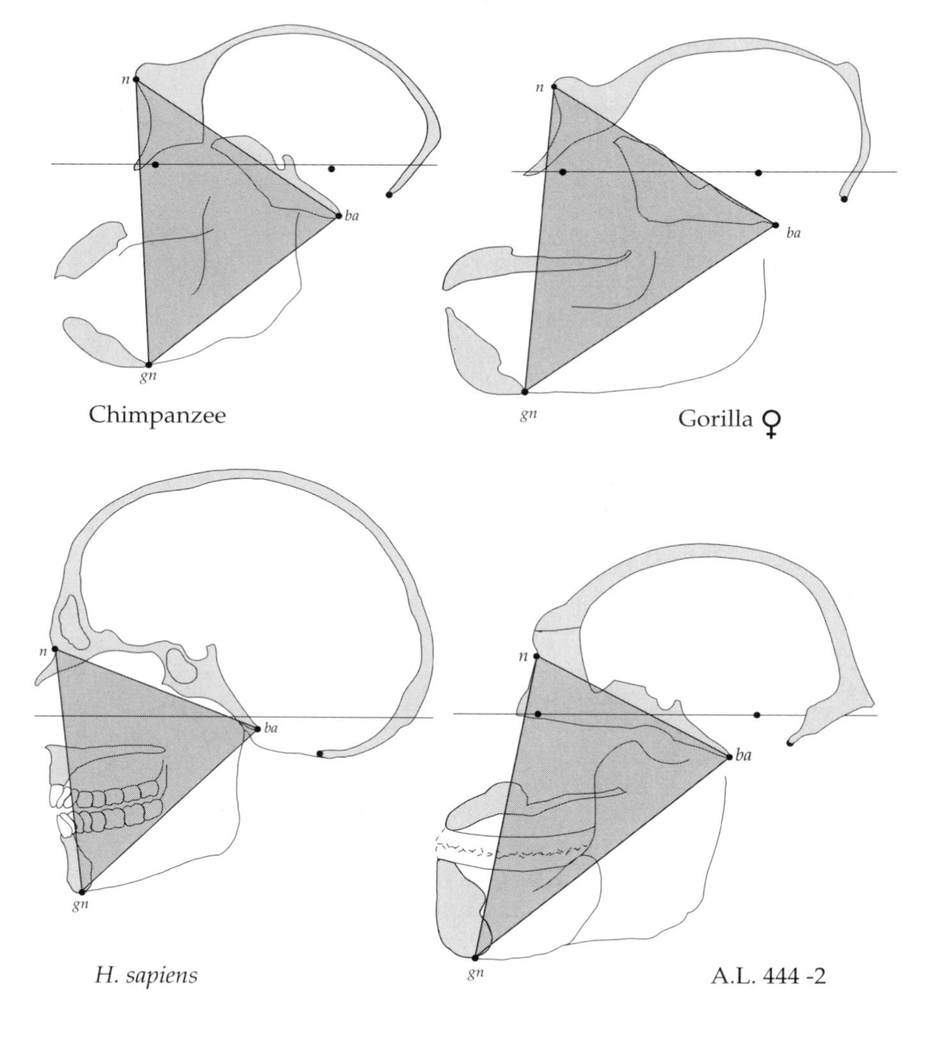

Figure 3.2 The facial triangle formed by nasion, basion, and gnathion. *Clockwise from left*: Midsagittal cross section of the skulls of a chimpanzee, female gorilla, A.L. 444-2, and modern *Homo sapiens*. The triangle in A.L. 444-2 is more skewed than the equilateral triangles in the other specimens.

Table 3.1 Measurements from the Face with Occluded Mandible

Specimen/ Sample		Nasion-Gnathion (50) (mm)	% of Biorbital Breadth	Gnathion–Basion (51) (mm)	% of Biorbital Breadth	Nasion–Basion (45) (mm)	% of Biorbital Breadth
Australopithecus afarensis							
A.L. 444-2		150	158	157	165	105	111
A.L. 417-1d		129	157	n/a	n/a	n/a	n/a
Homo sapiens	Mean (n = 10)	121	123	107	109	101	103
	Range	110–127	117–129	92–119	95–121	91–109	91–111
	SD	6	7	8	8	5	5
Gorilla gorilla female	Mean (n = 10)	140	136	144	140	125	121
	Range	130–146	124–142	135–149	131–147	111–142	106-138
	SD	4	3	4	3	10	9
Gorilla gorilla male	Mean (n = 10)	179	150	188	157	142	117
	Range	168-192	141-161	175-198	146-165	136-148	112-122
	SD	8	7	5	4	2	2
Pan troglodytes female	Mean (n = 10)	124	138	128	137	105	112
	Range	111–136	124–151	115–143	121–153	90–121	97–128
	SD	6	7	8	9	9	10
Pan troglodytes male	Mean (n = 10)	131	144	129	146	107	120
	Range	120–147	132–163	111–142	126–161	97–119	109–133
	SD	7	8	8	9	4	5

SD, standard deviation.
Numbers in parentheses in column headings refer to numbered measurements in Figures 3.7–3.9.

duces a skewed nasion–gnathion–basion triangle—that is, a triangle with unique proportions (Figure 3.2). In this triangle, the gnathion to nasion line is proportionately long and diagonal in relation to the FH. An approximately equilateral triangle is formed in modern humans, as well as in chimpanzees and female gorillas. (Male gorillas exhibit a tendency toward a somewhat skewed facial triangle.) Indeed, the discrepancy between the sides of the triangle in A.L. 444-2 is the greatest. The ratio between the shortest side (basion–nasion) and the longest one (gnathion–nasion) is the smallest in our entire sample. The basion–nasion side constitutes only 70% of the total facial height (gnathion–nasion) in the Hadar specimen, as opposed to a mean of 83% in modern humans; 79% and 89% in male and female gorillas, respectively; and 85% and 82% in male and female chimpanzees, respectively.

The similarities between facial triangles are not necessarily homologous, however; they may very well stem from a variety of factors. The anterior position of gnathion relative to the dental arcade (a configuration that results in a steep symphysis) in A.L. 444-2 not only contributes to the oblique orientation of the gnathion–nasion line but also reduces the horizontal distance between prosthion and this line. In the African apes, the posteriorly positioned gnathion (in addition to the degree of prognathism) increases the distance between this line and prosthion.

The mandible in A.L. 444-2 constitutes a greater percentage of facial height (orbitale to alveoloar plane) than in the great apes (Figure 3.3 and Table 3.2). Thus, the height of the corpus (measured in the coronal plane of orbitale) constitutes 67% of the facial height in the fossil.[1] The mean proportion in male chimpanzees is 54%; in females it is 57%; in male gorillas, 55%; and in female gorillas, 46%. In modern humans, the value is 78%. In modern humans, however, it is the proportionately diminutive upper face—as manifested by the small percentage of the biorbital distance that is occupied by the orbitale–alveolar plane distance—that dictates the relatively large index value (see the discussion in the section "Proportions of the Facial Mask").

The relative height of the mandibular corpus appears to be consistent in *A. afarensis*, as can be seen from measurements made on another, smaller (presumably female) specimen (A.L. 417-1) that consists of a maxilla and a mandible. In the latter specimen, the height of the corpus constitutes 69% of the orbitale–alveolar plane height. Thus, the height of the upper face and that of the mandibular corpus are slightly more disproportionate than in A.L. 444-2; this observation is of interest in light of the substantially smaller size of A.L. 417-1. In gorillas, which exhibit considerable sexual dimorphism, the situation is reversed. Corpus height in the female, constituting a mean of only 46% of the infraorbital facial height, is smaller than that of the male, in which it occupies 55% of that part of the face. Additional, independent, evidence of the great height of the mandibular corpus, as well as of its enormous breadth in *A. afarensis*, can be found in Table 3.3, where corpus

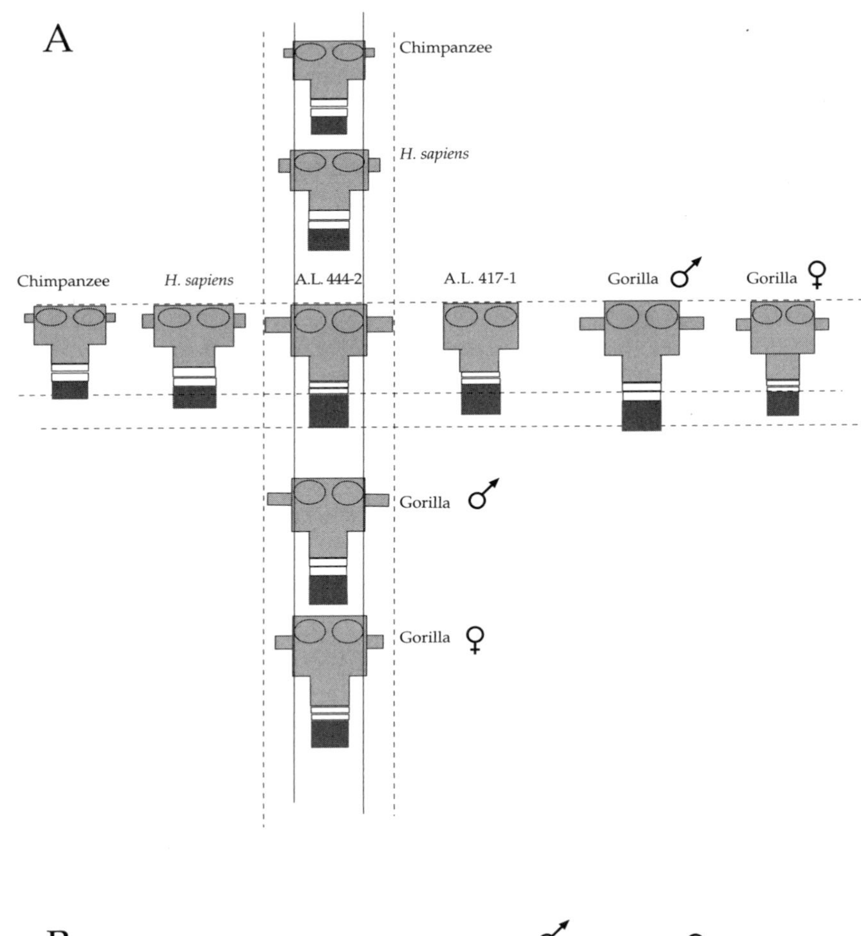

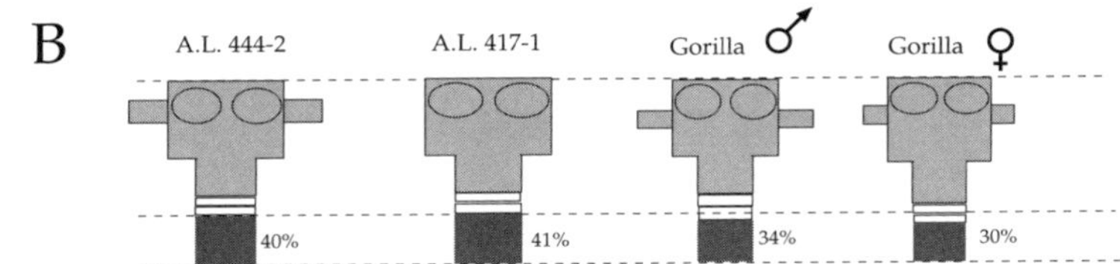

Figure 3.3 Proportions of the facial mask in A.L. 444-2 and other hominoids. (A) In the horizontal array the skulls are shown in proportion to A.L. 444-2's skull segments (dashed horizontal lines). In the vertical array the skulls are drawn to a fixed biorbital breadth dimension, represented by the two solid vertical lines. The two dashed vertical lines represent the maximum bizygomatic breadth of A.L. 444-2. (B) The skulls are drawn to the same total facial height, taken as the vertical distance between supraglabellar point and the inferior margin of the mandibular corpus in the coronal plane of the orbits. The distance between the lower two dashed lines represents the height of the A.L. 444-2 mandibular corpus (dark gray area).

height and breadth are compared with the orbital and biorbital breadth rather than with facial height.

It is clear, therefore, that because of its disproportionately deep corpus, the mandible of A.L. 444-2 contributes more to the total height of the face than does the mandible in any of the African apes. This is demonstrated in Figure 3.3, the lower graphic, where total facial height is drawn to the same scale for all specimens. Here, total facial height is measured from the supraglabella point to the lowest point on the corpus on the coronal plane of orbitale. Although its index of total facial height to width does not differ substantially from that of other hominoids in our sample (Table 3.2), the relatively tall face of most primates results from their modest upper facial breadth. In contrast, in A.L. 444-2, with its broad upper face, the relatively tall face is primarily the product of the very deep mandibular corpus (see the extent to which the zygomatic arch deviates laterally relative to the biorbital distance, as shown in Table 3.2 and in the vertical column of Figure 3.3). Thus, when total facial height is evaluated against the biorbital distance, A.L. 444-2 yields an index value that is high—176%—though not much different from that yielded by the narrower face of gorillas, precisely because the gorilla's relative corpus height is noticeably less than that of A.L. 444-2 and other *A. afarensis* specimens (Table 3.2 and the vertical column in Figure 3.3). In other words, it is only due to the deep corpus in the face in A.L. 444-2 that the facial height "overcomes" the great width to produce the appearance of a tall face.

Only two early hominin specimens in which the mandible occludes with the lower face—Sts. 52 (*A. africanus*)

Table 3.2 Measurements of Facial Height Including the Mandible

Specimen/Sample		Mandible Corpus Ht. at Coronal Plane of Orbitale (81) (mm)	Orbitoalveolar Ht. (58) (%)	Mandible Corpus Ht. (81)/ Orbitoalveolar Ht. (58) (%)	Total Facial Ht. (49) (mm)	Total Facial Ht. (49)/ Max. Bizygomatic Br. (22) (%)	Total Facial Ht. (49)/ Biorbital Br. (54) (%)
Australopithecus afarensis	Mean	40	59	68	150	100	169
A.L. 444-2		44	66	67	167	100	176
A.L. 417-1d		36	52	69	133	n/a	162
Homo sapiens	Mean (n = 10)	33	42	78	134	104	140
	Range	30–36	38–48	71–88	122–143	96–111	129–152
	SD	2	3	5	6	5	7
Gorilla gorilla female	Mean (n = 10)	32	61	46	158	107	167
	Range	21–41	51–69	31–61	150–169	97–120	154–179
	SD	7	3	10	3	3	4
Gorilla gorilla male	Mean (n = 10)	44	78	55	189	108	167
	Range	39–49	71–85	48–62	180–199	101–115	137–196
	SD	2	3	4	3	2	9
Pan troglodytes female	Mean (n = 10)	29	52	57	132	105	146
	Range	20–39	40–68	51–63	115–149	96–114	132–163
	SD	3	5	2	6	4	5
Pan troglodytes male	Mean (n = 10)	28	51	54	132	105	151
	Range	20–36	44–58	39–65	115–149	92–116	135–167
	SD	2	2	4	5	4	9
A. africanus							
Sts. 52[a]		26	52	50	n/a	n/a	n/a
A. robustus							
SK 12		40	61	66	n/a	n/a	n/a

Ht., height; Br., breadth; SD, standard deviation; n/a, not available.
[a]The M_3s in Sts. 52 are incompletely erupted.

and SK 12 (*A. robustus*)—permit the calculation of this ratio: 50% for Sts. 52 and 66% for SK 12. However, Sts. 52 is a subadult (its M3s are not yet in occlusion), and its corpus would likely have deepened to some degree with full adulthood. Nevertheless, a close approach to the index value for SK 12 is unlikely, and thus *A. afarensis* probably demonstrates a closer resemblance to the robust specimen than to the more generalized Sts. 52. Additional, fully adult mandibles and associated maxillae of various early hominin species are needed to confirm this.

The mandible's occlusion with the upper jaw reveals an additional rare morphological detail that sheds light on the angular relationship between the front of the palate and the mandibular symphysis (Figure 3.4). The almost horizontal orientation of the nasoalveolar clivus in A.L. 444-2 and other *A. afarensis* specimens renders the cross section of the clivus similar to that of the African apes. It forms an angle of 36° with the occlusal plane in A.L. 444-2 and 45° in A.L. 417-1, which is similar to the values in the great apes (Table 3.4). In contrast, the symphysis in the fossils is much taller and steeper than that of the apes, in which not only is the symphysis inclined but its lower part is also posteriorly extended as a shelf. The angle between the occlusal plane and the symphysis measures approximately 40° in the apes and 70° in A.L. 444-2 and A.L. 417-1. The unusual combination of a prognathic clivus and a steep symphysis in A.L. 444-2 results in an obtuse angle of 107° between these two elements. In A.L. 417-1 the angle is 115°.

In the African apes, the inclined symphysis and clivus form an acute angle that measures 83° in male and female chimpanzees and about 80° in the gorillas (Table 3.4); the occlusal plane almost exactly bisects this angle, producing angles of about 40°. The angle between the symphysis and clivus in our sample of modern humans is extremely obtuse, at 161°. Both arms of the angle are equally steep in reference to the occlusal plane and form angles of 82° and 79° with this plane.

Table 3.3 Relative Size of the Mandibular Corpus

Specimen/ Sample		Orbital Br. (68) (mm)	Mandible Corpus Ht. at Coronal Plane of Orbitale (81) (mm)	Corpus Ht. (81)/ Orbital Br. (68) (%)	Corpus Ht. (81) / Biorbital Br. (54) (%)	Mandible Corpus Br. at P4–M_1 (mm)	Corpus Br. at P4–M_1/ Orbital Br. (68) (%)	Corpus Br. at P4–M_1/ Biorbital Br. (54) (%)
Australopithecus afarensis	Mean	36	40	110	45	21	57	23
A.L. 444-2		40	44	110	46	23	57	24
A.L. 417-1d		34	36	106	43	17	50	20
A.L. 333 reconstruction		35	40	114	46	23	65	26
Homo sapiens	Mean (n = 10)	39	33	84	34	11	30	12
	Range	31–45	27–39	71–104	28–43	8–15	24–37	10–15
	SD	2	2	8	3	1	4	1
Gorilla gorilla female	Mean (n = 10)	40	33	82	33	17	43	17
	Range	35–44	27–38	61–101	25–43	13–21	32–54	11–23
	SD	2	3	12	4	2	6	2
Gorilla gorilla male	Mean (n = 10)	43	44	102	38	19	44	16
	Range	38–47	38–49	91–116	31–44	16–23	36–51	13–19
	SD	2	4	7	3	3	4	4
Pan troglodytes female	Mean (n = 10)	35	29	83	32	13	37	14
	Range	31-40	24-33	68-100	24-37	8-18	32-42	8-18
	SD	2	2	8	2	3	2	3
Pan troglodytes male	Mean (n = 10)	36	30	84	35	15	41	17
	Range	32–42	26–37	72–95	30–41	10–19	32–53	12–22
	SD	2	3	6	3	2	5	2

Ht., height; Br., breath; SD, standard deviation.

The Cranium: Lateral and Median Views

General Outline of the Cranium

A lateral view (Figures 3.5 and 3.6) of A.L. 444-2 reveals the outline of a relatively large cranium, with a distance from prosthion to opisthocranion, which coincides with inion, of 215 mm (Figure 3.7, measurement 42; Figures 3.7, 3.8, and 3.9 illustrate the measurements used in our study). The calvaria itself, from glabella to opisthocranion, is large, as well (167 mm) (Figure 3.7, measurement 1). These dimensions are substantially greater than those of our sample of chimpanzees and the well-known Sts. 5 cranium of *A. africanus*. The maximum cranial length in A.L. 444-2 is 119% of the length of that in Sts. 5, and the calvarial length is 114% of the value of that in Sts. 5. The

Figure 3.4 Midsagittal cross section of the mandibular symphysis and the nasoalveolar clivus. The horizontal line represents the occlusal plane.

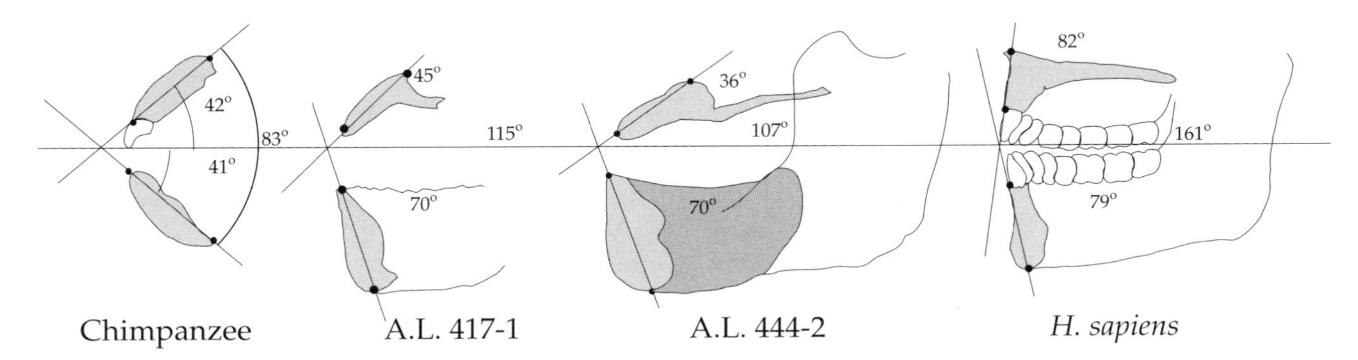

Table 3.4 Angulation of Nasoalveolar Clivus and Mandibular Symphysis (Degrees)

Specimen/Sample		Angle between Nasoalveolar Clivus and Symphysis	Angle between Clivus and Occlusal Plane	Angle between Symphysis and Occlusal Plane
Australopithecus afarensis	Mean	111	41	70
A.L. 444-2		107	36	70
A.L. 417-1d		115	45	70
Homo sapiens	Mean (n = 10)	161	82	79
	Range	138–186	63–99	68–92
	SD	12	11	7
Gorilla gorilla female	Mean (n = 10)	82	41	41
	Range	71–91	35–46	29–53
	SD	5	2	7
Gorilla gorilla male	Mean (n = 10)	77	37	41
	Range	67–86	27–51	33–48
	SD	5	7	3
Pan troglodytes female	Mean (n = 10)	83	40	43
	Range	73–91	31–49	35–49
	SD	4	4	3
Pan troglodytes male	Mean (n = 10)	83	43	40
	Range	71–94	33–52	31–49
	SD	5	4	5

general size of A.L. 444-2 approximates that of large female gorilla crania. Nevertheless, A.L. 444-2 and a female gorilla cranium are readily distinguishable because of the substantial differences in calvarial outline, facial size, degree of prognathism, location of inion, and the manner in which the face is hafted to the braincase.

The braincase of A.L. 444-2 displays a rounded profile that extends smoothly from glabella to inion. As a result of the calvaria's considerable length (which is much greater than its height), the midsagittal contour of the calotte is a long, flat arc that terminates quite abruptly at the posterior part of the braincase, where it plunges to-

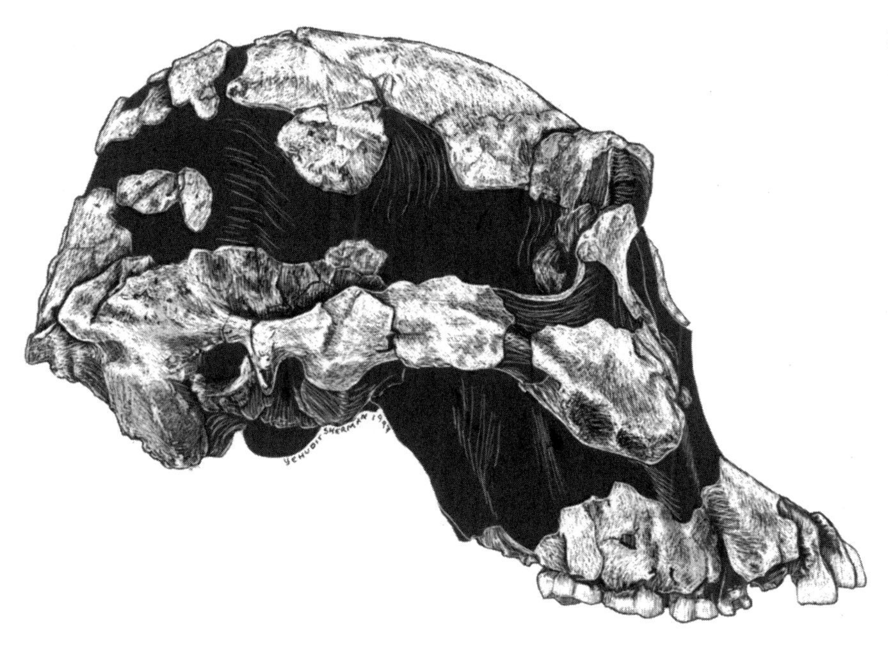

Figure 3.5 Artist's rendering of the reconstructed cranium of A.L. 444-2, right lateral view. See the picture gallery following page 122 for an enlarged view (75% natural size).

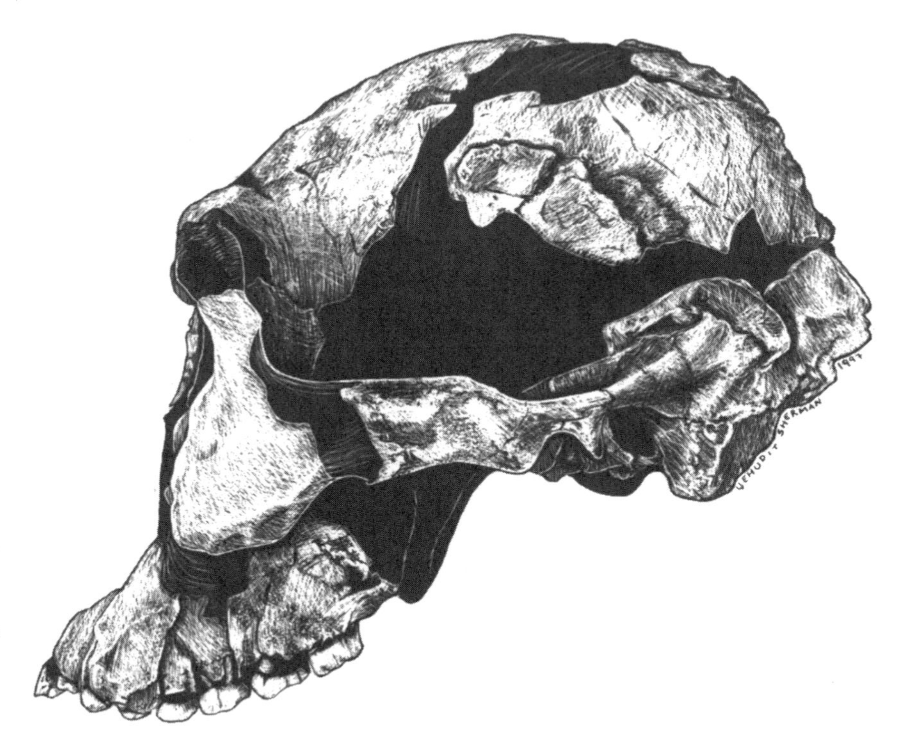

Figure 3.6 Artist's rendering of reconstructed cranium of A.L. 444-2, left lateral view. The left lateral view gives the impression that the skull is taller than does the right lateral view (Figure 3.5) because the left porion was artificially displaced inferiorly as a result of the crack at the base of the temporal squama. See the picture gallery following page 122 for enlarged view (75% natural size).

ward the occipital bone. There it turns into a straight, almost vertical, line that extends posteroinferiorly to inion. Because inion is situated so far below, close to the level of the FH (unlike the elevated inion in the great apes, for example), the course of this vertical section of the contour is quite long.

The calvaria's height to length ratio is smaller in A.L. 444-2 than that found in Sts. 5, for example (Table 3.5). Whereas in Sts. 5 the calvaria's height (with a vertical projected distance from basion to vertex of 102 mm; Figure 3.7, measurement 7) is 69% of its length, this ratio is 63% in A.L. 444-2. Two factors join to produce the high index in Sts. 5: its modest anteroposterior length, which is 12% smaller than that of A.L. 444-2, and its substantial vertical height, which is almost identical to that of A.L. 444-2 (only a 3% difference is observed). Smaller index values than those of A.L. 444-2 are found in *A. boisei*, two specimens of which yield a mean of 58%. Here the combination is reversed. On the one hand, the anteriorly protruding glabellar region adds to the length of the calvaria; on the other hand, the calvaria is low. A ratio of 59%, almost the same as in *A. boisei*, is calculated for *A. aethiopicus* cranium KNM-WT 17000.

Surprisingly, the index values for modern humans are not dramatically different from those found in Sts. 5. The mean index for modern humans is 75%. The similarity between Sts. 5 and modern humans is most likely the product of the different vertical positions of basion in these two species. Whereas in modern humans it lies close to the level of the FH, basion is located far beneath this line in Sts. 5. The outcome of these different positions is the relatively short vertical distance from basion to vertex in modern humans, compared to the distance in Sts. 5. (See the section "The Axis of the Reconstructed Cranial Base: The Inclination of Clivus".) Therefore, a more meaningful index of calvarial height relative to length is the ratio of the height of the vertex above the FH to the length of the calvaria (Table 3.5). Differences among fossil specimens and comparative samples highlight the relatively high calvaria of modern humans and the relatively low calvaria of *A. boisei* and *A. aethiopicus.* In A.L. 444-2 the index of this ratio is 50%, and in Sts. 5 it is 52%. The female gorilla sample yields an index of 47%, and the chimpanzees have 49%. Indeed, this index singles out modern humans, whose mean vertex height is 63% of the calvaria's length. With a mean index of 46%, the calvaria of *A. boisei* is quite low, but it is lower still in KNM-WT 17000 (44%).

The modest elevation of A.L. 444-2's midsagittal contour above the level of the orbit can also be expressed metrically by means of one of W. E. Le Gros Clark's (1950) well-known indices (Figure 3.10). The relevant index evaluates the portion of the calvarial height above the orbit and expresses it as a percentage of the total calvarial height (the height of the vertex).[2] In A.L. 444-2, 51% of the calvaria's height resides above the roof of the orbit, and this value is smaller than that in Sts. 5 (as can be ascertained through simple observation). In Sts. 5, the

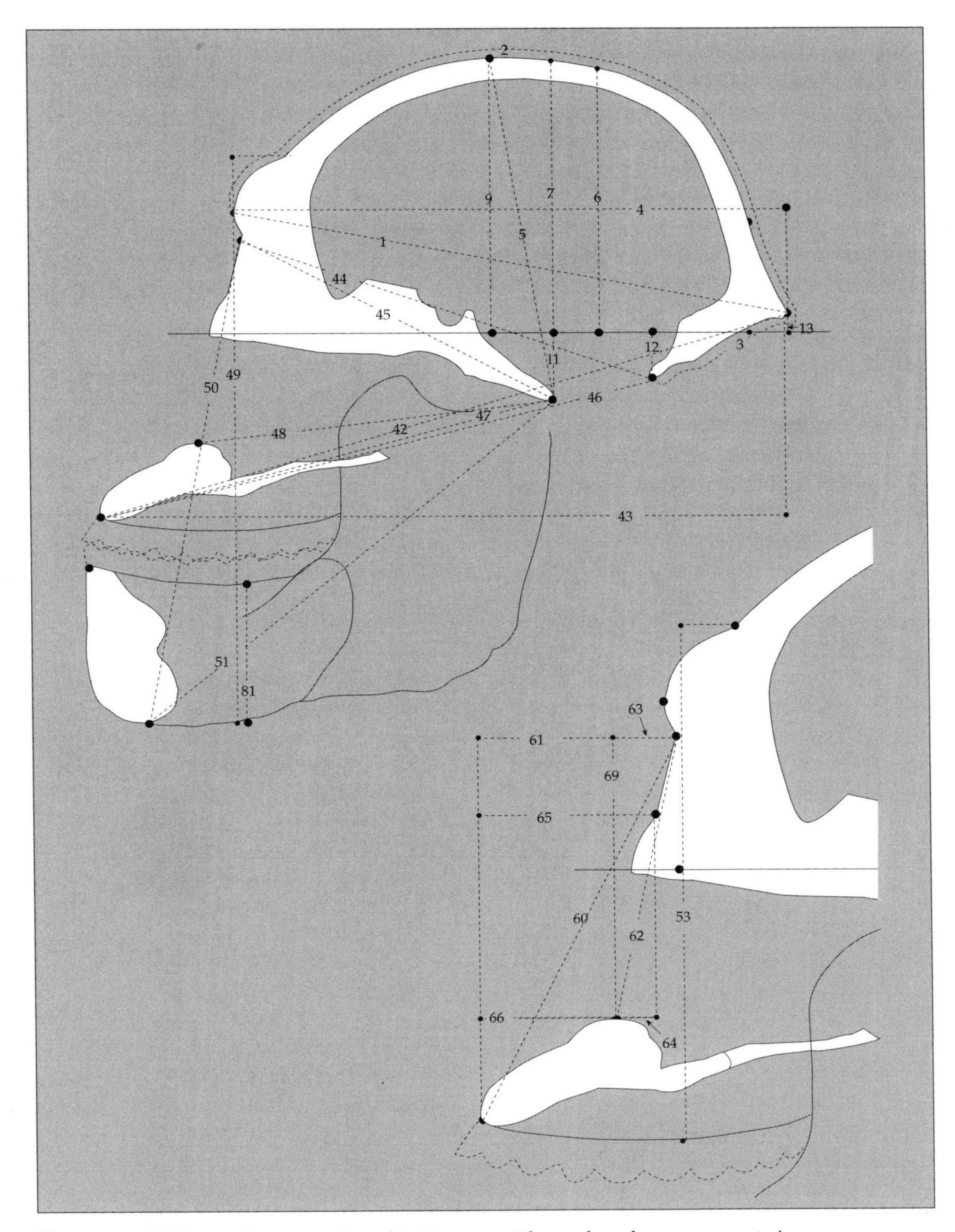

Figure 3.7 Midsagittal cross section of A.L. 444-2, with numbered measurements in the cranial midline. Measurement numbers refer to those defined in data tables.

index is 64%, which is even greater than the value of 59% yielded by the *H. habilis* specimen KNM-ER 1813.[3] The index in A.L. 444-2 is much smaller than that calculated for modern humans, where the mean is 70%, but it does not differ significantly from the female gorilla mean of 48% or the chimpanzee mean of 50%. Four *A. boisei* specimens provide a mean value of 50%; the index value for KNM-WT 17000 is again the lowest, at 46%.

As measured from glabella to opisthocranion, the midsagittal arc of the A.L. 444-2 calvaria is relatively large, at 235 mm. The chord between these points is, as mentioned, 167 mm. The arc/chord ratio, therefore, is 71%. As expected, this value, which indicates the relatively low magnitude of the calvaria's midsagittal curvature, is much greater in A.L. 444-2 than in modern humans, where it is 65%; however, it is almost iden-

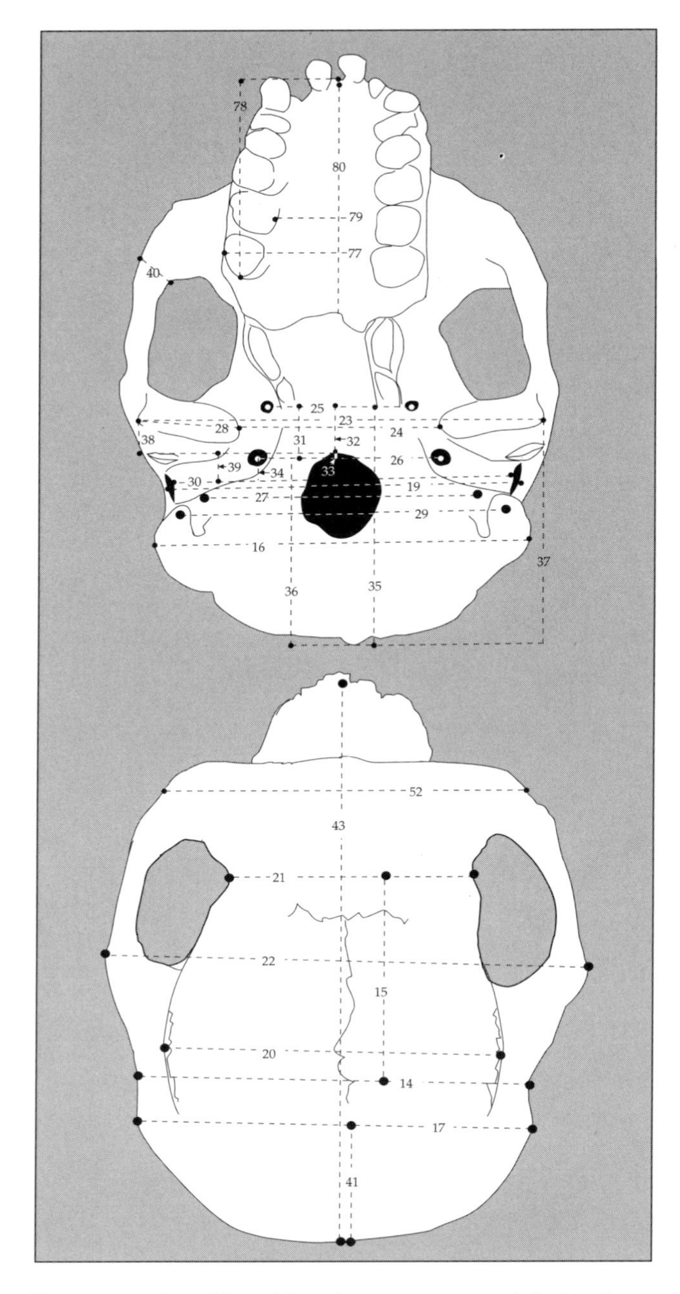

Figure 3.8 Breadth and length measurements of the basal (*top*) and superior (*bottom*) aspects of A.L. 444-2.

tical to the value measured in Sts. 5—72%—in spite of a substantial difference in size between the two skulls. The smaller-than-expected discrepancy between the values for the Hadar specimen and modern humans can be attributed to the position of opisthocranion in the latter, which does not coincide with that of inion. A separate point in modern humans, opisthocranion is located substantially superior to inion.

The arc/chord relationship with which we have compared A.L. 444-2 can be better represented graphically (Figure 3.11). By size-adjusting the maximum length of the calvaria in projection to that of A.L. 444-2, we are able not only to visually detect the differences in the index value but also to pinpoint their locations. In this manner we also eliminate the problem inherent in the varying position of opisthocranion from one hominoid species to the next.

Although these measurements and indices vary only slightly, the shapes they represent can differ substantially. The midsagittal outline of Sts. 5 forms part of a circle;

Figure 3.9 Breadth and height measurements of the facial (*top*) and posterior (*bottom*) aspects of A.L. 444-2.

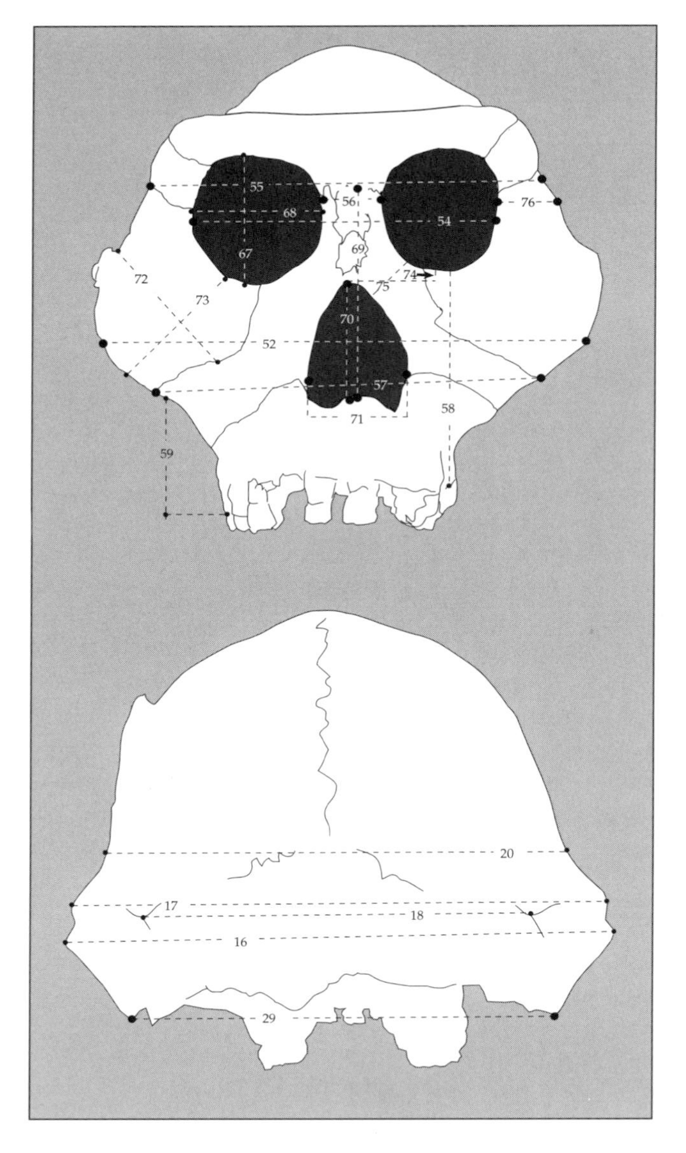

Table 3.5 Relative Height of the Calvaria (%)

Specimen/Sample		Height from Basion (7)/Length of Vault (1)	Height from FH (9)/Length of Vault (1)	Index of Vault Height above Orbits[a]
Australopithecus afarensis				
A.L. 444-2		63	50	51
Homo sapiens	Mean (n = 10)	75	63	70
	Range	70–81	57–71	66–73
	SD	3	4	3
Gorilla gorilla female[b]	Mean (n = 10)	64	47	48
	Range	57–71	42–52	45–51
	SD	4	2	2
Pan troglodytes female	Mean (n = 10)	64	49	51
	Range	56–69	44–55	42–60
	SD	3	3	4
Pan troglodytes male	Mean (n = 10)	64	49	49
	Range	56–75	42–58	39–58
	SD	5	4	6
A. africanus				
Sts. 5		69	52	64
H. habilis				
KNM-ER 1813		n/a	50	59
A. boisei	Mean	58	46	52
OH 5		55	45	52
KNM-ER 406		61	44	47
KNM-ER 13750		n/a	43	50
KNM-ER 23000		n/a	52	57
A. aethiopicus				
KNM-WT 17000		59	44	46

n/a, not available.

[a]Le Gros Clark's (1950) index FB/AB.

[b]Gorilla males were not measured due to the interference of the massive sagittal crest.

Figure 3.12 illustrates how this outline fits almost perfectly into the contour of the circle (for an explanation of the method, see the figure caption). However, the midsagittal outline of OH 5 consists of two straight lines with a short arc bridging them. One straight, relatively long segment extends from glabella to vertex (which is located on the posterior half of the calvarial arc), while the other segment descends steeply along the posterior end of the calvaria's midsagittal contour down to inion. A relatively short, curved segment at the top of the calvaria connects the two straight segments and fits into the circle. The rounded contour of the midsagittal outline of Sts. 5 is apparently a unique morphology that is consistent throughout *A. africanus*. It can be observed in the more fragmentary and deformed *A. africanus* crania Sts. 71 and Stw. 505, both of which are represented in Figure 3.12, and in the posterior section of MLD 37/38. Even the greatly deformed Stw. 13 hints at this morphology.

A similar midsagittal contour to that of OH 5 characterizes other *A. boisei* calvariae, although OH 5 stands out vis-à-vis the distinct segmentation of its contour. In KNM-ER 406, KNM-ER 732, KNM-ER 13750, and KNM-ER 23000, the glabellar region substantially deviates anteriorly from the outline of the circle, and the supraglabellar region is straight (and elevated relative to vertex). The posterior part of the calvaria in *A. boisei* deviates to different degrees from the circular outline, as well. In KNM-ER 406 and KNM-ER 23000, a rounded occipital contour follows the outline of the circle more faithfully than in OH 5, whose flatter occipital contour deviates the most toward the center of the circle.

In the African great apes, the outline of the braincase itself is rounded, forming an arc of an almost perfect circle. The contour of the supraorbital portion, however, extends sharply anterosuperiorly, deviating dramatically from the circular outline of the braincase. The contour of the supraorbital portion is clearly separated from the calvaria by a deep dip, which corresponds to the supratoral sulcus. In male gorillas, the paramedian section of the posterosuperiorly extended compound temporal/nuchal crest deviates notably from this rounded outline. Here, too, a marked depression in the contour

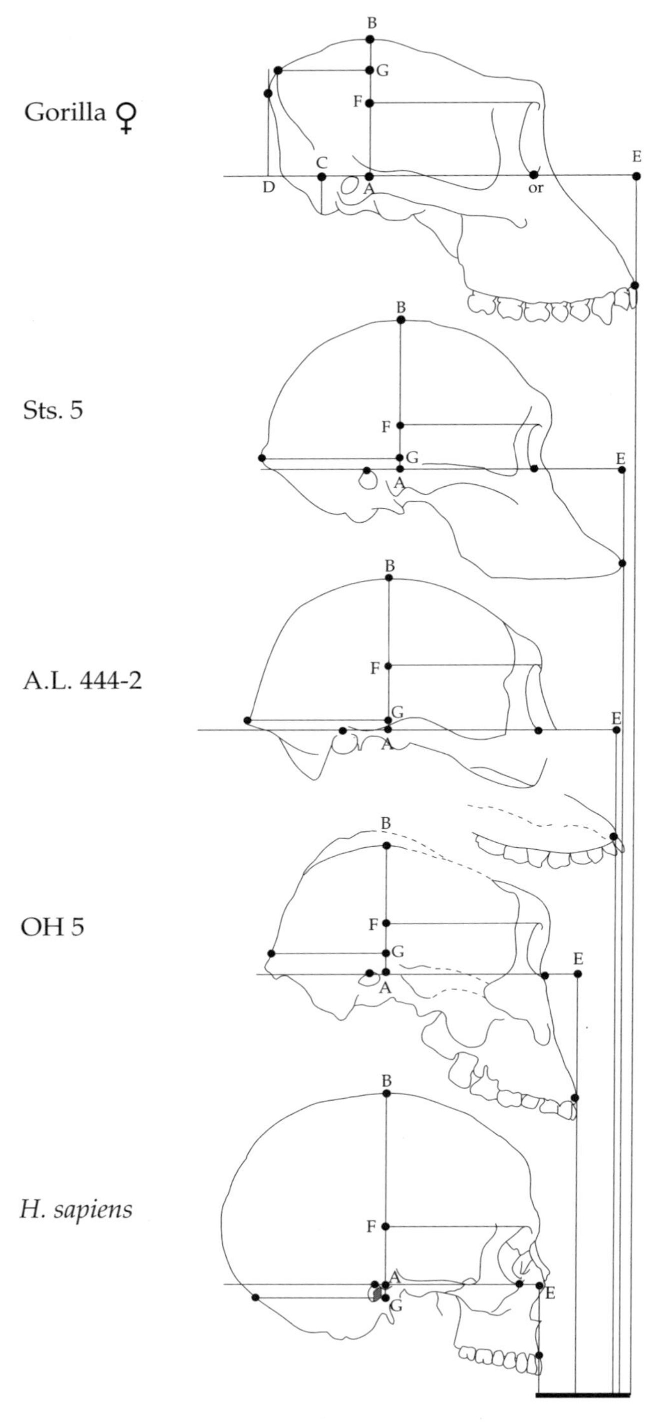

Figure 3.10 The cranial indices of W. E. Le Gros Clark. These indices express the height of the calvaria above the orbital roof (FB/AB) and the height of the nuchal arches relative to calvarial height (AG/AB). We have added the degree of prognathism, which expresses the horizontal distance to which prosthion extends anterior to sellion (indicated by the vertical line dropped from prosthion on each specimen). All crania are adjusted to the biorbital breadth of A.L. 444-2, and sellion is positioned on the same vertical line in all of the specimens. (Gorilla and OH 5 adapted from Tobias, 1967.)

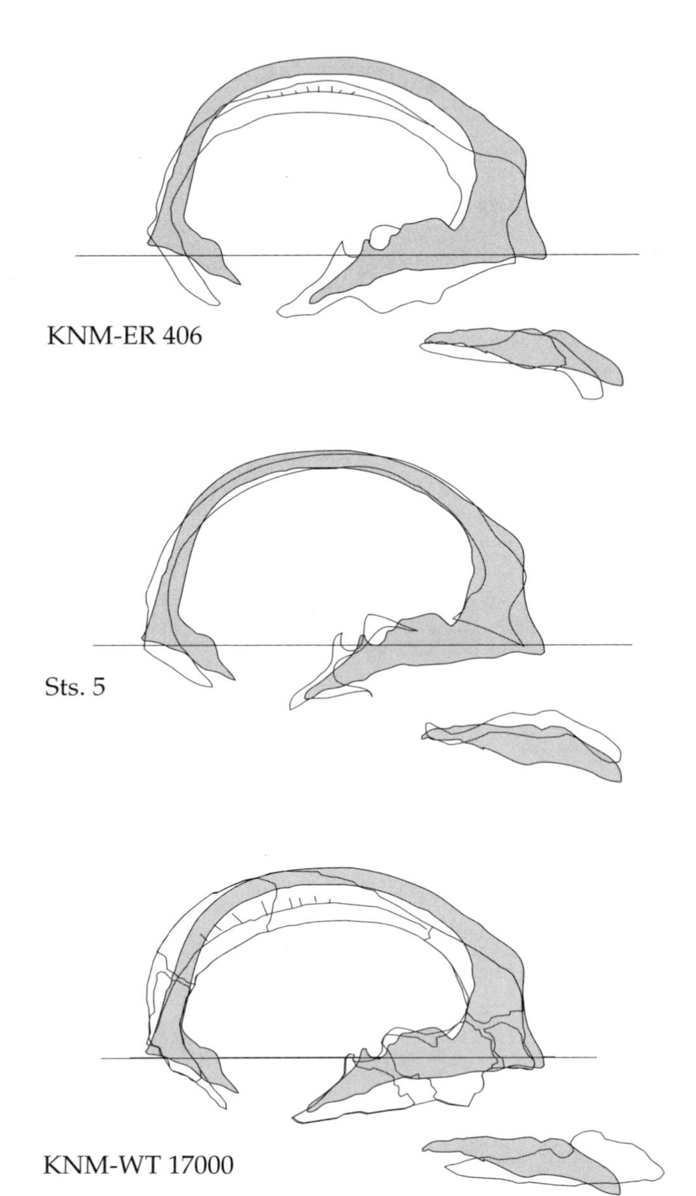

Figure 3.11 Calvarial height versus length in A.L. 444-2 and other hominins. Gray shading represents the midsagittal outline of A.L. 444-2. The crania are drawn to fit the glabella-opisthocranion distance of A.L. 444-2, and the Frankfurt Horizontal (FH) coincides in all of them. In comparison to A.L. 444-2, the KNM-ER 406 and KNM-WT 17000 calvariae are low, whereas that of Sts. 5 is as tall as the Hadar specimen's. (Cross sections of KNM-ER 406 and KNM-WT 17000 are based on CT images kindly made available to us by Dr. Fred Spoor.)

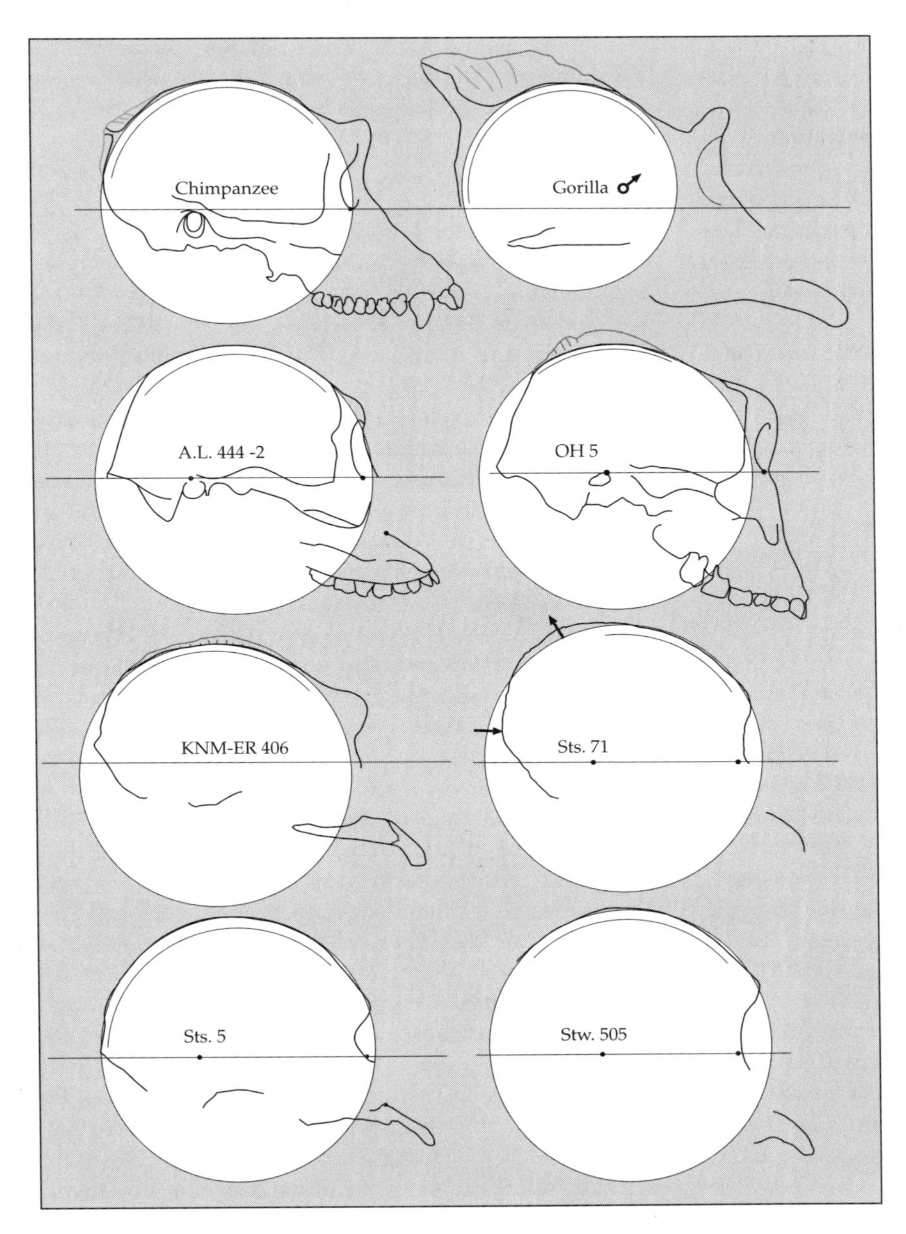

Figure 3.12 Position of the circular section of the midsagittal calvarial contour in A.L. 444-2 and other hominoids. The midsagittal contour of each calvaria is superimposed on a circle. The contours of the three *A. africanus* specimens fit the circle almost perfectly (the deviation seen in Sts. 71 is partly the outcome of postmortem deformation). The chimpanzee and male gorilla contours (as well as that of the female gorilla, not shown here) also follow the outline of the circle to a great extent, with only the exocranial structures—the supraorbital tori and the compound temporonuchal crests—deviating from it. In *A. boisei*, only a relatively small section of the contour fits the circle. This section is located posteriorly in relation to the calvaria's length; therefore, the anterior part of the midsagittal contour extends outside the circle, and the posterior part deviates toward the center of the circle. A.L. 444-2 is similar to the *A. boisei* configuration in this respect.

separates this element from the calvaria. The latter sulcus is usually situated higher topographically than the supratoral sulcus. This description is applicable primarily to male gorillas, whereas in female gorillas and chimpanzees of both sexes the posterior sulcus is weak or absent due to a much weaker nuchal crest.

Similar to Sts. 5, A.L. 444-2 has an almost completely circular calvarial contour. However, deviating from this rounded contour are both the occipital segment, which, straightening gradually as it descends, comes closer to the center of the circle, and the glabellar segment, which turns slightly outward (anterosuperiorly), away from the center of the circle. This deviation of the glabellar segment is significantly greater than the modest, almost nonexistent, deviation that marks the glabellar segment of the Sts. 5 and other *A. africanus* contours.

Whereas both the frontal and the parietal arcs contribute to the convex curvature of the calvarial outline in A.L. 444-2, the convex segment in OH 5 is primarily the product of the parietal's arc (compare with other *A. boisei* specimens, such as KNM-ER 406 and KNM-ER 13750). This configuration in *A. boisei* results from the concavity of the frontal squama's trigonum frontale. In both *A. boisei* and A.L. 444-2, the flattening of the occipital's midsagittal contour is related to the extent to which the ectocranial structures project posteroinferiorly, and to the size and orientation of the predominantly horizontal nuchal plane. In A.L. 444-2 and OH 5, the nuchal and occipital scales of the occipital bone meet to form an almost right angle, producing a sharp lip that marks the transition between one scale and the other. However, most *A. afarensis* and *A. boisei* specimens do not display this configuration. In those specimens, some of which are attributed to females and young adult individuals, the occipital midsagittal outline is rounded and the transition from one section of the bone (the occipital plane) to the other (the nuchal plane) is smooth and continuous (*A. afarensis*: A.L. 162-28, A.L. 224-9, A.L. 288-1a, A.L. 333-45, A.L. 439-1, A.L. 444-1, and KNM-ER 2602; *A. boisei*: KNM-ER 406, KNM-ER 407, KNM-ER 23000, L. 338y-6, and possibly Omo 323-1976-896). (See the section "Orientation of Segments in the Midsagittal Contour," including the discussion of the angular relationship between the different skull segments, and Chapter 5).

Position of Inion and Summit of the Arches of the Superior Nuchal Line in Relation to Calvarial Height

As in other hominins, but in contrast to the African apes, inion in A.L. 444-2 is low, close to the FH. It rises only 4.5 mm above this plane, and since the summit of the arch of the superior nuchal line lies 6 mm above the level of inion, this summit reaches a height of 10.5 mm above the FH. This value is similar to that of other early hominins, in which inion is always in the vicinity of the FH (Table 3.6).

To quantify the position of inion and the nuchal plane in relation to the height of the calvaria, we again use Le Gros Clark's system of indices (1950; see also Tobias, 1967) even though his measurements deal not with inion itself but with the summit of the superior nuchal line, which is usually above inion (Figure 3.10). Le Gros Clark expressed the height of this summit as a percentage of the total calvarial height above the FH. In A.L. 444-2, the index is 12%, which is similar to that of the other two specimens in our comparison: 11% in OH 5 and 10% in Sts. 5 (see Table 3.6 for additional data). A profound difference exists between A.L. 444-2 and other early hominins, on the one hand, and the African apes, on the other. A.L. 444-2 and the other early hominins show a much greater resemblance to modern humans than to the African apes. The index values in the latter are extremely high—female gorillas yield a mean index of 67%, while the mean value for males is as great as 108% (a value exceeding 100% indicates that the nuchal arches extend above the horizontal plane of vertex). In chimpanzees the values are lower but still far above those yielded by A.L. 444-2 and other early hominins: 47% the mean for the female chimpanzee sample and 51% the mean for the males. The low value in A.L. 444-2 conforms to the very acute angle of the nuchal plane in this specimen (as discussed in the section "Inclination of the Lower Occipital Scale: Inion–Opisthion"). In modern humans, as just mentioned, the values are much smaller, averaging −2% in our sample. The negative value indicates that the summits of the nuchal arches lie below the FH.

Proportions of the Calvaria's Segments

The arc length of the midsagittal contour of the A.L. 444-2 calvaria, from glabella to inion, is 236 mm and includes the following segments: from glabella to bregma (gl–br), 101 mm; from bregma to lambda (br–la), 106 mm; and from lambda to inion (la–i), 29 mm. Evaluating the arc length of the different segments (Figure 3.13; Tables 3.7 and 3.8) reveals that the parietal arc (br–la) constitutes the longest arc segment of the three, 45% of the total arc length of the calvaria. This proportion is larger than in any hominoid calvaria in our comparative sample. The parietal arc achieves this relative dimension primarily because of the short superior scale of the occipital squama (la–i), which comprises only 12% of the total arc. At 14%, the latter segment is also very short in chimpanzees. (Because of early suture fusion and crest formation in the rear of the calvaria, it is usually not possible to define these segments in gorilla crania.) However, the short upper occipital scale in chimpanzees does not result in a relatively long parietal arc, which is only 39% of the calvarial arc length in our sample. Instead, it is the relatively long frontal arc, the longest of all the segments (48% of the total) that influences these relationships. In light of the similar lambda–inion segment length in chimpanzees and A.L. 444-2, we note an essential difference: in the latter, the reduction of this segment's length stems from the low position of lambda, whereas in chimpanzees, the reduction results from the ascent of inion as part of the elevated nuchal crest.[4]

The parietal arc in Sts. 5 is intermediate between the short arc seen in chimpanzees and the long one in A.L. 444-2. In Sts. 5, the frontal arc constitutes only 38% of the total arc, but the upper scale of the occipital is relatively long, at 19%. The arrangement of calvarial arcs in Sts. 5 is essentially that seen in KNM-ER 1813 (Wood, 1991a) and is also present in other hominins. Note that the dramatically larger brain size in modern humans does not affect these relationships.

When the inion–opisthion segment is included in a measurement of the calvaria's arc length in the Hadar skull, the figure reaches 279 mm; however, inclusion of this segment does not greatly affect the comparative data (Figure 3.13 and Table 3.7). Specimen A.L. 444-2 still emerges with the proportionately shortest lambda–opisthion arc, which constitutes 25% of the total arc and is even shorter than that of chimpanzees (33%; see Table 3.8). These results are somewhat surprising in light of the fact that the distance between the posteriorly situated foramen magnum and the center of the cranial base in

Table 3.6 Height of the Nuchal Area

Specimen/Sample		Height of Inion Above FH[a] (13) (mm)	Nuchal Area Above FH (mm)	Height of Vertex (9) (mm)	Nuchal Area Height Index (%)[b]
Australopithecus afarensis	Mean	6	11	81	13
A.L. 444-2		5	11	86	12
A.L. 333 reconstruction		6	10	76	13
Homo sapiens	Mean (n = 10)	–8	–3	115	–2
	Range	–21–1	–8–3	109–127	–7–3
	SD	8	4	6	3
Gorilla gorilla female	Mean (n = 10)	48	48	72	67
	Range	40–56	40–56	68–75	56–82
	SD	7	7	3	11
Gorilla gorilla male	Mean (n = 10)	81	81	76	108
	Range	76–88	76–88	70–82	101–114
	SD	5	5	6	5
Pan troglodytes female	Mean (n = 10)	27	35	67	47
	Range	17–35	20–40	63–72	30–60
	SD	7	6	3	10
Pan troglodytes male	Mean (n = 10)	30	38	69	51
	Range	20–42	30–48	61–81	37–69
	SD	7	6	6	10
A. africanus	Mean	5	9	72.5	13
Sts. 5		3	7	73	10
Sts. 71		(11)	15	72	(20)
MLD 37/38		1	6	n/a	8
H. habilis					
KNM ER 1813		–6	–3	74	–4
A. boisei	Mean	–1	6	74	8
OH 5		–4	8	77	11
KNM-ER 406		0	5	71	7
KNM-ER 13750		–2	–2	71	–3
KNM-ER 23000		–2	10	82	12
KNM-ER 732		4	9	71	12
A. aethiopicus					
KNM-WT 17000		9	9	66	14

[a]Negative values indicate that inion is below the FH. Values in parentheses are estimates.
[b]Le Gros Clark's (1950) index AG/AB.

chimpanzees is greater than in hominins. Thus, the effect of the very short lambda–inion arc in A.L. 444-2 appears to be the determining factor in the modest length of this specimen's occipital arc (see Chapter 5). The ascent of inion in the apes overrides the shortening effect of the posterior position of opisthion and lends the nuchal scale length greater weight in calculating the proportions.

Orientation of Segments in the Midsagittal Contour

Inclination of the Frontal Chord

Because of the great morphological differences between the hominin and the African ape supraorbital region, the orientation of the anterior segment of the calvarial contour (nasion–bregma chord) in these groups does not readily lend itself to comparison. The varying vertical and horizontal position of nasion, the difference in the degree of anterior protrusion of the glabella, and the inconsistent presence of the supratoral sulcus necessitate substituting the supraglabella point for nasion in our measurements (Figure 3.14). Thus, the chord connecting bregma to supraglabella permits us to evaluate the inclination of the frontal squama, though it must kept in mind that the great topographical differences between the frontal of the great apes and that of the hominins—the presence of a deep supratoral sulcus in the apes—carries the bone surface in the latter far below the frontal chord.

In A.L. 444-2 (Figure 3.15) the angle between the frontal chord (as defined above) and the FH is relatively small at 27°, as compared to 44° in modern humans and 36° in

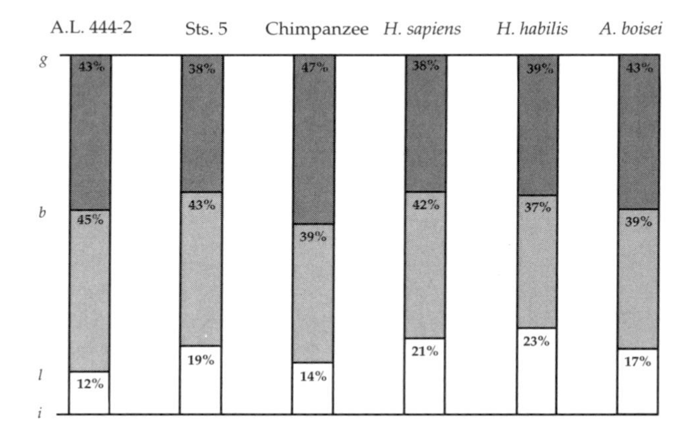

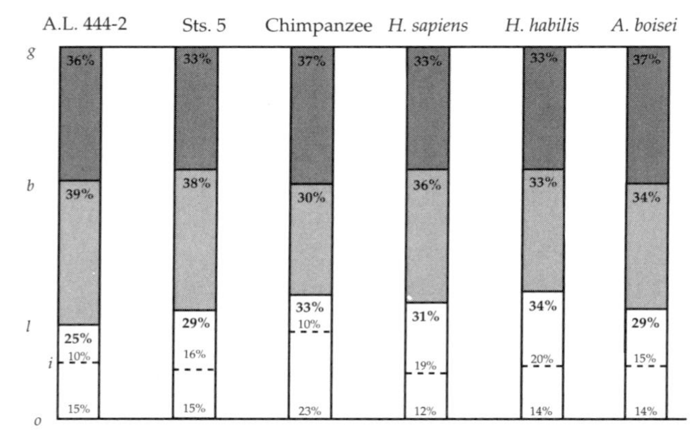

Figure 3.13 Proportions of the midsagittal arc segments of the calvaria in A.L. 444-2 and other hominoids. The midsagittal arc segments are shown in proportion to the total arc of the calvaria from glabella to inion (*upper*) and from glabella to opisthion (*lower*).

Sts. 5. The orientation of this chord in Sts. 5 appears to be somewhat steeper than what is seen in the less complete specimens Sts. 71 and Stw. 505 (Table 3.9). Its relatively high value in *A. africanus* (relative to A.L. 444-2) could readily have been inferred from Le Gros Clark's system of indices and from the almost complete overlap between the midsagittal cranial contour and the circle depicted in Figure 3.12. KNM-ER 1813, the sole representative of *H. habilis* in our comparative sample, exhibits a similar inclination, at 34°, to that of Sts. 5. Nevertheless, the lower value seen in A.L. 444-2 is significantly higher than that of our sample of female gorillas (mean 19°) but not far from that of chimpanzees (mean 24°). The mean value for four *A. boisei* specimens (OH 5, KNM-ER 406, KNM-ER 13750, and KNM-ER 23000) is 25°, which is essentially the same as the KNM-WT 17000 value and does not differ significantly from that of A.L. 444-2 (Table 3.9).

The combination of a relatively large frontal angle and a curved frontal arc that deviates substantially from the chord (a small chord–arc ratio) will result in a more anterior position of the frontal arc's summit relative to the calvarial length. Similarly, a small frontal angle and a flat arc are expected to result in a more posterior position of the frontal arc's summit relative to the calvarial length. In modern humans, for example, the coronal plane of the frontal arc's summit is close to the plane of the glabella (Figure 3.15). The horizontal distance between the coronal plane of the frontal arc's summit and that of the glabella constitutes only 12% of the total length of the calvaria, in contrast to 27% in A.L. 444-2. As expected, this distance constitutes a smaller segment (21%) of the length in Sts. 5, as is the case in KNM-ER 1813. However, because of the more complex topography in apes, these measurements are not comparable to those yielded by hominins. Nevertheless, female gorillas exhibit a mean value of 38%, similar to that found in male and female chimpanzees. In general, the flatter the frontal arc, the closer the summit comes to bregma. The summit is the farthest from bregma in modern humans. It falls at about the halfway point in Sts. 5 and KNM-ER 1813 and is closer to bregma in A.L. 444-2. In *A. robustus* and *A. boisei* the frontal squama is usually completely flat or concave; hence, the frontal arc's "summit" is a moot point here since the arc often mimics the chord.

Inclination of the Parietal Chord

The parietal chord (br–la) in A.L. 444-2 forms an angle of 35° with the FH (Figure 3.15). This measurement is in sharp contrast to the 14° angle that the chord forms in female gorillas and the 21° and 20° angles in male and female chimpanzees, respectively. In the apes, the low value stems primarily from the relatively elevated position of lambda, which is situated close to the vertical level of bregma. In A.L. 444-2, it is the low position of lambda (relatively closer to the level of the FH) that creates the larger angle of the parietal chord (and the great length of the parietal arc, as discussed). Although in modern humans the angular value (29°) is similar to that of A.L. 444-2, here it is achieved not by a low position of lambda (which in humans is relatively high in reference to the FH) but by a very high position of bregma. In contrast, the acute angle measured on KNM-ER 1813 (25°) and Sts. 5 (25°) is an outcome of a very high position of lambda relative to bregma. In *A. boisei*, the mean angle of four specimens (OH 5, KNM-ER 406, KNM-ER 13750, and KNM-ER 23000) is 27°, which, as in A.L. 444-2, is the product of the relatively low position of lambda.

An unusual configuration can be found in KNM-WT 17000. The very high inclination value (38°) stems from the very low position of lambda, which approximates the level of the FH. Lambda is so low in this specimen that it overrides the effect of the low position of bregma in determining the angular value.

Table 3.7 Length of the Vault Arc Segments (mm)

Specimen/Sample[a]		Glabella–Opisthion	Glabella–Inion	Glabella–Bregma	Bregma–Lambda	Lambda–Opisthion	Lambda–Inion	Inion–Opisthion
Australopithecus afarensis								
A.L. 444-2		279	236	101	106	70	29	41
Homo sapiens	Mean (*n* = 10)	364	318	119	132	112	67	46
	Range	331–385	290–342	111–130	106–154	105–120	55–78	35–53
	SD	18	18	6	13	5	7	6
Pan troglodytes female	Mean (*n* = 10)	210	161	82	61	70	20	50
	Range	193–221	150–171	72–89	55–65	57–85	14–25	43–60
	SD	12	9	7	5	12	5	7
Pan troglodytes male	Mean (*n* = 10)	212	166	75	68	68	23	46
	Range	203–219	160–177	69–82	64–73	60–77	21–26	34–56
	SD	6	7	5	3	6	2	7
A. africanus								
Sts. 5		242	212	80	92	70	39	31
H. habilis								
KNM-ER 1813		254	219	85	82	86	51	35
A. boisei	Mean	264	224	94	87	76	38	38
OH 5		280	234	n/a	n/a	83	37	46
KNM-ER 406		247	214	92	86	67	34	34
KNM-ER 13750		n/a	217	95	83	n/a	37	n/a
KNM-ER 23000		264	231	95	91	78	45	33

n/a, not available.

[a]Gorillas were not measured due to interference of the massive cranial crests.

Table 3.8 Contribution of the Arc Segments to Arc Length of the Vault (%)

Specimen/Sample[a]		Frontal Arc Glabella–Bregma (1)[b]	Frontal Arc Glabella–Bregma (2)	Parietal Arc Bregma–Lambda (1)	Parietal Arc Bregma–Lambda (2)	Occipital Arc Lambda–Opisthion (1)	Occipital Upper Scale Arc Lambda–Inion (1)	Occiptal Upper Scale Arc Lambda–Inion (2)	Occipital Lower Scale Arc Inion–Opisthion (1)
A. afarensis									
A.L 444-2		36	43	39	45	25	10	12	15
Homo sapiens	Mean (*n* = 10)	33	38	36	42	31	19	21	12
	Range	31–35	36–40	32–40	37–46	28–33	15–21	17–24	9–15
	SD	1	2	2	2	2	2	2	2
P. troglodytes female	Mean (*n* = 10)	39	51	29	38	33	9	13	23
	Range	37–42	48–55	25–33	34–43	29–38	7–11	9–13	21–27
	SD	2	3	3	4	4	2	3	3
P. troglodytes male	Mean (*n* = 10)	35	45	31	41	33	11	14	22
	Range	33–38	43–48	32–34	38–43	30–37	8–14	13–15	21–25
	SD	2	2	2	2	2	2	1	2
A. africanus									
Sts. 5		33	38	38	43	29	16	19	15
H. habilis									
KNM-ER 1813		33	39	33	37	34	20	23	14
A. boisei	Mean	37	43	35	39	29	15	17	14
OH 5		n/a	n/a	n/a	n/a	30	13	16	16
KNM-ER 406		37	43	35	40	27	14	16	14
KNM-ER 13750		n/a	44	n/a	38	n/a	n/a	17	n/a
KNM-ER 23000		36	41	34	39	30	17	19	13

n/a, not available.

[a]Gorillas were not measured due to interference of the massive cranial crests.

[b](1) = value relative to total arc (g–o); (2) = value relative to g–i arc.

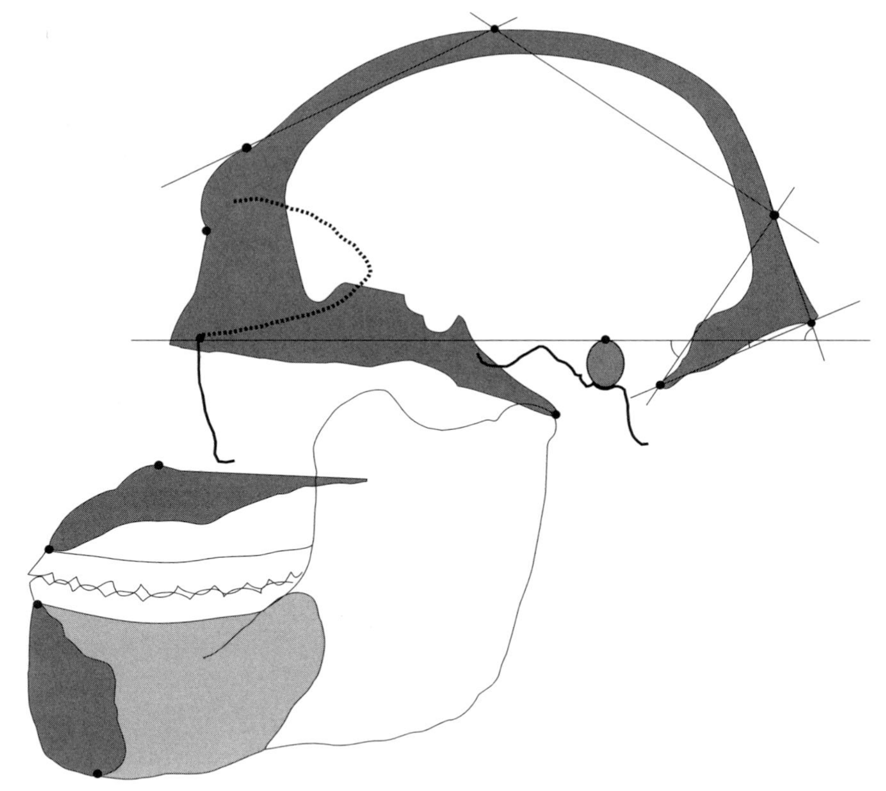

Figure 3.14 Midsagittal cross section of A.L. 444-2, displaying the major craniometric landmarks and the inclination of the major calvarial chords. The heavy dashed line represents the sagittal contour of the orbital cone in the sagittal plane of orbitale. The dark solid line below this represents the sagittal contour of the infraorbital region in the sagittal plane of orbitale. The dark solid line below the external auditory meatus (shaded oval) represents the sagittal cross section through the glenoid region at the mid-mandibular fossa level; the anterior part of this cross section indicates the outline of the articular eminence, and the posterior part indicates the anterior face of the pars mastoidea.

From drawings of the midsagittal cross sections (Figure 3.15) one can see that in modern humans, the frontal arc rises farther above the corresponding chord than the parietal arc does (in other words, the frontal arc has a greater curvature), whereas in A.L. 444-2 the situation is reversed: the parietal arc is more curved than the frontal arc. A more extreme manifestation of the latter relationship is found in *A. boisei.* In Sts. 5 (as well as in KNM-ER 1813) these two arcs seem to have a similar degree of curvature; thus, the midsagittal contour matches the contour of a circle (Figure 3.12), as mentioned earlier.

Inclination of the Occipital Sagittal Chord (Lambda–Opisthion)

The occipital sagittal chord forms a 122° angle (open anteriorly) with the FH in A.L. 444-2 (Figures 3.14 and 3.15). In modern humans, the angle is identical. However, in female gorillas, a mean angle of 94° is observed; in female chimpanzees, it is 98°, and in the males, it is 96°. The discrepancy between these apes and A.L. 444-2 and, indeed, between apes and hominins in general, is due to the posterior position of the apes' foramen magnum (i.e., opisthion), which dictates the steeper inclination of the chord. (See the discussion in the section "Anteroposterior Position of the Foramen Magnum in Relation to Length of the Calvaria".) However, another major factor contributes to the low values of this angle: the height of lambda. The obtuse angle measured on A.L. 444-2 stems from both the anterior position of foramen magnum and the low position of lambda (relative to the FH) on the posterior slope of the calvaria.

Because of the relatively elevated position of lambda in Sts. 5, we are left with the posterior location of opisthion as the sole explanation for the relatively small angle of the occipital chord (107°) in this specimen. The value of the angle in modern humans lies toward the other end of the spectrum (recall that this value is the same in A.L. 444-2). However, in modern humans, it is the extremely anterior location of opisthion that overrides the effect of lambda's high position. In KNM-ER 1813, the inclination of the chord is 112°, and in *A. boisei*, the mean of three specimens (KNM-ER 406, KNM-ER 23000, and OH 5) is 118°. Again, the similarity between OH 5 and A.L. 444-2 is great and for the same reason: the low position of lambda in OH 5 (as in A.L. 444-2) brings the angle to 125°. In KNM-WT 17000, where lambda descends to its lowest position among the hominins—almost coinciding with inion—the angle is the largest, as expected, at 132°.

Throughout our hominoid sample, the position of lambda dictates the orientation of both chords, lambda–opisthion (the occipital sagittal chord) and lambda–bregma (the sagittal chord of the parietal bone). As a result, a fixed relationship exists between these two chords in the form

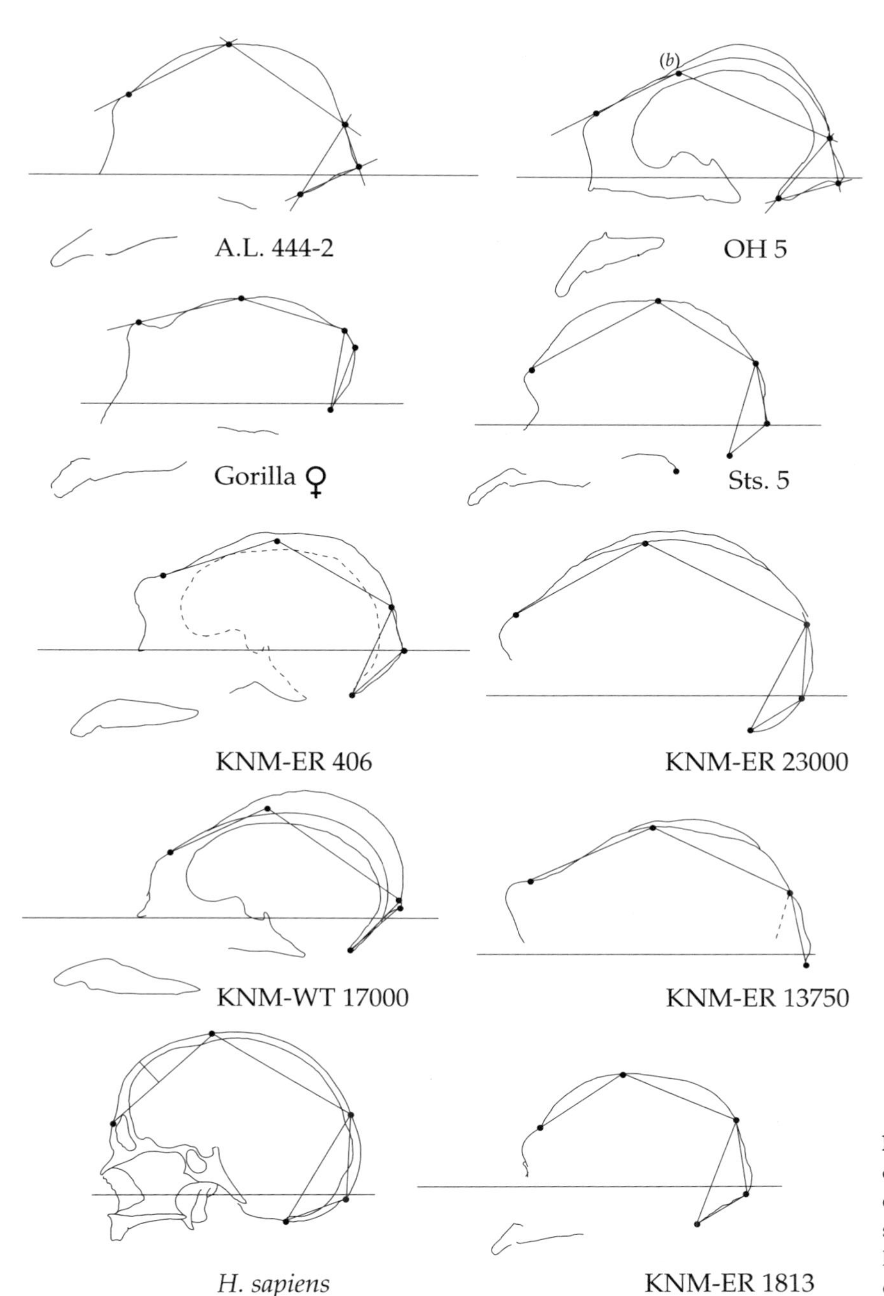

Figure 3.15 Inclination of the major calvarial chords in A.L. 444-2 and other hominoids. Clockwise from the supraglabellar point, the anthropometric landmarks are bregma, lambda, inion, and opisthion (see text for details). (Not to scale.)

of a relatively constant ±90° angle in all hominoids—from the apes, in which lambda's position is very high, to KNM-WT 17000, in which it is very low (Figure 3.16). This constant lambda angle attests to the fact that bregma and opisthion have a fixed position *relative to lambda*, which is very variable in its position. Specimen A.L. 444-2 fits this pattern perfectly, with a 90° angle between the chords (Figure 3.14).

The relationship between the occipital sagittal chord and arc is discussed in Chapter 5. However, for a given arc/chord index value, arcs of various shapes can be delineated. In A.L. 444-2, for example, the occipital "arc" takes the shape of a right angle and forms an almost perfect isosceles triangle with its chord (Figure 3.15). A line perpendicular to the triangle's base (the lambda–opisthion chord) extending from inion would divide the base into two almost equal parts. The section adjacent to lambda would be slightly shorter than the one adjacent to opisthion. This geometrical configuration is identical to that seen in OH 5 (Figure 3.15); however, in the latter, the lambda–opisthion chord falls entirely within the bone mass of the occipital, whereas in A.L. 444-2, the chord lies

Table 3.9 Inclination of the Vault's Chord Segments (Degrees)

Specimen/Sample		Frontal Chord (Supraglabella–Bregma)	Parietal Chord (Bregma–Lambda)	Occipital Chord (Lambda–Opisthion)	Occipital Upper Scale Chord (Lambda–Inion)	Occipital Lower Scale Chord (Inion–Opisthion)
Australopithecus afarensis	Mean	27	35	119	76	35
A.L. 444-2		27	35	122	70	25
A.L. 333 recon. (A.L. 333-45)		n/a	n/a	116	82	45
Homo sapiens	Mean (n = 10)	44	29	122	100	25
	Range	41–46	25–36	116–128	96–110	17–34
	SD	2	3	5	5	6
Gorilla gorilla female	Mean (n = 10)	19	14	94	18	79
	Range	17–20	7–20	86–102	11–25	69–88
	SD	2	7	8	7	10
Pan troglodytes female	Mean (n = 10)	24	20	98	45	71
	Range	21-30	14-23	92-102	30-57	65-82
	SD	4	4	4	11	8
Pan troglodytes male	Mean (n = 10)	24	21	96	54	67
	Range	17–30	16–23	88–101	50–57	59–76
	SD	4	3	6	3	7
A. africanus	Mean	32	26	110	82	36
Sts. 5		35	25	107	80	40
Sts. 71		32	n/a	n/a	n/a	n/a
MLD 37/38		n/a	26	113	84	31
Stw. 505		30	n/a	n/a	n/a	n/a
H. habilis						
KNM-ER 1813		34	25	112	80	30
A. boisei	Mean	25	27	118	84	30
OH 5		(29)[a]	(25)	125	72	16
KNM-ER 406		19	30	113	90	42
KNM-ER 13750		24	26	n/a	79	n/a
KNM-ER 23000		28	28	115	95	32
A. aethiopicus						
KNM-WT 17000		24	38	132	n/a	42

[a]Values in parentheses are estimates (±2°).

within the endocranial space, quite far from the internal bone table, perhaps as a result of the greater concavity of the endocranial surface (Figure 3.14). Other crania of both *A. afarensis* and *A. boisei* exhibit a rounded, convex arc from lambda to opisthion. As such, those specimens resemble other hominins and many chimpanzees and female gorillas. A rounded midsagittal occipital contour is also seen in *A. africanus* (Sts. 5, MLD 1, MLD 37/38, and Sts. 71).

The near joining of lambda and inion in KNM-WT 17000, the product of the unusually low position of lambda, results in an unusual juxtaposition of the occipital arc and chord and thus also in the formation of an extremely long, narrow triangle (lambda–inion–opisthion). The lambda–inion side is much shorter than the lambda–opisthion and inion–opisthion sides. A triangle with similarly long, narrow proportions characterizes great ape crania, too, but in these, the proximity of inion and lambda stems from the elevation of inion toward a high lambda (Figure 3.15). Thus, the specimens differ considerably in the orientation and location of the triangle in relation to the FH.

Inclination of the Chord of the Upper Occipital Scale (Lambda–Inion)

The upper occipital scale chord forms a 70° angle with the FH in A.L. 444-2. In modern humans, this angle exceeds the perpendicular, at a mean value of 100°. The great proximity of inion to lambda and the obliteration of lambda at a rather young age make this measurement inapplicable to most great ape specimens. The very fact that this measurement is inapplicable to the great apes demonstrates just how much the morphology of the occipital bone in A.L. 444-2 differs from that of the great apes. However, in chimpanzee specimens in which the measuring points are discernible (usually in young females), the angle produced by the very short upper occipital scale with the horizontal is 45° (female) and 54° (male). The long upper occipital scale

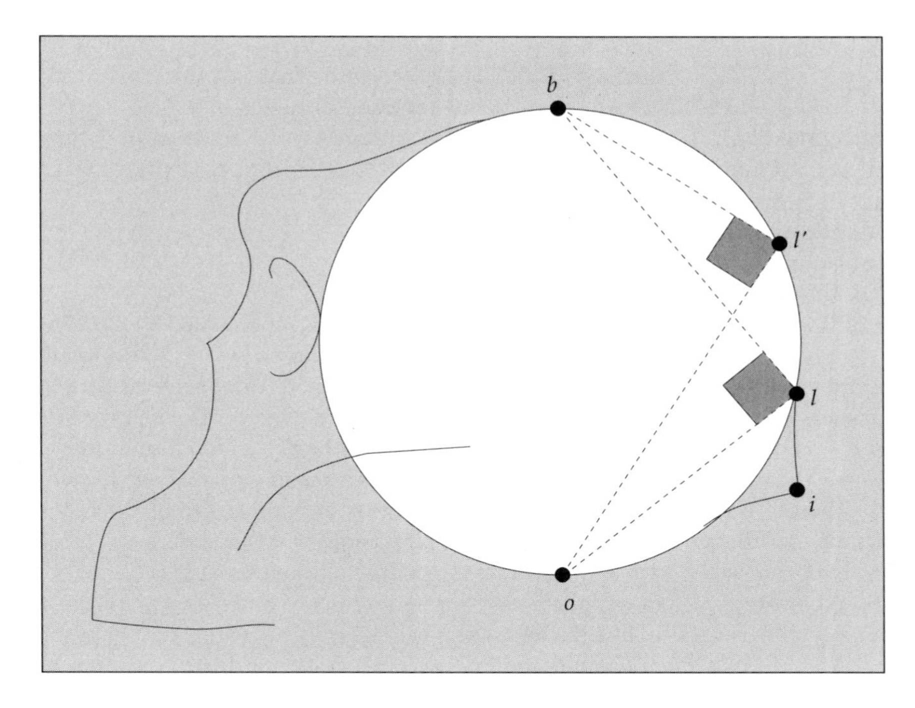

Figure 3.16 Consistency of the "lambda angle." The points shown in this representation of the posterior cranium are bregma (*b*), two positions of lambda (*l′* and *l*), inion (*i*), and opisthion (*o*). Note that although lambda changes position, the lambda angles *bl′o* and *blo* remain the same.

in Sts. 5 forms an 80° angle with the horizontal, and this is identical to the angle in KNM-ER 1813, which has an even longer upper scale. The mean angle in *A. boisei* (84°) is greater than that in A.L. 444-2. Note that the specimens with the more obtuse angles in this species (KNM-ER 23000, at 95°, and KNM-ER 406, at 90°) differ in their occipital topography from those with smaller angles (OH 5, at 72°, and perhaps Omo 323-1976-896, which, though fragmentary, appears to belong to this category). In the specimens of *A. boisei* with a more obtuse angle, the midsagittal cross section of the whole occipital is rounded, and inion does not provide a sharp demarcation of the transition between the lower and upper scales. OH 5 resembles A.L. 444-2 in its angulated midsagittal cross section and in the posteroinferior protrusion of inion from the endocranial contour of the cross section.

Inclination of the Lower Occipital Scale (Inion–Opisthion)

The chord of the lower occipital scale in A.L. 444-2 forms a rather small angle (25°) with the FH. In modern humans, this chord is identically inclined (Figures 3.14 and 3.15 and Table 3.9). In female gorillas, however, the value is 79°, and in chimpanzees, the nuchal scale is more inclined, at 71° (female) and 67° (male). This angle constitutes one of the most significant differences in cranial morphology between A.L. 444-2 and the great apes and, indeed, between hominins and African apes in general.

In A.L. 444-2, the value of the angle is smaller (in other words, the nuchal plane is more horizontal) than in many other hominins, including A.L. 333-45, in which the value is substantially greater, at 45° (Kimbel et al., 1984). This difference apparently stems partly from the degree of the posteroinferior protrusion of inion at the tip of the massive external occipital protuberance and to the varying position of opisthion (note the differences in the inclination of both the upper scale and the occipital chord in these two specimens; Table 3.9). In Sts. 5, the angle measures 40°, and in KNM-ER 1813, it is 30°. To a large extent, the discrepancy between these two specimens is due to the low position of inion, beneath FH, in KNM-ER 1813, as in modern humans, in which the small angle of the chord also results from the anterior position of the foramen magnum. Only one specimen of *A. boisei* (OH 5) demonstrates a similar anatomical configuration to that of A.L. 444-2. Here, too, inion is extended posteroinferiorly on a massive ectocranial structure, and hence the nuchal scale angle is only 16°. Two other specimens of *A. boisei*, however, exhibit rather high values for this angle: 42° in KNM-ER 406 and 32° in KNM-ER 23000. The broad range of values seen in *A. boisei* resembles the range of the two *A. afarensis* specimens A.L. 444-2 and A.L. 333-45 (see Chapter 5 for further discussion of this variation).

Again, the angle between the chords of the upper and lower scales in those specimens in which inion protrudes from the general outline of the calvaria (and thus causes a more horizontal orientation of the nuchal scale)—A.L. 444-2 and OH 5—is approximately 90°. As expected, this angle is larger in the specimens with a steeper nuchal scale.

Inclination of the Reconstructed Foramen Magnum (Basion–Opisthion)

A substantial part of the central axis of the cranial base from opisthion to nasion is missing in A.L. 444-2 (Figure 2.3 a and b). However, the inclination and orientation of the petrous elements of the temporal bones, the remains of the orbital roofs (the floor of the anterior cranial fossa), and the lateral part of the floors of the middle cranial fossae facilitated a reliable reconstruction of the missing parts of the central cranial base.

Based on this reconstruction, the length of the foramen magnum is 32 mm, and this line—basion to opisthion—forms an angle of +16° with the FH ("+" indicates that the angle opens posteriorly; "–" indicates an anteriorly open angle, i.e., a foramen that opens anteroinferiorly). The angle was most likely slightly smaller than this figure, but because of minor damage to the posterior margins of the foramen magnum, opisthion is elevated a few millimeters from what we infer to have been its true position.

The inclination of the foramen magnum in A.L. 444-2 is, at +16°, smaller than in the *A. africanus* Sts. 5 cranium (+20°)[5] but larger than the angle measured in MLD 37/38, which is approximately +12°. These values are greater than those measured in *A. boisei* (+7° in OH 5 and +9° in KNM-ER 406) and dramatically greater than in modern humans, where the mean is –8°. Hence, A.L. 444-2 and modern humans differ by 24° in this measurement (Table 3.10). The foramen magnum in our sample of gorillas is much more steeply inclined (+30° in females and +27° in males) than in the Hadar specimen and is slightly less so in chimpanzees (+20° in females and +18° in males).

The midsagittal cross section of the reconstructed center of the cranial base in A.L. 444-2 indicates that the descent of opisthion in relation to the FH is primarily responsible for the difference in the inclination of the foramen magnum between modern humans and A.L. 444-2 (Figure 3.17 and Table 3.10). Whereas basion lies an equal vertical distance below the FH in both taxa, the relative position of opisthion differs between them considerably. In *H. sapiens*, the mean opisthion position is 26 mm below the FH, and in A.L. 444-2, this distance is only 13 mm. The configuration observed in A.L. 444-2 resembles that of the African apes; in other words, the inclination of the foramen magnum in the latter should be attributed not to basion's distance from the FH but to the small vertical distance of opisthion from this plane (9 vs. 26 mm in female gorillas and 10 vs. 20 mm in male chimpanzees; see Table 3.10).

The means of opisthion–FH and basion–FH distances and the ratios between them in the Plio-Pleistocene hominin taxa (Table 3.10) do not differ significantly from the A.L. 444-2 measurements, although the two values are smaller in the Hadar hominin. That is, both basion and opisthion are situated closer to the FH in A.L. 444-2. It is worthwhile noting that the distance between opisthion and the FH (25 mm) in KNM-ER 1813 resembles that in modern humans.

Inclination of the Occlusal Plane

The occlusal plane in A.L. 444-2 is represented by the third right molar (the left one is displaced upward as a result of the deformation that influences the posterior part of the left side of the palate) and the left central incisor. The occlusal plane lies parallel to the FH. In modern humans, the occlusal plane exhibits a mean slope of –8° (downward and forward), whereas in gorillas, the plane slopes upward, reaching a mean angle of +7° in both sexes (Table 3.10). In contrast to gorillas, chimpanzees display an occlusal plane that slopes downward and forward, more so in females (–6°) than in males (–4°). The inclination of the occlusal plane in A.L. 444-2 does not differ from that of most other early hominins, where the occlusal or alveolar plane is parallel to or slopes slightly anteroinferiorly relative to the FH. It is noteworthy that the one *A. boisei* specimen that permits measurement of this angle yields a value of +12°, whereas in KNM-WT 17000, the (missing) occlusal plane was undoubtedly parallel to the FH or even sloped anterosuperiorly relative to it.

Axis of the Reconstructed Cranial Base: Inclination of the Clivus and Position of the Sella Turcica

The inclination of the medullary clivus in modern humans is rather steep, with the straight segment from basion to the base of the dorsum sellae forming an angle of 64° with the FH (Figure 3.17). In female gorillas, the comparable angle is 29° (and similar values are measured on chimpanzees; Table 3.10). The reconstructed clivus angle in A.L. 444-2 is 44°. Despite great variation in the inclination of the clivus in our sample (Table 3.10 and Figure 3.17), three distinct groups can be seen. The group with small angles consists of the African apes; early hominins are the second group, with medium-size angles; and the third group, modern *H. sapiens*, exhibits large angles. However, in some of the early hominins—OH 5 and MLD 37/38—the angle is also rather high.

The inclination of the clivus seems to have the greatest influence on the position of the sella turcica relative to the FH. Recall that the vertical distance between basion—the lower end of the clivus—and the FH is more or less constant among hominoid taxa. Thus, the steeper the clivus, the greater the elevation of the sella turcica will

Table 3.10 Inclination of the Cranial Base Segments

Specimen/Sample		Inclination of the Occlusal/Alveolar Plane[a] (degrees)	Inclination of Foramen Magnum[b] (degrees)	Inclination of the Medulary Clivus (degrees)	Vertical Distance Opisthion–FH (12) (mm)	Vertical Distance Basion–FH (11) (mm)
Australopithecus afarensis						
A.L. 444-2		−1	16	44	13	20
Homo sapiens	Mean (*n* = 10)	−8	−8	64	26	20
	Range	−12–0	−14–0	55–72	22–30	15–26
	SD	4	5	6	3	3
Gorilla gorilla female	Mean (*n* = 10)	7	30	29	9	26
	Range	0–14	23–35	25–33	3–17	23–29
	SD	6	5	3	6	2
Gorilla gorilla male	Mean (*n* = 10)	7	27	27	6	21
	Range	4–12	20–32	22–30	2–10	14-23
	SD	3	5	3	4	4
Pan troglodytes female	Mean (*n* = 10)	−6	20	29	11	19
	Range	−10–0	15–30	25–31	2–17	15–25
	SD	4	5	3	5	3
Pan troglodytes male	Mean (*n* = 10)	−4	18	32	10	20
	Range	−9–0	5–30	28–35	7–15	12–26
	SD	3	7	3	3	4
A. africanus	Mean	0	16	49	17	25
Sts. 5		0	20	49	17	28
Sts. 71		0	n/a	n/a	19	n/a
MLD 37/38[d]		n/a	12	n/a	16	21
H. habilis						
KNM-ER 1813		0	(0)[c]	n/a	25	(23)
A. boisei	Mean	12	10	51	18	23
OH 5		12	7	58	16	18
KNM-ER 406		n/a	13	43	20	27
KNM-ER 23000		n/a	n/a	n/a	19	n/a
A. aethiopicus						
KNM-WT 17000		(0)	n/a	47	20	25

[a]Negative values indicate inferior slope relative to FH. For specimens without teeth (Sts. 5, KNM-WT 17000) the alveolar plane is substituted.
[b]Negative values for foramen magnum inclination indicate anteroinferior orientation of the foramen.
[c]Values in parentheses are estimates.
[d]FH in MLD 37/38 based on Sts. 5; data are approximations.

be, as depicted schematically in Figure 3.18. Indeed, in modern humans, the extreme steepness of the medullary clivus is accompanied by a relatively large vertical distance between the FH and the sella turcica. In the African apes, in contrast, the inclined clivus results in a lower position of the sella turcica, which coincides with that of the FH. Similarly, the reconstructed sella turcica in A.L. 444-2 is located very close to the FH. Sella turcica is also observed at the level of FH in other early hominins, such as Sts. 5 and MLD 37/38. However, in OH 5, the sella turcica is relatively high, with about one-third of the clivus length situated above the level of the FH—as predicted from the steepness of the clivus in this specimen. The inclination of the clivus in another *A. boisei* specimen, KNM-ER 406, as well as in KNM-WT 17000, is associated with a lower position of the sella turcica, as expected (Figure 3.17).

A more traditional way of expressing cranial base flexion is through the angle formed between basion–sella turcica (prosphenion) line and the sella turcica–nasion line. Because the great variation in the position of nasion and the difficulty in identifying this point in fossils reduce the value of this method, we have not employed it here. However, for the record, the angle is 150° in the reconstructed base of A.L. 444-2.

In the anterior part of the midsagittal contour of the A.L. 444-2 cranial base, a deep olfactory pit (deep in relation to the orbits' ceilings) similar to that of the African apes has been reconstructed based on A.L. 58-22, a craniofacial fragment that shows this morphology (Kimbel et al.,

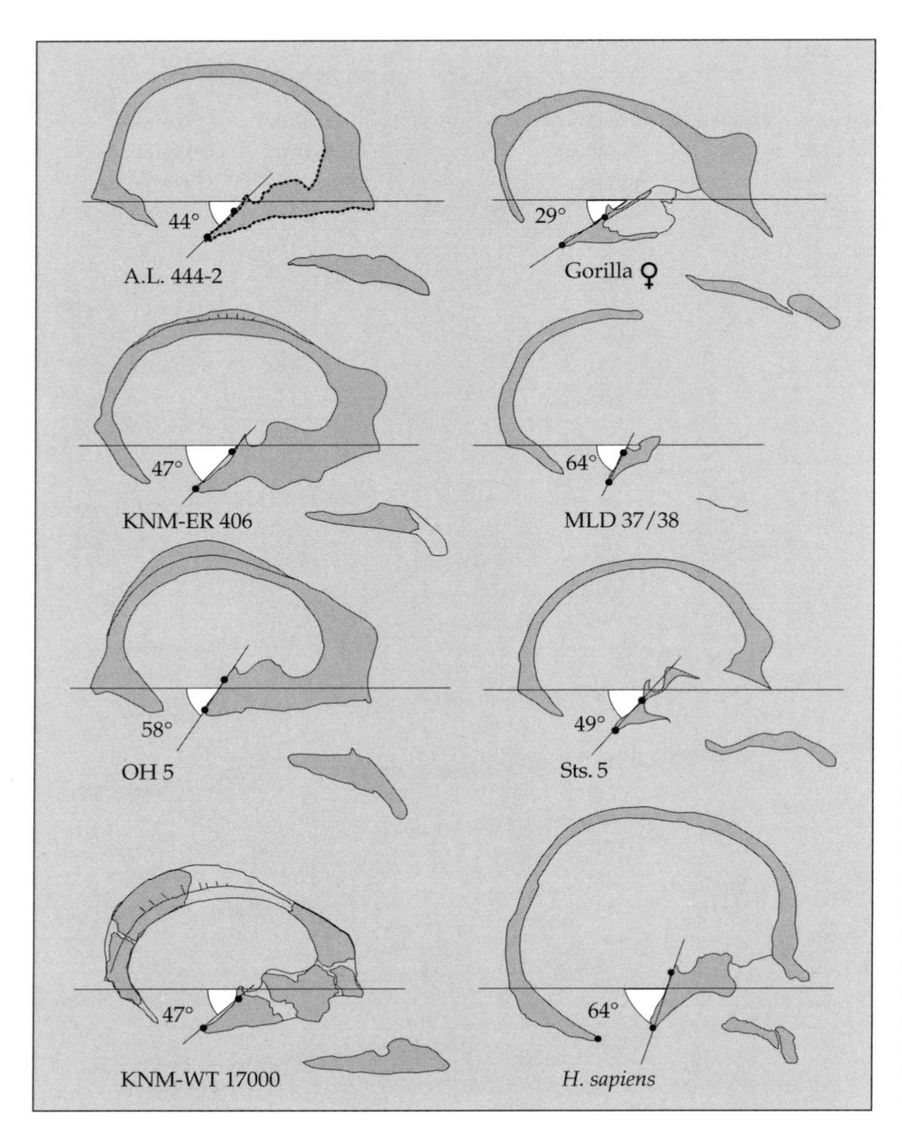

Figure 3.17 The medullary clivus and the position of the sella turcica relative to Frankfurt Horizontal (FH). In the early hominins, sella turcica lies at approximately the level of FH. In the apes, this structure is below FH, and in *Homo sapiens*, it is above. Note that the angle formed by the clivus with the FH is small in the apes and large in modern humans. In most of the early hominins, the size of the angle falls in between these values. MLD 37/38 orientation based on Sts. 5 FH. (Cross sections of KNM-ER 406, KNM-WT 17000, and MLD 37/38 are based on CT images kindly made available to us by Dr. Fred Spoor. The cross section of OH 5 is from Tobias, 1967.)

1982) (Figure 3.14). In this respect, *A. afarensis* resembles the great apes, as do other *Australopithecus* species.

Anteroposterior Position of the Foramen Magnum in Relation to Length of the Calvaria

Le Gros Clark's (1950) index of condylar position (see CD/CE in Figure 3.10) yields a value of 45% in A.L. 444-2. We provide this index value for the record, even though we believe that it does not express the true position of the condyles or the foramen magnum (which lies between D and E), as Le Gros Clark's calculations were affected by prognathism: that is, specimens that vary in the degree of prognathism yield different values although these specimens may exhibit the same distance between the condyles and opisthocranion.

To eliminate the effect of prognathism on the index of the foramen magnum's true position, we have adopted Weidenreich's method, in which the anteroposterior position of opisthion is expressed in relation to the length of the braincase—the horizontal distance between the coronal plane of the glabella and that of the most posterior point on the braincase[6] (which coincides with inion and opisthocranion in the Hadar cranium)[7] (Weidenreich, 1943: 131). In A.L. 444-2, the segment that stretches from opisthion to the most posterior point of the vault (horizontal distance) constitutes 24% of the length of the entire braincase (Figure 3.19 and Table 3.11). This value is smaller than that obtained on modern humans (31%) but is still much higher than those yielded by gorillas (females, 13%, and males, 7%) and chimpanzees (females, 14%, and males, 12%).[8] Even the 1984 *A. afarensis* com-

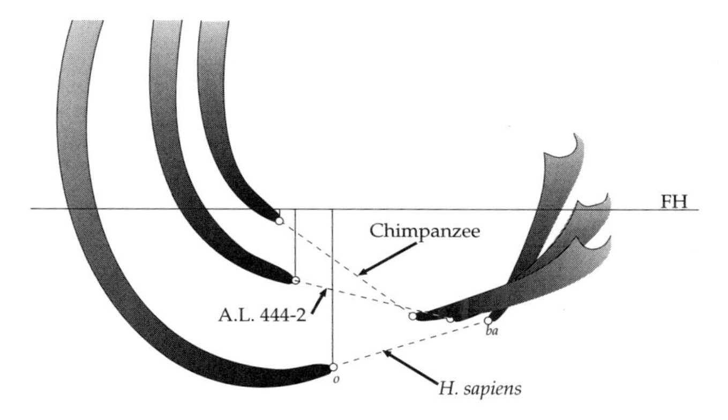

Figure 3.18 Schematic representation of the differences among hominoids in the position of opisthion and basion (and the resulting inclination of the foramen magnum), the inclination of the medullary clivus, and the position of sella turcica. In the apes, opisthion (*o*) is posteriorly located and elevated relative to the Frankfurt Horizontal (FH); basion (*ba*) is also posteriorly located and the clivus is only slightly inclined; the sella turcica lies below FH. In modern humans, opisthion is very low relative to the FH; basion is anteriorly located; the clivus is steep; and the sella turcica lies above FH. *A. afarensis* cranium A.L. 444-2 represents an intermediate configuration in respect to all these elements. The morphocline suggested here is a transition from the posteroinferior orientation of the foramen magnum in apes to its anteroinferior orientation in modern humans. This change occurs through a vertical descent of opisthion; basion undergoes no such change. The foramen magnum migrates anteriorly through an anterior shift of basion and opisthion (with basion remaining on the same horizontal plane). As basion advances anteriorly, the clivus gains a steeper inclination and the sella turcica ascends above the level of the FH. At the same time, the occipital squama moves posteriorly with the increase in brain volume.

posite reconstruction, with its very low values (perhaps also the effect of the extensive reconstruction of the entire frontal), yields an index that is 19% higher than that of the African apes. A.L. 444-2 is similar to other early hominins in this respect. The index in KNM-ER 1813 is 25%, and the mean for three *A. boisei* specimens (OH 5, KNM-ER 406, and KNM-ER 23000) is 21%. KNM-WT 17000 yields an approximate value of 21%.

Shape of the Facial Slope

A lateral view of the A.L. 444-2 cranium reveals a prognathic face with a midsagittal profile that is divided into well-defined segments. Below a modestly protruding glabellar region and the slightly depressed nasion area beneath the glabella, there are three distinct contours with clearly marked transitions between them (Figures 3.5 and 3.6): (1) a concave upper facial contour features a slightly recessed sellion borne by the nasal bones and a slightly elevated rhinion at its inferior termination; (2) below rhinion, the upper part of the concave middle facial contour rests on the lateral margin of the nasal aperture and the lower end on the canine jugum (the "nasocanine contour" of Kimbel et al., 1984); (3) the lowermost part of the facial contour is convex, resting on the nasoalveolar clivus (the "nasoalveolar contour" of Kimbel et al., 1984). This last segment continues smoothly into the moderately procumbent central incisors, which complete the convex profile.

This segmented facial profile is shared by many other primates and thus can be considered the plesiomorphic arrangement. In contrast, many *A. africanus* specimens exhibit a single, straight facial contour from sellion to the alveolar plane (i.e., the nasocanine and nasoalveolar contours are merged in lateral view; Kimbel et al., 1984). The straight contour is due to the predominating influence of the straight anterior pillars on *A. africanus* midfacial morphology and of the flat nasoalveolar clivus "stretched" between the lower ends of the pillars (that is, the clivus is on the same plane as the pillars) (Rak, 1983). A straight or slightly concave, unsegmented facial profile from sellion to prosthion characterizes *A. robustus* and *A. boisei*, although in these taxa the anterior position of the

Figure 3.19 Position of the foramen magnum relative to the maximum projected length of the calvaria in A.L. 444-2 and other hominoids. The dashed bar indicates the calvarial length; the solid line indicates the distance between the most posterior point on the calvaria (in projection) and opisthion (*o*).

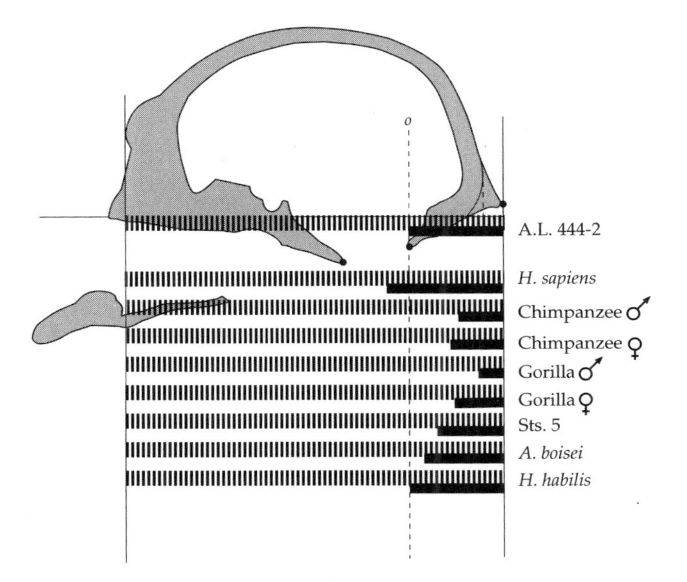

Table 3.11 Relative Position of Foramen Magnum

Specimen/Sample		Projected Vault Length (4) (mm)	Horizontal Distance Opisthion to Posteriormost Point on Vault[a] (mm)	Index of Foramen Magnum Position[b] (%)
Australopithecus afarensis	Mean	158	35	22
A.L. 444-2		165	40	24
A.L. 333 reconstruction		(151)[c]	29	(19)
Homo sapiens	Mean (n = 10)	182	56	31
	Range	158–194	45–62	28–34
	SD	11	6	2
Gorilla gorilla female	Mean (n = 10)	156	20	13
	Range	147–171	12–27	8–16
	SD	11	6	4
Gorilla gorilla male	Mean (n = 10)	168	11	7
	Range	166–170	9–13	4–10
	SD	2	2	2
Pan troglodytes female	Mean (n = 10)	136	19	14
	Range	128–143	15–22	11–16
	SD	5	3	2
Pan troglodytes female	Mean (n = 10)	138	17	12
	Range	126–147	9–24	7–18
	SD	7	4	3
A. africanus	Mean	135	27	20
Sts. 5		143	26	19
Sts. 71		(127)	(27)	(21)
H. habilis				
KNM-ER 1813		146	36	25
A. boisei	Mean	161	33	21
OH 5		167	40	24
KNM-ER 406		158	28	18
KNM-ER 23000		159	32	20
A. aethiopicus				
KNM-WT 17000		150	(32)	(21)

[a]Not necessarily opisthocranion.
[b]Index calculated as opisthion to posterior vault/proj. vault length × 100.
[c]Values in parentheses are estimates.

peripheral parts of the face obscures the central part of the face in lateral view (this is the classical definition of the "dished" face in these robust *Australopithecus* species). Here the clivus remains prognathic but is depressed relative to the more lateral facial elements, a configuration that produces what Rak (1983) coined the "nasoalveolar gutter." Despite the extensive prognathism characterizing KNM-WT 17000, the peripheral part of the face—the zygomatic bones—bears the anterior profile. The entire sunken central portion of the face is hidden in a lateral view.

The concave, segmented facial profile of A.L. 444-2 lies considerably posterior to the glabella-prosthion line (Figure 3.20). The point where the gap between these two lines is greatest falls close to the level of the inferior margin of the nasal aperture. In other words, this is the deepest point in the facial profile. Although part of this gap could be attributable to slight postmortem deformation that resulted in an inward sinking of the central part of the face, the fact that we see the same topography in A.L. 417-1d allows us to discount the effect of deformation.

Although much of the central portion of the face of *A. garhi* is missing, the preserved area and the manner in which it has been reconstructed (Asfaw et al., 1999) suggest that the specimen exhibits the plesiomorphic facial contour. Furthermore, the detail of the specimen's profile seems to resemble that of A.L. 444-2 in that the deepest point of the face, as we have inferred from the illustration in Asfaw et al. (1999), is located toward the bottom of the profile.

In comparison to A.L. 444-2, the deepest part of the facial contour in chimpanzees and gorillas is located higher up on the superior part of the face, just beneath glabella. The elevated depression is the consequence of two distinguishing features of the African great ape face: total facial prognathism is greater than in A.L. 444-2, so

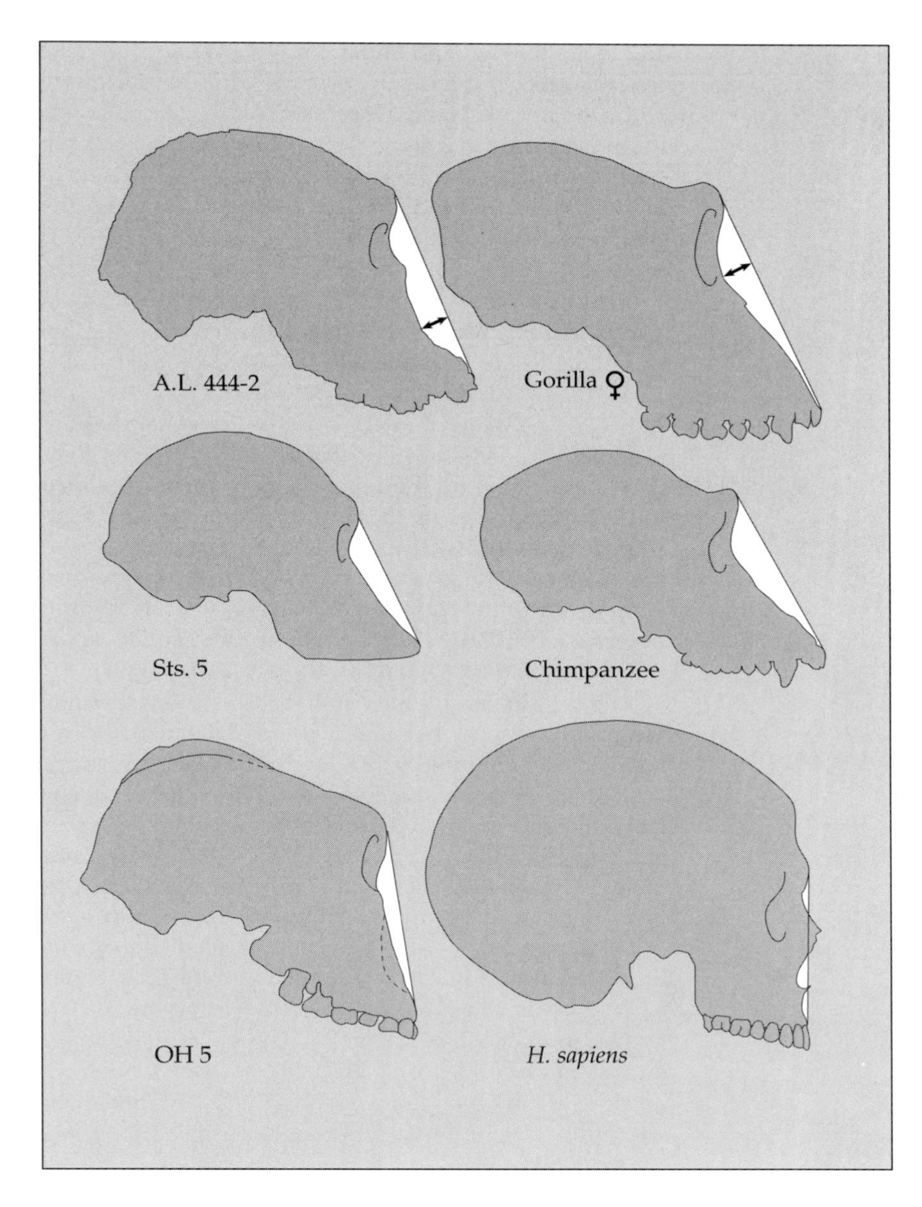

Figure 3.20 Midsagittal outline of the face relative to the glabella–prosthion line in A.L. 444-2 and other hominoids. Note that contrary to the configuration in gorillas, chimpanzees, and Sts. 5, the deepest part of the outline in A.L. 444-2 is low, at the level of the nasal sill.

the level at which the facial contour begins its antero-inferior projection is much higher (in the vicinity of sellion), and in the African apes glabella juts out considerably past the level of sellion.

The extreme retreat of prosthion in modern humans brings about a dramatically anterior extension of the nasal bones and the anterior nasal spine. In other words, rhinion is situated considerably anterior to the glabella–prosthion line, hence producing a jagged profile.

Degree of Prognathism and Relative Length of the Dental Arcade

Prognathism can be expressed metrically in several ways. Some methods evaluate the degree to which the front of the dental arcade projects beyond the coronal plane of the facial frame (i.e., the horizontal distance between them). Angular methods evaluate the degree of facial slope as an angle formed with the FH or the occlusal plane (which often nearly parallel one another). Each of these methods has its advantages and disadvantages. In the first, the values are absolute, and there is a need to correct for size; in the second, a similar horizontal distance between the front of the palate and the facial frame (for example, nasion) can yield different degrees of angular prognathism as the vertical distance between these structures varies.

In this study we rely on what Rak (1983) has termed the "index of palate projection relative to sellion." This index (Figure 3.21) expresses the percentage of the palate's length that projects anterior to the coronal plane of sellion.

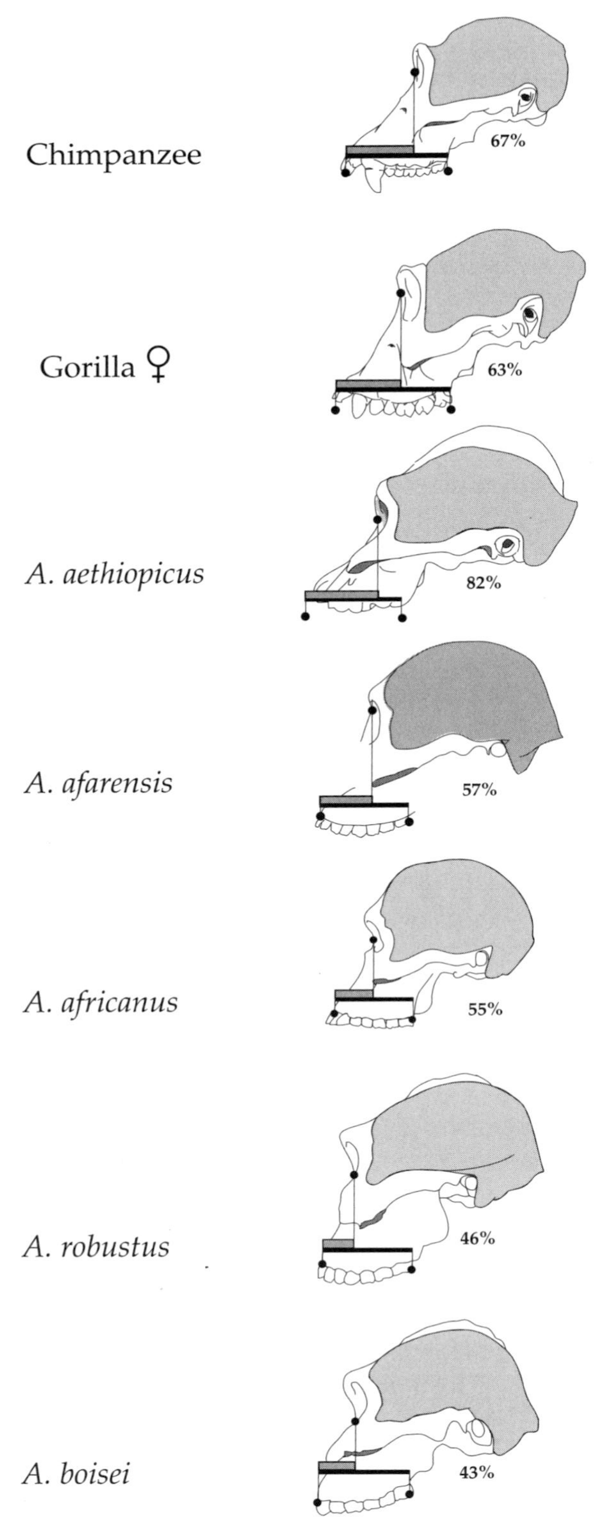

Figure 3.21 Prognathism expressed as the percentage of palate length extending anterior to the coronal plane of sellion (Rak's 1983 Index of Prognathism). The figure for *A. afarensis* is the mean for A.L. 444-2 (55%) and A.L. 417-1 (59%). Other species' values are based on the samples enumerated in Table 3.13.

In female gorillas and chimpanzees, 63% of the palate length and 67% thereof, respectively, lie anterior to the coronal plane of sellion, whereas in A.L. 444-2, only 55% extends beyond this plane. This value is the same as the mean for *A. africanus* (55%, *n* = 2). However, while there is a substantial difference between A.L. 444-2 and the highly prognathic Sts. 5 (68%), the Hadar specimen is much more prognathic than Sts. 71 (43%).

As mentioned, the fact that the occlusal plane almost parallels the FH helps us evaluate the relationship between the palate and sellion in less complete specimens. In another adult *A. afarensis* specimen, A.L. 417-1d, the index is 59%, which is very close to the value for A.L. 444-2. It turns out, however, that when we add less complete specimens—for example, Sts. 52a with its estimated index value of 40%—to the *A. africanus* sample, Sts. 5 appears to be exceptionally prognathic. Even more fragmentary specimens, such as TM 1511, MLD 6/23, and Sts. 17, seem to ally with Sts. 71 rather than Sts. 5 in this respect. In terms of this index of prognathism, we should expect *A. afarensis* and *A. africanus* to overlap substantially.

Only a single specimen of *A. robustus*, SK 48, permits the computation of this index. In this specimen, 46% of the palate length projects anterior to the coronal plane of sellion. Nevertheless, here, too, less complete specimens leave the impression of a modest protrusion of the palate. These include SKW 11, SK 12, SK 13/14, SK 46, and SK 83. The single Kromdraai specimen, TM 1517a, appears to be more prognathic than those from Swartkrans.

Three specimens of *A. boisei* permit calculation of this index; the mean value is 43%, for OH 5, KNM-ER 406, and KNM-ER 732 (on the reconstruction necessary for the latter two specimens, see Rak, 1983). This value is the smallest among the species of *Australopithecus*. Even though the mean value is small, there is substantial variation in it, with individual index values ranging from 35% (OH 5) to 53% (KNM-ER 406) (on the variation characterizing the masticatory system of *A. boisei*, see Rak, 1988). On the other end of the hominin spectrum stands KNM-WT 17000. The degree to which the palate protrudes in this specimen—the index is a remarkable 82%—exceeds even the index of the African great apes. The short vertical distance between sellion and the alveolar plane creates an extremely prognathic face in this cranium (see the following discussion of the angular prognathism).

In modern humans, of course, the index values are very low (15%). In KNM-ER 1813, the only *H. habilis* cranium that permits this measurement, the index value (40%) is also quite low, although the low value in *Homo* is achieved in a rather different manner from that of the robust *Australopithecus* species. In the robust species, the low value is due in part to the posterior retraction of the entire palate, which remains highly prognathic subnasally, whereas in *Homo*, it is due to the shrinkage of the palate and the prominence of the nasal bridge, which bears sellion (see Figure 3.22; Rak, 1983).

These index values agree with the values obtained via more traditional methods that involve angular measures of prognathism (Table 3.12). The facial angle, formed by the sellion–prosthion line and FH, is 65° in A.L. 444-2. (We use sellion instead of nasion due to great variation in the position of the latter landmark in hominoids.) This value is greater than in the African apes; the angle is 59° and 54° for female and male chimpanzees, and 57° and 52° for female and male gorillas, respectively. In both ape species, the males, with their smaller values, are more prognathic than the females.

The female specimen A.L. 417-1d displays a facial angle of 60°, which is similar to that of A.L. 444-2. Several less complete maxillae, however, such as A.L. 200-1a, A.L. 413-1, and A.L. 486-1, provide the impression of greater prognathism than do these two specimens, whereas A.L. 333-1 and A.L. 333-2 appear to be less prognathic.

In *A. africanus*, the mean value is 61°; although the sellion–prosthion angle for Sts. 5 is similar to that of the African apes (53°), the facial angle for Sts. 71 is identical to that of A.L. 444-2 (65°). This difference was already noted in the discussion of the relationship between the anterior projection of the palate relative to the facial frame. Here, too, in the less complete specimens of the hypodigm, the alveolar plane substitutes for the FH. In Sts. 52a, the angle between the occlusal plane and that of the face is 63°, while in our reconstruction of MLD 6/23 it is 65°. This is the same impression—that of a relatively obtuse angle—given by some less complete specimens of this species, such as TM 1511, TM 1512, and Sts. 17.

One specimen of *A. robustus*, SK 48, and two of *A. boisei*, OH 5 and KNM-ER 406, permit the measurement of this angle. SK 48 and OH 5 exhibit a rather steep face, with angles of 70° and 72°, respectively. At 60°, the angle in KNM-ER 406 differs substantially. This difference seems to indicate a rather wide range of variation in the "dish-faced" robust hominins, as also suggested by the other methods described above. Indeed, despite the unavailability of this angle in the majority of *A. robustus* and *A. boisei* specimens, superficial examination of less complete specimens contributes to this impression (Rak, 1983, 1988; Kimbel and White, 1988b).

With an angle of 41°, KNM-WT 17000 is highly prognathic. This specimen's extreme angular prognathism is produced by a combination of the anterior position of prosthion and a very short vertical distance between sellion and the alveolar plane—in other words, a vertically low face.

In modern humans, the dental arcade is short, constituting 47% of the biorbital distance (Rak, 1983). For comparison's sake, we measured a male baboon's dental arcade, which constitutes 119% of the biorbital distance—an extreme example from the other end of the spectrum.

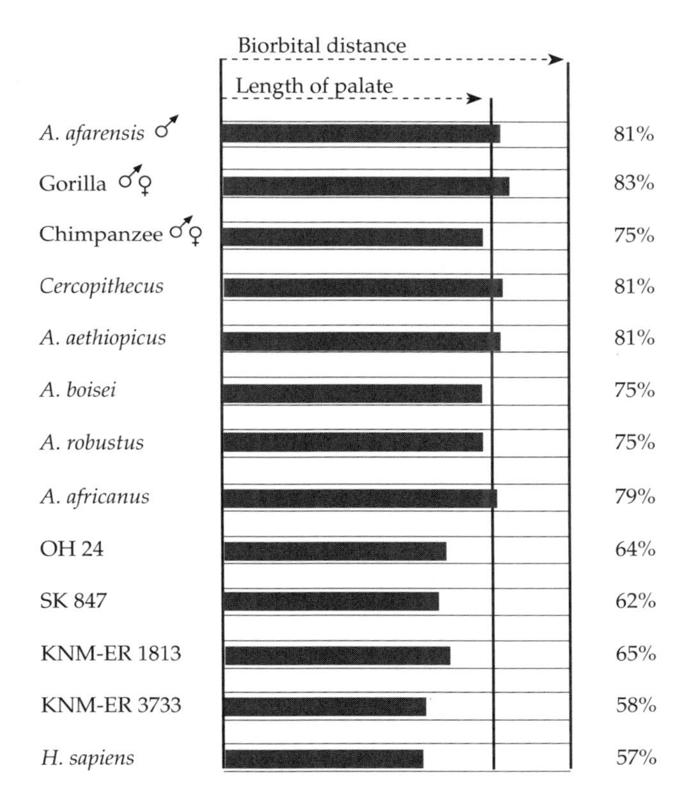

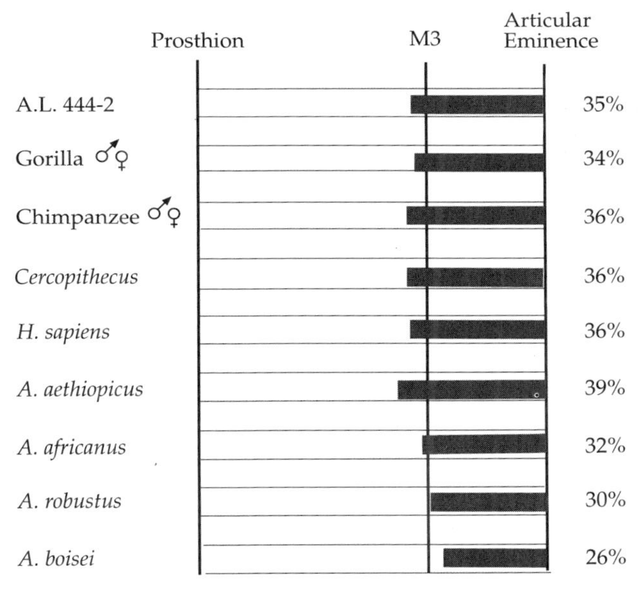

Figure 3.22 Graph expressing the relative length of the palate in A.L. 444-2 and other catarrhine primates. Palate length is shown relative to the biorbital distance (*above*) and to the horizontal distance between prosthion and the articular eminence (*below*).

Table 3.12 Measures of Prognathism

Specimen/Sample		Index of Palate Protrusion[a] (%)	Sellion–Prosthion Angle (x°)	Sellion–Nasospinale Angle[b] (x°)	Nasospinale–Prosthion Angle (x°)	Sellion–Rhinion Angle (x°)	Rhinion–Nasospinale Angle (x°)	Palatal Length (78)/ Biorbital Br. (54) (%)
Australopithecus afarensis	Mean	57	63	72	41	65	80	81
A.L. 444-2		55	65	76	39	74	85	81
A.L. 417-1d		59	60	68	42	56	74	n/a
Homo sapiens	Mean (n = 10)	15	84	82	90	54	100	57
	Range	10–26	78–90	78–87	74–104	45–60	93–105	53–60
	SD	8	4	3	10	6	5	3
Gorilla gorilla female	Mean (n = 10)	63	57	64	43	59	70	83
	Range	55–67	50–62	60–68	30–47	56–66	59–75	78–94
	SD	4	4	3	7	3	5	7
Gorilla gorilla male	Mean (n = 10)	71	52	62	34	57	68	82
	Range	64–79	45–57	53–68	23–43	45–63	62–75	78–85
	SD	5	4	4	6	6	5	3
Pan troglodytes female	Mean (n = 10)	67	59	68	45	72	63	71
	Range	58–72	50–66	59–75	37–51	64–79	51–75	65–75
	SD	5	5	6	5	5	8	4
Pan troglodytes male	Mean (n = 10)	68	54	64	39	68	59	79
	Range	61–77	47–63	56–74	29–47	61–73	48–77	70–88
	SD	6	6	5	6	4	9	6
A. africanus	Mean	54	61	68	48	68	67	80
Sts. 5		68	53	66	37	58	70	73
Sts. 71		43	65	72	47	72	72	86
Sts. 52a		(52)[c]	(63)	68	57	67	67	n/a
Stw. 505		n/a	(58)	66	50	74	60	n/a
MLD 6		n/a	(65)	n/a	n/a	n/a	n/a	n/a
Homo habilis								
KNM-ER 1813		40	65	76	47	n/a	n/a	65
OH 24		n/a	n/a	n/a	n/a	n/a	n/a	64
A. robustus								
SK 48		46	70	90	50	n/a	(97)	69
TM 1517		n/a	n/a	n/a	n/a	n/a	n/a	82
A. boisei	Mean	43	66	83	48	82	84	78
OH 5		35	72	88	55	82	90	79
KNM-ER 406		53	60	78	40	80	77	71
KNM-ER 732		(41)	n/a	n/a	n/a	n/a	n/a	(83)
A. aethiopicus								
KNM-WT 17000		82	41	58	22	59	55	81

n/a, not available.
[a]Index after Rak (1983).
[b]Or center of nasal sill (in *A. boisei*, *A. robustus*, and *A. aethiopicus*).
[c]Values in parentheses are estimates.

Figure 3.22 demonstrates that the ratio in various *Australopithecus* species, ranging from 75% to 79%, resembles that found in other hominoids. Specimen A.L. 444-2 falls within this group. Its dental arcade constitutes 81% of the biorbital distance, a value that is similar to that of gorillas and identical to that calculated for KNM-WT 17000. In *H. habilis* (KNM-ER 1813 and OH 24), in contrast, the dental arcade is reduced to about 64%–65% of the biorbital distance.

The magnitude of prognathism can be the outcome of various factors: the length of the dental arcade (as exemplified by *H. sapiens* and the baboon described above), the arcade's position relative to the braincase (as in *A. boisei*, described below), or a combination of the two. Thus, given the observations that we have summed up in Figure 3.22, we can conclude that the orthognathism that characterizes most robust *Australopithecus* crania is achieved not through a shortening of the dental arcade (as is the case in both early *Homo* and *H. sapiens*) but through a *retreat* of the whole palate toward the coronal plane of the articular eminences, as previously suggested by Rak (1983). For example, *A. boisei* exhibits the most specialized configuration: the distance between the M3 and the articular eminence is only 26% of the distance between the articular eminence and prosthion. In A.L. 444-2, the comparable value is 35%, which is similar to that of chimpanzees, gorillas, and many other primates. But there is considerable variation in *A. boisei* (Rak, 1988); in the more prognathic KNM-ER 406, for instance, the entire palate is located farther away from the articular eminence than it is in the orthognathic OH 5. One can clearly see in Figure 3.22 that KNM-WT 17000 exhibits the greatest distance between the M3 and the articular eminence. This distance constitutes 39% of the prosthion–articular eminence length. (For further discussion of this topic, see the section "Relationship between Elements of the Masticatory System.")

One puzzling issue at this point is how to accommodate the fact that gorillas and chimpanzees have a much greater degree of angular prognathism (the angle formed between the sellion–prosthion line and FH) than does A.L. 444-2, but still share with the Hadar cranium the same relative palate length (as compared to biorbital width) and position (as judged by the M^3–articular eminence length) (see above and Figure 3.22). The solution to this seeming paradox may lie in the greater proximity of the facial frame itself (comprising the nasal skeleton, the orbits, and their rims) to the coronal planes of porion and the articular eminence in the great apes than in the Hadar cranium. In our sample of female and male chimpanzees, the horizontal distance between orbitale and porion constitutes 84% and 89%, respectively, of the biorbital distance; in gorillas, the figures are 89% for females and 92% for males. The figure in A.L. 444-2 is larger, 115%. For a hominin comparison, the value calculated for Sts. 5 is 100%, and for OH 5 it is 95%. (In Figure 3.23, in which A.L. 444-2, a chimpanzee cranium, and Sts. 5 have been scaled to the same orbitale–porion horizontal distance, the Hadar skull is unrealistically small, so that, indeed, the porion–orbitale distance in that specimen is very long.) These data indicate that the great ape facial frame, but not the palate, lies closer to porion than in hominins. Thus, the snout of the great ape cranium protrudes to a much greater extent relative to the facial frame than in A.L. 444-2.

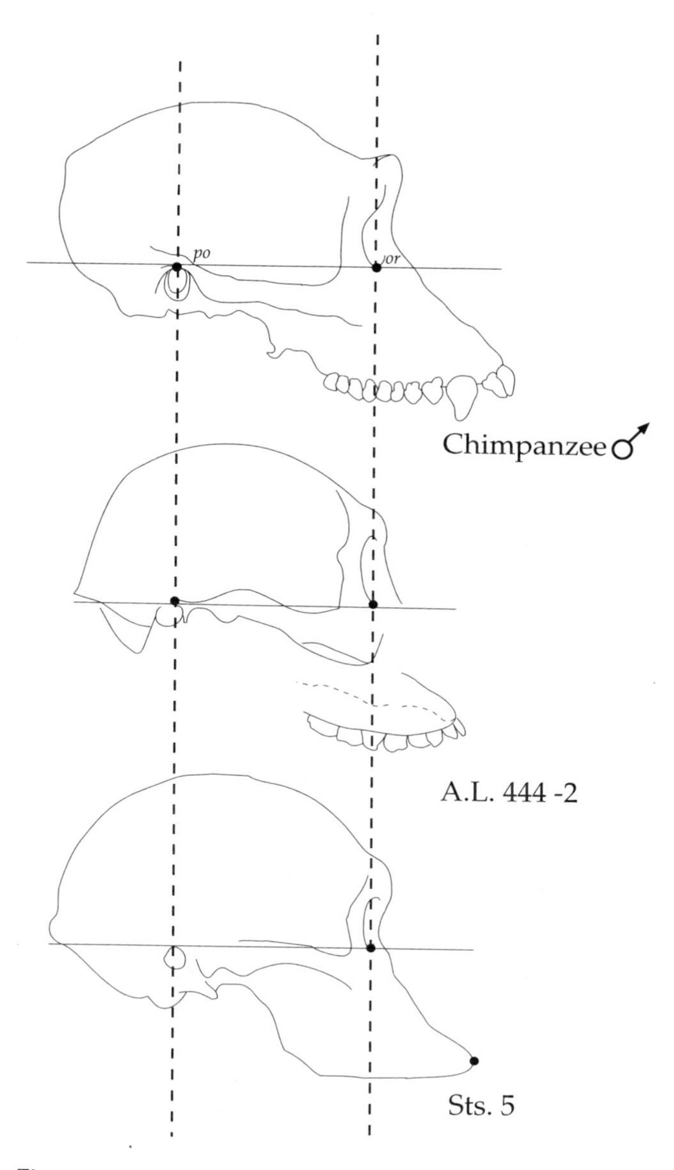

Figure 3.23 The porion–orbitale distance in a chimpanzee, A.L. 444-2, and Sts. 5. All three crania are scaled to the porion–orbitale distance of A.L. 444-2. Note that with these adjustments for size, the A.L. 444-2 cranium appears smaller than that of the chimpanzee and Sts. 5 because the porion–orbitale distance in A.L. 444-2 is actually greater.

Prognathism of the Facial Segments

The prognathism of the different segments of the face in A.L. 444-2 are expressed in the following measurements: the upper segment, sellion to nasospinale, yields an angle of 76° with the FH; the lower segment, nasospinale to prosthion, forms an angle of 39°. In modern humans, the mean angle of the upper segment as measured in our sample is 82°. The inclination of the nasoalveolar clivus, the lower segment of the facial profile, is 90° in our human sample. From these latter values we can see that the differences in the lower segment's inclination and length are what contribute most significantly to the difference in the degree of *total* facial prognathism—the sellion–prosthion angle—between modern humans and A.L. 444-2 (Figure 3.24).

In the female gorilla sample, the mean value for the upper segment is 64° and for the clivus, it is 43°; males

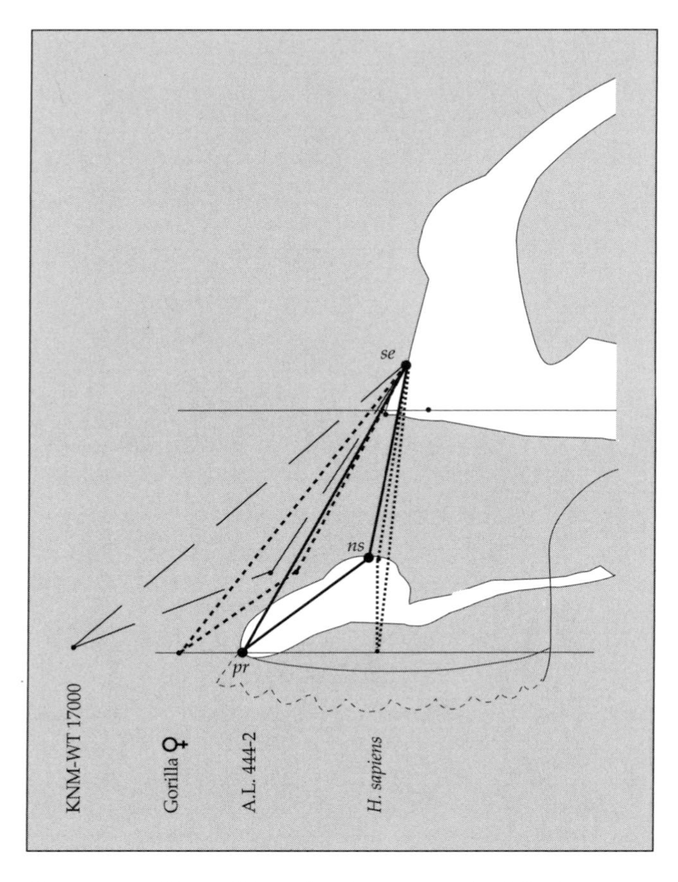

Figure 3.24 Prognathism of various segments of the face (sellion–prosthion, sellion–nasospinale, and nasospinale–prosthion) in A.L. 444-2 and other hominoids. The outlines are drawn to fit the vertical projected sellion–prosthion distance of A.L. 444-2.

exhibit smaller values, especially those representing the inclination of the clivus. Chimpanzees present a similar picture, with mean segment angles of 68° and 45°, respectively, in the females; again, males yield lower values, which indicate a greater inclination of the segments, particularly of the clivus. The difference between the angles of the nasoalveolar clivus in A.L. 444-2 and gorillas, for example, stems primarily from the anterior position of nasospinale in the latter (Figure 3.24). Recall the anterior position of the entire palate in the apes, as represented by the index of palate protrusion.

From Table 3.12, which presents the values of prognathism in hominins, we can conclude that the variation in the *A. africanus* values is primarily the product of variation in the length and inclination of the clivus. See, for example, the values for Sts. 5 and Sts. 52a, in which the inclination of the sellion–nasospinale line is almost identical, but the difference in the inclination of the clivus is 20°. Very high values characterize the inclination of the sellion–nasospinale line in *A. robustus* and *A. boisei* (in these species, this line runs from sellion to the center of the nasal sill, the homologous site for nasospinale), with some specimens exhibiting values even greater than those yielded by modern humans. These very high values of the upper segment in some robust *Australopithecus* specimens override the effect of a clivus that is not only inclined but also very long, and thus they bring about orthognathism of the whole face. Orthognathism in *A. robustus* and *A. boisei* is clearly the outcome of the spatial relationship between sellion and prosthion, while the nasoalveolar clivus remains strongly inclined.

In KNM-WT 17000 the corresponding values are 58° for the upper facial segment and a very small 22° for the lower. The degree of prognathism of the upper segment in this hominin is even greater than that seen in male gorillas. The lower segment, the nasoalveolar clivus, is also considerably more prognathic in KNM-WT 17000, with an angle that is approximately half that of the apes (Figure 3.24). As a result, total facial prognathism in this specimen is much greater than in gorillas—rendering KNM-WT 17000 the only hominin to exhibit such a configuration.

The inclination of the nasal bones is not necessarily related to the degree of total facial prognathism or to prognathism of the nasoalveolar clivus.[9] That is, the "prognathism" of the nasal profile (the inclination of the sellion–rhinion line in reference to FH) is often unrelated to the prognathism of the face as a whole (the inclination of the sellion–prosthion line). In A.L. 444-2, the inclination of the nasal profile is 74°. Modern humans exhibit an angle of 54°; in female gorillas, the profile is inclined at 59°, and in the males, at 57°. The segment is much steeper in chimpanzees, as shown in Table 3.12.

Since nasion is often high enough to coincide with glabella in the African apes—especially gorillas—sellion is deeply recessed and the contour of the nasal bones assumes a long, concave shape, unlike the rather straight profile in A.L. 444-2 and A.L. 417-1d.

A. africanus displays a rather steep sellion–rhinion profile, which is very similar to that of A.L. 444-2. Indeed, the only exception to the *A. africanus* configuration is the prognathic Sts. 5, which exhibits the smallest angle, at 58°, of all the *A. africanus* specimens on which the angle can be measured (Table 3.12).

The shrinking of the human facial skeleton and the resulting posterior retreat of prosthion apparently do not influence the inclination of the sellion–rhinion segment; hence, the nasal bones emerge as the most dramatically protruding part of the facial profile in modern humans. In OH 5, in contrast, orthognathism is associated with an almost vertical orientation of the nasal bones, and rhinion is the deepest topographical point in the central face (Table 3.12). This anatomy is similar to that of *A. robustus* and was undoubtedly present in a number of incomplete specimens, including SK 11, SK 12, SK 13, SK 46, and SK 83.

The substantial facial prognathism in gorillas is also expressed in the inclination of the actual plane of the nasal aperture. Nasospinale lies nearly 10 mm anterior to the coronal plane of rhinion. Rhinion and nasospinale form an angle of 70° in the females and 68° in the males (Table 3.12). Chimpanzees yield lower values: 63° in the females and 59° in the males. The horizontal distance between rhinion and nasospinale in A.L. 444-2 is only 3 mm. Hence, the rhinion–nasospinale line is steeper than in the apes and forms an angle of 85° with the FH. Nevertheless, the considerable length and prognathism of the nasal bones in the African great apes, particularly in gorillas, act together to reduce the slope of the nasal opening and to extend its superior end substantially away from the coronal plane of the orbits. In A.L. 444-2, in contrast, the nasal opening is steep because of the position of both rhinion and nasospinale and lies near the coronal plane of the orbits. (For further discussion, see the section "The Nasal Aperture: Its Shape, Inclination, and Position in the Facial Mask.)

A somewhat similar morphology to that found in the African great apes is seen in prognathic hominins, such as Sts. 5 and KNM-WT 17000 (despite the fragmentary state of the inferior parts of the latter specimen's nasal bones). In relatively orthognathic hominins, such as OH 5, SK 12, and SK 48, the plane of the "formal" nasal aperture, from rhinion to nasospinale (or, rather, to the center of the nasal sill), is vertical but recessed relative to the facial elements that constitute the medial parts of the face. Hence, a sagittal wall forms on each side of the nasal opening, and a "subnasal gutter" spans the area between the bottom parts of these two walls. The corridor formed by the walls leads to the nasal cavity itself (Rak, 1983). (For further discussion, see the section "The Nasal Aperture: Its Shape, Inclination, and Position in the Facial Mask.") In modern humans, at least in our western Asiatic sample, rhinion is located substantially anterior to nasospinale at a distance of 8 mm. Hence, the angle of the rhinion–nasospinale line is greater than 90°, with a mean of 100°.

In gorillas, chimpanzees, and many other prognathic primates, the snout is connected topographically to the infraorbital plates via sagittally oriented bony walls that stretch from the medial part of the infraorbital region to the protruding palate (Figure 3.25). These walls in the gorilla and chimpanzee are extensive in lateral view because of the magnitude of prognathism, on the one hand, and the infraorbital region's posteroinferior retreat from the inferior orbital margin, on the other. In A.L. 444-2, the infraorbital region is essentially vertically oriented, but since the snout is less prognathic, the lateral walls are not as extensive as in the great apes. In *A. africanus*, as can be seen in specimens TM 1511, Sts. 5, Sts. 71, Stw. 505, and MLD 6, the infraorbital region slopes downward and forward; hence, despite the substantial prognathism, the lateral walls are not dominant at all. Although it is very prognathic, KNM-WT 17000 displays only a small trace of the lateral wall of the snout, as the infraorbital region also slopes anteriorly to a considerable degree. In *A. robustus* and *A. boisei*, no sign of the wall can be found. The absence of the canine fossa's medial walls in *A. robustus* and *A. boisei* is self-explanatory: the inferior portion of the infraorbital bone surface slopes so anteriorly that it reaches the coronal plane of prosthion.

Relationship between Elements of the Masticatory System

A system of measurements and indices was developed by Rak (1983) to quantify the geometric relationships within the masticatory system. In essence, the method evaluates the extent to which the dental arcade retreats posteriorly beneath the braincase, on the one hand, and the extent to which the masticatory muscles, primarily the masseter, move anteriorly away from the temporomandibular joint on the other. The indices developed for this method are as follows:

- The first index, shown in Figure 3.26, column A, evaluates the distance to which the dental arcade extends anteriorly past the coronal plane of the deepest point in the sagittal facial contour (sellion). This anteriorly protruding section of the dental arcade is expressed as a percentage of the entire arcade length. (The index is discussed above, in the section "Degree of Prognathism and Relative Length of the Dental Arcade.")
- The second index, shown in Figure 3.26, column B, measures the portion of the dental arcade that protrudes anteriorly beyond the coronal plane of the zygomatic tubercle (the most anterior point of the masseter muscle's origin). This anteriorly protruding section of the dental arcade is expressed as a percentage of the entire arcade length.
- The third index, shown in Figure 3.26, column C, expresses the horizontal distance to which the masseter origin extends anteriorly beyond the deepest point of the face (sellion). This distance (from the articular eminence to the zygomatic tubercle) is expressed as a percentage of the horizontal distance between the articular eminence and sellion.
- The fourth index, shown in Figure 3.26, column D, expresses the proximity of M^3 (the posterior end of the dental arcade) to the articular eminence. This distance is expressed as a percentage of the horizontal distance between the articular eminence and sellion.
- The fifth index, shown in Figure 3.26, column E, summarizes the "efficiency" of the whole mastica-

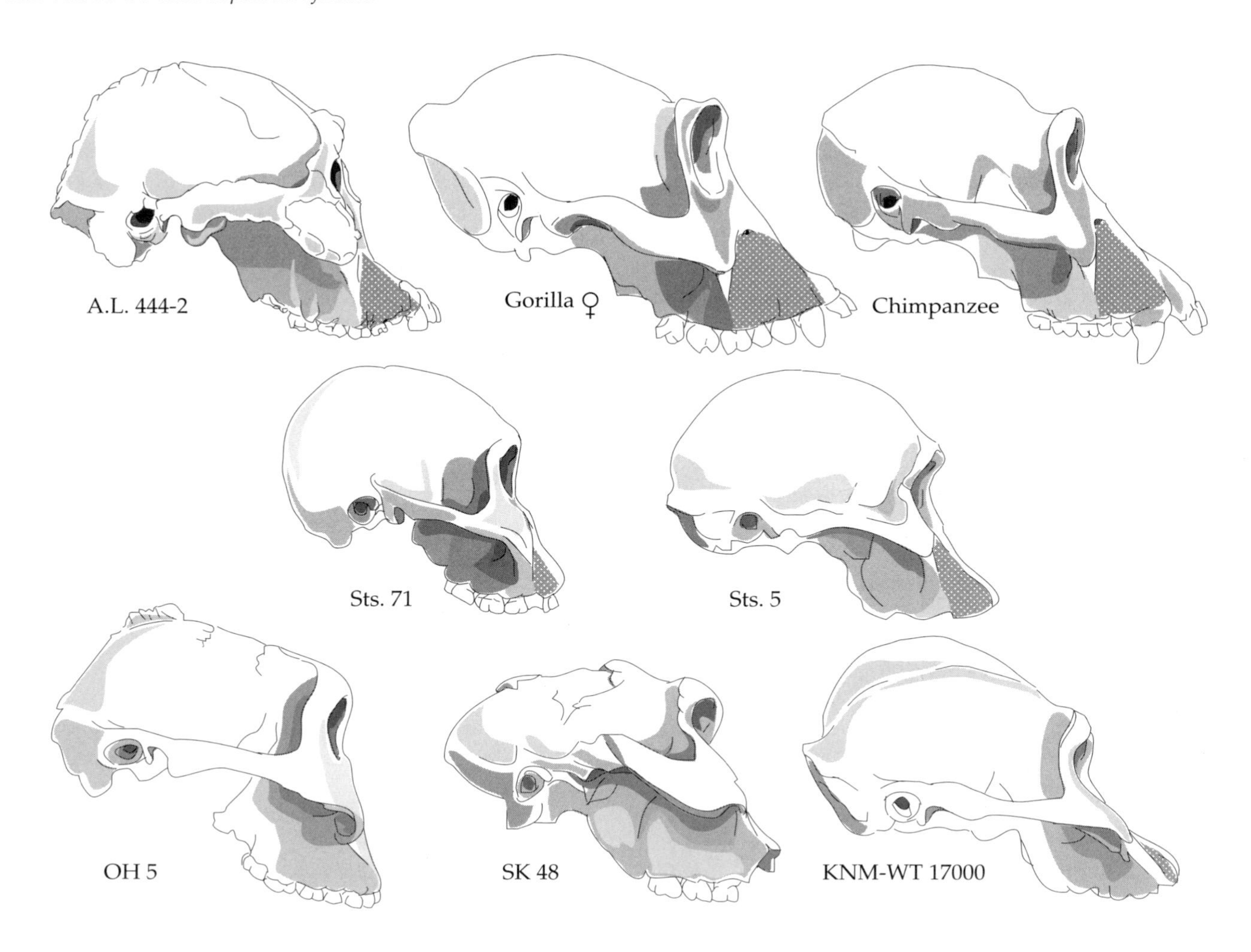

Figure 3.25 The medial wall of the canine fossa in A.L. 444-2 and other hominoids. The extent of the canine fossa, which forms the lateral wall of the snout, is indicated by the dotted area. The wall extends from the infraorbital region and the zygomaticoalveolar crest to the anterior outline of the face.

tory system. Coined the "index of overlap" (Rak, 1983), it measures the extent to which the dental arcade length overlaps the distance between the articular eminence and the zygomatic tubercle. This overlapping segment is expressed as a percentage of the total length of the masticatory system—the horizontal distance between the articular eminence and prosthion.

The gorilla and the chimpanzee represent the generalized configuration, as their index of overlap values resembles those of other primates (Figure 3.26 and Table 3.13). Modern humans, whose index value is 29%, are also part of this generalized group. However, the shrinkage of the palate's length in modern humans introduces some incompatibility to the comparison. (If the length of a gorilla's palate were substituted for the length of the shrunken human palate, the value in modern humans would be reduced to 21%.)

A. afarensis, as represented by A.L. 444-2 (the only specimen that permits measurements of this sort) exhibits the most generalized masticatory system of any *Australopithecus* species. The value of the index of overlap in this specimen is 26%. As expected, all the components that contribute to this index are generalized, as well (Table 3.13). In all of the indices, A.L. 444-2 resembles *A. africanus* but is somewhat more generalized. Until the discovery of A.L. 444-2, *A. africanus* had displayed the most generalized masticatory system of any *Australopithecus* species. The indices of A.L. 444-2 fall between those of *A. africanus* and the African apes but are much closer to the former.

Specimen KNM-WT 17000 presents an interesting case. Even though it is extremely prognathic, its masseter

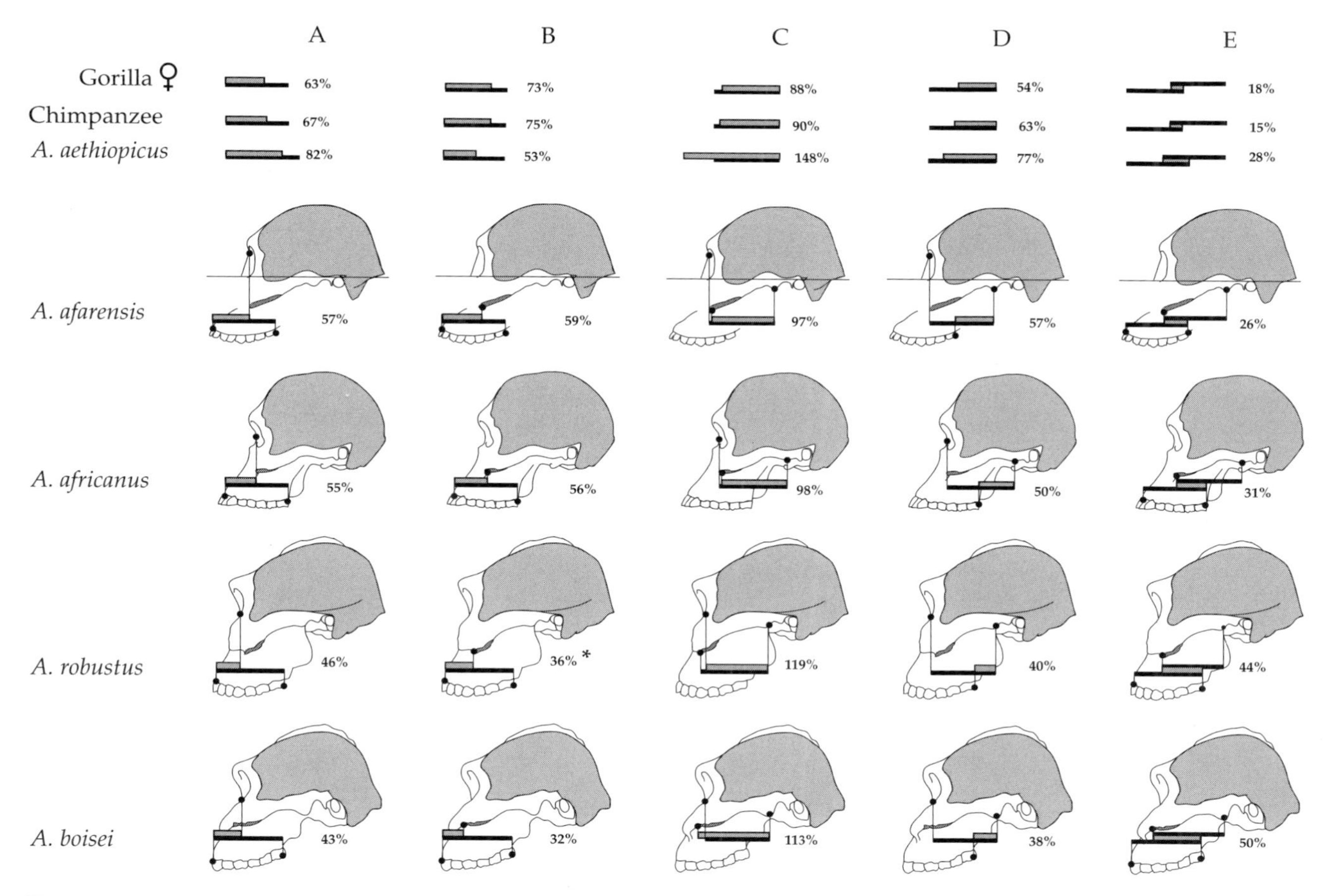

Figure 3.26 Rak's (1983) indices expressing the configuration of the masticatory system in A.L. 444-2 and other hominoids. The letters at the top refer to those in Table 3.13 as follows: A, the degree of the palate's protrusion anterior to sellion; B, the degree of the palate's protrusion anterior to the zygomatic tubercle; C, the degree of the masseter's forward extension relative to the distance between sellion and the articular eminence; D, the position of M³ relative to the distance between sellion and the articular eminence; E, the degree of overlap between the palate length and the articular eminence–zygomatic tubercle distance. The values for *A. afarensis* are based on A.L. 444-2, except for palate protrusion, which is the mean for A.L. 444-2 and A.L. 417-1. All other species' values are based on samples enumerated in Table 3.13. The asterisked value is a correction of the value stated in the original publication. (Adapted from Rak, 1983.)

seems to "overcome" the prognathism through the substantially anterior location of its insertion on the extended zygomatic bone. Still, the index of overlap is only 28%.

The two antagonistic shifts reach their extreme in *A. boisei*. In this species the masseter origin is extended maximally forward, and the dental arcade exhibits the greatest posterior retraction of all the hominins. Here, 50% of the total length of the masticatory system—the horizontal distance between the articular eminence and prosthion—is occupied by the portion of the palate length that overlaps with the distance between the articular eminence and the most anterior part of the masseter. By themselves the values constituting the overlapping index are rather variable in *A. boisei*, but their interaction yields a relatively narrow range of variation in the index, and thus a consistently derived masticatory configuration (Table 3.13). For further discussion of the variation in the masticatory system within *A. boisei*, see Rak (1988).

The Zygomatic Arch

The zygomatic arch of A.L. 444-2 is long, massive, and rugosely marked by the masseter muscle's origin. Its length, from porion to the point where it is attached to the body of the zygomatic bone, is 84 mm.

The vertical depth and the shape of the arch as seen in lateral view vary significantly from species to species. In A.L. 444-2, the arch is extremely deep vertically, mea-

Table 3.13 Indices of the Masticatory Apparatus (%)

Specimen/Sample		A Palate Protrusion Anterior to Sellion	B Palate Protrusion Anterior to Masseter	C Position of Anterior Part of Masseter	D M3 Position	E Overlapping Index
Australopithecus afarensis	Mean	57	54			
A.L. 444-2		55	59	97	57	26
A.L. 417-1d		59	48	n/a	n/a	n/a
Gorilla gorilla female	Mean (*n* = 6)	63	73	88	54	18
	Range	58–66	65–83	79–97	48–61	11–24
	SD	4	5	8	7	4
Gorilla gorilla male	Mean (*n* = 6)	70	76	94	66	15
	Range	64–78	72–84	82–106	57–76	8–17
	SD	5	7	9	8	6
	Pooled mean (*n* = 12)	67	75	91	60	17
	Range (*n* =12)	58–78	65–85	79–106	48–76	8–24
	SD	5	6	9	8	5
Pan troglodytes female	Mean (*n* = 6)	67	77	87	63	14
	Range	58–72	72–81	79–96	56–68	11–18
	SD	5	3	6	4	3
Pan troglodytes male	Mean (*n* = 6)	68	74	93	63	17
	Range	61–77	70–83	88–102	57–74	10–18
	SD	6	5	7	6	3
	Pooled mean (*n* = 12)	67	75	90	63	15
	Range (*n* = 12)	58–77	70–83	79–102	56–74	10–18
	SD	6	4	7	5	3
A. africanus	Mean	55	56	98	50	31
Sts. 5		68	65	103	66	22
Sts. 71		43	47	93	35	40
A. robustus	Mean	46	36	119	40	44
SK 48		46	31	119	40	50
TM 1517		n/a	42	n/a	n/a	37
SK 52		n/a	34	n/a	n/a	45
A. boisei	Mean	43	32	113	38	50
OH 5		35	28	110	31	57
KNM-ER 406		53	26	130	49	50
KNM-ER 732		41	41	100	36	43
A. aethiopicus						
KNM-WT 17000		82	53	148	77	28

From Rak (1983).

suring between 22 and 24 mm at the midpoint of its length (in its present, deformed condition, the left arch is deeper than the right). This dimension is only 13 mm in modern humans, 17 mm in female gorillas, and 21 mm in male gorillas. It is the smallest in chimpanzees, at 11 mm.

Two structures contribute to the shape and the substantial vertical depth of the zygomatic arch in A.L. 444-2. Both structures lie anterior to the relatively delicate posterior root of the arch, which measures only 8 mm vertically (immediately anterior to the postglenoid process; Figure 3.5). One is the very prominent articular tubercle in the arch's lower margin, and the other is a pronounced rise in the superior margin of the arch. Although these structures are not situated exactly opposite each other on the arch, together they demarcate the beginning of its increase in vertical depth. From these structures forward, the arch gains in depth continuously until it reaches the body of the zygomatic bone. In A.L. 444-2, the great depth of the arch contrasts with the thinness of its posterior root.

The configuration is different in *A. boisei*. Surprisingly, the zygomatic arch is not especially deep, measuring about 13 or 14 mm in specimens OH 5, KNM-ER 406, and KNM-ER 13750. Although the articular tubercle is massive and protrudes considerably downward, there is no convexity on the superior margin of the arch. Instead, this margin runs in a straight line anteriorly to the body of the zygomatic bone. In *A. robustus* (e.g., SK 46, SK 48, SK 83, and TM 1517) a strongly convex hump breaks the otherwise straight contour of the superior margin of the arch. However, as in gorillas, but unlike in A.L. 444-2, this hump drops in height anteriorly.

Even though the zygomatic arch in A.L. 444-2 is slightly convex on the sagittal plane, its superior margin is still at the level of FH (Figure 3.27). As in modern humans and other hominins, the superior margin of the anterior root of the zygomatic arch is situated at the level of the orbital floor (orbitale). This configuration contrasts sharply with that of both *A. boisei*, as seen in specimens KNM-ER 406, KNM-ER 13750, and possibly OH 5, and *A. aethiopicus* specimen KNM-WT 17000. In all of these, the superior margin of the anterior root of the arch reaches a higher level than the floor of the orbit. The superior margin at the anterior end of the arch in the African apes is situated substantially below the level of orbitale, a disparity that results in the anteroinferior inclination of the zygomatic arch. This description is particularly applicable to gorillas (Figure 3.27), in which the superior margin of the zygomatic arch extends anteriorly as a horizontal line and then drops anteroinferiorly to create the characteristic hump. The topographical drop in chimpanzees occurs immediately anterior to the posterior root of the zygomatic arch and thus forms no hump. As a result, the contour of the zygomatic arch itself is smoother and more horizontally inclined than in gorillas—that is, a long section of the arch is more aligned with the FH but is situated substantially lower.

Although the site where the superior margin of the arch merges with the body of the zygomatic bone is reconstructed in A.L. 444-2, there is little doubt that the upper margin of the arch formed a right angle with the frontal process of the zygomatic bone. A similar angle is found in gorillas and chimpanzees, although the actual position of the angle, as just mentioned, is considerably below the level of orbitale. *A. boisei* typically displays an obtuse angle (actually, a curved line) between these two processes, as the outcome of the extreme broadening of their roots. In KNM-WT 17000, a right angle is observed between the two processes, with a distinct transition from one process to the other.

Root of the Zygomatic Process of the Maxilla

In A.L. 444-2 the zygomatic process emerges from the body of the maxilla above the middle of the dental arcade's length. The process takes the form of a swollen structure whose cross section runs in a smooth, continuous curve from its temporal surface to its facial surface. A sagittal cross section of the root of the zygomatic process (Figure 3.28) helps clarify this morphology. The excavated canine fossa at the front of the zygomatic process gives the anterior margin of the cross section an indented appearance in A.L. 444-2. Its posterior, or temporal, surface is rounded. This uneven appearance of the cross section is typical of primates with a deeply excavated canine fossa. Many *A. afarensis* specimens (A.L 199-1, A.L 200-1a, A.L 333-1, A.L 417-1d, A.L 427-1, A.L 442-1, and A.L 486-1) bear the remains of the zygomatic process and closely match the cross section described above. The lowest point on the root of the zygomatic process of the maxilla in A.L. 444-2, as in the other *A. afarensis* specimens, is located at the coronal level of M^1. This position reflects the relatively posterior origin of the process in this species, as discussed

Figure 3.27 The inclination and position of the zygomatic arch in A.L. 444-2, *A. boisei*, and African apes. The anterior root of the zygomatic arch is below the orbital floor in the female gorilla and the chimpanzee; approximately level with the floor in A.L. 444-2, as in *A. africanus* and *Homo*; and above it in *A. boisei*. Note the difference between gorillas and chimpanzees in the orientation of the zygomatic arch relative to the Frankfurt Horizontal (FH).

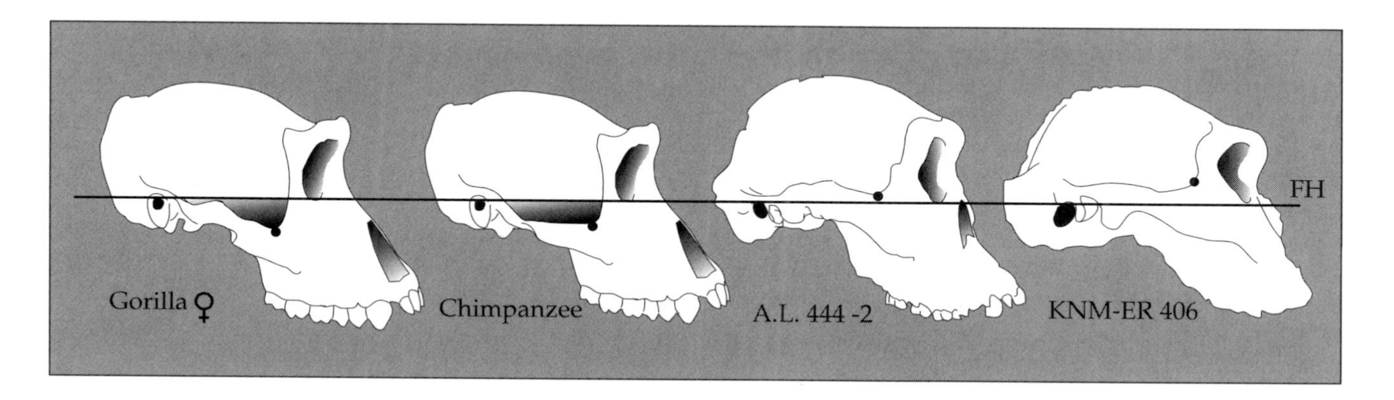

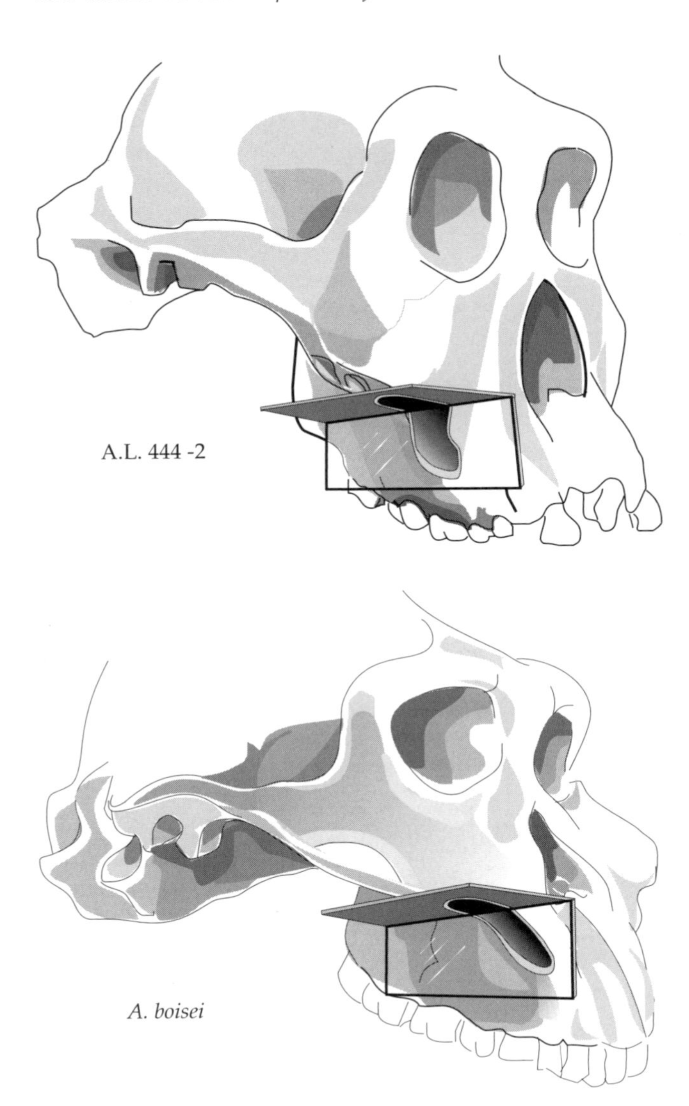

Figure 3.28 Sagittal and transverse sections through the root of the zygomatic process of the maxilla in A.L. 444-2 and *A. boisei.* The sagittal part of the cross section is vertical in A.L. 444-2. In *A. boisei,* the sagittal part of the cross section is inclined, forming an integral part of the visorlike structure commonly found in the infraorbital region in this species.

in the section "Relationship between Elements of the Masticatory System."

A substantial morphological difference can be observed between *A. afarensis* and *A. boisei,* which is particularly significant in light of the similar size of their facial skeletons and massiveness of their facial components (Figure 3.28). The anatomical configuration of the root of the zygomatic process of the maxilla in *A. boisei* has been described in Rak (1983) as follows:

> The zygomatic process of the maxilla in *A. boisei* is most unusual. The manner in which it emerges from the body of the maxilla is an apparent adjustment to the visorlike shape of the infraorbital region in the more lateral part of the face. The process is thin anteroposteriorly, assuming the shape of an obliquely oriented plate, with its inferior part (the root of the zygomaticoalveolar crest) reaching forward and its temporal surface facing almost inferiorly. As the posterior aspect of the visor, this surface is concave, unlike the posterior surface of the inflated zygomatic process (with an unconfined sinus) that characterizes the generalized morphology, including that of *A. robustus.* The morphology of the zygomatic process of the maxilla of *A. boisei* can be seen clearly in specimens KNM-ER 406 and KNM-ER 732; it is so distinctive that even the minute remains of the process that are found on the eroded palates of KNM-ER 405 and KNM-ER 733 are sufficient to suggest it. Reconstruction of the more lateral part in the latter specimens, which can be achieved through extrapolation of the remains of the process, will undoubtedly result in a visor-shaped infraorbital region. Furthermore, on the basis of the orientation of the remains of the process, the visor of KNM-ER 405 can be assumed to be the most extended and upward flaring of all the known *A. boisei* specimens. Specimen OH-5 has very little of this particular morphology of the zygomatic process, which is in accordance with the modest flare of its visorlike infraorbital region. (p. 56)

The face of A.L. 444-2, despite the similarity in size to that of large *A. boisei* specimens, demonstrates the generalized hominin configuration. Similarly, hominins with a more generalized face than that of *A. boisei*—including modern humans; *H. habilis* (KNM-ER 1813, OH 62, Stw. 53); and other early *Homo* specimens, such as SK 847 and KNM-ER 1470—share with A.L. 444-2 the primitive shape of the cross section of the maxillary zygomatic process. *A. africanus* and even *A. robustus* also exhibit this common anatomy, as in the following examples: *A. africanus* specimens Sts. 5, Sts. 17, Sts. 71, and Stw. 505; and *A. robustus* specimens TM 1517, SK 46, and SK 48, in which the process, though generalized, is also extensively pneumatized and swollen. However, in KNM-WT 17000, the cross section resembles that of *A. boisei*; in other words, the shape, orientation, and anterior position of the inferior edge of the root of the process are all suited to a visorlike infraorbital region.

The Lateral Orbital Region

In A.L. 444-2, the lateral orbital region (consisting of the frontal process of the zygomatic bone and the zygomatic process of the frontal bone) lies almost entirely on one anteriorly facing plane. Only the marginal process, which protrudes from the posterolateral edge of the frontal process, swings around and faces laterally. Superiorly, the lateral orbital region extends anteriorly to join the (reconstructed) bulbous lateral extremity of the supraorbital torus, thus assuming a slightly concave appearance in lateral view. In contrast, the lateral contour in modern humans coincides with the rim of the orbital margin itself because the surface of the zygomatic bone's frontal process faces laterally instead of anteriorly.

In its anteriorly directed frontal process, the A.L. 444-2 zygomatic bone resembles that of gorillas. Similarly, the upper part of this process in gorillas is slightly concave, as it is the product of the bulbous lateral end of the supraorbital torus. Chimpanzees are usually characterized by a flat frontal process that faces anterolaterally. Its sharp medial margin serves as a clear border between the orbital cavity and the surface of the lateral orbital region. This configuration is reminiscent of the comparable site in *Homo*.

In *A. africanus*, the surface of the frontal process also faces forward. It is flat transversely but sharply angulated on the sagittal plane. The best examples of this morphology are Sts. 5 and Sts. 17. It is also seen in Sts. 71, in which the lateral orbital margin is concave but the anterior bone surface faces more laterally. This morphology is undoubtedly variable.

The lateral orbital region in *A. boisei* is nearly flat on both the sagittal and transverse planes. The surface slopes anteriorly to adjust to the topography of the visor-shaped zygomatic bone and faces slightly laterally. This topography and orientation are present in every *A. boisei* specimen in which the relevant anatomy is preserved, such as OH 5, KNM-ER 406, KNM-ER 732, and KNM-ER 13750. The sample is rather small in *A. robustus*; the morphology can be observed only in SK 48, SK 52, and TM 1517. In these three specimens, the process is narrower and more laterally facing than in *A. boisei*, and, as such, it also differs from that of *A. africanus*.

Specimen KNM-WT 17000 is unique among the hominins—and, indeed, among all the primates—in the anatomical configuration of this region. Although very little remains, we can observe its reconstructed state. In the absence of a defined orbital rim demarcating the lateral margin of the orbital cavity, the transverse cross section assumes a convex appearance because its bone surface passes insensibly from the deep part of the orbital cavity to the lateral margin of the frontal process.

When examined in a lateral view, the slightly concave sagittal contour of the lateral orbital region—combined with the similarly curved, midsagittal profile of the upper face seen in A.L. 444-2—presents a configuration that is typical of many other primates. However, a quite different morphology is observed in Sts. 5 (Rak, 1983), in which two sharply angulated contours coincide in a lateral view. As a result, the facial mask looks folded along a horizontal axis that stretches from the recessed center of one lateral orbital region to that of the other. A less pronounced version of this "folded" morphology is observed in specimens Sts. 17 and Sts. 71, and is probably also present in Stw. 505, although that specimen lacks much of the lateral part of the face. In the *A. boisei* specimens OH 5, KNM-ER 406, KNM-ER 732, and KNM-ER 13750 and the *A. robustus* specimen SK 48, a different configuration is evident. A massive, midsagittally convex glabellar region bulges anteriorly, contrasting sharply in lateral view with the concave contour of the lateral orbital margin. In KNM-WT 17000, the morphology is similar to that of *A. robustus* and *A. boisei*, even though its relatively delicate glabellar region is not as anteriorly prominent as is common in these two taxa.

Shape and Height of the Squamosal Suture

Only the superior segment of the squama of the left temporal bone is preserved on A.L. 444-2; the entire base of the squama is missing. From what does remain, we can infer that the squama was highly arched, reaching a great height on the braincase. The shape of the arch can be confirmed on A.L. 333-45 (as well as on the juvenile A.L. 333-105), although in A.L. 444-2 the squama seems to be more elevated. In this respect, the temporal squama of *A. afarensis* resembles that of modern humans and other hominins and differs considerably from that seen in the African apes (Figure 3.29; see also the discussion of this issue by Weidenreich, 1943). The path of the squamosal suture in the apes is straight and situated low on the braincase. At its summit, the height of the suture constitutes only 39% of the calvaria's height in female gorillas and 37% in males; in female chimpanzees, this value is 29% and in males, 32%. The suture in modern humans ascends rather far, occupying 43% of the calvaria's height. Note that the height of the braincase acts to reduce the index value in humans.

In A.L. 444-2, the impression that the squama is very tall may stem partially from deformation of the cranial base. This deformation produced a gap between the squama and its base and thus caused the left porion to descend about 9 mm relative to the right porion. (See Figure 2.3a, where the gap can be seen between fragments 2 and 4; see also Figure 2.5.) If the gap is excluded from the sutural height calculations, the distance to which the suture ascends above the FH is 56 mm, which constitutes 67% of the calvaria's height.

The index value in *H. sapiens* is not comparable to the early hominin data because of the dramatic increase in the modern human calvarial height. The relatively low *A.*

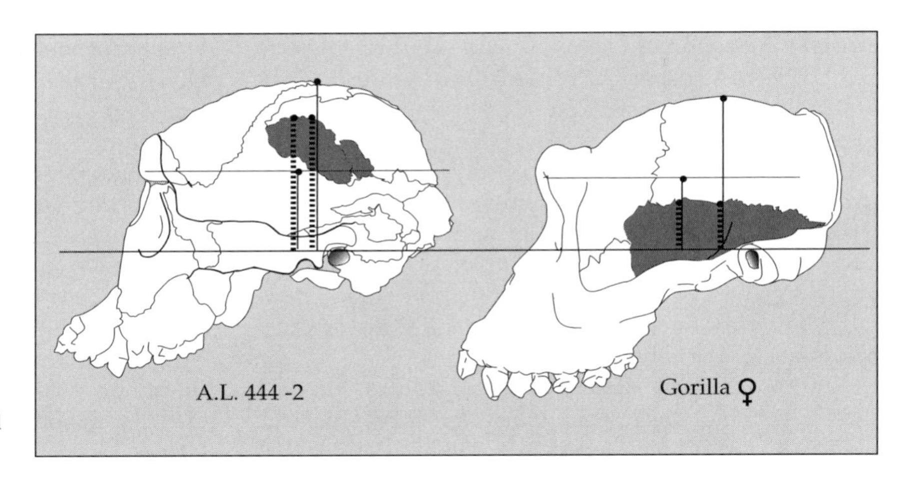

Figure 3.29 Differences in the height of the squamosal suture in A.L. 444-2 and a female gorilla. Two pairs of lines (one dotted and one solid) are displayed on each specimen. The pair on the left represents the height of the squamosal suture relative to the height of the orbital roof. The pair on the right represents the height of the suture relative to the calvarial height.

boisei skull displays the opposite effect on the index calculation. By expressing sutural height relative to the orbital roof, we can eliminate the effect of calvarial height variation. As can be seen in Table 3.14 and Figure 3.29, the apex of the suture attains a height that is only 74% of the orbital roof's position above the FH in gorillas and 64% in chimpanzees, whereas in modern humans, the value is 136%. In other words, the squamosal suture in humans is substantially higher than the orbital roof. In A.L. 444-2, the suture is even more elevated, at 143% of the orbital roof's height—a value that is higher not only than the mean of four *A. africanus* specimens, at 106%, but also than the *A. boisei* mean of 132%. The latter species is known for the extensive overlapping of its temporal squama with the parietal (Rak, 1978). (This index eliminates the effect of the above-mentioned gap running along the base of the temporal squama in A.L. 444-2, as the gap is a factor in both the numerator and denominator.)

In spite of its height, the squamosal suture in A.L. 444-2 lacks the extensive striae parietalis that characterize the skull of *A. boisei* (Rak, 1978). Indeed, in the Hadar cranium, the squama overlaps the parietal bone only slightly, in contrast to the extensive overlap observed in *A. boisei.* In the *A. boisei* specimen KNM-ER 23000, in which the exposure of the entire area of the suture permits the pertinent measurements, the extent of overlap is immense. Extending 28 mm (arc measurement) along the coronal arc of the parietal, the overlap constitutes 52% of the remaining uncovered coronal parietal arc. This value in OH 5 is 36%, and in L.338y-6 (a juvenile), it is 53%. The exact extent of overlap in A.L. 444-2 cannot be measured, but undoubtedly it was considerably smaller than it was in *A. boisei.*

A comparison of the coronal cross section of the human calvaria with that of A.L. 444-2 (Figure 3.30) clearly reveals that the evolutionary increase in cranial capacity is accompanied by an allometric change in size of the parietal bone while the height of the squama remains constant. In other words, the increase in the size of the braincase in modern humans is achieved not through a proportional increase in the area of both the temporal squama and the parietal but, rather, through a dramatic expansion of only the parietal.

The Cranium: Vertical View

General Outline of the Cranium

In superior view (Figure 3.31), the external outline of the cranium—which is delineated by the lateral aspect of the zygomatic arches and thus includes the temporal foramina—forms a near circle. The lateral flair of the zygomatic arches—which deviate considerably laterally from their posterior roots and converge anteriorly from their widest point toward the midline via a smooth, somewhat flattened arc—lends the contour its circular appearance. An additional factor that contributes to this shape is the arches' smooth merging with the lateral ends of the supraorbital torus by way of the zygomatic bones. Only a slight, almost indistinguishable, angle demarcates the transition between the lateral contour of the arches and the peripheral face. In fact, this outline of the cranium is identical to that in the basal view (which is described in further detail in the section "The Cranium: Basal View").

Unlike their configuration in A.L. 444-2, the zygomatic arches in gorillas, much as in chimpanzees, first deviate laterally and then straighten out on the sagittal plane before converging slightly toward the midline. The arches then abruptly turn medially, forming a distinct angle with the body of the zygomatic bones. Consequently, the configuration renders the outline of the gorilla cranium more rectangular in a superior view. The contour in chimpanzees is somewhat more rounded and thus bears a greater resemblance to that of A.L. 444-2.

Table 3.14 Height of the Squamosal Suture

Specimen/Sample		Maximum Suture Ht. (mm)	Index of Squamosal Suture/Vault Ht.[c] (%)	Index of Squamosal Suture/Ht. of Orbital Roof[c] (%)
Australopithecus afarensis				
A.L. 444-2		56	67	143
Homo sapiens	Mean (*n* = 10)	49	43	136
	Range	43–55	38–49	119–158
	SD	4	3	14
Gorilla gorilla female	Mean (*n* = 10)	27	39	74
	Range	21–31	36–41	70–78
	SD	3	3	4
Gorilla gorilla male	Mean (*n* = 10)	29	37	75
	Range	23–33	32–42	55–89
	SD	5	5	16
Pan troglodytes female	Mean (*n* = 10)	19	29	63
	Range	16–22	26–33	61–67
	SD	2	3	3
Pan troglodytes male	Mean (*n* = 10)	21	32	65
	Range	18–23	26–35	56–74
	SD	2	3	7
A. africanus	Mean	35	46	106
Sts. 5		34	45	110
Sts. 71		39	53	121
Stw. 505		35	43	85
MLD 37/38[a]		33	44	106
A. robustus				
TM 1517a		37	n/a	n/a
A. boisei	Mean	47	64	132
OH 5		44	59	119
KNM-ER 406		52	74	148
KNM-ER 13750		42	60	120
KNM-ER 23000[b]		49	61	140
A. aethiopicus				
KNM-WT 17000		39	62	111

Ht., height.
[a]The index of squamosal suture height relative to orbital roof is combined with the face of Sts. 5.
[b]Orbitale in this specimen is based on KNM-ER 406.
[c]See Figure 3.29.

The anterior outline of the supraorbital region in A.L. 444-2 is essentially a straight line, giving the rounded cranial contour a truncated appearance in front. The (partially destroyed) glabellar region creates a modest anterior protrusion in the midline. Upon closer observation, the supraorbital outline is seen to first retreat as it passes laterally from glabella and then, in the more peripheral area, to advance anteriorly until its lateral ends reach a more anterior coronal plane than the more medial parts. (See the description of the frontal bone and comparative hominin data in Chapter 5.) In contrast, the supraorbital torus in the African apes usually exhibits a transversely convex supraorbital contour that continues to retreat as it moves laterally from the midline. In this respect, A.L. 444-2 resembles chimpanzees, many of which have a straighter torus outline, rather than gorillas.

It appears, therefore, that in gorillas, the straight, nearly parallel lateral sides of the cranial contour—the zygomatic arches—merge in front with the arched outline of the supraorbital torus, whereas A.L. 444-2 exhibits the reverse: the curved sides of the cranium merge with a torus whose outline appears to be straight or perhaps even slightly concave in a superior view. The snout (the maxilla) in this view protrudes sharply anteriorly beyond the straight anterior outline of the supraorbital torus. The lateral wall of the snout forms a right angle with the outline of the torus. In the more prognathic gorilla, and to a lesser extent in chimpanzees, the snout is carried so far

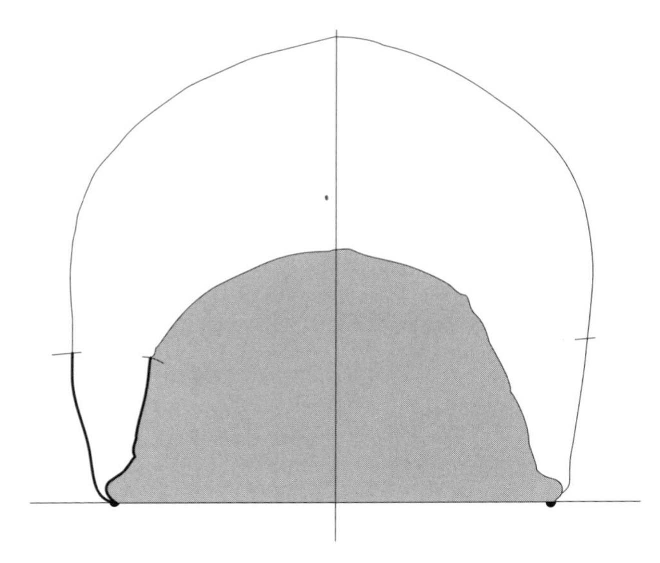

Figure 3.30 Coronal cross section of the calvaria at the level of porion in A.L. 444-2 (shaded outline) and modern humans (unshaded outline). The two crania are drawn to a fixed biporial breadth based on A.L. 444-2. Note that whereas the height of the temporal squama (heavy line) is almost identical in the two, the modern human calvaria is almost twice as high as that of A.L. 444-2. Thus, this difference in height stems solely from the increase in the length of the parietal arc in humans.

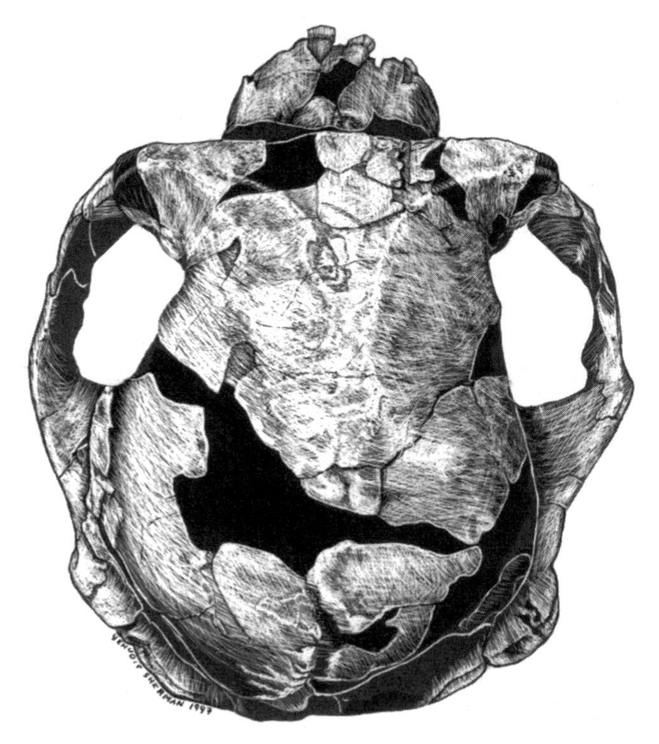

Figure 3.31 Artist's rendition of the superior view of the A.L. 444-2 reconstruction. See the picture gallery following page 122 for an enlarged view (75% natural size).

forward that part of the midface—rhinion, which is located substantially anteriorly, and the transverse buttress (Rak, 1983), which extends toward rhinion—is exposed in this view. In contrast, the entire central portion of the face is hidden by the torus in a superior view of A.L. 444-2, as in other early hominins (Figure 3.32).

The relatively modest magnitude of prognathism (in comparison with the great apes), on the one hand, and the lateral deviation of the zygomatic arches, on the other, clearly influence the "shape" index of the entire Hadar cranium. At 167 mm, the maximum bizygomatic breadth constitutes 81% of the total cranial length in horizontal projection. In comparison, other primates yield a substantially lower ratio: 69% in our sample of modern humans, 71% in chimpanzees, and 75% in female gorillas.[10] The similarity between modern *H. sapiens* and chimpanzee values underscores the caution one must exercise when evaluating the results of the shape index. In chimpanzees, prognathism influences the total cranial length, whereas in the orthognathic *H. sapiens*, the length of the cranium is the result of the dramatic extension of the braincase posterior to the biporion plane. The index is more significant, therefore, when primates with a similar endocranial capacity are compared. The value of 71% for Sts. 5, which is not far from the chimpanzee values, is still much lower than the ratio in A.L. 444-2. However, the *A. boisei* specimen KNM-ER 406 (which, with its slight reconstruction, is one of the two *A. boisei* crania that permit this measurement) has a substantially higher index value, at 93%, than that of *A. afarensis*. The ratio calculated for OH 5 is 88%, still higher than the index in A.L. 444-2 (as is the index for the extensively reconstructed cranium of the female KNM-ER 732, at 87%). Given the great similarity between *A. boisei* and the Hadar cranium in their braincase length and bizygomatic breadth, the larger shape index in the former is apparently the result of its much smaller degree of prognathism and the resulting reduction in total cranial length. The morphological reasons for the similarity in the bizygomatic breadths of these crania are discussed later in the sections "Outline of the Calvaria" and "The Facial Mask."

Specimen KNM-WT 17000 represents the other extreme of this configuration, with a total cranial length of 219 mm, which is longer than in any of the other early hominins, including A.L. 444-2. This length is the outcome of immense prognathism. However, the extensive prognathism, combined with the somewhat narrower maximum breadth, results in an index value (73%) that, as expected, is lower than in the Hadar specimen and approximately the same as in the African apes.

Shape and Dimensions of the Temporal Foramen

In *A. boisei*, the disparity between the extremely constricted postorbital region and the substantial lateral ex-

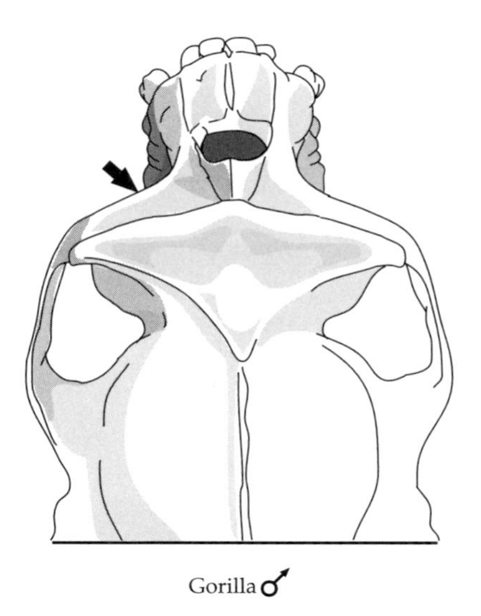

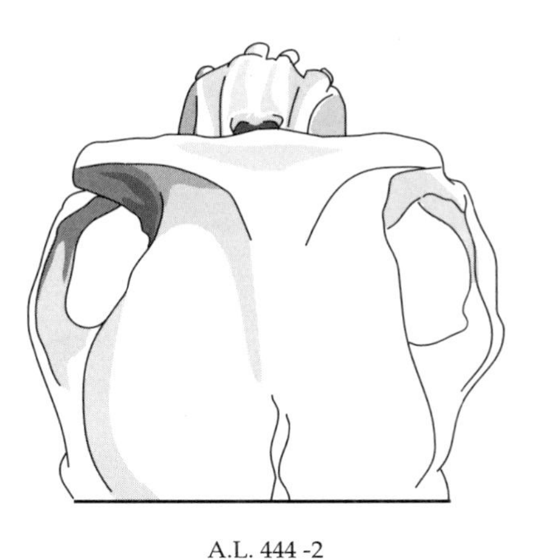

Figure 3.32 Superior view of A.L. 444-2 and a male gorilla cranium. In the gorilla much more of the face is exposed in this view than in A.L. 444-2. In the gorilla the "transverse buttress" (Rak, 1983) of the face is in full view (arrow). This structure is not present in the face of A.L. 444-2, though it is hinted at in some other *A. afarensis* faces (e.g., A.L. 333-1).

tension of the zygomatic arches produces huge temporal foramina. When the width across the postorbital constriction is expressed as a percentage of the maximum bizygomatic distance, the mean index value calculated for five *A. boisei* specimens (KNM-ER 406, KNM-ER 732, KNM-ER 13750, KNM-ER 23000, and OH 5) is 39%, whereas it is 46% in A.L. 444-2. Because the mean absolute values of bizygomatic breadth in *A. boisei* and the *A. afarensis* specimen are the same, the basis for the difference in the index values is a difference in postorbital width. (See the following discussion about the bizygomatic breadth relative to the biporial saddle breadth and the section "The Facial Mask," where the bizygomatic breadth is expressed as a percentage of the biorbital breadth.) Indeed, at 77 mm, the postorbital width in the Hadar cranium is about 16% greater than the mean for *A. boisei* (Figure 3.33; for further discussion, see Chapter 5).

In KNM-WT 17000, this index value is identical to that of the *A. boisei* mean, 39%, but the configurations responsible for these values are slightly different. The temporal foramen in KNM-WT 17000 is wide; however, its size is most likely the product of an extremely constricted frontal, the deep medial excavation of the temporal fossa. Indeed, although the lateral flare of the zygomatic arches is not unusual (they constitute only 122% of the biporial saddle breadth, as opposed to the *A. boisei* mean of 130%; see below), the amount of the total cranial (i.e., bizygomatic) width occupied by the postorbital constriction is the same as the *A. boisei* mean (Table 3.15).

The mean index in our sample of chimpanzees is 56% for females and 53% for males; and in the more size-dimorphic gorilla sample, the female and male values are also close, at 45% for females and 41% for males. *A. africanus* yields a value of 49% (the mean of Sts. 5 and Sts. 71), and in modern humans the value is 74%.

The anteroposterior dimension of the temporal foramen in A.L. 444-2 is longer than its width, and thus the foramen resembles that of many hominins and other primates. In sharp contrast is the unique foraminal shape in *A. boisei,* in which the mediolateral width usually exceeds the length. Clearly, this difference in shape is related to the extreme lateral deviation of the zygomatic arch and the medial excavation of the temporal fossa in *A. boisei.* The much broader postorbital region in A.L. 444-2 laterally "displaces" the temporal foramen while maintaining the smaller size and anteroposteriorly elongated shape encountered in all other hominoids. (For a more detailed description of the temporal foramen in a basal view, see the section "The Temporal Foramen and Adjacent Anatomical Structures.")

Swinging from its posterior root to the point where it merges with the visorlike zygomatic bone, the zygomatic arch in *A. boisei* crania forms part of a circle (Figure 3.34). The most lateral point on the arch (zygion) is located approximately at the midpoint of the arch's length, or perhaps even farther forward, whereas in A.L. 444-2, as in many other primates, zygion tends to be located closer to the posterior root of the arch. This is also the case in KNM-WT 17000.

The magnitude of lateral flair of the zygomatic arches can best be expressed in relation to the distance from one posterior root to the other (the porial saddle). As expected, this lateral deviation is maximal in *A. boisei,* where the bizygomatic breadth is 130% of the biporial saddle breadth ($n = 4$). In A.L. 444-2, the index is slightly lower, at 127%; at KNM-WT 17000's 122%, it is lower yet and finally reaches 118% in Sts. 5. Modern humans yield a very low index value of 107%. In female gorillas, it is 120%, and in the males, 121%, whereas the chimpanzee index is 108% and 114% for females and males, respectively.

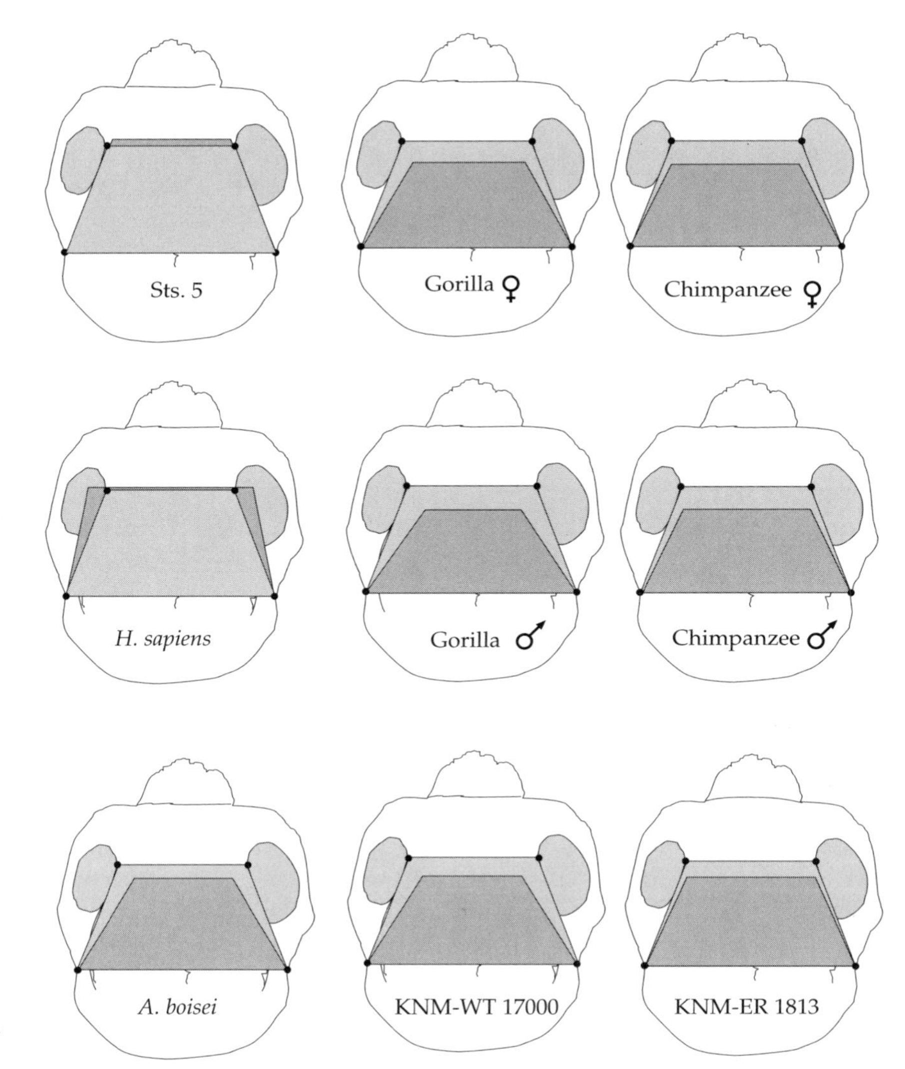

Figure 3.33 Relationship between the postorbital constriction, the breadth between the porial saddles, and the anteroposterior distance between them in A.L. 444-2 and other hominoids. This figure is based on the data in Table 3.15. The four points on each trapezoid represent the configuration in A.L. 444-2. The skulls are adjusted to the same biporial distance, which explains why Sts. 5, with its narrow cranial base, yields such a long anteroposterior dimension here.

The zygomatic arch in *A. boisei*, as can be discerned in superior view, twists about its long axis. At the posterior root, the medial aspect of the arch faces upward; at the anterior root, it is the lateral aspect that faces upward. Therefore, the anterior part of the arch appears as a wide, horizontal bone surface that extends laterally far from the orbital rim (Figure 3.34). In A.L. 444-2 the configuration of the arch is more generalized. The bone surface of the anterior root is almost vertical and, in comparison to *A. boisei*, is close to the lateral orbital margin. *A. robustus* and *A. aethiopicus* are more generalized than *A. boisei* in this regard.

An effect of the deeply excavated temporal fossa is the isolation of the supraorbital region as a distinct element, the supraorbital bar (Clarke's [1977] "costa supraorbitalis"). In A.L. 444-2, as expected from the more modest postorbital constriction, the two bars are not as long or as confined as in crania of *A. boisei* or other robust species of *Australopithecus*, including KNM-WT 17000. (For further discussion, see Chapter 5.)

Outline of the Calvaria

The effect of plastic deformation on the calvaria is clearly visible in a superior view of A.L. 444-2 Figure 3.31). As noted, this deformation affords the left side of the braincase a slightly more bulging contour than the right. Although the left side deviates farther laterally from the midsagittal plane, the elongated oval shape of the braincase is still evident. The ratio of the maximum width of the calvaria (122 mm, measured at the base of the temporal squamae) to calvarial length (glabella–opisthocranion) is 73% (Table 3.15).

This "cephalic" index can obviously yield identical values for different calvarial shapes. Therefore, the shape of the cranial vault is more meaningfully expressed as the

Table 3.15 Cranial Size and Shape in Superior View

Specimen/Sample		Projected Cranial Length (43)[a] (mm)	Maximum Bizygomatic Br. (22) (mm)	Cranial "Shape" Index (22/43) (%)	Calvarial Shape Index (20)/(1)[b] (%)	Postorbital Constriction (21) (mm)	Biporial Saddle Br. (14) (mm)	Postorbital Constriction/ Bizygomatic Br. (21/22) (%)	Index of Zygomatic Flare Relative to Biporial Saddle Br. (22/14) (%)	Anteropost. Distance (15) between the Plane of Constriction (21) and Biporial Saddle (14) (mm)	Tapering Index (21/14) (%)
Australopithecus afarensis											
A.L. 444-2		207	167	81	73	77	132	46	127	82	58
Homo sapiens	Mean (n = 10)	187	128	69	66	95	120	74	107	64	79
	Range	157–198	116–136	63–74	61–70	90–103	112–127	71–78	104–113	60–73	76–86
	SD	14	7	4	3	4	5	2	3	4	3
Gorilla gorilla female	Mean (n = 10)	199	148	75	71	66	123	45	120	54	54
	Range	190–207	142–153	71–78	66–77	61–72	117–129	40–51	114–126	48–69	47–60
	SD	5	2	3	3	4	4	3	2	5	3
Gorilla gorilla male	Mean (n = 10)	n/a	175	n/a	n/a	71	144	41	121	64	49
	Range	n/a	165–181	n/a	n/a	66–76	134–154	29–46	115–129	55–76	43–54
	SD	n/a	4	n/a	n/a	3	5	4	3	5	2
Pan troglodytes female	Mean (n = 10)	176	125	71	75	70	115	56	108	47	61
	Range	168–187	120–132	65–77	71–79	65–74	110–123	50–62	105–111	41–51	54–68
	SD	6	3	4	2	2	4	3	1	3	4
Pan troglodytes male	Mean (n = 10)	183	130	72	77	70	115	53	114	48	61
	Range	170–198	122–138	67–78	71–83	67–74	102–131	48–58	104–125	40–57	56–64
	SD	8	7	2	4	2	8	3	5	5	3
A. africanus	Mean	175	122	71	68	60	106	49	118	60	61
Sts. 5		175	127	71	68	64	107	50	118	60	61
Sts. 71		n/a	116	n/a	n/a	56	105	48	n/a	n/a	n/a
H. habilis											
KNM-ER 1813		166	n/a	n/a	71	69	113	n/a	n/a	42	61
A. robustus	Mean	n/a	141		(63)	71	120	48	n/a	n/a	
SK 48		n/a	148	n/a	(63)	71	120	48	n/a	n/a	n/a
TM 1517		n/a	134	n/a	n/a		n/a	n/a	n/a	n/a	n/a
A. boisei	Mean	193	172	91	72	66	133	39	130	61	54
OH 5		192	168	88	67	65	140	39	120	64	46
KNM-ER 406		193	180	93	72	62	133	34	135	63	67
KNM-ER 13750		n/a	180	n/a	n/a	69	128	38	141	63	53
KNM-ER 23000		n/a	160	n/a	78	66	130	41	123	55	52
KNM-ER 732[c]		(165)	(144)	(87)	(62)	61	(110)	42	(131)	(58)	(55)
A. aethiopicus											
KNM-WT 17000		219	160	73	69	62	131	39	122	57	47

[a]Sometimes nasospinale and/or rhinion protrude further forward in *H. sapiens* than prosthion.
[b]In *H. sapiens* sometimes maximum cranial breadth is larger than breadth at the base of temporal squama.
[c]Values for KNM-ER 732 are not figured in the mean for *A. boisei*.
Br., breadth
Values in parentheses are estimates.

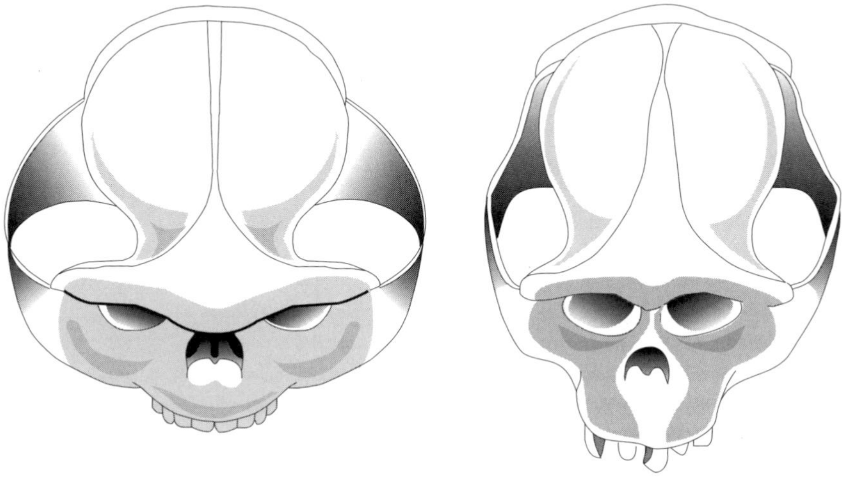

A. boisei

A.L. 444 -2

Figure 3.34 Schematic representation of the differences in the zygomatic arches of A.L. 444-2 and *A. boisei*. In *A. boisei* the zygomatic arch twists about its long axis such that the temporal surface of its posterior root faces superiorly, and of its anterior root, inferiorly. In A.L. 444-2, as in most other primates, the temporal surface of the anterior root faces predominantly posteriorly.

ratio of the postorbital breadth to the maximum breadth of the vault. In his study of OH 5, Tobias (1967) calculated the latter index and referred to the morphology as the "posteroanterior tapering of the cranial vault." The postorbital constriction in the Hadar specimen constitutes 58% of the maximum calvarial width, which here spans the porial saddles.[11] In *A. boisei*, the mean value of four specimens is somewhat smaller, at 54%. Sts. 5 yields a value of 61%, which is similar to the value calculated for our sample of chimpanzees (61% for both females and males). In gorillas, the index is lower, at a mean of 54% for females and only 49% for males. In our modern human sample, the mean is 79%. KNM-WT 17000 exhibits one of the most extreme hominin values, at 47%. Recall that the value for the *A. aethiopicus* cranium is primarily the result of a highly constricted postorbital region, as discussed earlier. These numerical differences between the taxa are not as dramatic as the obvious differences shown by simple visual inspection; in superior view, the calvaria of A.L. 444-2 does *not* resemble that of *A. boisei*, in spite of similar tapering index values.

The anteroposterior distance between the coronal planes of the postorbital and maximum vault breadths is an important factor in producing the shape of the braincase that is not taken into account by the quantitative expressions of shape discussed thus far. As the anteroposterior distance between these planes diminishes, the shape of the calvaria changes, although the breadth measurements may remain the same. It seems that the narrower the postorbital constriction, the farther the orbital cones "force" it posteriorly. Indeed, what provides the *A. boisei* braincase with its unusual teardrop appearance is (a) the considerable maximum width at its posterior end, (b) the narrow postorbital constriction at the anterior end, and (c) the relatively short distance between the coronal planes on which these two widths reside. As described in the section "General Outline of the Cranium in Basal View," the flatness of the posterior occipital outline in *A. boisei* further contributes to the teardrop shape of the calvaria. The more ovoid shape of the cranial vault in the Hadar specimen resembles the shape in *A. africanus* and, even more so, in chimpanzees. These similarities result primarily from the combination of a relatively high ratio of the postorbital constriction to the biporial saddle breadth, as well as the large anteroposterior distance between these two breadths. This is summarized graphically in Figure 3.33, where all crania are scaled to the biporal saddle breadth of A.L. 444-2.

To evaluate the effect of each of the three components on the shape of the vault, we plotted the percentage of the biporial saddle breadth occupied by the width across the postorbital constriction against the relative anteroposterior distance (relative to the biorbital breadth) between these two breadth measurements (Figure 3.35). In this way, the unique positions of A.L. 444-2 and modern humans in the graph are clearly discernible. It is primarily the enormous anteroposterior distance between the two breadth measurements in A.L. 444-2 that isolates this specimen, whereas the ratio between these two measurements in *H. sapiens*—the outcome of the wide postorbital constriction—is what sets this taxon apart.

The Cranium: Frontal View

The Facial Mask

Outline of the Facial Mask

The outline of the facial mask in A.L. 444-2 is an inferosuperiorly asymmetric hexagon (Figure 3.36). The nearly

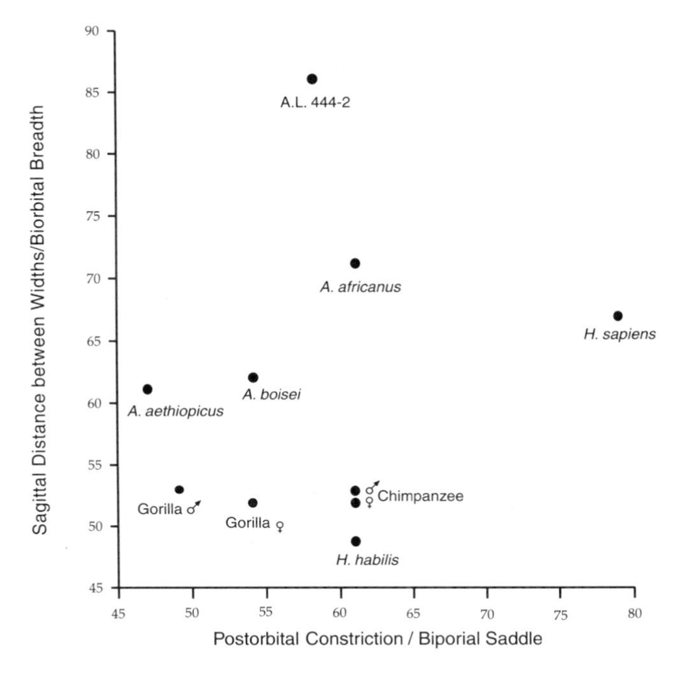

Figure 3.35 Bivariate scattergram of calvarial shape. The percentage of the biporial saddle breadth occupied by the breadth across postorbital constriction (x-axis) is plotted against the anteroposterior distance (relative to the biorbital distance) between these two breadth measurements. Note the isolated positions of A.L. 444-2 and modern humans in the graph. The large anteroposterior distance between the two breadth measurements isolates A.L. 444-2, whereas the small difference in magnitude between these two measurements sets *H. sapiens* apart. Among the fossil samples plotted here, only *A. boisei* is represented by more than one specimen.

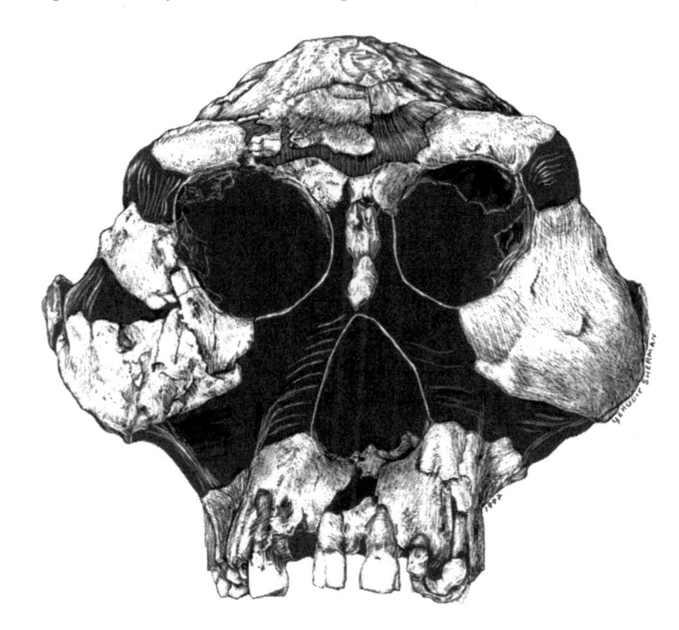

Figure 3.36 Artist's rendering of the anterior view of the A.L. 444-2 reconstruction. See the picture gallery following page 122 for an enlarged view (75% natural size).

straight horizontal line formed by the bilaterally extended supraorbital region constitutes the upper border of this hexagonal outline, while the occlusal plane constitutes the shorter, lower border. As it moves away from the midline, the horizontal supraorbital contour joins the lateral margin of the facial profile in a well-defined corner. Although this part of the facial mask is reconstructed, there is little doubt that, when the mask is examined in an anterior view, the junction between the supraorbital element and the lateral margin of the orbit was marked by a distinctly squared-off transition.

Three factors combine to render the upper part of the A.L. 444-2 facial contour similar to that of gorillas (and to a lesser extent, chimpanzees): the supraorbital element's horizontal orientation; the well-defined corner at the lateral extremity; and the overall homogeneity of the supraorbital element, though characterized by a modest lateral vertical thickening. There is no perceptible division into individual arches. In no other cranium of *Australopithecus* do we see this combination of features, with the probable exception of the Belohdelie frontal bone. (For a detailed comparative description of the supraorbital region, see Chapter 5.) The supratoral sulcus, which separates the supraorbital torus from the frontal squama in the African apes and is responsible for the major differences in frontal topography, is hidden in this view and thus does not diminish the great similarity discussed here.

In A.L. 444-2, the temporal process of zygomatic bone—the anterior root of the zygomatic arch—deviates considerably laterally from a vertical line dropped from the lateral end of the supraorbital torus. The mainly coronal orientation of the zygomatic's frontal process and its morphological accommodation to the lateral flair of the temporal process result in an extremely broad base of the frontal process. This morphology provides the lateral margins of the facial mask with an upwardly tapered appearance in A.L. 444-2—the upper half of the superoinferiorly asymmetric hexagon mentioned above. Gorillas, particularly males, also exhibit a substantial lateral deviation of the anterior root of the zygomatic arch. However, this deviation does not produce an upwardly tapered facial contour, at least not to the extent seen in A.L. 444-2. The outline of the lateral part of the upper facial mask is more or less vertical, the outcome of the straight, vertical lateral margin of the frontal process. In A.L. 444-2, the great width of the frontal process's base is achieved through an equal deviation of the contour medially and laterally from the imaginary vertical line (Figure 3.37). This morphological configuration is equivalent to that seen in another *A. afarensis* specimen, A.L. 333-1 (Figure 3.38), and is also the pattern in other *Australopithecus* crania (Rak, 1983).

However, the topography of the body of the zygomatic bone differs among *Australopithecus* species and

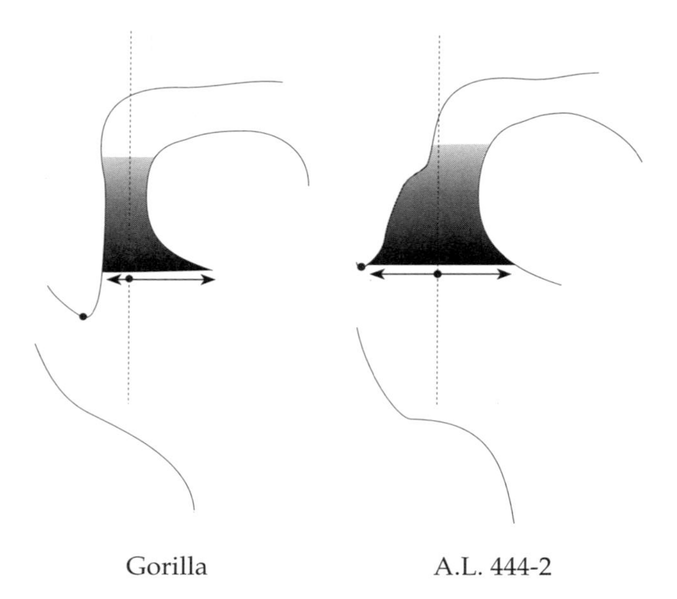

Figure 3.37 The root of the frontal process of the zygomatic bone in the gorilla and A.L. 444-2. In A.L. 444-2 the substantial width of the root of the frontal process is the result of both a lateral and a medial deviation of its margins.

influences the facial profile differently in each of them. In *A. africanus*, the large "zygomatic prominence" (Rak, 1983) protrudes both anteriorly and laterally, bearing on its surface the lateral contour of the face in an anterior view. The zygomatic prominence is also responsible for the lateral termination of the facial contour in a rather sharply defined point (Figure 3.39). This morphology can be seen clearly on Sts. 5, Sts. 52a, and Sts. 71. In A.L. 444-2, the most peripheral part of the face (the lateral corner of the hexagon) is not distinctly angled as in *A. africanus* but forms a wide curve in the lateral outline of the face (Figure 3.39).

In *A. robustus* specimens SK 48, SK 52, and TM 1517, the entire lateral portion of the body of the zygomatic bone, including the base of the wide temporal process, acts as a corner that demarcates the transition between the anteriorly and laterally facing parts of the bone (this also holds true in less complete or more deformed specimens, such as SK 46 and SK 83). This configuration provides the contour of the facial mask with a square outline. A similar topography can be seen in modern humans and many other primates. In *A. boisei* the upper lateral outline of the facial mask descends inferolaterally from the supraorbital torus for a great distance, merges smoothly with the lateral part of the zygomatic bone (which in this species assumes a visorlike shape), and finally terminates abruptly at the elevated inferior margin of the "visor" (Figure 3.39). In some *A. boisei* specimens, such as OH 5, KNM-ER 406, KNM-ER 732, and KNM-ER 13750, the oblique lateral outline of the facial mask is continuous with the inclined lateral segment of the supraorbital element. This configuration obliterates the distinct transition between the supraorbital element and the lateral outline of the face, and, as a result, the upper half of the facial hexagon is smooth rather than angular (that is, segmented), as in A.L. 444-2 (see Rak, 1983). The KNM-WT 17000 cranium appears similar to those of *A. boisei* in the outline of its facial mask.

In spite of the considerable damage to the zygomaticoalveolar crest in A.L. 444-2—the inferolateral side of the hexagon—there is no doubt that the crest was curved and its lateral half, inferiorly directed, formed a marked submalar incisure. This shape can be deduced from the actual remains of the crest on the left side of the face and the reconstruction of the crest on the right. The shape of the reconstruction is based on the bony remains and the constellation of the adjacent bony parts. Indeed, the zygomaticoalveolar crest is moderately to strongly curved in every adult and juvenile specimen of *A. afarensis* with the relevant area preserved (for example, A.L. 333-1, A.L. 333-86, A.L. 333-105, A.L. 413-1, A.L. 417-1d, and A.L. 427-1).

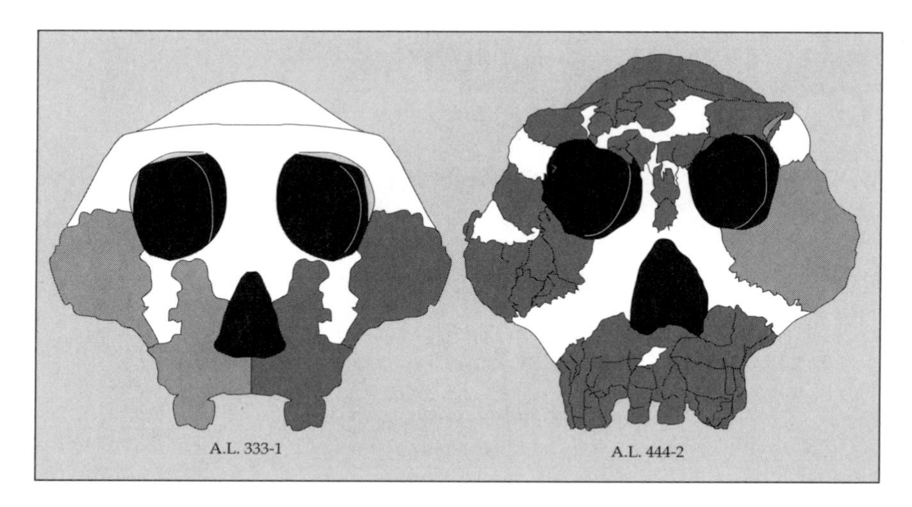

Figure 3.38 Outline of the facial mask of A.L. 333-1 and A.L. 444-2. The facial mask of A.L. 333-1 has been reconstructed; the right side is a mirror image of the original left side. Note the great width of the root of the frontal process of the zygomatic bone in both specimens. This anatomy of the process provides both facial masks with a laterally tapering outline.

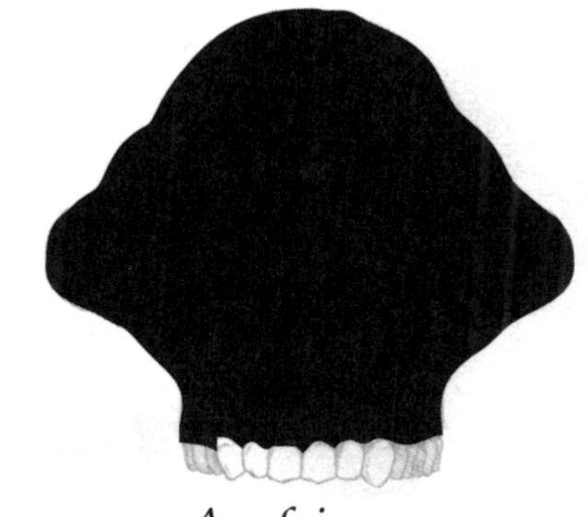

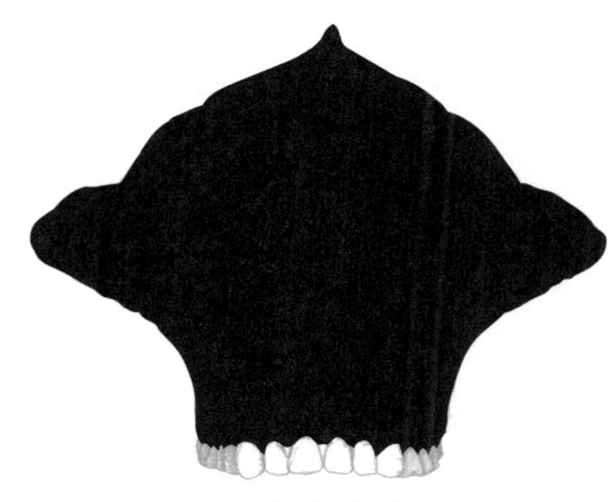

Figure 3.39 Comparison of an idealized facial mask in *A. africanus*, *A. boisei*, and A.L. 444-2.

Found in many primates, including *Homo* and most chimpanzees and gorillas, the submalar incisure thus represents the plesiomorphic configuration (Clarke, 1977; Rak, 1983; Kimbel et al., 1984). It differs from the condition in *A. africanus* (including the Taung juvenile), *A. aethiopicus*, *A. robustus*, and *A. boisei*.[12] In these species, the crest's lower margin never curves to face inferiorly but instead follows a straight, diagonal path from its origin on the maxilla to its termination at the lateral end of the zygomatic bone. (For further discussion of the shape of the facial mask, see the section "Proportions of the Facial Mask.")

Measurements of Facial Height

As expected, the facial height and width measurements of A.L. 444-2 reflect the skull's large size overall. Obviously, however, it is the facial topography and the proportions between the height and width measurements that determine the shape of the face. These are discussed in detail below.

The traditional facial height measurement from the (reconstructed) nasion to prosthion is 100 mm, which is similar to the mean (102 mm) for the female gorilla sample. However, because of the differences in the horizontal displacement of prosthion in specimens with varying degrees of prognathism, the distance between nasion and prosthion does not express the actual *vertical* facial height. We thus prefer to use the vertical distance between the alveolar plane and the plane of the supraglabellar region as an expression of facial height (Figure 3.7, measurement 53). In this manner, we also avoid the issue of the extreme variation in the position of nasion and the common problem of identifying this point in fossils. When the facial height is measured in this way, it is 113 mm in A.L. 444-2, which is slightly higher than the female gorilla mean, 108 mm. The facial height of Sts. 5, according to this method, is 86 mm (76% of the A.L. 444-2 value), and of OH 5, it is 117 mm (104% of the A.L. 444-2 value). See Table 3.16 for additional height measurements and comparative data.

Measurements of Facial Breadth

The biorbital distance is of special relevance, as we use it to represent the overall size of the skull.[13] In A.L. 444-2, the biorbital distance is 95 mm, which is only slightly less than in female gorillas in which the mean measurement is 108% of the A.L. 444-2 value. In Sts. 5, this distance is substantially shorter, at 85 mm—or 89% of A.L. 444-2's value—whereas in OH 5 it is 100 mm, or 105% of A.L. 444-2. See Table 3.17 for additional breadth measurements and comparative data.

Table 3.16 Facial Height Measurements (mm)

Specimen/Sample		Nasion–Prosthion (60)	Total Vertical Facial Ht.: Supraglabella–Alveolar Plane (53)	Orbitovalveolar Ht. (58)
Australopithecus afarensis	Mean	87	101	55
A.L. 444-2		100	113	66
A.L. 417-1d		(74)	88	50
A.L. 333 reconstruction		n/a	n/a	48
Homo sapiens	Mean (n = 10)	73	86	41
	Range	66–79	77–96	35–45
	SD	4	5	3
Gorilla gorilla female	Mean (n = 10)	102	108	63
	Range	96–109	103–113	58–69
	SD	3	2	3
Gorilla gorilla male	Mean (n = 10)	120	129	77
	Range	110–131	120–138	70–84
	SD	8	4	4
Pan troglodytes female	Mean (n = 10)	88	93	51
	Range	81–99	83–102	42–60
	SD	5	6	5
Pan troglodytes male	Mean (n = 10)	93	96	53
	Range	82–106	88–104	46–62
	SD	8	5	4
A. africanus	Mean	78	95	53
Sts. 5		77	86	51
Sts. 71		73	88	47
Stw. 505		90	110	62
Sts. 52a		71	n/a	53
H. habilis	Mean	67	84	47
KNM-ER 1813		66	86	47
OH 24		67	82	46
A. robustus	Mean			57
SK 48		76	97	61
TM 1517		n/a	n/a	52
A. boisei	Mean	101	113	69
OH 5		107	117	75
KNM-ER 406		(94)	(108)	62
KNM-ER 732[a]		(82)	91	47
A. aethiopicus				
KNM-WT 17000		88	88	53

Ht., height.
n/a, not available. Values in parentheses are estimates.
[a]Values for KNM-ER 732 are not figured in *A. boisei* means.

Proportions of the Facial Mask

Ratio of the biorbital distance to the bizygomatic breadth on the coronal plane of the orbits. The upward tapering of the face, discussed earlier in reference to the outline of the facial mask, can be quantified by means of a simple index (Figure 3.40, 1: b/a). The bizygomatic breadth on the coronal plane of the orbits measures 142 mm in A.L. 444-2. Expressing this distance as a percentage of the biorbital breadth yields an index value of 149%, which is considerably higher than in many other primates. Although the values for gorillas, at 138% for males and 139% for females, are lower than in A.L. 444-2, they exceed the values yielded by other primates. For example, the mean index for male chimpanzees is 123% (not significantly higher than the *Cercopithecus aethiops* value of 121%), for female chimpanzees, 114%, and for modern humans, a relatively low 118%.

Table 3.17 Facial Breadth Measurements (mm)

Specimen/Sample		Biorbital Breadth (54)	Maximum Bizygomatic Breadth (22)	Maximum Bizygomatic Breadth in Orbital Plane (52)	External Palatal Breadth (77)
Australopithecus afarensis	Mean	89	157	139	72
A.L. 444-2		95	167	142	82
A.L. 417-1d		83	n/a	n/a	61
A.L. reconstruction (A.L. 333-1)		87	147	136	67
Homo sapiens	Mean (n = 10)	95	129	113	64
	Range	85–101	117–136	107–120	59–68
	SD	5	7	5	4
Gorilla gorilla female	Mean (n = 10)	103	146	135	68
	Range	92–109	140–153	125–145	61–75
	SD	6	4	7	5
Gorilla gorilla male	Mean (n = 10)	121	178	162	74
	Range	115–126	165–190	152–175	69–78
	SD	2	8	6	2
Pan troglodytes female	Mean (n = 10)	91	122	109	60
	Range	86–96	113–129	103–116	53–68
	SD	2	5	3	3
Pan troglodytes male	Mean (n = 10)	89	129	112	60
	Range	82–101	119–139	100–127	52–69
	SD	5	6	7	5
A. africanus	Mean	82	122	115	69
Sts. 5		85	127	120	68
Sts. 71		79	116	110	70
Homo habilis	Mean	86			65
KNM-ER 1813		85	n/a	n/a	63
OH 24		86	n/a	n/a	66
A. robustus	Mean	92	141	133	70
SK 48		100	148	138	69
TM 1517		84	134	128	70
A. boisei	Mean	103	176	166	79
OH 5		100	168	165	81
KNM-ER 406		103	180	172	77
KNM-ER 13750		105	180	162	n/a
KNM-ER 732[a]		(88)	(144)	(127)	n/a
A. aethiopicus					
KNM-WT 17000		94	159	149	79

[a]Values for KNM-ER 732 are not figured in *A. boisei* means.
Values in parentheses are estimates.

Other *Australopithecus* species values approach or exceed that observed in A.L. 444-2. Two *A. robustus* specimens (SK 48 and TM 1517) yield a mean of 145%, while 140% is the mean for two *A. africanus* specimens (Sts. 5 and Sts. 71). *A. boisei* is the most extreme case, with a mean index of 158% (for OH 5, KNM-ER 406, and KNM-ER 13750, and the female KNM-ER 732). Note in Table 3.18 that the index in one of the *A. boisei* males (KNM-ER 406) reaches 167%! In KNM-WT 17000, the index is also high, at 159%. The very high values in *A. boisei* and *A. aethiopicus* apparently stem from a combination of two factors: the size and anterior encroachment of the temporalis (which push zygion anteriorly and thus place the anterior root of the zygomatic arch farther away from the orbit) and the transformation of the laterally flaring zygomatic bone into a visorlike structure.

Taken at face value, such indices can be misleading with regard to the true shape of the facial mask; the upward tapering of the face—the inclination of the upper lateral profile of the facial hexagon—is the outcome not only of the disparity between the bizygomatic distance (on the coronal plane of the orbits) and the biorbital dis-

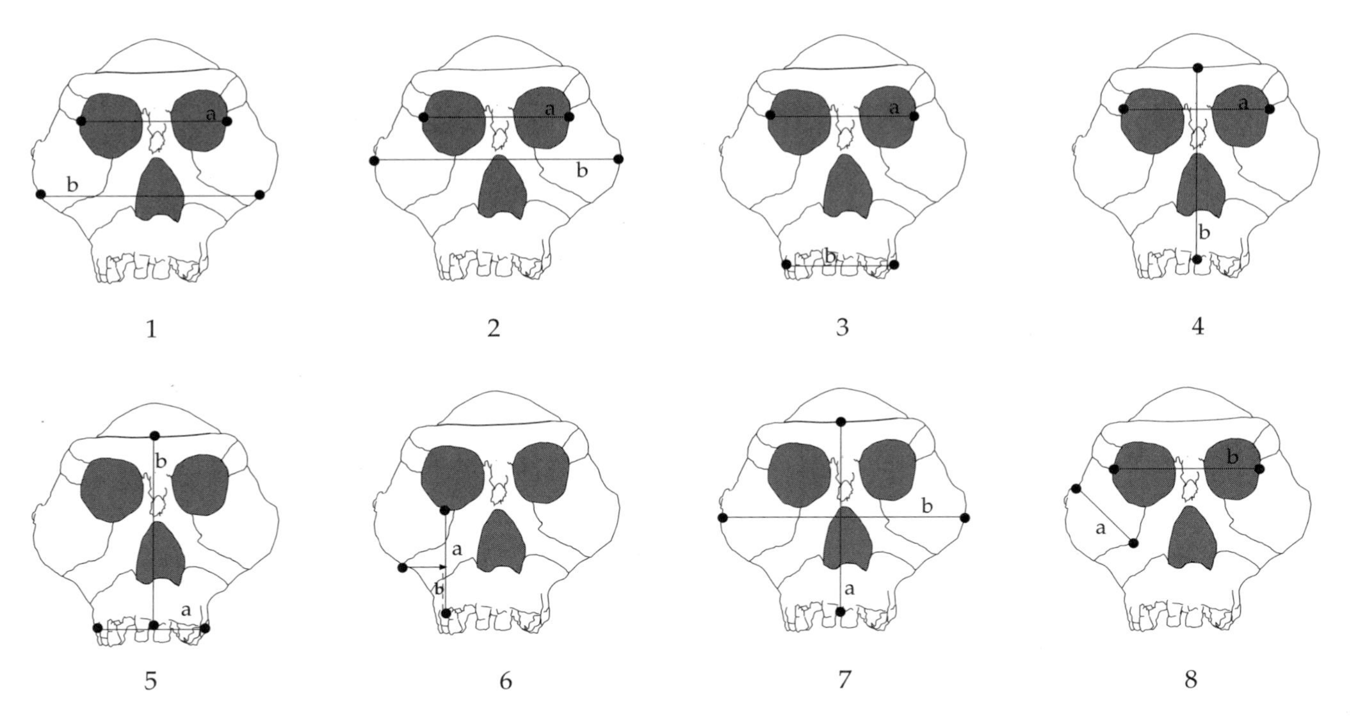

Figure 3.40 Proportions of the facial mask of A.L. 444-2. The numbers and letters correspond to those in the text, in the section "Proportions of the Facial Mask."

tance, but also of the vertical distance separating these two chords. When this distance is small, the maximum width of the face is closer to the supraorbital torus. In gorillas, for example, the very low temporal process of the zygomatic bone carries the maximum facial width at the coronal plane of the orbits downward, far below the inferior orbital margin (Figure 3.41). As a result, the vertical distance between the two chords is considerable, which lends a square appearance to the facial contour, despite an index that is higher than in other primates (recall the index value of 138% in male gorillas as compared to 123% in male chimpanzees). In contrast, the vertical proximity between these two chords in some *A. boisei* specimens achieves the opposite effect (see especially KNM-ER 406 and KNM-ER 13750)—a much stronger tapering of the upper facial contour than what would be expected from the index value itself. Here, the superolateral flaring of the inferior margin of the visorlike zygomatic bone is what brings the bizygomatic chord closer to the biorbital chord (and the supraorbital torus), thus reducing the vertical distance between them. The specimen that displays this configuration to the most extreme degree is KNM-WT 17000 (*A. aethiopicus*), in which the *inferior* margin of the temporal process of the zygomatic bone is situated almost at the same horizontal level as the orbital floor. Such a configuration results in an extremely tapered outline of the upper face.

The substantial vertical distance between the two facial chords causes the upper part of the A.L. 444-2 facial mask to appear more square than that of *A. africanus*, despite the latter's lower index value. The small vertical distance between the two horizontal chords, which provides the more robust australopiths with an extremely tapered upper facial outline, has the opposite effect on the lower part of their facial contour. There, the resulting large vertical distance between the occlusal plane and the chord of maximum facial width tends to make the lower sides of the hexagon steeper. Also contributing to this steepness is the straight outline of the zygomaticoalveolar crest (Figure 3.41).

In A.L. 444-2, the relatively small distance between the chord of facial width and the occlusal plane tends to flatten the lower, rather than the upper, part of the hexagonal facial mask. The curved zygomaticoalveolar crests in this specimen add to the less inclined appearance of the lower sides of the facial mask. This morphology is also very well expressed in A.L. 333-1. In comparison to *A. boisei* and *A. aethiopicus*, *A. africanus* and *A. robustus* tend toward this more generalized facial configuration, in which the distance between the two facial width chords is relatively large.

Ratio of maximum facial breadth (maximum bizygomatic breadth) to the biorbital distance. In the discussion of the superior view, the degree of lateral flare of the zygomatic

Table 3.18 Proportions of the Face (%)

Specimen/Sample		1[a] Bizygomatic Br. at Orbital Plane (52)/ Biorbital Br. (54)	2 Max. Bizygomatic Br. (22)/ Biorbital Br. (54)	3 Palatal Br. (77)/ Biorbital Br. (54)	4 Total Facial Ht. (53)/ Biorbital Br. (54)	5 Palate Br. (77)/ Total Facial Ht. (53)	6 Masseteric Ht. (59)/ Orbitoalveolar Ht. (58)	7 Total Facial Ht. (53)/Max. Bizygomatic Br. (22)	8 Zygomatic Br. (72)/ Biorbital Br. (54)
Australopithecus afarensis	Mean	152	173	76	111	74	52	67	46
A.L. 444-2		149	176	82	119	76	49	68	51
A.L. 417-1d		n/a	n/a	73	(108)	(74)	(57)	n/a	n/a
A.L. reconstruction (A.L. 333-1)		(154)	(169)	(74)	(107)	(73)	51	65	40
Homo sapiens	Mean (n = 10)	118	135	66	91	74	44	67	26
	Range	113–127	130–143	62–73	79–103	63–82	39–51	59–79	24–29
	SD	4	4	3	7	6	4	6	2
Gorilla gorilla female	Mean (n = 10)	139	151	68	113	62	45	74	28
	Range	131–149	147–156	61–74	104–121	55–68	38–54	70–78	21–32
	SD	4	2	3	5	3	5	2	3
Gorilla gorilla male	Mean (n = 10)	138	154	63	114	56	50	73	27
	Range	126–147	140–167	54–69	104–129	48–63	42–57	69–78	21–33
	SD	6	8	4	6	4	3	2	3
Pan troglodytes female	Mean (n = 10)	114	138	71	101	69	48	75	22
	Range	110–120	132–141	67–74	95–107	61–75	42–54	70–80	17–27
	SD	3	2	2	4	3	3	3	3
Pan troglodytes male	Mean (n = 10)	123	144	71	107	67	48	74	22
	Range	114–134	131–155	60–79	99–116	59–74	40-54	68–79	19–26
	SD	5	7	6	6	5	4	3	2
A. africanus	Mean	140	148	85	106	80	63	72	29
Sts. 5		141	149	80	101	79	63	68	28
Sts. 71		139	147	89	111	80	66	76	30
MLD 6		n/a	n/a	n/a	n/a	n/a	65	n/a	n/a
Sts. 52a		n/a	n/a	n/a	n/a	n/a	60	n/a	n/a
TM 1511		n/a	n/a	n/a	n/a	n/a	(58)	n/a	n/a
H. habilis	Mean			76	98				
KNM-ER 1813		n/a	n/a	74	101	n/a	n/a	n/a	(22)
OH 24		n/a	n/a	78	95	n/a	n/a	n/a	n/a
A. robustus	Mean	145	154	76			62		33
SK 48		138	148	69	97	n/a	63	66	33
TM 1517		152	159	83	n/a	n/a	61	n/a	(33)
A. boisei	Mean	158	169	78	111	70	59	64	36
OH 5		165	168	81	121	69	48	70	36
KNM-ER 406		167	175	75	108	71	(64)	60	42
KNM-ER 13750		154	171	n/a	n/a	n/a	n/a	n/a	31
KNM-ER 732		144	(163)	n/a	105	n/a	64	63	n/a
A. aethiopicus									
KNM-WT 17000		159	169	84	94	90	60	55	40

[a]Numbers correspond to Figure 3.40.
Ht., height; Br., breadth; Max., maximum; n/a, not available.
Values in parentheses are estimates.

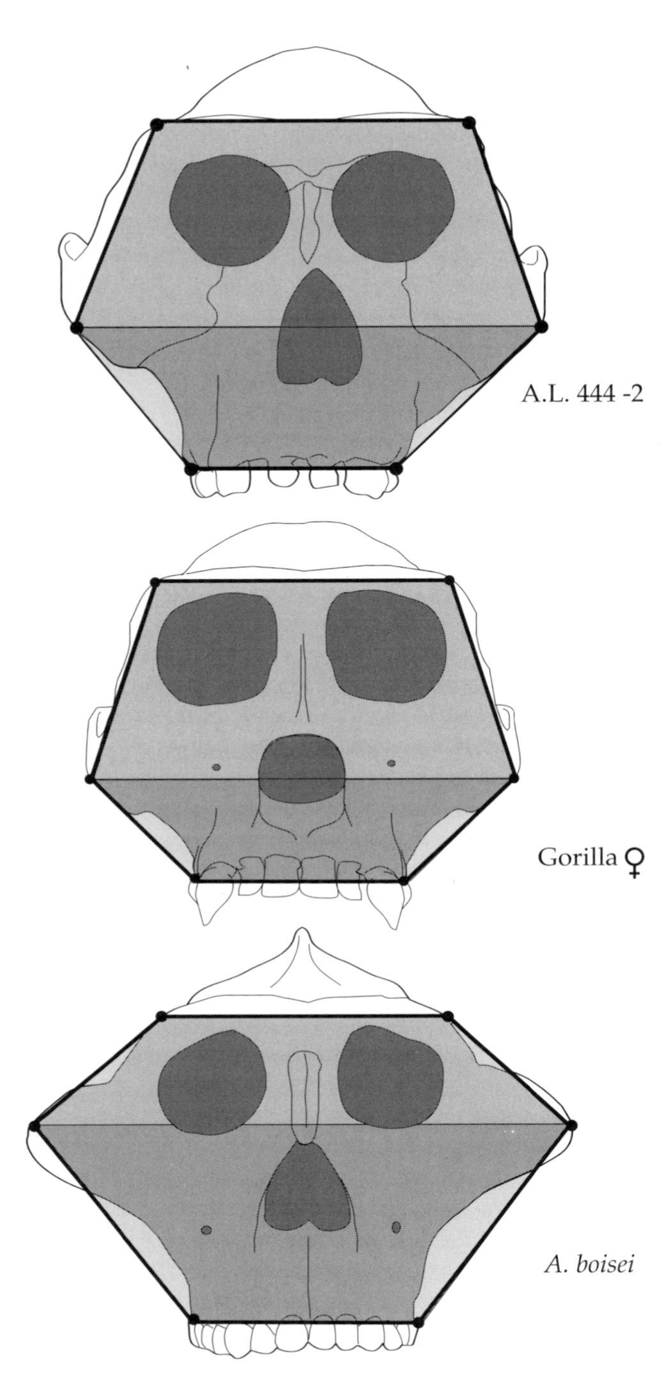

Figure 3.41 The hexagonal shape of the facial mask in A.L. 444-2, a female gorilla, and *A. boisei*. This schematic view of the facial masks emphasizes their different proportions. In *A. boisei*, the lower half of the facial hexagon is deep and the upper part is shallow, whereas the female gorilla and A.L. 444-2 exhibit the reverse proportions. All are scaled to the biorbital distance of A.L. 444-2.

arches is expressed relative to the minimum breadth of the cranial base—that is, the distance between the porial saddles. The substantial lateral deviation of the zygomatic arches is also noticeable when the maximum bizygomatic breadth is compared to the biorbital distance (Figure 3.40, 2: b/a). The discrepancy between these two measurements naturally influences the width of the face at the coronal plane of the orbits and thus also the shape of the facial mask. In A.L. 444-2, the maximum bizygomatic distance (167 mm) constitutes 176% of the biorbital distance, which is, indeed, a very high value compared to most other primates. In our female gorilla sample, for instance, the mean index is 151%, and in the males, 154%—still far from the Hadar value. Smaller values are obtained for chimpanzees (females, 138%, and males, 144%) and other primates (for example, a sample of four *Cercopithecus* specimens yields a value of 144%). Modern *H. sapiens* exhibits a modest flare, at 135%. Two *A. africanus* specimens, Sts. 5 and Sts. 71, have a mean "flaring" index of 148%, which is less than the mean index of 154% for two *A. robustus* specimens (SK 48 and TM 1517).

The high ratio of A.L. 444-2 places it in the same area of the range occupied by *A. boisei*. In fact, the Hadar specimen's ratio is even greater, though negligibly so, than the highest value of the *A. boisei* sample—the 175% of KNM-ER 406. The mean index for three male *A. boisei* specimens (KNM-ER 406, KNM-ER 13750, and OH 5) is 171%. KNM-WT 17000 yields a value of 169%. It is important to note that the large bizygomatic distance in A.L. 444-2 does not persist anteriorly, as it does not reach the extreme values of the bizygomatic breadth on the coronal plane of the orbits in *A. boisei* (see above). The transformation of the infraorbital region into a visorlike structure in *A. boisei* further increases the width at the coronal level of the orbits.

Ratio of palatal width to the biorbital distance. The palate (*external* dimensions at the coronal plane of the orbits or of M^2) of A.L. 444-2 is broad in relation to the biorbital distance (Figure 3.40, 3: b/a), especially when viewed in comparison to gorillas and, to a lesser extent, chimpanzees. In A.L. 444-2, the external width of the palate constitutes 82% of the biorbital distance, whereas the index in chimpanzees is 71%. (Compare with *Cercopithecus*, where the mean index value of four individuals is 72%.) In gorillas, as just mentioned, the index values are even smaller: 68% in the females and 63% in the males. The addition of two other *A. afarensis* specimens that permit the calculation of this index, the 1984 composite reconstruction (A.L. 333-1) and A.L. 417-1d, lowers the mean index for the species. Nevertheless, the mean value of 76% still indicates that the palate is wider externally than in the African apes, particularly gorillas.

In general, a wide palate characterizes all the australopiths and, indeed, most hominins, although considerable variation within species is observed. A value of 78% is the mean obtained from two *A. boisei* specimens (OH 5 and KNM-ER 406). In KNM-WT 17000, the extremely wide palate constitutes 84% of the biorbital distance. Another rather high mean (85%) is obtained from *A. africanus* specimens Sts. 5 and Sts. 71. Similarly, two specimens of *A. robustus* that differ greatly in this index (69% for SK 48 and 83% for TM 1517) yield a mean of 76%. Modern humans, in contrast, have an externally narrow palate and thus a low index value of 66%.

The configuration in chimpanzees apparently represents the generalized condition, as other primates yield similar values. We can conclude that the shrinkage of the masticatory system relative to skull size (the biorbital distance) in modern humans is what produces their comparatively low value. The *H. habilis* specimen KNM-ER 1813, with an index of 76%, does not yet display the narrow palate of modern *H. sapiens.*

Ratio of total facial height to biorbital breadth. The face of A.L. 444-2 is relatively tall; total facial height constitutes 119% of the biorbital distance (Figure 3.40, 4: b/a). (Recall that the height dimension employed here is the vertical distance from the supraglabellar plane to the alveolar plane.) Two other (reconstructed) specimens of *A. afarensis* yield somewhat lower values (108% for A.L. 417-1d and 107% for the 1984 composite reconstruction). The ratio in A.L. 444-2 is similar to that calculated for our gorilla sample (113% for females and 114% for males), whereas chimpanzee values (101% for females and 107% for males) are close to the values for Old World monkeys with a generalized face (for example 100% is the mean for a sample of four *Cercopithecus* specimens). In *A. africanus*, the mean for two specimens is 106%. As in A.L. 444-2 and male gorillas, the face of *A. boisei* is tall relative to the biorbital distance. The mean value for two males (OH 5 and KNM-ER 406, with its reconstructed facial height) is 115%, whereas in the female KNM-ER 732, the index is lower, at 105%. Cranium SK 48 of *A. robustus* and the single *A. aethiopicus* specimen, KNM-WT 17000, have relatively short faces, and thus yield the lowest index values in *Australopithecus* (97% and 94%), respectively. Such low index values mean that the facial height measurement is less than the biorbital distance. The large facial breadth measured at the plane of the orbits of KNM-WT 17000 enhances the apparent shortness of its face.

In modern *H. sapiens*, the diminished masticatory system results in reduced facial breadth and height. The modest facial height relative to the biorbital distance thus yields a very low mean index of 91%. Cranium KNM-ER 1813, representing *H. habilis*, seems to adhere to the more generalized pattern. Its height measurement, slightly greater than the biorbital distance, yields an index of 101%. The extensively reconstructed OH 24 yields an index of 95%.

Ratio of palatal width to total facial height. In the preceding two sections, we evaluated the (external) width of the palate and the height of the face, each in relation to the biorbital width, and observed that the values of both ratios are higher in A.L. 444-2 than in other primates. The ratio of external palatal width to total facial height is clearly grounded in those indices. This ratio in the Hadar specimen is 76% (Figure 3.40, 5: a/b). The index values estimated for the Hadar composite reconstruction and for A.L. 417-1d are very similar (73% and 74%, respectively). This value in A.L. 444-2 is greater than the mean value (70%) calculated for two *A. boisei* specimens (KNM-ER 406 and OH 5). Since their palatal width is almost the same as in the Hadar specimen, the lower index value in *A. boisei* must stem from its greater facial height dimension, as is quite clear from the more complete—less extensively reconstructed—specimen, OH 5. In KNM-WT 17000, in contrast, the value is high, at 90%. Here, the index is the product of a very wide palate and vertically short face (the facial height measurement in this specimen is almost the same as the biorbital distance). The mean value for two *A. africanus* specimens is 80%. In modern humans, the index is 74%, which is similar to that of A.L. 444-2. In humans, however, the value results from a modest palatal width and facial height, as was ascertained from their low ratio to the biorbital distance.

At the other end of the comparative range we find the gorilla sample (particularly the males). Having already observed that the external palatal arch is very narrow in male gorillas while the face is extremely tall, we are not surprised to find a low index value, at 56%. Indeed, much of the facial appearance specific to male gorillas stems directly from this combination. Chimpanzees exhibit less extreme facial proportions, although their index values are still quite low (67% for the males and 69% for the females).

The pattern that emerges in hominins, therefore, is a rather wide palate, especially in terms of facial height, with specimen A.L. 444-2 falling in the middle of the hominin range. The proportions found in female gorillas and chimpanzees probably represent the generalized configuration; identical index values were also measured on *Cercopithecus*, which yields a mean of 67% ($n = 4$).

Ratio of the height of the masseter origin to the height of the lower face. The height of the zygomatic tubercle above the alveolar margin of the maxilla constitutes 49% of the orbitoalveolar height in A.L. 444-2 (Figure 3.40, 6, b/a; Table 3.18). The tubercle, which marks the anterior limit of the masseter muscle origin on the inferior margin of the zygomatic arch, resides at a similar relative level of facial height in A.L. 333-1 (51%) and the reconstructed A.L. 417-1d (57%), the only other adult *A. afarensis* specimens on which both measurements can be made (see Kimbel et al., 1984). Gorillas, chimpanzees, and *Homo* share with *A. afarensis* the vertical position of the anterior masseter origin approximating the midway point of orbitoalveolar height, which may be considered the plesiomorphic condition for hominins. All other species of *Australopithecus* for which this relationship is known show a much higher relative position of the anterior masseter origin, with an index ranging from 59% to 63% (Kimbel et al., 1984). Adult specimens of *A. africanus* align with those of robust *Australopithecus* species in this regard. The higher index values in *A. africanus* and robust *Australopithecus* species reflect not only the increased height of the masseter origin but also its more lateral position relative to the muscle's insertion on the mandibular ramus. This position maximizes the vertical components of the masseter muscle's line of action on both the coronal and sagittal planes (see the discussion in Rak, 1983: 114–116; and Kimbel et al., 1984: 346–349).

Ratio of total facial height to the maximum facial breadth (*the facial shape index*). The facial shape index in A.L. 444-2 is 68%. This value is the product of a substantial total facial height combined with an immense bizygomatic breadth, which tend to balance each other out (Figure 3.40, 7: a/b). To gain insight into the significance of the shape index, one needs to evaluate each of its components independently (see indices 2 and 4 in Table 3.18), as an extremely broad face might dictate the index in one individual while a very short face will dictate it in another.[14] In other words, two such individuals could end up with the same index value.

A very similar index value to that of A.L. 444-2 is calculated for the Hadar composite reconstruction (65%). These values for *A. afarensis* do not differ substantially from the 67% calculated for modern humans. In the latter, however, the value is produced by a short face and a modest bizygomatic breadth. The index value for A.L. 444-2 resembles that of the other *Australopithecus* specimens. The mean value for two *A. africanus* crania (Sts. 5 and Sts. 71) is 72%, while SK 48, the only *A. robustus* specimen permitting these measurements, yields an index of 66%. The mean value for three *A. boisei* specimens (KNM-ER 406, KNM-ER 732, and OH 5) is 64%. At 55%, the value for KNM-WT 17000 constitutes a noticeable deviation, which confirms the specimen's visual appearance—a very low, broad face.

The African apes, particularly chimpanzees, have higher index values. Chimpanzees yield an index of 75%; in gorillas, the rather large lateral deviation of the zygomatic arches overrides the effect of the tall face and thus produces an index value that is almost the same as in chimpanzees, 74%.

Ratio of the zygomatic bone's anterior surface area to the surface area of the facial mask. The combination of the immense absolute size of A.L. 444-2's zygomatic bone, its anteriorly facing lateral section (the base of the frontal and temporal processes) and the extremely broad root of its frontal process renders the zygomatic bone a prominent component of the facial mask. Because of the size of the zygomatic bone's surface, the zygomatic process of the maxilla is less dominant in the makeup of the facial mask—or, more specifically, of its infraorbital region. The ratio of the zygomatic bone's "width" to the overall size of the facial mask appears to be the largest among the primates. The bone's width constitutes 51% of the biorbital breadth (Figure 3.40, 8: a/b), a value that is far greater than the 28% in female gorillas and the 27% in male gorillas. In chimpanzees, the width of the zygomatic bone produces an even lower value, at 22%. A similar value—23%—is obtained from the sample of four *Cercopithecus* specimens. Orangutans, it is worth noting, display an extremely narrow zygomatic bone, which constitutes only 22% of the biorbital breadth ($n = 5$). Given the already narrow orbits and interorbital breadth of the orangutan, this distance is, indeed, quite small. Modern humans do not differ, either, from what appears to be the generalized configuration, with a ratio calculated at 26%.

Among the hominins, only *A. boisei* yields a value anywhere near that of A.L. 444-2. The index calculated for KNM-ER 406 is 42%; for OH 5, with its partially reconstructed zygomatic bone, it is 36%. KNM-WT 17000 yields a value of 40%. In *A. robustus*, the only two specimens in which this ratio can be calculated yield an index of 33%, and in the two specimens of *A. africanus* that permit these measurements (Sts. 5 and Sts. 71) the value drops even farther, to 29%.

Summary of the facial proportions of A.L. 444-2. The facial proportions of A.L. 444-2 are summarized as follows:

- The breadth of the face at the coronal plane of the orbits is very large, substantially greater than in gorillas. Though less than the breadth measurement of *A. boisei*, that of A.L. 444-2 is not far from it.
- The location of the facial breadth on the coronal plane of the orbits is low relative to the height of the facial hexagon and thus makes the upper half of the hexagon taller than the flattened lower half. Reminiscent of the more generalized face (in chimpanzees and gorillas), the proportions of the hexagon in A.L. 444-2 are quite different from those seen in *A. boisei* and KNM-WT 17000, in which they are reversed. In the latter specimens,

the lower half of the hexagon is taller than the upper half.

- The palate width, as measured between the external faces of the alveolar processes, is relatively large, especially when compared to that of male gorillas, which have an unusually narrow palate. It seems that all the Plio-Pleistocene hominins have relatively wide palates, wider than what appears to be the generalized configuration, as seen in chimpanzees and female gorillas. In modern humans, the palate is externally narrow, the product of the general shrinkage of the dental arcade and the alveolar processes supporting it.
- The lateral deviation of the zygomatic arches relative to the orbits is immense, even exceeding (though slightly) that of the most extreme *A. boisei* specimen. However, this large bizygomatic breadth does not persist anteriorly to the level of the coronal plane of the orbits, where *A. boisei* exhibits the greater width.
- The face is tall relative to the biorbital distance, as in male gorillas and male *A. boisei* specimens.
- An externally very wide palate united with a tall face distinguishes A.L. 444-2 from male gorillas, on the one hand, and KNM-WT 17000, on the other. Male gorillas gain their characteristic appearance from the combination of a narrow palate and a tall face. The unique facial appearance of KNM-WT 17000 derives from its very wide palate and short face.
- The relative position of the zygomatic tubercle is low, the plesiomorphic configuration, and differs substantially from what is found in the robust *Australopithecus* species and *A. africanus*, in which the origin of the anterior part of the masseter is relatively high on the face.
- The facial shape index is the product of a tall face combined with a large bizygomatic breadth. This proportion is not significantly different from that seen in other *Australopithecus* species and modern humans. The only notable exception is *A. aethiopicus*, in which the combination of a short face and a large bizygomatic breadth produces a very low index. The facial shape index in A.L. 444-2 also differs from that of the African apes, especially chimpanzees, in which the modest bizygomatic breadth reduces the index value.
- The area of the zygomatic bone surface is immense, constituting a large percentage of the facial mask's surface area. A.L. 444-2 is unique in this configuration. Only *A. boisei* comes close to the index value. Because of the size of the zygomatic bone in A.L. 444-2, the zygomatic process of the maxilla appears less dominant in the makeup of the facial mask than in other hominins. In the apes, this process constitutes a substantially larger percentage of the composition of the infraorbital part of the face.

Facial Topography

The upper central part of the face in A.L. 444-2 is missing and has therefore been reconstructed. However, the adjacent structures—the inferior margin of the nasal aperture, the lower part of the aperture's lateral margin, the maxillary process of the zygomatic bone, the zygomatic process of the maxilla, the "borrowed" glabellar segment from specimen A.L. 438-1b, and even the remains of the slightly concave nasal bones—leave little doubt as to the anatomic and topographic accuracy of the reconstructed portion.

The Central Part of the Face

The fundamental aspects of the Hadar specimen's facial morphology and topography are shared with many other primates. The contention that *A. afarensis* has a generalized face was made as early as 1980, although the evidence at that time consisted only of fragments from several individuals (Rak, 1983).

The lateral margins of the nasal aperture are sharp and extend anteromedially to the canine juga. In this respect, A.L. 444-2 is consistent with every other *A. afarensis* specimen that bears the remains of these elements (A.L. 199-1, A.L. 200-1a, A.L. 333-1, A.L. 333-2, A.L. 413-1, A.L. 417-1d, A.L. 427-1, A.L. 442-1, A.L. 486-1). Topographically, the bone surface of the midface gradually advances anteromedially from the level of the peripheral face and terminates as the thin, sharp-edged plate at the lateral margins of the nasal opening. Thus, the transverse cross section of the face shows no trace of the central concavity (or "dish" face) that characterizes robust *Australopithecus* species (Figures 3.42 and 3.43).

This topography in A.L. 444-2 also differs considerably from the configuration of the *A. africanus* face. There, the zygomatic prominence, which is located at the roots of the temporal and frontal processes of the zygomatic bone and in some specimens (Sts. 5, Sts. 71, Stw. 370, and probably Sts. 17, Sts. 63, TM 1511, and MLD 6) is the most prominent structure of the facial mask, protrudes anterolaterally. A by-product of this protrusion is the topographical dip of the central part of the face relative to the peripheral part. In fact, it is this orientation of the transverse cross section of the infraorbital bone surface—which moves anteriorly the farther it gets from the midline—that forces the reconstruction of a prominence in the specimens that are missing the more lateral part of the face, as in MLD 6 and Sts. 17. Other specimens (such as Sts. 52a

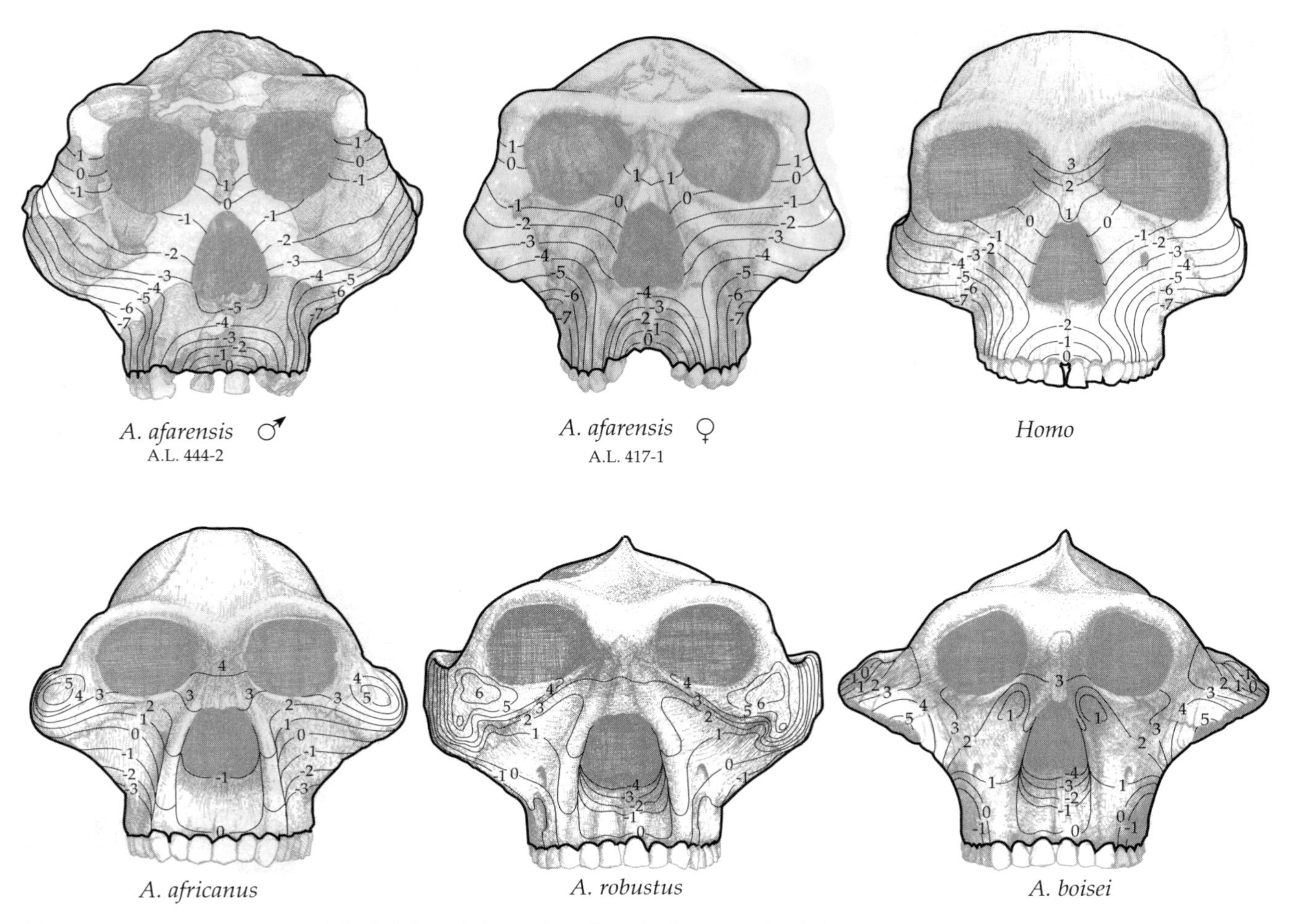

Figure 3.42 The topography of the facial mask in *A. afarensis* and other hominins. The *A. afarensis* specimens are A.L. 444-2 and A.L. 417-1 (in which the supraorbital region and frontal squama are hypothetical). As illustrated here, *Homo* represents a generalized facial mask (that of *H. habilis*). The specimens of *Homo*, *A. africanus*, *A. robustus*, and *A. boisei* are idealized composites and have been modified from Rak (1983).

and probably TM 1512) exhibit a less derived morphology, with the peripheral face remaining in a more posterior plane than the central nasal region. In such specimens, we expect the prominence to protrude less.

The lateral nasal margins of *A. africanus* are almost always demarcated by distinct anterior pillars, whose medial aspect is rounded in a transverse cross section (Rak, 1983). As a result, the nasal aperture in *A. africanus* usually lacks the sharp, paper-thin lateral margins seen in the *A. afarensis* face. The anterior pillars are sharply defined all along their lateral aspect by what has been termed the "maxillary furrow" (Rak, 1983), a distinct groove that sets the pillars apart from the more peripheral part of the face. In essence, a modification of the canine fossa that characterizes *A. afarensis* and the great apes, the maxillary furrow represents a derived character state in the *Australopithecus* facial morphocline. In this morphocline, the primitive canine fossa undergoes a transformation into the "maxillary furrow" of *A. africanus*, which then becomes the inferiorly located vestigial "maxillary fossula" of *A. robustus* (and *A. aethiopicus*) and eventually disappears altogether in the most specialized state, that of *A. boisei* (Rak, 1983) (Figure 3.44).

The face of A.L. 444-2 displays a primitive, hollowed canine fossa on each side of the snout, just as in all the other *A. afarensis* specimens in which the region is preserved. (On the right, the canine fossa is completely preserved and intact, whereas on the left, the plastically deformed region and the displacement of the snout make it barely noticeable.) Lying at the junction of the protruding snout and the posteroinferiorly receding infraorbital plate, the fossa is created by the topographical discrepancy between these two structures. In some *A. afarensis* specimens, the canine fossa is deeply hollowed, lending the maxilla (in anterior

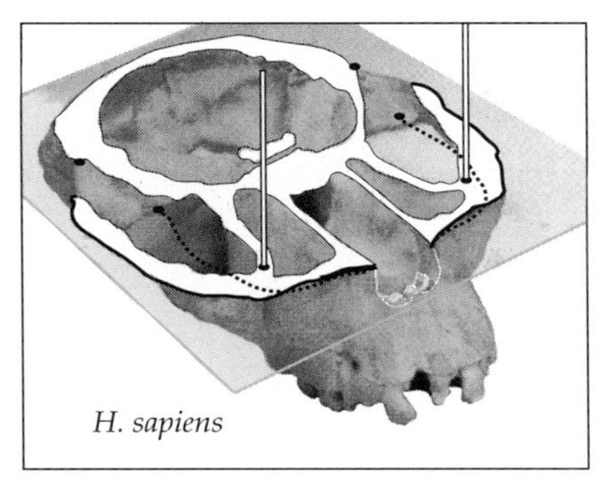

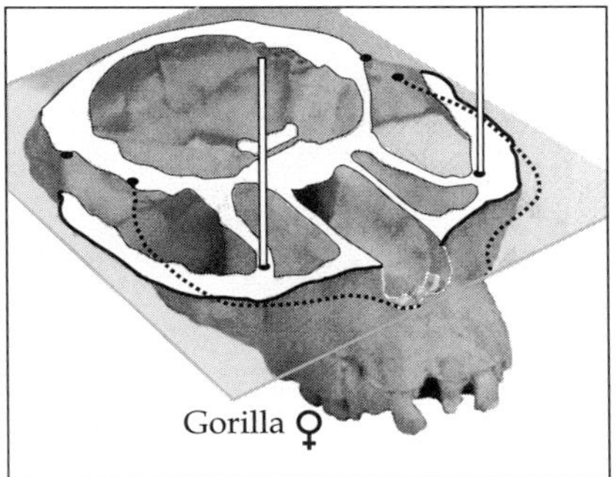

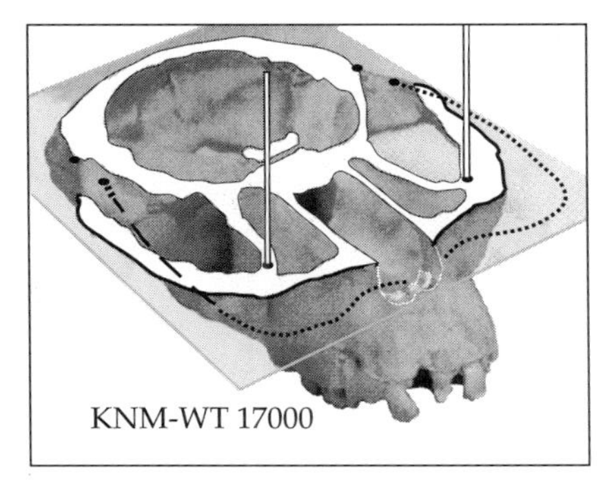

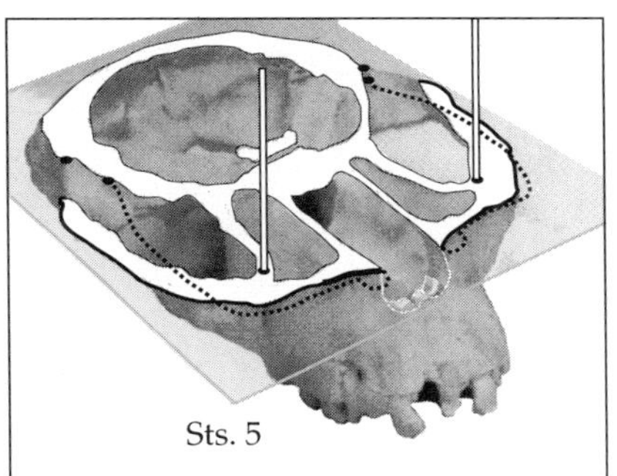

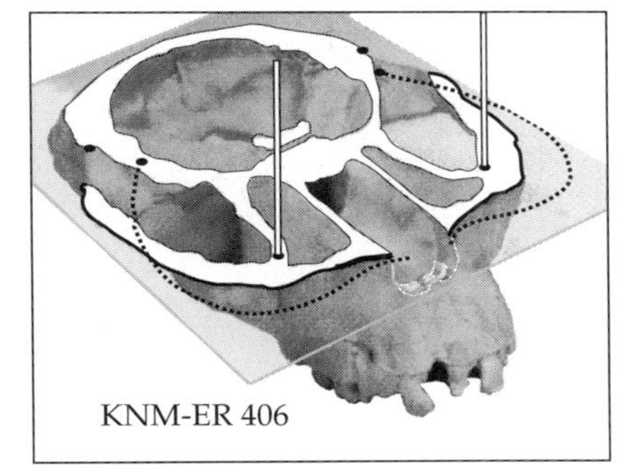

Figure 3.43 Contour of the transverse cross section through the middle of the infraorbital region in A.L. 444-2 and other hominoids. The actual cross section of A.L. 444-2 is depicted. The other specimens are represented by the contour of the section that extends from the lateral margin of the nasal aperture to the porial saddle (black dots). This outline is superimposed on the A.L. 444-2 cross section. All the contours have been adjusted to fit the biorbital breadth of A.L. 444-2 (the two vertical lines). Note the horizontal anteroposterior differences in the position of the porial saddle in A.L. 444-2 and the other hominoids, particularly modern humans.

view) a pinched appearance (A.L. 333-1, A.L. 413-1, A.L. 427-1); in others (probable females) it is a shallow depression (A.L. 200-1a, A.L. 417-1d, A.L. 442-1).

In gorillas, chimpanzees, and other nonhominin primates, a long, vertically (sagittally) oriented bone plate extends anteriorly from the medial part of the infraorbital region as the lateral wall of the snout. In these prognathic primates, this bone plate bears the majority of the canine fossa surface, which also impinges to a much smaller extent on the anterior-facing surface of the maxilla's zygomatic process. Contrasting with this morphology is the human configuration, in which the walls connecting the reduced snout to the facial plane are so short anteroposteriorly that the canine fossa is all but confined to the zygomatic process. The comparatively modest protrusion of the palate (compared to the great apes) in A.L. 444-2 reduces the anteroposterior length of the lateral walls of the snout, and the anterior face of the zygomatic process contributes proportionally more to the formation of the canine fossa. In other specimens of *A. afarensis*—for example, the facial fragments A.L. 333-1 and A.L. 427-1—the sagittally oriented wall of the canine fossa is more dominant, a sign that these specimens were more prognathic than A.L. 444-2. However, A.L. 417-1d more closely resembles A.L. 444-2 in these respects.

The Peripheral Infraorbital Region

The infraorbital bone plate consists of a single plane in A.L. 444-2. There is no trace of the "transverse buttress" seen in gorillas, in which it is situated above the infraorbital foramen and forms a distinct ridge that extends medially from the body of the zygomatic bone toward the upper part of the snout (Figure 3.32) (Rak, 1983). The buttress clearly demarcates the upper margin of the canine fossa and divides the infraorbital region, particularly its medial part, into two distinct planes. The superior plane passes anteroinferiorly from the inferior orbital

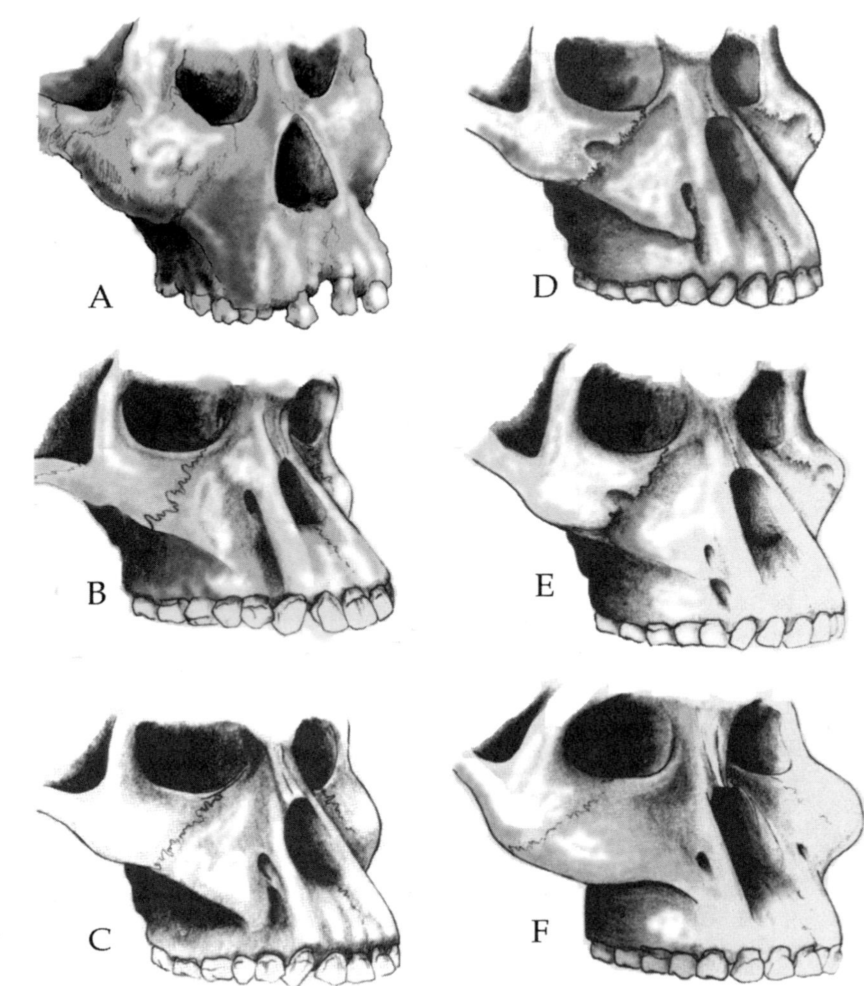

Figure 3.44 The facial structure of *A. afarensis* and other *Australopithecus* species. The drawings depict variation in facial structure within and among early hominin species: A, *A. afarensis*, based on A.L. 444-2; B, *A. africanus*, based on Sts. 5; C, *A. africanus*, based on Sts. 71; D, *A. robustus*, based on SK 46 and TM 1517; E, *A. robustus*, based on SK 12 and SK 48; F, *A. boisei*, based on KNM-ER 406 and OH 5. (Adapted from Rak, 1983.)

margin to the buttress, and the lower slants postero-inferiorly from the buttress to form the floor (posterior wall) of the canine fossa. If the face of A.L. 444-2 did have a transverse buttress—recall that the medial part of the infraorbital region is reconstructed—it would probably have been modest in size.

In another *A. afarensis* specimen, A.L. 417-1d, the fully preserved region does not display any signs of such a buttress. However, a third specimen, A.L. 333-1, exhibits a deeply excavated canine fossa along with a small but clearly discernable transverse buttress that marks the upper boundary of the fossa (Rak, 1983). These configurations indicate that the presence of a transverse buttress is a variable character in *A. afarensis* and that its expression corresponds to the variably deep canine fossa and degree of prognathism.

The single-planed infraorbital plate in A.L. 444-2 descends from the inferior orbital margin and recedes somewhat posteriorly (Figure 3.14). The infraorbital bone plate's inferior margin—the zygomaticoalveolar crest—lies slightly posterior to the coronal plane of the orbit. In contrast, the inferior margin of the infraorbital region in robust *Australopithecus* species and in *A. africanus* is situated anterior to the orbital plane. This configuration reflects the anterior advancement of the masseter muscle origin, part of the derived morphology of the robust *Australopithecus* masticatory system. In *A. boisei*, representing the extreme end of this morphocline, the infraorbital bone plate extends so far forward that it loses the conventional suspensory bracing of the basic facial frame (and, as explained above, also obliterates the anterior pillars and the fossulae still seen in *A. robustus*). As a result of this peculiar configuration, a structure that can support itself architecturally is necessary—hence the modification of the more lateral bone surface of the infraorbital region into a visorlike form (Rak, 1983). Even though A.L. 444-2 resembles some of the largest *A. boisei* specimens in size and rugosity of the muscular attachment areas, its infraorbital morphology is still the most generalized of any *Australopithecus* species. (See also the section "Relationship between Elements of the Masticatory System.")

In *A. boisei*, the entire facial surface of the zygomatic bone (the visor) faces anterosuperiorly and is completely smooth. To illustrate the nature of the visorlike structure and compare it with the comparable region in A.L. 444-2, we are aided by Figure 3.45 in which three virtual plates are translucent and of equal size. Each plate represents a cross section in a different plane. A diagonal cross section (plate A), oriented 45° to the sagittal plane, extends anterolaterally from the root of the frontal process of the zygomatic bone. In *A. boisei*, this cross section reveals an extensive concave surface that is the product of the elevated inferior edge of the zygomatic bone, which is situated almost as high as the level of the inferior orbital margins. The concave surface extends all the way to the inferoanterior edge of the zygomatic bone. Perpendicular to plate A (45° to the sagittal plane) lies the cross section represented by plate B, which reveals a *convex* contour. A horizontal cross section beneath the level of the inferior orbital margin (plate C) exposes a smoothly convex arc from the infraorbital bone plate to the zygomatic arch in *A. boisei*. Note that part of the visor's inferior margin (its lateral part) is elevated to a position above the plane of the transverse cross section. It is the combination of the contours of these three planes that elucidates the visorlike structure of the *A. boisei* infraorbital region.

A convex contour in a horizontal cross section (plate C) is characteristic of the zygomatic bone in all hominoid faces. However, unlike *A. boisei*, the other hominoids—including *A. afarensis* (represented by specimen A.L. 444-2, in the upper row in Figure 3.45), *A. africanus*,

Figure 3.45 The generalized lateral infraorbital region in A.L. 444-2 and the "visorlike" region in *A. boisei*. The three sections emphasize differences between the lateral infraorbital regions of A.L. 444-2 and *A. boisei*. Plane A, passing vertically through the middle of the root of the frontal process of the zygomatic bone, forms a 45° angle with both the sagittal and the coronal planes. In A.L. 444-2, plane B, vertical and tangent to the lateralmost point on the zygomatic bone's surface, forms a 90° angle with plane A; the mediolateral location of plane B in *A. boisei* is equal to the horizontal distance between this plane and the lateral orbital margin in A.L. 444-2. Plane C is horizontal and is located at mid-infraorbital height level. See text for discussion.

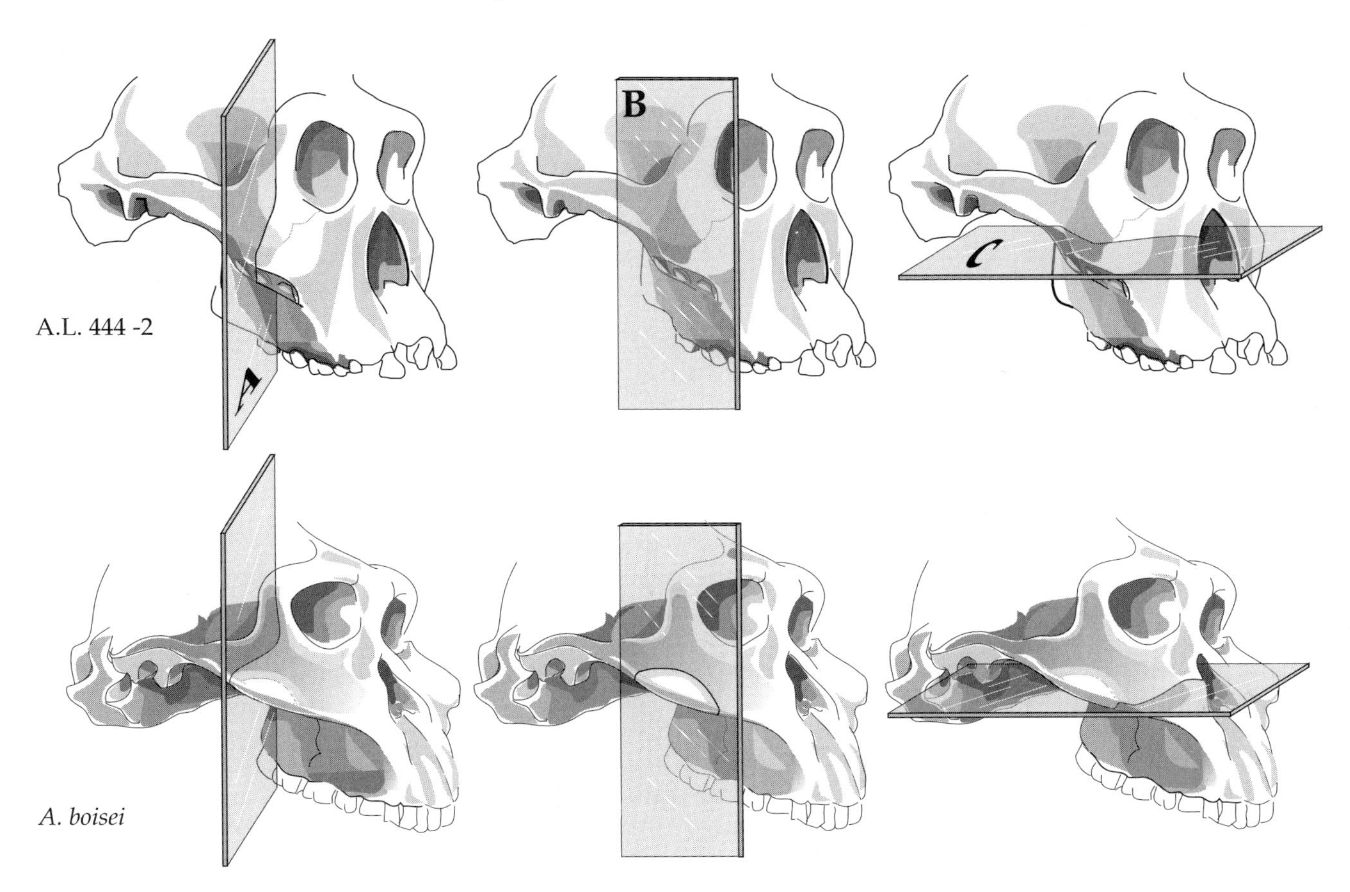

and *A. robustus*—exhibit a more clearly segmented (almost angulated) contour—that is, one with a convexity localized at a distinct "corner" of the zygomatic bone, where the bone's anteriorly directed facial surface joins the laterally facing temporal process.

The diagonal cross section (plate A) through the zygomatic bone of A.L. 444-2 exposes a steep, nearly vertical contour, slightly concave superiorly near the frontal process of the zygomatic bone, but convex over the inferior half of the bone. As a result of the verticality of this bone surface, which lies entirely behind plate B, the latter plate fails to intersect the zygomatic bone at all.

The presence of a visor in the face of *A. boisei* is variable (Rak, 1983, 1988) and results from a particular interplay among the components of the masticatory system (see the section "Relationship between Elements of the Masticatory System"). In some *A. boisei* specimens, a retraction of the dental arcade reduces the need for an anteriorly extended masseter muscle origin and a concomitant transformation of the infraorbital region into a visor (see OH 5, for example). In more prognathic individuals, similar index values are achieved through the anterior extension of the masseter origin, which brings about the same degree of "efficiency" of the masticatory system (for example, in KNM-ER 406). Indeed, though OH 5 exhibits a modest visor and KNM-ER 406 an extensively developed one, the "index of overlap" is similar in the two specimens, at 57% and 50%, respectively.

As early as 1983 one of us wrote:

> It is noteworthy that *A. boisei* is not the only primate to have the infraorbital region transformed into a visorlike structure. A fossil monkey, *Theropithecus brumpti*, apparently was confronted with the same problem. As the masseteric origin became more anteriorly displaced beyond the coronal plane of the orbital openings, the bone bearing the muscle was left there, too, without a structure from which to be suspended. Hence, as in *A. boisei*, the bone in *T. brumpti* obtained a visorlike shape. Specimen L.345-287 demonstrates this in a most extreme form. . . . Although the visor does exist in other specimens of this species, it is less pronounced [in some, it is absent altogether]. Just as the relatively large sample of this fossil monkey from the Omo reveals a wide range of variation in the development of the visor (differences related to individual age, sex, robusticity, and, particularly, the degree of forward extension of the zygomatic bone are easily recognized), a similar distribution *could be expected* from a more extensive sample of *A. boisei* specimens. Even in the small available sample, substantial differences exist between the specimens: the visor in KNM-ER 406 is most developed and in OH-5 and KNM-ER 732, much less so. OH 5 is a young individual, approximately 16 years old (Tobias, 1967), and KNM-ER 732 is a female. As mentioned previously, fragments of other specimens indicate a more developed visor than in KNM-ER 406 (for instance, KNM-ER 733 and KNM-ER 405). (Rak, 1983: 99; emphasis added by the authors)

In light of the range of variation demonstrated by the visorlike structure in *A. boisei* and other primates, we find it surprising that the claimed absence of a visor in the face of a new *A. boisei* specimen (KGA10-525) from Konso (Suwa et al., 1997) caused such consternation, leading to the following far-reaching conclusion: "In particular, [the] lack of the visorlike facial morphology contradicts Rak's dichotomization of East and South African robust *Australopithecus*" (pp. 491–492). Had these researchers attempted to calculate the index of overlap for this specimen, whose state of preservation would apparently permit the necessary measurements, we might understand why this *A. boisei* specimen has no visor.

The KNM-WT 17000 cranium of *A. aethiopicus* is again a revealing case in regard to the visor. The specimen is extremely prognathic; the protrusion of its palate is immense, extending far beyond sellion, with an index value of 82% (Figure 3.26, Index A). To maintain anatomical and functional integrity—that is, to keep up with the extension of the palate—the masseter origin advanced an incredible distance anteriorly. Index C of the masticatory system (Figure 3.26) is 148%! The index of overlap (Index E in Figure 3.26, at 28%) achieved through this advancement does not reach the values of *A. boisei* but still is almost double the value yielded by the less prognathic chimpanzees. Indeed, such an extreme anterior advancement of the masseter origin—the inferior margin of the infraorbital region—necessitates the transformation of the zygomatic bone into a visor for precisely the same reason as in *A. boisei* and *T. brumpti*, that is, the need for a self-supporting structure.

The Subnasal Region

In spite of damage and distortion to the maxilla in A.L. 444-2, the nasoalveolar clivus closely resembles the corresponding region in many other *A. afarensis* specimens (A.L. 199-1, A.L. 200-1a, A.L. 333-1, A.L. 333-2, A.L. 413-1, A.L. 417-1d, A.L. 427-1,[15] A.L. 442-1, and A.L. 486-1). The enlarged sample of *A. afarensis* maxillae, including that of A.L. 444-2, corroborates earlier descriptions of this species' subnasal topography as apelike (Rak, 1983; Ward and Kimbel, 1983). Because of this resemblance to the great apes and other primates, this topography is interpreted here as plesiomorphic. The midsagittal cross section of the nasoalveolar clivus in A.L. 444-2 juts forward as a broad, convex arch, and the clivus in its entirety is situated in front of the central facial plane. The horizontal cross section at the level of the alveolar margin forms a parabola, as do more superior cross sections at midclivus and beneath the nasal sill. This primitive topography can also be seen in *A. garhi* (Asfaw et al., 1999) and *A. anamensis* (M. Leakey et al., 1995).

This topography contrasts sharply with what is found in robust *Australopithecus* specimens, in which a large section of the mainly flat bone surface of the clivus (and the nasal aperture itself) is recessed behind the level of the more peripheral part of the facial surface. The horizontal cross section of the maxilla at the level of the alveolar margin is more rectangular because of the rearrangement of the anterior teeth in a straight line at the front of the robust *Australopithecus* palate (Robinson, 1956).

In *A. africanus*, the surface of the clivus is also frequently flat and at the same topographic level as the prominent anterior pillars. Sometimes it is slightly sunken between them (contrast Sts. 5 and Stw. 73 with TM 1512 and Sts. 52a). Indeed, the clivus has been described as a flat surface that is "stretched" between the lower thirds of the pillars (Rak, 1983). The rectangular shape of the palate's anterior part is the product of this configuration, with the anterior pillars forming the corners of the rectangle just above the level of the alveolar margin.

The two prominent anterior pillars in the face of *A. africanus*, the flat clivus between them, and the two distinct maxillary furrows lateral to the pillars form a well-defined topographic unit, referred to as the "nasoalveolar triangular frame" (Rak, 1983). This single-plane frame is situated farther forward than the more peripheral face (see Figure 3.43, which illustrates the midfacial topography in a transverse cross section, and also Figures 3.44 and 3.46). However, in the absence of these pillars, furrows, and a flat clivus, the face of *A. afarensis* does not exhibit a nasoalveolar triangular frame.

The retraction of the dental arcade while the anterior pillars are still present at the anterior ends of the two maxillae on each side of the clivus results in the formation of the nasoalveolar gutter in the face of *A. robustus* (Rak, 1983). In this configuration, the upper end of the clivus (the nasal sill) lies between two sagittally oriented "walls" that stand lateral to it (forming a "corridor" that leads to the actual nasal opening). Consequently, nasospinale (which marks the true opening of the nasal cavity) is situated behind the more peripheral facial surface. This morphology is also found in *A. boisei*, although the presence of the anterior pillars was to a large extent obliterated through further specialization (Rak, 1983).

Confirmation of this structural interpretation of the gutter's formation—the relationship between the retraction of the palate and the appearance of the gutter—can

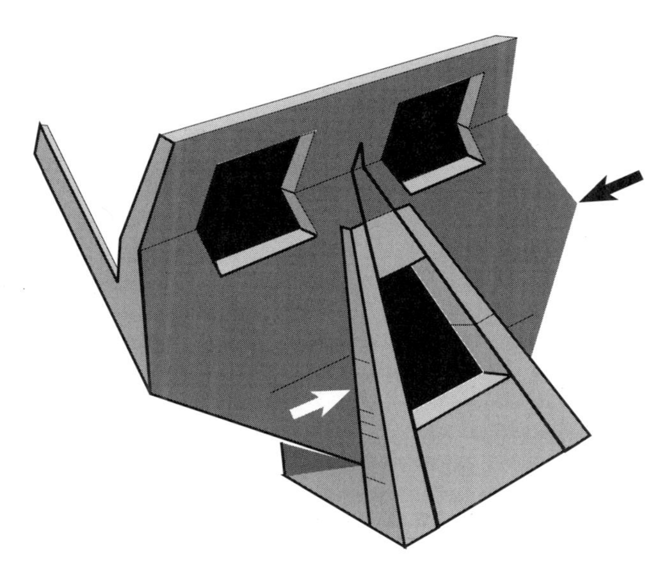

Figure 3.46 Schematic representation of the common facial structure in *A. africanus*. The nasoalveolar triangular frame (light gray area), the maxillary furrow and the zygomatic prominence (arrows) are common elements in the distinctive facial configuration of *A. africanus*.

be found in the topography of the relevant region in KNM-WT 17000. In this specimen, the palate extends anteriorly, which makes the specimen the most prognathic hominin, as described earlier. Although flat on the sagittal plane and concave on the coronal plane, the clivus does not take the form of a true nasoalveolar gutter, as it is situated much farther anteriorly than the more peripheral parts of the face, with nasospinale lying on the same plane as the anterior surfaces of the zygomatic processes of the maxillae. In other words, no defined corridor of lateral walls leads to the nasal opening (compare KNM-WT 17000 and SK 12, for example).

The inferior edges of the sloping, anteriorly extended infraorbital bone plates (the zygomatic processes of the maxillae) are located at the same level and almost the same orientation as the clivus itself, thus enhancing the appearance of the immense width of the clivus in KNM-WT 17000.

The Interorbital Region, the Glabellar Region, and the Nasal Bones

Most of the glabellar region is damaged or missing in A.L. 444-2. As we have already noted, the interorbital region has been reconstructed based on the morphology of the A.L. 438-1b frontal fragment. The latter specimen, associated with a partial mandible and upper limb elements (Kimbel et al., 1994) is clearly from a large male individual. In terms of size and morphology, it makes a perfect fit with the medial curvature that demarcates the orbital roof and the superomedial corners of the orbits of A.L. 444-2 and has thus been incorporated in the reconstruction of that specimen (see fragment f in Figure 2.2 A). This composite furnishes us with the only view of the glabellar region's anatomy in an adult *A. afarensis* cranium (the region is preserved in the juvenile cranium A.L. 333-105).

On the basis of the frontal fragment of A.L. 438-1b and the reconstructed orbits of A.L. 444-2, we may conclude that the interorbital region of the latter is very narrow. Its width is no more than 19 mm, which constitutes only 48% of the orbital breadth (Table 3.19).[16] Although the interorbital width is rather variable within species,[17] the narrowness of the region in *A. afarensis* is corroborated by specimen A.L. 417-1d, with an index of 46%. In addition, when compared with the Taung juvenile, A.L. 333-105 also demonstrates this anatomy. These index values are smaller than those of the African apes, in which the interorbital width constitutes 69% of the orbital breadth in female chimpanzees and 55% in the males. In male gorillas, the interorbital distance is quite wide, at 79% of the orbital breadth, whereas in the females the figure is 56%. In orangutans, in contrast, which are known for their extremely narrow interorbital region, the distance between the orbits is only 31% of the orbital breadth ($n = 2$

Table 3.19 Measurements and Proportions of the Orbital Region

Specimen/Sample		Interorbital Br. (56) (mm)	Orbital Br. (68) (mm)	Orbital Ht. (67) (mm)	Interorbital Br. (56)/ Orbital Br. (68) (%)	Orbit Ht. (67)/ Orbit Br. (68) (%)
Australopithecus afarensis	Mean	18	38	34	47	90
A.L. 444-2		19	40	37	48	93
A.L. 417-1d		16	35	(30)[a]	46	(86)
Homo sapiens	Mean (n = 10)	26	39	34	66	87
	Range	23-28	34-41	31-36	62-72	78-103
	SD	2	2	2	4	7
Gorilla gorilla female	Mean (n =10)	21	38	38	56	99
	Range	17–25	35–41	33–43	45–66	92–104
	SD	2	3	3	7	3
Gorilla gorilla male	Mean (n = 10)	32	41	38	79	92
	Range	26–39	35–47	33–44	69–89	82–109
	SD	3	3	2	7	7
Pan troglodytes female	Mean (n = 10)	24	35	32	69	92
	Range	20–27	30–39	25–39	65–74	83–105
	SD	2	3	3	2	7
Pan troglodytes male	Mean (n = 10)	20	37	33	55	91
	Range	17–23	33–42	26–39	50–61	81–99
	SD	2	2	4	3	6
Pongo pygmaeus female	Mean (n = 2)	11	35	38	31	108
Pongo pygmaeus male	Mean (n = 2)	12	35	41	34	117
A. africanus	Mean	22	33	33	66	99
Sts. 5		19	32	30	59	94
Sts. 71		20	30	30	67	100
Stw. 505		27	38	39	71	103
A. robustus						
SK 48		25	37	32	68	86
A. boisei	Mean	28	37	35	74	94
OH 5		31	34	34	91	100
KNM-ER 406		26	40	35	65	88
KNM-ER 23000		28	n/a	n/a	n/a	n/a
Omo 323-1976-896		27	n/a	n/a	n/a	n/a
KNM-ER 732		21	32	31	66	97
A. aethiopicus						
KNM-WT 17000		22	34	37	65	109

Ht., height; Br., breadth; n/a, not available.
[a]Based on reconstructed skull.

males), which is an impressively small value given that the orbits are so narrow as well.

The relative interorbital width in A.L. 444-2 is smaller than in any other *Australopithecus* cranium. In *A. africanus*, the mean value of three specimens (Sts. 5, Sts. 71, and Stw. 505) is 66%, and a single *A. robustus* specimen, SK 48, yields an index of 68%. The mean of three *A. boisei* specimens is 74%. Note that even in this small sample, the range of values is great, from 65% (KNM-ER 406) to 91% (OH 5), with the female KNM-ER 732 yielding an index of 66%. Similarly, the value for specimen KNM-WT 17000 is 65%.

According to the A.L. 438-1b glabellar fragment, the glabellar region in *A. afarensis* is not as massive as in robust *Australopithecus* specimens. The medial part of the thin, lightly constructed supraorbital elements attaches abruptly to the glabellar region, giving the superomedial orbital corner a well-defined angle. This configuration, along with the modest interorbital distance, whose width is vertically constant almost up to the glabella itself, leaves the glabellar region with a confined area that is "squeezed" between the superomedial corners of the orbits. The glabellar region does not bulge forward to overhang nasion and sellion, as it does in some other *Australopithecus* crania. This anatomical configuration resembles that of female gorillas, though there are some differences that are discussed below. In most male gorillas, the interorbital distance is great, and glabella, which usually coincides with nasion, overhangs sellion.

In A.L. 444-2, and in hominins in general (Clarke, 1977), the glabellar point is situated lower than the superior orbital margins. In contrast, it is elevated above the level of the orbits in gorillas, especially in males, and in some chimpanzees, although the descent of the medial part of the supraorbital torus tends to carry the glabella down in many chimpanzee skulls. In *A. boisei*, although the medial part of the supraorbital element usually does not descend, glabella is located lower than the superior orbital margins. In *A. robustus* and KNM-WT 17000, a combination of an inflated interorbital region (more so in *A. robustus*) and arched supraorbital elements that descend toward the midline carries glabella inferiorly.

The Orbits and Their Shape

Only the right orbit of A.L. 444-2 survives, although its medial wall and the medial third of its lower margin have been reconstructed. Despite the unnatural, distortion-related shape of the orbital contours, it is apparent that the orbits are both tall and narrow. The superolateral corners are sharply angulated, with a more acute angle on the right than on the left because of the above-mentioned superior displacement of the right supraorbital element's lateral end. In contrast, the inferolateral corners are blunt.

The right orbit has a roughly egglike shape, which is attributable in large part to the superolateral displacement of the zygomatic bone fragment that bears the inferolateral corner of the orbit along the frontal process of the zygomatic (Figure 3.36). Restoring this fragment to its proper place would render the orbital contour more square by straightening and elevating the inferior orbital margin and reestablishing the correct position of the inferolateral corner of the orbit.

In its present, deformed state, the right orbit measures 40 mm in width and 39 mm in height. These chords are perpendicular to each other, and the transverse one is parallel to the FH. In other words, the transverse axis of the orbit does not dip inferolaterally as has been described in reference to some *A. boisei* crania (Rak, 1983). Returning the displaced zygomatic fragment to its original site does not influence the width measurement, but it does reduce the height measurement to 37 mm. With the suggested correction, the shape index is 93%, which resembles that of gorillas and chimpanzees. In those groups, especially in female gorillas, the orbits are also square, the height and width chords are perpendicular to each other, and the width chord is parallel to the FH. In chimpanzees, the orbits tend to be a bit more rounded. The supraorbital torus follows the curved path of the superior orbital margins. In Sts. 5 rectangular orbits, whose height is smaller than their width, bear well-defined orbital corners. However, this specimen might be an exception for *A. africanus*, as other specimens, such as Sts. 71, Taung, and perhaps also Sts. 17 and Stw. 505, display squarish orbits that are relatively tall in comparison with their width (Table 3.19). In both the rectangular and the square shapes, the breadth axis is horizontal. In specimen SK 48, the sole representative of *A. robustus*, the orbits are also rectangular in shape. The ill-defined orbital margins of KNM-WT 17000 make it difficult to ascertain the orbital shape, a situation that is unique among the hominins. In modern humans, the shape of the orbit shows a wide range of variation, as do the orbital index values.

The little that is left of the inferior orbital margin and the floor of the orbital cavity in A.L. 444-2 suggests that the orbital floor is not recessed below the level of the inferior margins of the orbital rim itself. This topography contrasts with that of most gorillas and chimpanzees, in which the orbital floor has an excavated appearance. A.L. 444-2 resembles modern humans, whose elevated orbital floor almost reaches the level of the inferior orbital margins.

The lateral segment of the inferior orbital margin in A.L. 444-2 is sharp, in contrast to the blunt, ill-defined margin in *A. robustus* (SK 46, SK 48, SK 79, SK 83, and TM 1517) and *A. aethiopicus* (KNM-WT 17000). On the one hand, in these two taxa the transition between the floor of the orbit and the facial surface of the zygomatic bone is inconspicuous, and therefore the lateral part of the inferior orbital margin appears as a flat, anteriorly extended shelf (Rak, 1983). On the other hand, *A. boisei* (KNM-ER 406, KNM-ER 732, KNM-ER 733, and OH 5) exhibits a sharp sill that demarcates the edge of the orbital floor, similar to that of A.L. 444-2. In *A. africanus*, the orbital floor, especially its lateral portion, drops dramatically to a substantial depth, immediately behind the inferior orbital margin. This topography can be observed on Sts. 52a, Sts. 71, and Taung and is suggestive of the chimpanzee orbit.

The Nasal Aperture: Its Shape, Inclination, and Position in the Facial Mask

The morphology of the inferior margin of the nasal aperture and of the lower parts of its lateral margins is described in the discussion of the nasal sill, the subnasal region, and the lateral margins of the nasal cavity in the section "Facial Topography" and in Chapter 5. The maxillary fragments bear the sole surviving sections of the nasal aperture's circumference and permit reconstruction of the remainder of the nasal opening. Although the nasal bones survive to a large extent, their inferior extremities are missing, necessitating reconstruction of the superior margin of the nasal aperture. Furthermore, the nasal bones are "floating," as they have no contact with the frontal bone.

In its reconstructed state, the nasal aperture forms a tall triangle. This reconstruction is based on the nasal aperture of four *A. afarensis* specimens (A.L. 333-1, A.L. 417-1d, A.L. 486-1, and the juvenile A.L. 333-105) and on the fact that the maximum width of the nasal aperture undoubtedly lies very close to the nasal sill in A.L. 444-2 and all the other specimens in which portions of the aperture's lateral margins are preserved. Plastic deformation influencing primarily the left lower part of the face leaves the reconstructed nasal aperture slightly asymmetric; on the left side, the lateral margins deviate more laterally than on the right (Figure 3.36).

The maximum nasal aperture width, based on the distance between the undeformed right side and the midline, is 25 mm, measured slightly above the nasal sill. Nasal aperture height (rhinion to nasospinale) is estimated at 37 mm, yielding a shape index of 68%. Of special interest is A.L. 417-1d, whose nasal aperture is relatively complete. Although the nasal bones are missing, the clearly visible remains of the nasomaxillary suture indicate the position of the aperture's apex. With a breadth of 23 mm and a height of 27 mm, the triangular opening yields an index of 85%, indicating a proportionally wider aperture than in A.L. 444-2. The almost complete anterior nasal opening of another maxillary specimen, A.L. 486-1, with a breadth of 22 mm, an estimated height of 27 mm, and an index of 81%, reveals a similar nasal shape. The mean "shape" index in *A. afarensis*, therefore, is 78%. This index is substantially lower than those yielded by the African apes (104% for female gorillas and 111% for male gorillas; 100% for female chimpanzees and 93% for male chimpanzees).

In both absolute and relative terms the nasal opening of *A. afarensis* is the narrowest of any *Australopithecus* species yet described. For example, 101% is the mean index of five *A. africanus* specimens, of which Sts. 5, at 108%, is the most extreme. A value greater than 100% indicates, of course, a nasal aperture that is wider than it is tall. The only specimen of *A. robustus* that permits this index to be calculated is SK 12, which yields 104%, and the only intact *A. boisei* specimen, OH 5, provides an index of 97%. In modern humans, the range of variation is known to be great (see the data compiled by Tobias, 1967).

This "shape" index does not adequately express the actual shape of the nasal opening. It is, rather, the position of the maximum breadth of the opening relative to its height that counts. Along with other *A. afarensis* specimens, A.L. 444-2 differs substantially from gorillas, in which the maximum width is located high above the nasal sill, almost at the midpoint of the aperture's height (probably the primitive morphology, as it is shared by many other primates). Because of this configuration, the shape approximates a wide oval and sometimes even a square with rounded corners, whereas in *A. afarensis*, the nasal opening is triangular. Although, as mentioned earlier, the nasal aperture in chimpanzees is wide in proportion to its height (the width of the aperture is often greater than its height), its general shape closely resembles that of *A. afarensis*. What contributes to this similarity is not only the location of the widest part of the nasal opening close to its inferior margins, providing the aperture's triangular shape, but also the topographic relationship of the nasal opening to the nasoalveolar clivus and to the infraorbital plate lateral to it. Johanson and White drew attention to the similarity in these morphologies as early as 1979, and it has since been discussed by others (Rak, 1983; Kimbel et al., 1984).

The width of the nasal aperture in *A. africanus* and the presence of anterior pillars as the lateral margins of the nasal opening distinguish this taxon from *A. afarensis*. In this respect, therefore, a greater resemblance exists between *A. afarensis* and *H. habilis* (KNM-ER 1813 and OH 24) than between *A. afarensis* and *A. africanus*.

The narrowness of the nose of A.L. 444-2 can be inferred independently when the breadth of the nasal opening is compared with the breadth of the orbit. The maximum breadth of the nasal opening constitutes 63% of the orbital breadth, and in A.L. 417-1d, 70%. A wider nose characterizes both chimpanzees and gorillas; the index values in chimpanzees are 68% for females and 69% for males, and in female and male gorillas, it is 80% and 89%, respectively. The *A. afarensis* mean is lower than any of the other australopith values, with species' means ranging between 76% and 80% (Table 3.20).

Table 3.20 Measurements and Proportions of the Nasal Region

Specimen/Sample		Total Height of Nose na–ns (69) (mm)	Height of Nasal Aperture rh–ns (70) (mm)	Maximum Breadth of Nasal Aperture (71) (mm)	"Shape" Index of Nasal Aperture (71)/(70) (%)	Index of Total Nasal Height (70)/(69) (%)	Nasal Breadth (71)/Orbital Breadth (68) (%)	Minimum Distance Nasal Aperture to Orbit (75) (mm)	Index of Minimum Distance (75)/ Orbital Breadth (68) (%)
Australopithecus afarensis	Mean	58	29	23	78	55	67	14	38
A.L. 444-2		67	37	25	68	55	63	16	40
A.L. 417-1d		49	27	23	85	55	70	12	36
A.L. 486-1		n/a	(27)	22	81	n/a	n/a	n/a	n/a
A.L. reconstruction (A.L. 333-1)		n/a	26	(21)	(81)	n/a	n/a	n/a	n/a
Homo sapiens	Mean (n = 10)	50	31	26	84	62	64	15	37
	Range	44–53	25–34	24–29	73–97	57–67	60–71	13–19	33–39
	SD	3	3	2	10	3	4	2	2
Gorilla gorilla female	Mean (n = 10)	83	31	32	104	37	80	26	65
	Range	77–88	26–34	27–36	94–113	33–41	72–87	20–31	50–80
	SD	3	2	3	10	2	4	4	11
Gorilla gorilla male	Mean (n = 10)	83	33	36	111	39	89	33	81
	Range	78–87	27–38	29–44	103–120	35–43	76–103	28–41	73–93
	SD	3	3	4	4	2	9	2	5
Pan troglodytes female	Mean (n = 10)	58	25	25	100	44	68	19	53
	Range	52–64	20–31	19–22	82–115	37–55	58–80	16–21	48–59
	SD	3	3	3	8	5	7	2	3
Pan troglodytes male	Mean (n = 10)	58	27	25	93	47	69	17	47
	Range	51–67	22–32	19–32	72–115	42–57	54–85	11–23	32–60
	SD	4	3	4	12	5	10	4	8
A. africanus	Mean	51	25	25	101	50	80	14	49
Sts. 5		47	24	26	108	51	80	16	48
Sts. 71		51	24	25	104	51	83	15	50
Sts. 52a		48	23	25	107	48	n/a	14	n/a
Stw. 505		62	28	29	104	45	(76)	19	(50)
MLD 6/23		47	25	20	80	53	n/a	8	n/a
A.robustus									
SK 12		n/a	24	25	104	n/a	n/a	(20)	n/a
A. boisei	Mean	62	(31)	29	(93)	(50)	(78)	(20)	(55)
OH 5		66	30	29	97	45	85	22	65
KNM-ER 406		58	(32)	28	(88)	(55)	70	(18)	(45)
A. aethiopicus									
KNM-WT 17000		57	(26)	26	100	(46)	76	(18)	(53)

Values in parentheses are estimates.

The height of the nasal opening in *A. afarensis* constitutes a relatively great proportion of the total nasal height. In both A.L. 444-2 and A.L. 417-1d, the height of the actual nasal opening spans 55% of the distance between nasion (reconstructed in both specimens) and nasospinale. In other words, the length of the nasal bones (chord measurement) is smaller than the height of the opening. Only in the sample of modern humans, with a mean index of 62%, are greater values found—that is, proportionally shorter nasal bones. In chimpanzees and gorillas, the nasal opening is less than half the total nasal height (with index values ranging from 37% to 47%). The considerably long nasal bones are the product of the high position of nasion and the very low position of rhinion, which, indeed, lowers the height of the nasal opening itself. In *A. africanus*, the length of the nasal bones is almost the same as the height of the nasal opening, as can be seen from the mean value of 50% for five specimens. A value of 45%, representing proportionally longer nasal bones, is found in OH 5, the only other *Australopithecus* specimen that permits this measurement.

The nasal aperture is situated relatively high in the facial mask of A.L. 444-2. In its reconstructed state, the aperture's upper end lies at the same level as the inferior orbital margins. Confirmation of the high position of the *A. afarensis* nasal aperture is found in A.L. 417-1d and apparently also in A.L. 333-1. This configuration contrasts with that of gorillas, on the one hand, in which the superior end of the nasal aperture is typically located well below the level of orbitale, and of modern humans, on the other, in which the nasal aperture extends above the level of the inferior orbital margins. In this regard, A.L. 444-2 more closely resembles the condition seen in chimpanzees.

In other hominins the picture is variable. The nasal aperture in *A. boisei* (OH 5 and KNM-ER 406) lies below the level of the orbital floor, as in *A. robustus* (SK 12, SK 48, and SK 83) and *A. aethiopicus*. A similar configuration is found in *A. africanus*, as seen in Sts. 71, Stw. 13, and Stw. 505, whereas in Sts. 5, as in A.L 444-2, the nasal aperture is located slightly higher, closer to the level of the inferior margins of the orbits. On those specimens of early *Homo* that permit such observations, such as KNM-ER 1813 and SK 847, the top of the nasal aperture is at the level of the orbital floor or slightly above it.

Variation in the position of the nasal opening relative to the orbits is also manifested in the minimum distance between the openings of these two cavities (Table 3.20 and Figure 3.9, measurement 75). As expected from the position of the nasal opening, the minimum distance is small in *A. afarensis* (16 mm for A.L. 444-2 and 12 mm for A.L. 417-1d), even smaller when corrected for size (against the orbital width, 40% and 36%, respectively). These values are substantially smaller than the gorilla means (65% for females and 81% for males) and also smaller than the chimpanzee means (53% for females and 47% for males). Similarly, the *A. afarensis* values are smaller than those obtained from other australopiths on which the measurements can be conducted (Table 3.20).

However, the values yielded by this configuration reflect a more complex reality, as the minimum distance between the nasal opening and the orbit is the product of not only the vertical distance between them but also the horizontal (anteroposterior) distance. For example, in gorillas, particularly males (Table 3.20), the minimum distance is immense, encompassing the huge vertical distance between these openings (the nasal opening, as described above, lies far below the level of the inferior orbital margin) and the horizontal distance between them (the upper part of the nasal opening is located substantially anterior to the coronal plane of the orbital opening). In modern humans, despite the position of the nose above the level of the inferior orbital margin, the distance is greater than expected. Here, it is the horizontal distance between the protruding upper end of the nasal opening (rhinion) and the orbit that constitutes the main component of the distance between these two cavities. The relatively small dimension in *A. afarensis*, therefore, stems both from the elevated position of the nasal opening and from the small horizontal distance between the aperture's upper part and the coronal level of the orbit, as seen particularly in A.L. 444-2 and A.L. 417-1d. In A.L. 333-1, a greater horizontal distance exists between the upper part of the nasal aperture and the orbit. Indeed, in this specimen, the horizontal distance necessitates (as in gorillas) a transverse buttress, albeit a modest one, which is absent in the two other specimens. (See the section "Facial Topography.") In other *Australopithecus* species, including *A. africanus*, the flatness of the face—especially in the center of the facial mask—results in a very small horizontal distance between the orbit and the nasal opening; this distance is almost negligible in the robust taxa.

The horizontal distance between the openings of orbital and nasal cavities directly influences the orientation of the plane of the nasal aperture (the prognathism of the rhinion–nasospinale segment; see the section "Prognathism of the Facial Segments"). For example, were it not for the considerable horizontal distance between the openings in the gorilla face, the plane of the nasal aperture would be much more inclined. In other words, the anterior location of rhinion outweighs the influence of the nasal sill's extremely anterior position in this very prognathic species. The orthognathic australopith species—*A. robustus* (as exemplified by SK 12 and SK 48), *A. boisei* (as exemplified by OH 5), and some specimens of *A. africanus* (for example, Sts. 71)—exhibit a rather steep (vertical) plane of the *actual* nasal aperture. In these species, the posterior position of the nasal sill neutralizes the effect of a very posterior rhinion (located close to the coronal level of the orbit). The plane of the opening in the more prognathic Sts. 5, therefore, is more inclined. Specimen KNM-WT 17000 is the most extreme example of an inclined nasal aperture, which is the product of an anteriorly protruding palate and a sunken upper portion of the nasal opening (although rhinion is not preserved in this specimen). The other extreme is the modern human configuration, in which the situation is reversed. The palate is greatly diminished in size and hence orthognathic, and rhinion protrudes so far that the plane of the nose is not only vertical but also retreats inferoposteriorly from rhinion in many instances. In A.L. 444-2, as in A.L. 417-1d and A.L. 333-1, the plane of the nasal opening is rather steep, apparently because of the specimens' mild degree of upper facial prognathism, especially in comparison to chimpanzees and gorillas, and their slightly protruding rhinion.

A closer examination of the lateral nasal margins provides additional detail to the description of the facial contour in lateral view (see the section "Shape of the Facial Slope") and reveals that they demonstrate considerable variation. In modern humans, the lateral nasal margins appear as sharp crests at the anterior ends of bony plates that are almost on the sagittal plane. These crests protrude anteriorly from the more posterior peripheral facial plane. Concave in a sagittal plane, the margins retreat a great distance inferoposteriorly from rhinion. Then they advance anteroinferiorly to nasospinale. As a result of this configuration, the nasal opening is three-dimensional. In all the australopiths, including *A. afarensis*, the *actual* nasal opening is on one plane. A similar situation exists in chimpanzees and gorillas, although in the latter the topography of the lateral nasal margin is rather complicated (see Chapter 5). *H. habilis* (KNM-ER 1813 and OH 24) adheres to this morphology, whereas SK 847 (*H. erectus*) already exhibits the curving of the plane of the nasal opening, as readily observed in a lateral view, though not to the degree seen in modern humans.

The Cranium: Occipital View

Occipital View and the Coronal Section of the Calvaria

Despite deformation that slightly flattens the right lateral wall of the braincase and consequently deflects the left wall such that it bulges faintly, a posterior view of A.L. 444-2 clearly reveals the typical bell shape that characterizes many other *Australopithecus* crania, as well as those of the great apes (Figure 3.47). As such, the maximum width of the braincase is situated at a very low level. Although in A.L. 444-2 the calvaria is bell shaped, its lateral walls rise relatively steeply until, in their upper part, they curve abruptly toward the midline. Consequently,

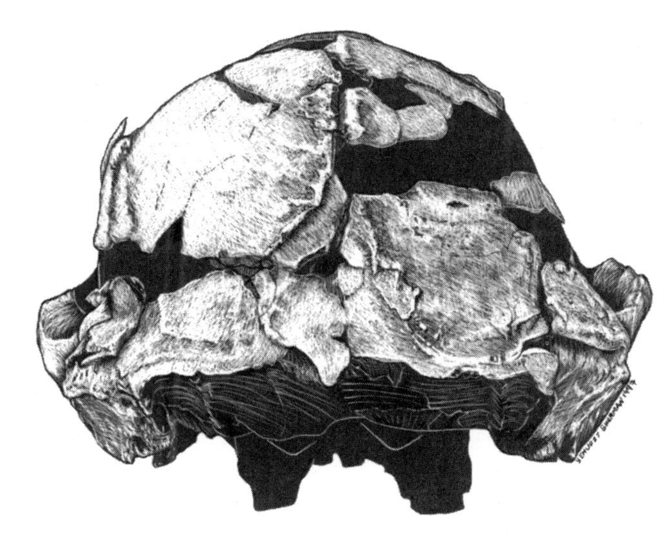

Figure 3.47 Artist's rendition of the posterior view of the A.L. 444-2 reconstruction. See the picture gallery following page 122 for an enlarged view (75% natural size).

in a posterior view, this steep-walled bell is also rather tall. The calvaria's summit itself consists of two slightly elevated peaks, one on each side of the midline, faint expressions of the two temporal lines. A compound sagittal crest lies posterior to the coronal plane of the calvarial outline, on the posterior slope of the braincase.

The coronal contour's maximum breadth is measured across both the supramastoid crests and the equally prominent mastoid crests. A slight indentation in the contour at this site is an expression of the shallow supramastoid sulcus that separates these crests from one another. The contour tapers quickly inferomedially, riding in an almost straight line along the lateral aspect of each of the mastoid processes toward their tip—the lowest points of the coronal outline (see Chapter 5 and Figure 3.48).

The coronal contour's shape index (or the ratio of the calvarial height on the coronal plane of porion to the minimum breadth of the cranial base—that is, the distance between the porial saddles;[18] Table 3.21) is 64% in A.L. 444-2, which is greater than the value for chimpanzees (59% for both females and males) and female gorillas (55%). The difference between A.L. 444-2 and the female gorilla sample diminishes when we add the index value (55%) of the other *A. afarensis* specimen on which these measurements can be performed, A.L. 333-45. The mean value for these two *A. afarensis* specimens thus is 60%.[19]

An examination of this shape index in Table 3.21 reveals that the *A. afarensis* value, though similar to the values for the female gorilla and chimpanzee samples, is lower than those for *A. africanus* and *H. habilis*. The mean index for four *A. africanus* specimens (MLD 37/38, Sts. 5, Sts. 71, and Stw. 505) is 68%, and 70% is the value calcu-

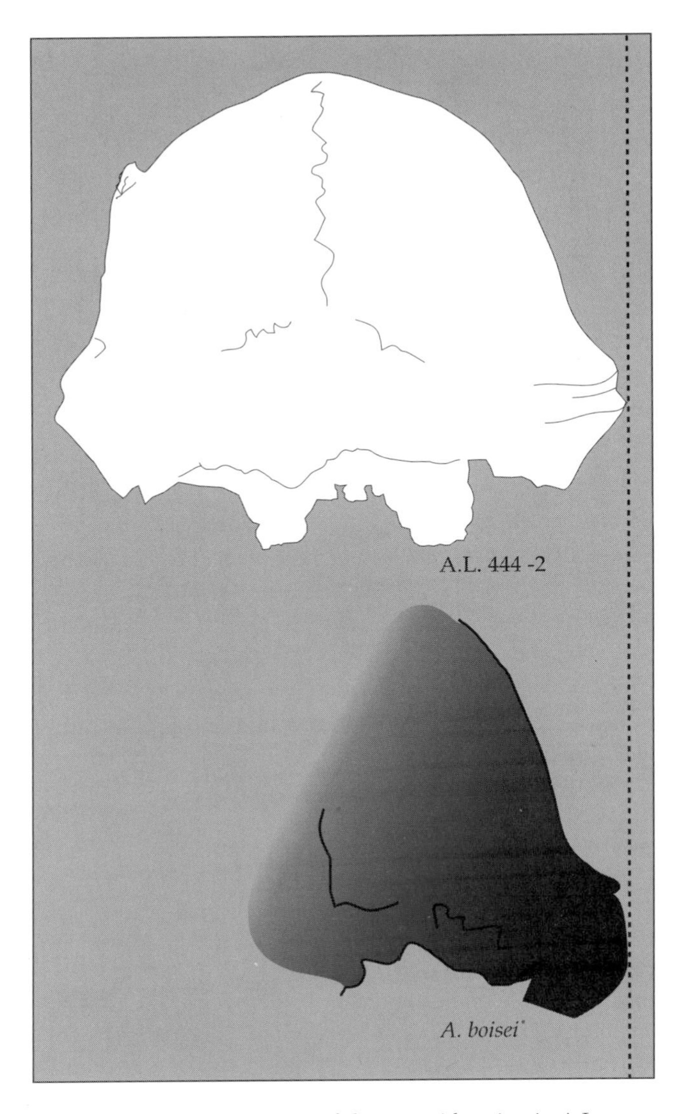

Figure 3.48 Lateral contour of the mastoid region in A.L. 444-2 and *A. boisei*. In *A. boisei* the lateral contour of the mastoid is characteristically rounded and the lateralmost point is situated low on the bone. In A.L. 444-2, as in other *A. afarensis* mastoids, the lateral contour is straight and oriented inferomedially; the lateralmost point is higher, approximating the supramastoid crest.

lated for *H. habilis* specimen KNM-ER 1813. However, *A. boisei* demonstrates somewhat lower values than *A. afarensis*, with a mean of 56% for five specimens (OH 5, KNM-ER 406, KNM-ER 732, KNM-ER 13750, and KNM-ER 23000). With an index of 49%, KNM-WT 17000 has the lowest relative calvarial height among known *Australopithecus* crania.

It is interesting to note that in these constellations, the calvarial height, rather than the width, plays the primary role in the determination of the index value, as illustrated by a comparison of A.L. 444-2 with KNM-ER 406 and KNM-WT 17000. In these three specimens, the breadth of the cranial base (as measured here) is about the same; thus, disparities in their index values can be attributed to the effect of differing calvarial heights (see also Figure 3.11, where the cranial length of these specimens has been drawn to fit that of A.L. 444-2).

When the height and breadth of the calvaria are expressed as a percentage of the biorbital distance (Table 3.22), the narrowness of the cranial base is clearly seen to be responsible for the high index values in Sts. 5, where the cranial base constitutes a smaller percentage of the biorbital distance than in A.L. 444-2, and in KNM-ER 1813. The effect of this narrowness on the index value helps explain the huge discrepancy shown in Figure 3.49 (and discussed below) between the coronal contour of the calvaria of A.L. 444-2 and that of Sts. 5 and KNM-ER 1813. This discrepancy is the result of our stretching the breadth of the narrow-based crania to fit the breadth of the wide-based A.L. 444-2 cranium. By the same token, one could demonstrate how narrow the Sts. 5 and KNM-ER 1813 crania are relative to their height by reducing their height to match that of A.L. 444-2.

Figure 3.49 is essentially a graphic expression of the traditional chord/arc relationship. We prefer this representation because it permits a visual identification of the ways in which A.L. 444-2 differs from various other crania; a numeric calculation of the chord/arc ratio does not highlight the actual morphological differences between the crania. For example, the graphic depiction draws attention to the great similarity between the coronal sections of A.L. 333-45, A.L. 444-2, and female gorillas. (Male gorillas, in contrast, display a bell-shaped outline that is squatter and has flatter lateral walls that incline medially from the broadest point of the braincase toward the base of the elevated sagittal crest. This configuration gives the coronal section an almost triangular appearance.) Similarly, we can see that the shape of the coronal contour in chimpanzees tends to be somewhat more bulbous than that of A.L. 444-2. With the adjustment of the cranial base breadth to that of A.L. 444-2, differences can easily be perceived between the coronal outlines of *A. afarensis*, *A. africanus* (Sts. 5, MLD 37/38), and *H. habilis* (KNM-ER 1813). The method also underscores the differences between the outlines of *A. afarensis* and the robust *Australopithecus* crania (KNM-ER 406 and KNM-WT 17000). Indeed, KNM-ER 1813 and Sts. 5 exhibit much taller, steeper-walled coronal outlines than does A.L. 444-2, whereas *A. boisei* and *A. aethiopicus* display a much lower, flatter outline. If the great apes represent the primitive configuration, we must view the height and shape of the braincase of *A. boisei* and *A. aethiopicus*, on the one hand, and *A. africanus*

Table 3.21 Shape of the Cranial Vault in Posterior View

Specimen/Sample		Height of the Apex (6) (%)	Biporial Saddle Breadth (14) (mm)	Height of Apex (6)/Biporial Saddle Breadth (14) (%)
Australopithecus afarensis	Mean	78	132	60
A.L. 444-2		84	132	64
A.L. reconstruction (A.L. 333-45)		72	131	55
Homo sapiens	Mean (n = 10)	115	120	96
	Range	108–126	112–127	88–101
	SD	6	5	4
Gorilla gorilla female	Mean (n = 10)	67	123	55
	Range	60–73	117–128	47–62
	SD	5	3	5
Gorilla gorilla male	Mean (n = 10)	n/a[a]	144	n/a
	Range	n/a	138–150	n/a
	SD	n/a	4	n/a
Pan troglodytes female	Mean (n = 10)	70	115	59
	Range	63–78	110–123	52–71
	SD	5	4	5
Pan troglodytes male	Mean (n = 10)	68	115	59
	Range	62–74	102–130	50–70
	SD	4	8	6
A. africanus	Mean	76	112	68
Sts. 5		74	107	69
Sts. 71		75	105	71
MLD 37/38		74	108	69
Stw. 505		81	(126)	64
H. habilis				
KNM-ER 1813		78	113	70
A. boisei	Mean	74	128	57
OH 5		76	140	54
KNM-ER 406		70	133	53
KNM-ER 13750		69	128	55
KNM-ER 23000		81	130	62
KNM-ER 732 female		69	(110)	63
A. aethiopicus				
KNM-WT 17000		64	131	49

[a]Gorilla males were not measured due to the interference of the sagittal crest.
n/a, not available.
Values in parentheses are estimates.

and *Homo*, on the other, as differently derived configurations. *A. afarensis* demonstrates the plesiomorphic condition vis-à-vis this morphology.

The Nuchal Region

The nuchal region in A.L. 444-2 is almost entirely concealed in a posterior view, as can be anticipated from the previous discussion of this region in lateral view and from the indices expressing the region's position and inclination (see the sections "Position of Inion and Summit of the Arches of the Superior Nuchal Line in Relation to Calvarial Height" and "Inclination of the Lower Occipital Scale (Inion–Opisthion)"), which demonstrate that the nuchal region is tucked underneath the calvaria. The vertically suspended, thin nuchal crest itself contributes to the concealment of the region in this view.

Not every *A. afarensis* specimen displays this configuration. For example, in A.L. 162-28, A.L. 224-9, A.L. 288-1a, A.L. 333-45, A.L. 439-1, A.L. 444-1, and KNM-ER 2602, the lower scale is steeper and thus more exposed in a posterior view, and the nuchal crest is not well enough developed to aid in hiding the scale. In fact, when compared to these other *A. afarensis* crania, A.L. 444-2 is a rarity. (For a

Table 3.22 Relative Calvarial Breadth and Height (%)

Specimen/Sample		Biporial Saddle Br. (14)/Biorbital Br. (54)	Ht. of Apex (6)/Biorbital Br. (54)
Australopithecus afarensis	Mean	145	85
A.L. 444-2		139	88
A.L. 333 reconstruction		(150)	(83)
Homo sapiens	Mean (n = 10)	126	121
	Range	118–135	110–132
	SD	6	8
Gorilla gorilla female	Mean (n = 10)	127	68
	Range	124–136	64–79
	SD	8	5
Gorilla gorilla male	Mean (n = 10)	120	n/a[a]
	Range	114–132	n/a
	SD	9	n/a
Pan troglodytes female	Mean (n = 10)	127	77
	Range	118–136	69–84
	SD	7	4
Pan troglodytes male	Mean (n = 10)	128	78
	Range	116–142	71–89
	SD	7	5
Other hominins			
Sts. 5		126	87
KMN-ER 1813		125	92
OH 5		140	76
KNM-WT 17000		139	68

[a]Gorilla males were not measured due to the interference of the sagittal crest.
Ht., height; Br., breadth. n/a, not available.

more detailed discussion, see the section "General Outline of the Cranium" and Chapter 5.) The contrast between A.L. 444-2 and the other *A. afarensis* specimens in the exposure of the nuchal area is also what distinguishes OH 5 (in which the nuchal region is concealed) from other *A. boisei* specimens. Thus, most of the *A. afarensis* and *A. boisei* specimens share this more generalized configuration with *A. africanus* and early *Homo*. In the great apes, the nuchal crest juts out posterosuperiorly from the braincase, and as developed as it may be in male gorillas, it never obscures the nuchal area in a posterior view.

The low position of the nuchal crest in A.L. 444-2—the line of transition between the cranial base and the rest of the braincase—and the crest's extended lip create an illusion of a taller braincase in a posterior view. Conversely, the great apes' nuchal crest, which is elevated far above the FH, produces the opposite effect. There, it appears as if the braincase is much lower than in A.L. 444-2. Indeed, as discussed earlier, the comparative metrics and graphics reveal a great similarity between these taxa in absolute height, as well as in the shape of the coronal section. Further detail of the topography of the nuchal region follows.

The Cranium: Basal View

General Outline of the Cranium in Basal View

As a result of postmortem deformation, the palate is not properly aligned with the calvarial midline in A.L. 444-2. Following the course of the palate posteriorly from its anterior end, we see that it deviates gradually away from the midline toward the right side. Whereas prosthion lies on the midline, the rear of the palate is displaced 15 mm to the right of its correct anatomical position (Figure 3.50). (For a detailed discussion of the shape of the palate, see Chapter 5.)

If we disregard the palate, the basal outline of the cranium forms almost a perfect circle. Beginning anteriorly at the zygomatic process of the maxilla on the right side, the contour runs over the zygomatic arches in what is an essentially smooth curve and continues along the cranial base until it completes the circle at the zygomatic process of the maxilla on the left side. The most lateral point on the contour (zygion) is located on the posterior root of the zygomatic arch, anterolateral to the articular tubercle. From this point, the contour gradually converges both anteriorly and posteriorly toward the midline.

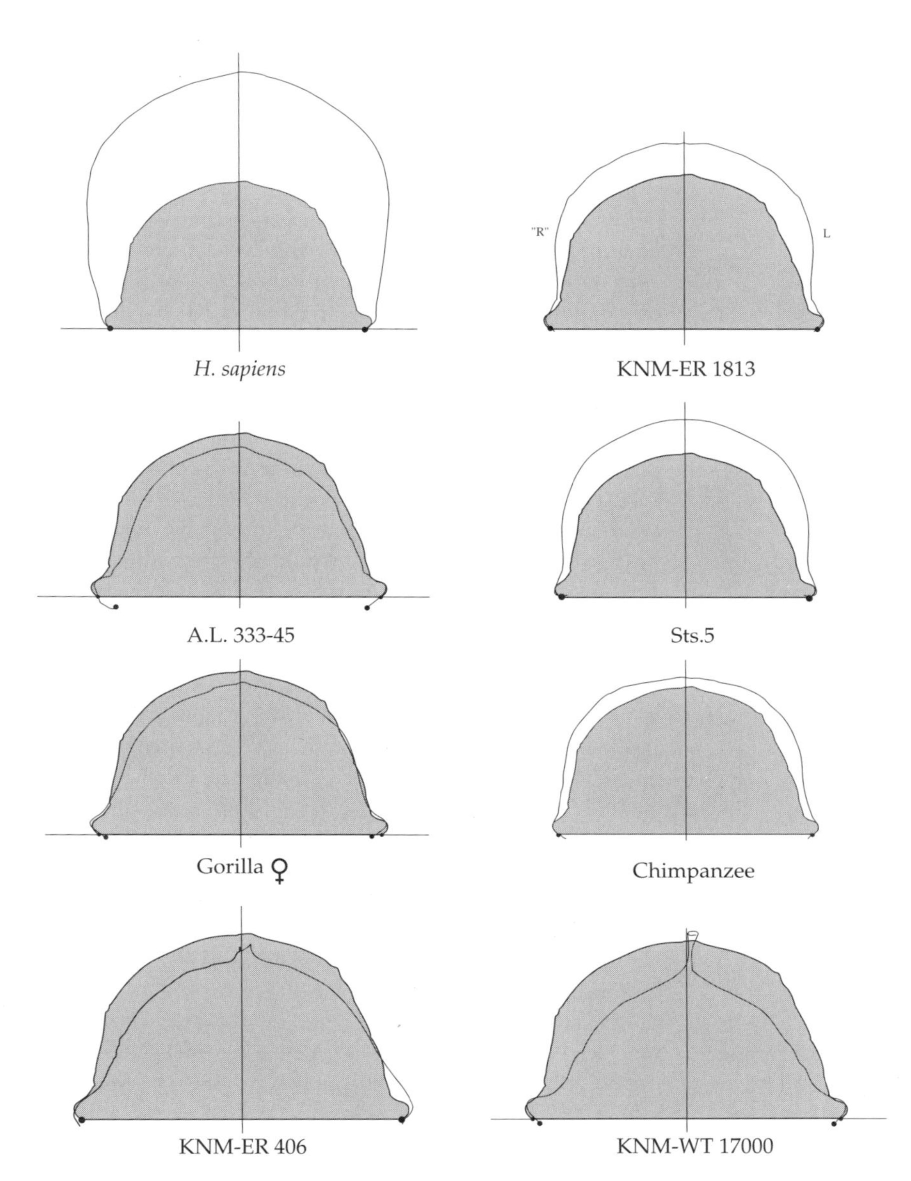

Figure 3.49 Coronal cross section of the calvaria through porion in A.L. 444-2 and other hominoids. All specimens have been scaled to the biporial saddle breadth of A.L. 444-2 (shaded area). In the KNM-ER 1813 outline, the right side is a mirror image of the less deformed left side. Note the similarity in the outlines of *A. afarensis* and the female gorilla, the much higher outline in *A. africanus* specimen Sts. 5, and the lower, sloping outlines of *A. boisei* (KNM-ER 406) and *A. aethiopicus* (KNM-WT 17000).

The lateral outline of the mastoid crest is situated posteromedial to the sagittal plane of zygion and in this location contributes to the circularity of the basal contour. The postglenoid process bulges laterally from the concave contour of the porial saddle because of the great lateral extension of the process beyond the plane of the external auditory meatus. Of the two structures that border the saddle—the posterior root of the zygomatic arch and the mastoid process—the root of the arch is situated farther laterally. Therefore, the porial saddle in A.L. 444-2 is uneven, with its deepest point located closer to the mastoid process, approximately 10 mm posterior to porion itself. In chimpanzees and female gorillas, the two structures adjoining the porial saddle extend laterally to a similar degree; a line drawn between the summits of the two structures is almost parallel to the sagittal plane (2°–4°) (Figure 3.51 and Table 3.23). Thus, the deepest point in the saddle is located at the midpoint between the mastoid crest and the root of the zygomatic arch. In this respect, the morphology in these apes resembles that of other hominins, such as KNM-ER 1813 and modern humans. In Sts. 5 this line forms a greater angle. With a more laterally protruding posterior root of the zygomatic arch, male gorillas show a somewhat greater resemblance to A.L. 444-2 than do female gorillas.[20]

The contour of the cranial base in chimpanzees and female gorillas exhibits a more rectangular shape, in contrast to the more circular contour in A.L. 444-2. The rectangular appearance stems from the more sagittal orientation of the zygomatic arches, the more clearly defined transition between the arches and the body of the zygomatic bones (in the form of a distinct angle, particu-

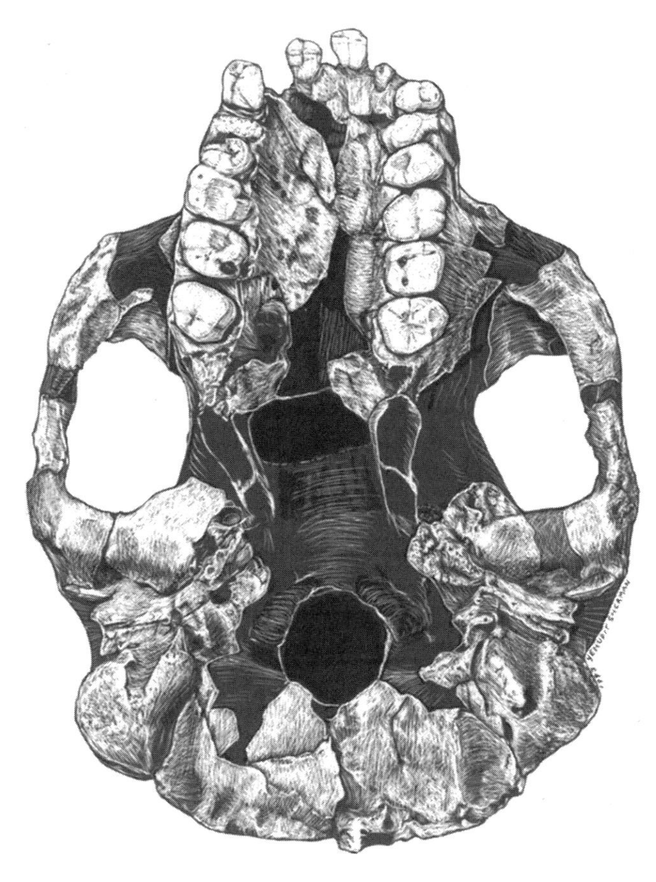

Figure 3.50 Artist's rendering of the basal view of A.L. 444-2 reconstruction. See the picture gallery following page 122 for an enlarged view (75% natural size).

larly in gorillas), and the alignment of the mastoid crests and the zygomatic arches on the same sagittal plane, as described above. This rectangular outline characterizes the crania of other hominins. A notable exception is *A. boisei*, in which the immense maximum bizygomatic breadth, coupled with the forward location of the two zygion points relative to the cranium's anteroposterior length (these points are closer to the coronal level of the orbits), gives the cranial outline a heartlike shape. It is the considerable horizontal distance between the sagittal plane of the mastoid crest and that of zygion, as well as the distance between these two points' coronal planes, that produces this unusual contour (Figure 3.51). Consequently, relative to the distance between the two structures defining the saddle, the deepest point in the *A. boisei* porial saddle is located even more posteriorly than in the Hadar cranium. This configuration is found frequently in *A. boisei* and is particularly extreme in specimens such as KNM-ER 406 and KNM-ER 13750. *A. aethiopicus* and *A. robustus* (see, for example, TM 1517) demonstrate the more generalized rectangular outline. In KNM-WT 17000, as in many other primates, including A.L. 444-2, zygion lies on the posterior part of the zygomatic arch, near the articular eminence on the root of the arch. Hadar cranium A.L. 444-2, with its rounded basal contour, falls between the generalized rectangular outline and the heart-shaped outline of *A. boisei* crania.

The posterior cranial base contour (between the left and right mastoid crests) is a broad, flat arch in A.L. 444-2. The ratio of the arc length that connects the most laterally extended portions of the mastoid crests to the chord distance between these two points is 85%. (Note that the arc length referred to here is that of the contour in a basal view and not the minimum arc length between the asteria.) This high value in A.L. 444-2 means that the most posterior part of the contour is close to the coronal plane of the mastoid crests. By inference, we can state that a small horizontal distance separates the coronal plane of porion and the most posterior point of the vault outline. Female gorillas and chimpanzees yield lower ratios, at 73% and 78%, respectively. Similarly, in other hominins, the chord constitutes a relatively small percentage of the arc. In Sts. 5, the index is 78%, and in KNM-ER 1813, it is 73%. In modern *H. sapiens*, however, it is as low as 59%. *A. boisei* exhibits a flat contour, similar to that of A.L. 444-2; the mean value of two specimens (KNM-ER 406 and KNM-ER 13750) is 84%.

In this discussion we have limited ourselves to only a few sample values of the chord/arc ratio because of the effects that the irregularity of the occipital bone surface and the intra- and interspecific variation in the anatomy and vertical position of the nuchal crest have on these measurements. To reduce this "noise" in the index value, we employ a more straightforward method than the chord/arc ratio to quantify the degree to which the posterior end of the contour extends beyond the coronal level of the mastoid crests. To this end we express the horizontal distance between the coronal plane of the mastoid crests and the most posterior part of the cranial base contour as a percentage of the distance between the supramastoid crests. Both A.L. 444-2 and A.L. 333-45 yield a value of 28% (Table 3.23 and Figure 3.52). This value is smaller by far than that calculated for modern humans (58%). The *H. habilis* cranium KNM-ER 1813 also yields a high value, 45%. Values of 35% and 34%, respectively, are calculated for Sts. 5 and MLD 37/38, which do not differ significantly from the 32% yielded by female chimpanzees and the 34% yielded by the males, nor from the female gorilla value of 34%. The elevated nuchal crest in male gorillas, and even in some of the females, makes it difficult to perform the measurement and hence renders a comparison inapplicable.[21]

Much smaller values characterize *A. boisei*. Three specimens (KNM-ER 406, KNM-ER 23000, and OH5) yield a mean of 25%. (The anatomy of less complete *A. boisei* specimens, such as KNM-ER 407 and Omo 323-1976-896, also suggest a small value.) The flat posterior

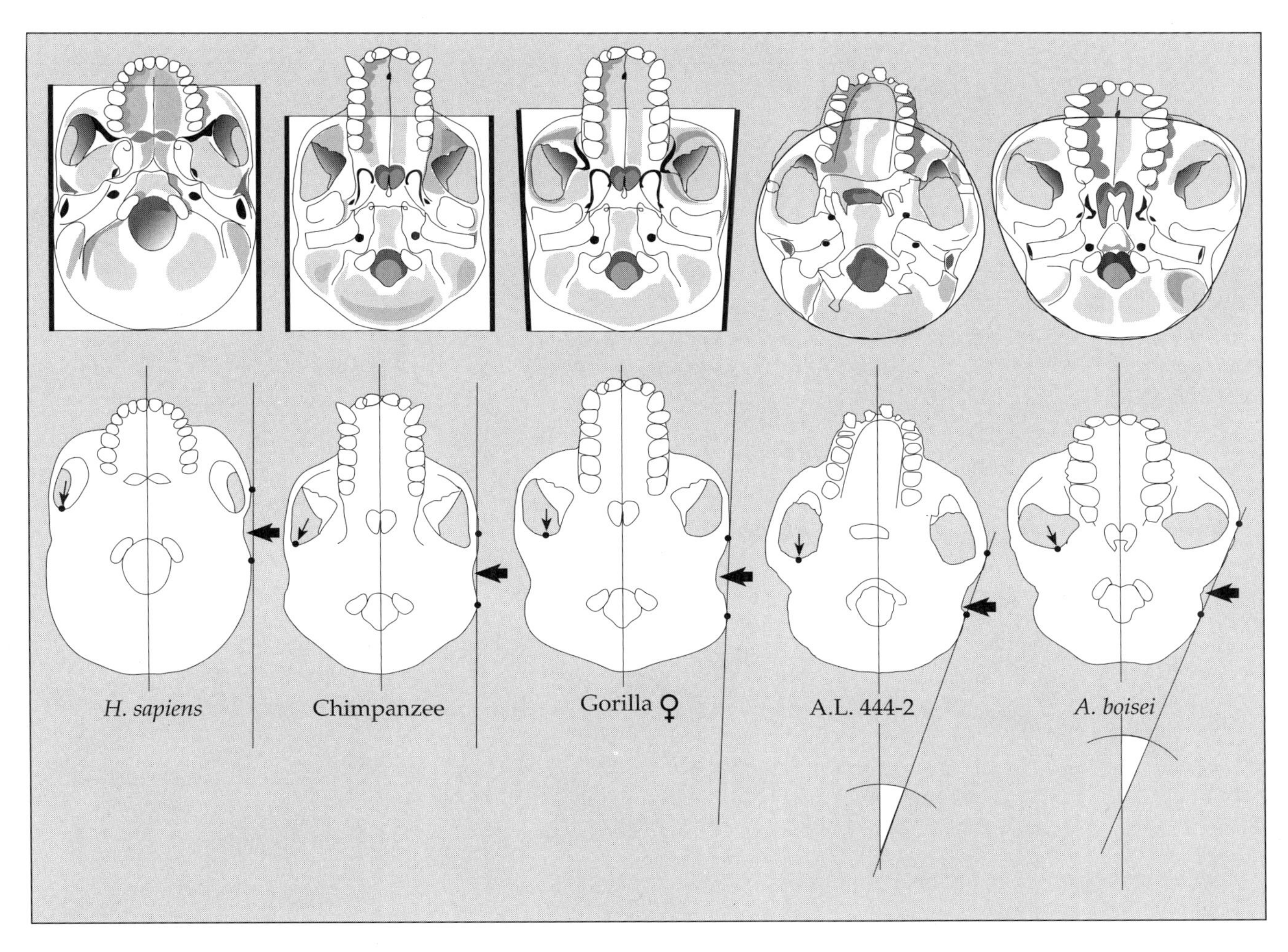

Figure 3.51 Outline of the basal view of A.L. 444-2 and other hominoid crania. Note the differences between the rounded outline in A.L. 444-2 and the squarer one in humans, chimpanzees, and female gorillas. The *A. boisei* outline (modeled on KNM-ER 406) is triangular with rounded corners. The large arrow indicates the deepest part of the porial saddle. The small arrow in the temporal foramen indicates the most posterior part of the foramen.

section of the *A. boisei* basal contour contributes, along with other factors, to the teardrop appearance of the braincase in a vertical view (see the section "Outline of the Calvaria"). An equally low value of 25% is calculated for KNM-WT 17000. Although the *A. afarensis* sample is small and the material is fragmentary, both A.L. 444-2 and A.L. 333-45 are closer to *A. boisei* and *A. aethiopicus* than are any other taxa (no adult *A. robustus* specimen permits the calculation of this index). This shortness of the nuchal plane may explain the greater relief of muscle origin scars and bony septa on the nuchal plane of many of the larger *A. afarensis* and robust *Australopithecus* crania.

In contrast to the smooth, convex arch of the posterior cranial base contour in A.L. 444-2, several *A. boisei* specimens exhibit a posteriorly extended lip about 30 mm lateral to inion, which gives their posterior contour a strongly segmented, trapezoidal shape (Figure 3.53). This feature is especially well developed in specimens KNM-ER 406, KNM-ER 13750, Omo 323-1976-896, and KGA 10-525 (as seen in Suwa et al., 1997), but it is less marked in OH 5 and KNM-ER 407. It is also exhibited in a most extreme fashion by *A. aethiopicus* (KNM-WT 17000). In *A. africanus*, the smooth, rounded posterior contour is similar to that of most *A. afarensis* crania, including KNM-ER 2602, A.L. 162-28, A.L. 288-1, A.L. 333-45, and A.L. 439-1. The fragmentary nature of the adult *A. robustus* sample does not permit observation of this anatomy.

These differences between the generalized pattern exhibited by *A. afarensis* and the unique configuration seen in *A. boisei* indicate that the orientation of the fan of

Table 3.23 Metrics of the Posterior Cranial Base Outline

Specimen/Sample		Max. Breadth of Cranial Base (17) (Bisupramastoid) (mm)	Max. Post. Extension of Cranial Contour from Coronal Plane of Supramastoids (41) (mm)	Post. Extension (41)/ Bisupramastoid Br. (17) (%)	Angle of Lateral Aspect of Cranial Base to Midline (degrees)[a,b]
Australopithecus afarensis	Mean	135	38	28	12
A.L. 444-2		139	39	28	12[c]
A.L. reconstruction (A.L. 333-45)		130	36	28	n/a
Homo sapiens	Mean (n = 10)	131	76	58	−1
	Range	124–139	64–89	50–64	−2–1
	SD	6	8	4	1
Gorilla gorilla female	Mean (n = 10)	136	46	34	2
	Range	129–143	38–55	29–40	0–3
	SD	4	5	3	1
Pan troglodytes female	Mean (n = 10)	124	39	32	4
	Range	117–129	35–46	27–37	0–8
	SD	3	3	2	2
Pan troglodytes male	Mean (n = 10)	121	41	34	3
	Range	111–132	35–47	29–40	0–8
	SD	6	4	3	3
A. africanus	Mean	113	39	35	7
Sts. 5		108	38	35	7[c]
MLD 37/38		117	40	34	
H. habilis					
KNM-ER 1813		111	50	45	n/a
A. boisei	Mean	133	34	25	15
OH 5		133	35	26	14
KNM-ER 406		138	42	30	15
KNM-ER 23000		128	25	19	17
A. aethiopicus					
KNM-WT 17000		135	34	25	12

[a]A negative value indicates that the cranial base is wider than the zygomatic arch.
[b]Refer to figure 3.51.
[c]The figure for A.L. 444-2 and Sts. 5 is the average of the two sides.
Max., maximum; Post., posterior. n/a = not available.

the temporalis muscle, the hypertrophied section of the fan, and the relationship of the fan to the nuchal muscles were not the same in these two taxa. (For a more detailed discussion, see the section "Pattern of Cranial Cresting.")

The Temporal Foramen and Adjacent Anatomical Structures

The general shape of the temporal foramen was discussed in the section "The Cranium: Vertical View." In the basal view, additional detail is revealed that is specific to this view. The various primate taxa in our comparative sample exhibit substantial differences in the outline of the posterior section of the temporal foramen. In chimpanzees, the posterior end of the foramen is a deep notch found posterolaterally, near the posterior root of the zygomatic arch. The presence of this notch gives the posterior margin of the foramen an asymmetric skewed appearance and a triangular shape to the foramen itself (Figures 3.51 and 3.54). In gorillas, in contrast, the outline of the foramen's posterior border is symmetrically concave, with the most posterior point of the outline situated equidistant from the zygomatic arch and the lateral wall of the braincase. Chimpanzees apparently represent the primitive configuration, as it is shared with many other primates, such as baboons, macaques, and gibbons.

In modern humans, the medial wall of the temporal fossa (the wall of the braincase) is very close to the straight zygomatic arch. As a result of this proximity, the temporal foramen assumes a long, narrow, oval shape. Its posterior border, therefore, is rounded and constricted by the aforementioned structures and thus is also narrow and

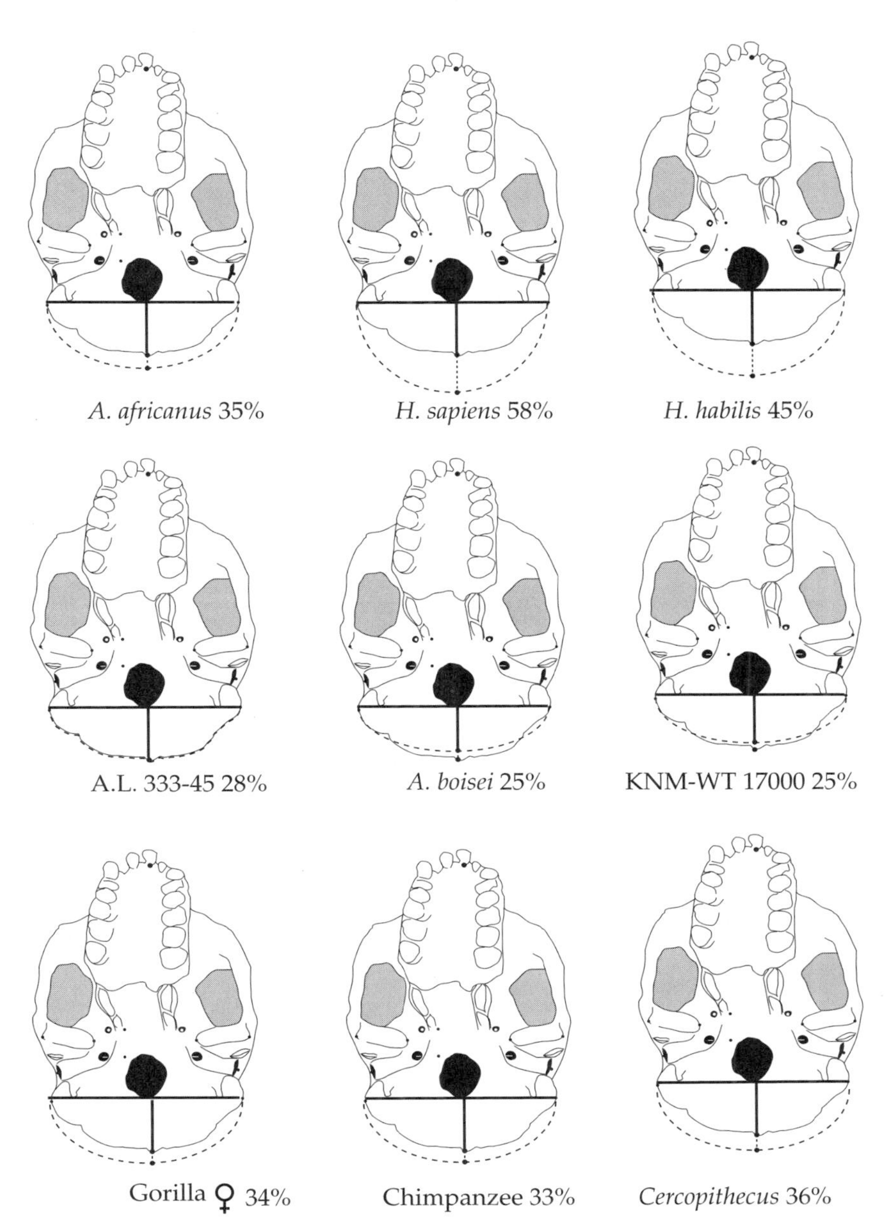

Figure 3.52 Posterior extension of the calvaria beyond the bisupramastoid line in A.L. 444-2 and other hominoids. A.L. 444-2 is represented by a drawing of the basal view of the cranium. The bisupramastoid lines of the compared hominoids have been superimposed on the Hadar outline and drawn to fit its bisupramastoid width. The values express the ratio between the extent to which the back of the cranium protrudes posteriorly from the bisupramastoid line (in mm) and the length of this line. In each comparison the schematic dashed outline represents the degree to which the compared hominoid differs from A.L. 444-2. Note the similarity between *A. afarensis* and robust *Australopithecus*.

symmetrical. Clearly, A.L. 444-2 is unlike chimpanzees in this respect and is more similar to gorillas and *Homo*. The outline of the posterior end of the foramen in A.L. 444-2 is symmetrically concave, and its most posterior point situated equidistant from the lateral and medial margins of the foramen. This anatomy, readily discernible on the right side (the side less affected by postmortem deformation) provides the foramen with a more or less oval shape, symmetrical along its axis. The only other *A. afarensis* specimen on which this region is complete enough for observation is the juvenile A.L. 333-105. Here, the symmetrical configuration of the posterior margin of the temporal foramen is confirmed and, indeed, is quite different from that displayed by chimpanzees of the same dental age. In both *A. afarensis* and chimpanzees, the morphology of the posterior end of the temporal foramen in juveniles is indistinguishable from that of the respective adults.

Even though the arrangement of the relevant area in *A. africanus* resembles that of a chimpanzee at first glance (at least as seen in the best preserved specimen, Sts. 5), a closer look reveals substantial differences between them. As in chimpanzees, the most posterior part of the foramen's contour is located adjacent to the posterior root of the zygomatic arch in *A. africanus*. However, this section is not angled as in the chimpanzee but, rather, forms a well-defined, indented notch in the posterolateral outline of the foramen. Undoubtedly a similar indentation existed

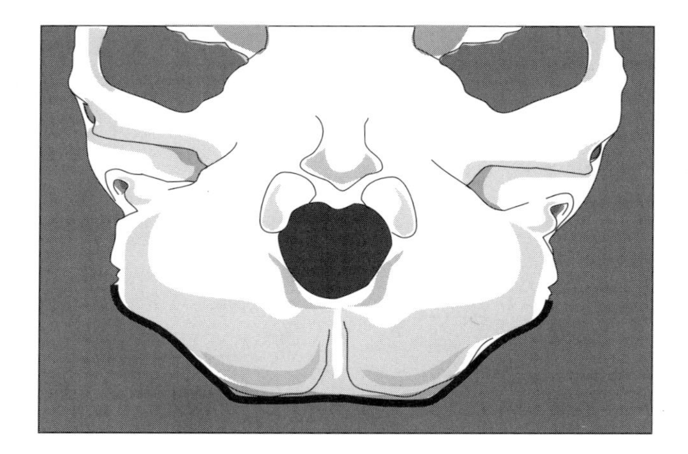

A. boisei

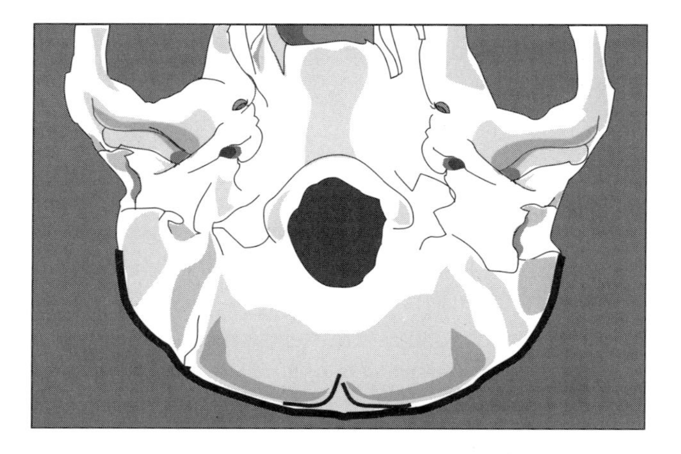

A.L. 444 -2

Figure 3.53 Outline of the posterior margin of the calvaria in A.L. 444-2 and *A. boisei*. The outline in A.L. 444-2 is rounded, whereas in *A. boisei* (modeled on KNM-ER 406) it is segmented. This lends the *A. boisei* occipital outline a trapezoidal shape.

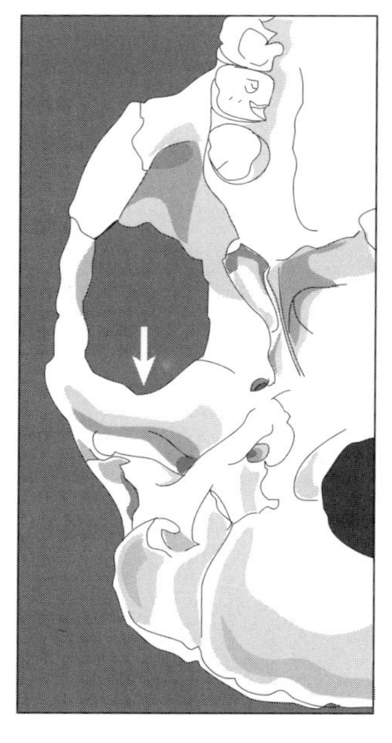

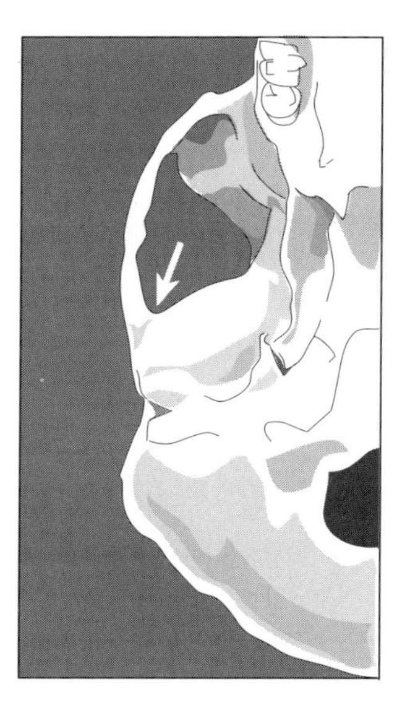

A.L. 444 -2 Chimpanzee

Figure 3.54 Morphology of the posterior margin of the temporal foramen in A.L. 444-2 and a chimpanzee. The most posterior part of the foramen (indicated by the arrow) is located at the lateral end of the posterior margin, near the posterior root of the zygomatic arch, in the chimpanzee and in the middle of the posterior margin in A.L. 444-2.

in MLD 37/38, as well as in Sts. 19 (assigned to *Homo* by Kimbel and Rak, 1993).

Another difference between the configurations seen in chimpanzees and *A. africanus* (and Sts. 19) is that the medial part of the articular surface of the temporomandibular joint (as determined by the path of the synovial line) in chimpanzees extends considerably anterior to the posterior border of the foramen, whereas in *A. africanus*, the entire articular surface remains behind the coronal plane of the foramen's posterior margin. In this respect, chimpanzees differ from all the hominoids discussed thus far but resemble other primates, such as gibbons, vervets, and macaques; two *A. afarensis* specimens (A.L. 166-9 and A.L. 333-45) also appear to be more similar to chimpanzees than is A.L. 444-2.

The crania of *A. boisei* exhibit a different shape of the temporal foramen and thus also of its posterior margin. The deepest, most posterior point of the foramen is located posteromedially because of the long, diagonally (anterolateral–posteromedial) oriented articular eminence and the mediolaterally restricted preglenoid plane (see the section "Inclination of the Articular Eminence"). This morphology can be seen in OH 5, KNM-ER 406, KNM-ER 13750, and probably KNM-ER 23000. The foramen in *A. boisei*, at least in the more extreme examples, assumes the shape of a wide triangle whose mediolateral perpendicular height is greater than its base. In KNM-WT 17000, the most posterior point of the foramen margin is situated relatively medially (although an intrusion of the prominent infratemporal crest into the foramen creates an irregularity in the outline; see Figure 3.55), but the foramen is longer anteroposteriorly than it is wide, as it is also in *A. robustus* specimen TM 1517.

In A.L. 444-2, the structure that forms the anterior margin of the temporal foramen—the inferior border of the zygomatic bone—is extremely thick (for further detail, see Chapter 5). This structure's massivity stands in

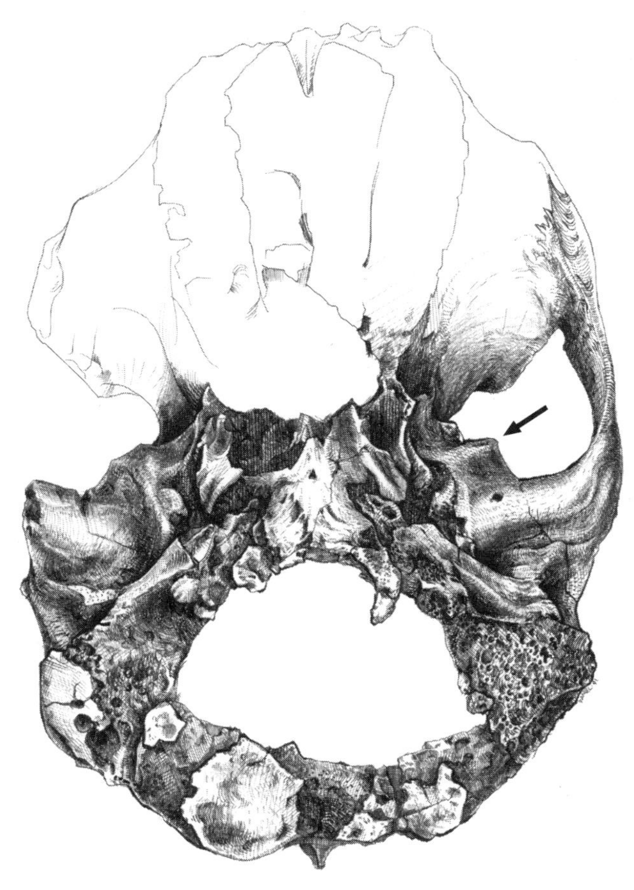

Figure 3.55 Artist's rendering of the basal view of *A. aethiopicus* cranium KNM-WT 17000. The preglenoid plane intrudes into the space of the temporal foramen (arrow).

great contrast, not only to the extraordinarily delicate region in chimpanzees but also to the more massive region in gorillas. Not one gorilla, including even the most robust in our sample, approaches the degree of massivity seen in A.L. 444-2. However, we can conclude from the data in Table 3.24, which compares the width of the masseteric scar to the biorbital breadth and to the height of the zygomatic bone, that a relatively large masseteric scar generally characterizes the hominins.

General Topography of the Cranial Base

Examination of the base of African ape crania reveals that the surface is convex on both the coronal and sagittal planes in chimpanzees and gorillas (the convexity on the coronal plane is greater in female gorillas than in the males). As a result, the whole surface assumes a puffy appearance, with the occipital condyles and foramen magnum occupying the most elevated[22] area in this topography.

As can be clearly observed in a posterior view, the convex coronal outline extends continuously from one supramastoid crest to the other (Figure 3.56). The occipital condyles crown the summit of the arch, and the mastoid processes constitute the lateral ends of its slope. Aligned with this contour are the so-called paramastoid processes (with which the transverse process of the atlas vertebra articulates), the mastoid tips, and the lateral aspect of the mastoid processes. None of these structures deviates from the smooth curvature of the arch (the dotted line in the figure). Male gorillas sometimes diverge from this pattern. In such cases, which occasionally include large male chimpanzees as well, the dominant mastoid processes in the periphery protrude somewhat from the coronal contour.

The convex sagittal outline of the cranial base in African apes is the product of two sloping surfaces. One extends anterosuperiorly from the foramen magnum at the summit of the arch to the posterior nasal choanae and the base of the pterygoid processes (this slope is essentially the roof of the nasopharynx). The steeper slope, comprising the nuchal plane of the occipital squama, extends from the foramen magnum to the nuchal crest. As a result of the dramatic ascent of the crest, the sagittal outline of the nuchal region is extremely long and convex.

In contrast to its arched appearance in chimpanzees and female gorillas, the coronal outline of the region in A.L. 444-2 is flat and topographically depressed between the mastoid processes in the lateral periphery (Figure 3.56). In the center of the depressed bone surface lies the foramen magnum.

The midsagittal outline of the skull base is slightly convex, almost flat, especially when compared to that of the great apes. Such flatness is undoubtedly the product of a low inion and the horizontal inclination of the nuchal plane, as discussed in the section "Inclination of the Lower Occipital Scale (Inion–Opisthion)."

Despite the differences between the *A. afarensis* nuchal regions that are reasonably preserved—those of A.L. 444-2 and of specimens in which the midsagittal cross section is more rounded (such as A.L. 162-28 and A.L. 439-1)—a closer examination reveals that A.L. 444-2 simply represents a more extreme version of the same morphology. In A.L. 333-45, for example, the coronal outline of the region stretching between the mastoids is rather flat (Figure 3.56).

Cranium OH 5, which exhibits the most completely preserved cranial base in the *A. boisei* hypodigm, resembles A.L. 444-2 in this region. The cranial base of OH 5 is flat on both the sagittal and coronal planes, and the nuchal region is topographically reccessed in relation to the mastoid processes. However, within the *A. boisei* hypodigm, considerable variation in the topography of

Table 3.24 Thickness of the Massetric Scar

Specimen/Sample		Inferior Zygomatic Margin Thickness (40) (mm)	Inferior Margin Thickness (40)/ Zygomatic Height (73) (%)	Inferior Margin Thickness (40)/ Biorbital Breadth (54) (%)
Australopithecus afarensis	Mean	14	36	15
A.L. 444-2		17	39	18
A.L. 333-1		10	32	11
Homo sapiens	Mean (*n* = 10)	8	34	8
	Range	6–10	25–50	7–11
	SD	1	7	1
Gorilla gorilla female	Mean (*n* = 10)	8	21	9
	Range	6–9	15–25	8–9
	SD	1	4	1
Gorilla gorilla male	Mean (*n* = 10)	7	16	6
	Range	4–9	11–21	5–8
	SD	2	4	2
Pan troglodytes female	Mean (*n* = 10)	5	19	6
	Range	2–9	10–30	3–10
	SD	3	9	4
Pan troglodytes male	Mean (*n* = 10)	4	14	4
	Range	3–6	11–20	3–6
	SD	1	4	1
A. africanus	Mean	9	38	11
Sts. 5		7	26	8
Sts. 71		11	41	14
Sts. 52a		10	48	n/a
A. robustus	Mean	15	48	16
SK 48		13	43	13
TM 1517		16	53	19
A. boisei	Mean	15	40	15
OH 5		15	38	15
KNM-ER 406		16	43	16
KNM-ER 13750		14	39	13
KNM-ER 732 female		11	38	13
A. aethiopicus				
KNM-WT 17000		16	44	17

the cranial base can be observed. Here, too, much of the topographical variation stems from the degree of horizontality of the nuchal region on the sagittal plane and the size of the mastoid processes on the coronal plane (see the section "Inclination of the Lower Occipital Scale [Inion–Opisthion]"). Just as A.L. 333-45 and A.L. 439-1 lie at the opposite end of the range of variation from A.L. 444-2 in the *A. afarensis* sample, the *A. boisei* specimens KNM-ER 406, KNM-ER 407, and KNM-ER 23000 are at the opposite end of the range from OH 5.

Not much remains of the nuchal region of KNM-WT 17000, at least not of its external bone table. Nevertheless, the topography of the skull base has a flat appearance on both the coronal and sagittal planes. The nuchal region itself gives the impression of having an extensive surface area. The posteriorly extended lip of the nuchal crest and the flat posterior face of the mastoid processes, which lie almost at the level of the nuchal plane itself, together produce this flat, extensive appearance.

Topographically, the *A. africanus* cranial base exhibits few distinguishing characteristics. It is convex on the sagittal plane, almost flat on the coronal plane, and somewhat puffy in appearance. In contrast to *A. afarensis* and *A. boisei*, the very modest mastoid processes in *A. africanus* have a negligible role in demarcating the periphery of the nuchal region, as seen in specimens MLD 37/38 and Sts. 5. A similar topography, brought about by the same factors, is observed in *H. habilis*, in specimens such as KNM-ER 1813, OH 24, and Stw. 53, as well as in modern humans. In the latter, however, the medial aspect of the

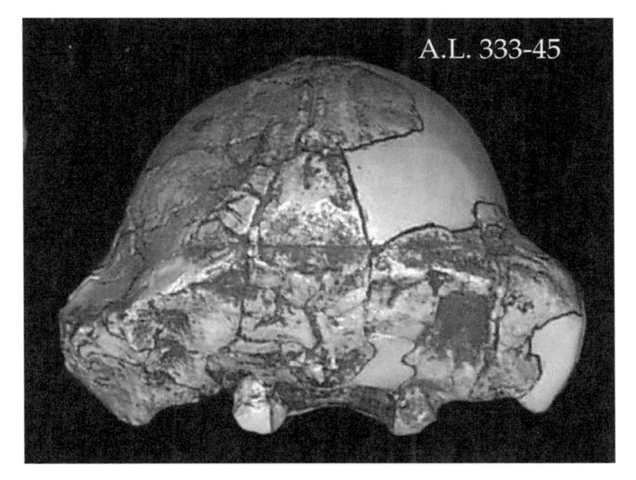

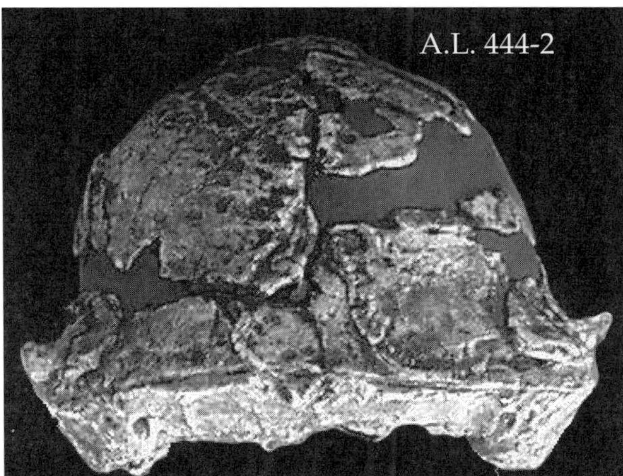

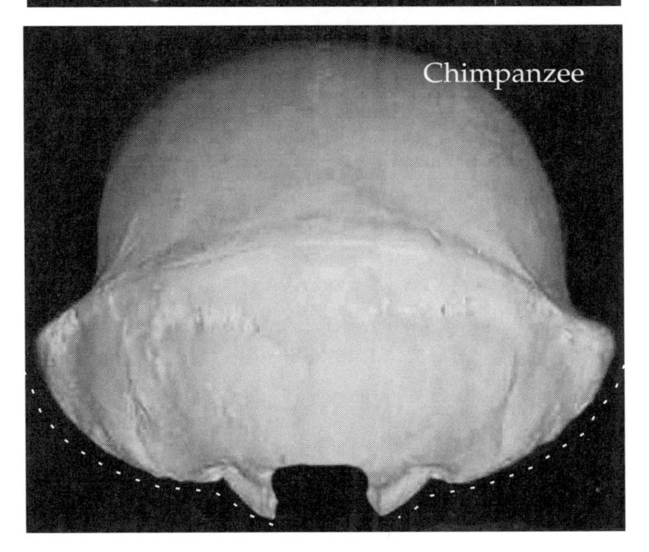

Figure 3.56 Coronal outline of the cranial base in a chimpanzee and *A. afarensis*. Compare the continuously arched outline (white dotted line) from the topographically elevated occipital condyles to the supramastoid crest in chimpanzees to the straight, recessed outline between the two mastoid tips in the *A. afarensis* specimens. In the latter, the outline of the lateral aspect of the mastoid is separated from the rest of the cranial base by the mastoid tips.

prominent mastoid processes, which protrude inferiorly below the nuchal plane, forms a distinct wall that demarcates the lateral ends of the nuchal region.

Metrics and Angular Relationships

Anteroposterior Measurements

As compared to the great apes, A.L. 444-2, like all other hominins, exhibits a relatively anteriorly situated foramen magnum and shortened anterior cranial base. The forward location of the foramen clearly affects the position of the surrounding elements of the cranial base. Because this difference in position occurs along the anteroposterior axis of the cranial base, its effect on the surrounding structures is more noticeable here than along the mediolateral axis. Unlike the very long anteroposterior segment between the foramen magnum and the posterior nasal opening in the great apes, this region in A.L. 444-2 and other hominins, including modern *H. sapiens*, is short and confined (as also appears to be the case in what survives of that region in *Ard. ramidus* [White et al., 1994, 1995]). This difference is expressed quantitatively by a very small horizontal distance between the coronal planes of basion and foramen ovale (or the root of the pterygoid process). Consequently, the basioccipital element, which bridges these structures, is also very short (Table 3.25). The estimated basioccipital length is 20 mm in A.L. 444-2.[23] With a length measurement of 19 mm, the basioccipital element that survives with specimen A.L. 417-1 confirms the small basioccipital length in *A. afarensis*. In *A. afarensis* basioccipital length constitutes approximately 22% of the biorbital breadth (the mean of two specimens). These values are much smaller than those in the African great apes, both absolutely and in proportion to the overall size of the cranium (as judged by the biorbital breadth). Basioccipital length constitutes 28% of the biorbital distance in female chimpanzees and 29% in males, and reaches 30% in female gorillas and 33% in males (Table 3.25).

The length of the basioccipital element can be measured in three *A. africanus* specimens. In Sts. 5 it is longer than in *A. afarensis*, occupying 29% of the biorbital breadth. Shorter than the element in Sts. 5, that of MLD 37/38 probably constitutes a smaller percentage of the (unavailable) biorbital distance, as the two crania are otherwise approximately the same size. Biorbital breadth cannot be measured on Stw. 187a; however, our data indicate that the absolute length of its basioccipital (17.5 mm) is among the smallest of the known hominins. Only OH 24 has a shorter basioccipital element, which may be due in part to crushing (see Tobias, 1991). The mean of two *A. boisei* specimens (OH 5 and KNM-ER 406) is 22%, the same as the index in our modern *H. sapiens* sample (22%). However, KNM-WT 17000 has a relatively long

Table 3.25 Absolute and Relative Measures of Cranial Base Length

Specimen/Sample		Basioccipital Length (mm)	Basioccipital Length/ Biorbital Br. (54) (%)	CC-FO (31) (mm)	CC-FO (31) /Biorbital Br. (54) (%)	Tympanic Length (CC-TP) (mm)	Tympanic Length/ Biorbital Br. (54) (%)	Petrous Length (CC-PA) (mm)	Petrous Length/ Biorbital Br. (54) (%)
Australopithecus afarensis	Mean	20	22	22	21	31	37	18	19
A.L. 444-2		20	21	20	21	35	37	18	19
A.L. 417-1		19	23	n/a	n/a	n/a	n/a	n/a	n/a
A.L. 333-45		n/a	n/a	24	n/a	27	n/a	17	n/a
Homo sapiens	Mean (*n* = 10)	21	22	18	19	23	24	20	21
	Range	18–24	20–26	14–22	16–22	19–25	19–27	17–23	18–23
	SD	1	1	2	2	2	3	2	2
Gorilla gorilla female	Mean (*n* = 10)	29	30	27	27	39	39	28	28
	Range	24–35	24–35	22–33	22–33	31–45	31–44	25–30	25–32
	SD	2	3	3	3	4	4	2	2
Gorilla gorilla male	Mean (*n* = 10)	37	33	35	30	48	42	31	27
	Range	33–41	27–39	29–40	25–35	43–52	37–47	24–38	23–32
	SD	2	3	4	3	3	3	3	3
Pan troglodytes female	Mean (*n* = 10)	26	28	21	23	34	37	25	27
	Range	23–30	22–35	17–26	19–27	30–39	31–42	21–30	22–31
	SD	2	3	3	2	2	3	2	2
Pan troglodytes male	Mean (*n* = 10)	28	29	24	25	38	40	25	26
	Range	22–32	22–36	18–29	20–31	31–45	36–48	21–29	22–30
	SD	3	4	3	3	4	4	2	3
A. africanus	Mean	21	29	22	26	25	27	24	28
Sts. 5		25	29	22	26	23	27	24	28
MLD 37/38		21	n/a	21	25	26	(30)	24	(28)
Stw. 187		17	n/a	n/a	n/a	n/a	n/a	n/a	n/a
Early *Homo*	Mean	19	(14)	15	16	24	29	21	26
Sts. 19		19	n/a	17	n/a	22	n/a	20	n/a
OH 24		(12)	(14)	15	17	30	35	20	23
KNM-ER 1813		n/a	n/a	13	15	19	22	24	28
A. boisei	Mean	22	22	19	18	32	35	26	27
OH 5		20	20	19	19	36	36	24	24
KNM-ER 406		25	24	18	17	34	33	30	29
KNM-ER 407		21	n/a	(14)	n/a	25	n/a	23	n/a
A. aethiopicus									
KNM-WT 17000		25	27	20	23	28	30	22	23

Values in parentheses are estimates.

Ht., height; Br., breadth; n/a, not available; CC, carotid canal; FO, foramen ovale; TP, lateral tympanic point; PA, petrous apex.

basioccipital element, which constitutes 27% of the biorbital breadth.

A result of the foramen magnum's anterior position in A.L. 444-2 is that basion advances past the coronal plane of the external auditory meati. More than one-third of the reconstructed foramen's length extends past the biporial line. In this respect, the Hadar cranium seems to be the most extreme when compared not only to the African great apes, whose foramen lies well behind this line, but also to other hominins, including modern humans, in which basion is situated near or at the level of the biporial line. In robust *Australopithecus* species, the foramen magnum is more anteriorly situated than in modern humans. A.L. 444-2 resembles the robust specimens KNM-ER 406, KNM-ER 407, and SK 47, with the foramen magnum extending anterior to the bicarotid canal line, and thus well past the biporial line (Dean and Wood, 1982). However, a resemblance is also seen between the robust specimens and some crania of early *Homo*, in which basion is situated anterior to the biporial line, as in OH 24 (again, this may be influenced by distortion). Nevertheless, unlike the configuration in A.L. 444-2, basion in early *Homo* does not reach the bicarotid canal line. In *Ard. ramidus* (White et al., 1994), too, basion appears to approximate this line.

Graphic representation is helpful in evaluating these relationships, which are depicted in Figure 3.57. The illustrations are based on precise metric data and have been adapted in part and with modification from Dean and Wood (1982). They show that the proximity of basion to the biporial line and to the bicarotid canal line is not always correlated. Only in *A. afarensis*, *A. robustus*, and *A. boisei* does the anterior margin of the foramen magnum lie both anterior to the former and in the coronal plane of the latter.

The shortness of the cranial base (anterior to the foramen magnum) in A.L. 444-2 results in a reduction of the anteroposterior length of the quadrilateral formed between the bicarotid line and the line that connects the foramina ovale. In general, this quadrilateral is substantially shorter in hominins than in the great apes, as expected. Since the position of the foramina ovale on the anteroposterior axis appears to be fixed in relation to the adjacent structures (such as the spheno-occipital synchondrosis and the base of the lateral pterygoid plates), the reduction in the anteroposterior length of the quadrilateral is influenced mostly by the anterior migration of the carotid canals (Figure 3.57).

Other anteroposterior length measurements of the cranial base were discussed in detail in the discussion of the lateral view of the skull and are not repeated here. However, they are intimately connected to the basal aspect of the cranium and express significant anatomical differences between taxa vis-à-vis the shape of the cranial base. Among the measurements that profoundly affect the shape of the cranial base and exhibit variation between and within taxa we can mention the following:

- The horizontal distance between the coronal plane of M^3 and the articular eminence (see the section "Relationship between Elements of the Masticatory System")
- The position of the masseter muscle's origin—that is, the zygomaticoalveolar crest—relative to the anteroposterior length of the cranium and to the length of the dental arcade (see the section "Relationship between Elements of the Masticatory System")
- The length of the dental arcade and its position relative to the braincase (see the section "Degree of Prognathism and Relative Length of the Dental Arcade")
- The anteroposterior position of the foramen magnum relative to the length of the braincase (see the section "Anteroposterior Position of the Foramen Magnum in Relation to the Length of the Calvaria")

The differences in these measurements are manifested in a variety of morphologies. In the great apes, the posterior position of the foramen magnum, on the one hand, and the anterior position of the entire dental arcade, on the other, bring about a voluminous space at the center of the cranial base between the posterior nasal choanae and the foramen magnum. The similar measurements in chimpanzees and Sts. 5 are accompanied by similarly spacious cranial base areas. Furthermore, the metric differences between specimens of *A. africanus*, especially between Sts. 5 and Sts. 71, are neatly translated into the differences one would expect in their cranial bases. (Although the base of Sts. 71 is far less well preserved than that of Sts. 5, one can still detect the differences.) The small amount of prognathism in the robust *Australopithecus* species (especially *A. boisei*) is achieved, as discussed, through the posterior retraction of the whole dental arcade. Combined with the anterior position of the foramen magnum, this posterior retraction results in a reduction of the volume in the center of the cranial base. Here, too, the size of this space varies greatly, depending on the anteroposterior position of the dental arcade (compare the position of M^3 in KNM-ER 406 and OH 5).

A basal view reveals that *A. boisei* and hominoids equipped with a more plesiomorphic masticatory system differ in the relationship between the zygomaticoalveolar crest and the length of the dental arcade, as one would expect from the metrics (discussed in the section "Relationship between Elements of the Masticatory System"). In *A. boisei*, the zygomatic process of the maxilla flares out laterally from the most anterior part of the dental arcade,

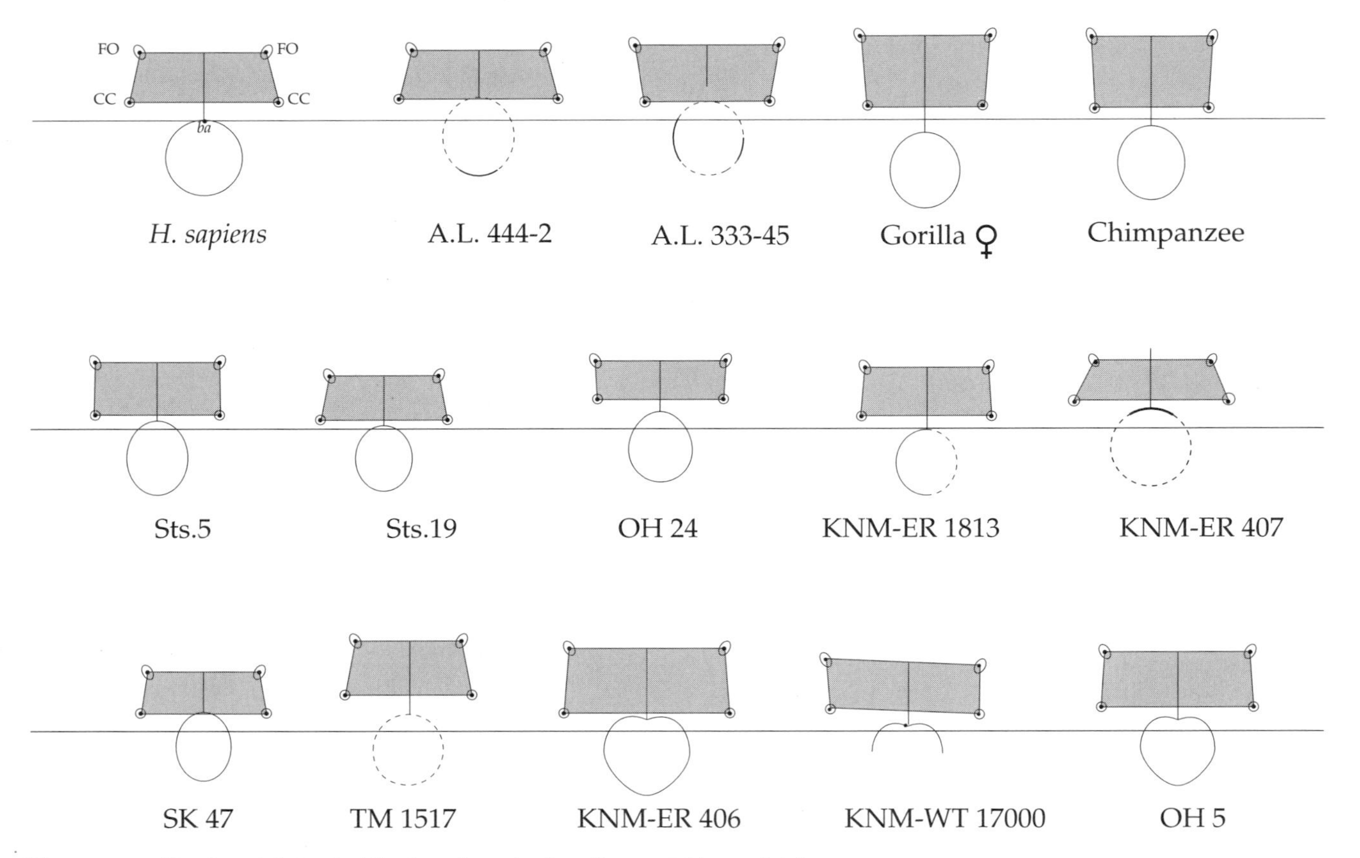

Figure 3.57 The "quadrilateral of the four foramina" on the cranial base of A.L. 444-2 and other hominoids. This schematic view (based on Dean and Wood, 1982) depicts the variations in the shape of the quadrilateral and its relationship to the foramen magnum and the bitympanic line (horizontal line in the figure). Note the following: (1) In the apes, the quadrilateral is wider anteriorly and narrower posteriorly, whereas in the hominins, the proportions are reversed. (2) In the apes, the quadrilateral is tall and narrow, and in the hominins, it is shorter and broader. (3) The anterior end of the foramen magnum (basion) lies substantially posterior to the bitympanic line in the apes but touches the line or lies anterior to it in the hominins. (4) In A.L. 444-2, the foramen magnum encroaches on the quadrilateral (that is, it passes the bicarotid line); this proximity can also be observed in some robust *Australopithecus* specimens. FO = foramen ovale; CC = center of carotid foramen.

whereas in chimpanzees and the more generalized hominins it emerges from the posterior half of the dental arcade. Cranium KNM-WT 17000 presents an interesting combination of the two configurations. On the one hand, the extremely anterior position of the dental arcade acts to create an immense horizontal distance between the palate and the foramen magnum. On the other hand, the zygomaticoalveolar crest flares laterally from the most anterior part of the dental arcade. In modern humans, as well as in early specimens of *Homo*, all the components of the masticatory system are diminished (rather than retracted posteriorly), as already discussed. Prosthion, therefore, extends only slightly anterior to the basal contour of the cranium, but the zygomaticoalveolar crest still exhibits a plesiomorphic arrangement in its posterior position relative to the reduced length of the dental arcade. It is the shrinkage of the masticatory system that reduces the *absolute* horizontal distance between M^3 and the articular eminence and, along with the relatively anterior position of the foramen magnum (compared to that of the great apes), brings about a tightening of the nasopharyngeal space.

The morphology of the cranial base in A.L. 444-2 conforms perfectly to the metrics of both prognathism and the configuration of the plesiomorphic masticatory system. A large part of the dental arcade extends beyond the contour of the cranial base. Thus, the zygomatic processes emerge from the middle of the dental arcade length, and the coronal plane of M^3 lies a substantial horizontal distance away from the coronal plane of the articular eminence. Nevertheless, the considerably anterior position of the foramen magnum is one of the major factors responsible for the constriction of the space in the center of the skull base. The picture, however, is apparently more complex, given the intra- and intertaxic variation in the distance between the facial frame (the coronal plane of orbitale) and the braincase (the coronal plane of porion), as discussed in the section "Degree of Prognathism and Relative Length of the Dental Arcade"). We can reasonably assume that, in turn, variation in this distance introduces variation in the size of the space of the nasopharynx, regardless of the effect wielded by the foramen magnum's position or the degree of retraction of the dental arcade.

Breadth Measurements

In contrast to the great taxonomic diversity in the position of anatomical structures along the anteroposterior axis of the cranial base, the juxtaposition of the various structures and their location relative to the breadth of the base (whatever its size) appear to be relatively conservative. If detected at all, such variation is minor. Any change in the breadth of the cranial base corresponds to changes in the width of the structures and landmarks within the cranial base itself. This means that the widening of the base does not consist merely of an addition to the lateral periphery that leaves the central structures untouched; rather, the entire area stretching between the two porial saddles participates proportionally in the broadening of the base. (An analogy can be drawn with a balloon that has markings on it. When the balloon is inflated, the markings drift away from each other while maintaining their relative positions.) For example, the distance between the foramina ovale (Table 3.26) constitutes the same percentage of cranial base breadth (biporial saddle breadth) in both the very narrow base of *A. africanus* and the extremely broad base of *A. boisei* (44%; Table 3.27). (For a discussion of the breadth of the braincase relative to the biorbital distance in these two taxa, see the section "The Occipital View and the Coronal Section of the Calvaria" and Table 3.22.)

A similar relationship between the biforamen ovale distance and the biporial saddle breadth characterizes *A. afarensis*: the distance between the foramina ovale constitutes 42% of the cranial base breadth (the mean of A.L. 333-45 and A.L. 444-2). In contrast, the African apes, particularly male gorillas, exhibit a narrow biforamen ovale width, but without a parallel reduction of cranial base breadth so that the biforamen ovale distance occupies only 36% of the biporial saddle breadth. (For ratios of these measurements to the biorbital distance—our yardstick—see Table 3.27.) This reduction in the biforamen ovale distance in the male gorilla corresponds to a narrowing of other cranial base elements, including the dental arcade, particularly its posterior portion (see the section "Proportions of the Facial Mask" and Table 3.27). Female gorillas, like male chimpanzees, yield values that are slightly larger, though not significantly so. In hominins, including modern humans, the values are similar to that of A.L. 444-2.

The distance between the two carotid foramina is also fairly constant among hominins, despite variation in cranial base breadth. In A.L. 444-2, the bicarotid distance constitutes 45% of the cranial base breadth, and in A.L. 333-45, it is 39%. The mean for *A. afarensis* is 42%, a proportion similar to that of other early hominins and modern humans but higher than that of the African apes (Table 3.27). Small as they might appear, these differences between the hominins and the apes are still structurally significant, as they are apparently related to the forward encroachment of the foramen magnum in hominins (see the section "Relationship between Breadth and Length Measurements; Angular Measurements").

Among the hominins, the relationship between the bi-entoglenoid and cranial base breadth is also more or less fixed. Again, despite the great differences that Sts. 5 and OH 5 exhibit in both the absolute and relative width of the base, the ratios are almost the same, at 61% in Sts. 5 and 64% in OH 5. Similarly, 61% is the value calculated for both A.L. 444-2 (59% for A.L. 333-45) and modern humans. The mean of four *A. boisei* specimens is 64%, while the *A. aethiopicus* value is 61%. Note, however, that early *Homo* yields a somewhat lower value. Again, we find that in respect to the bi-entoglenoid and cranial base breadth relationship, the African apes deviate from the hominin norm.

One noticeable exception in the width measurements of the hominins is the distance between the articular tubercles (the lateral ends of the articular eminences), whose ratio to the width of the cranial base varies substantially among taxa. Whereas in modern humans the distance between the tubercles constitutes 97% of the breadth of the base—that is, the tubercles are located farther medially than the porial saddles—the distance in *Australopithecus* crania is considerably larger than the breadth of the base. The ratio in A.L. 444-2 is 114%, which is less than in most specimens of *A. boisei*, where the disproportion is particularly conspicuous in the larger specimens, such as KNM-ER 406 (117%), KNM-ER 13750 (121%), and KNM-ER 23000 (120%) (Table 3.27). Other *Australopithecus* species show less lateral displacement of the articular tubercles; both *A. aethiopicus* and *A. africanus* have ratios of 110%. Early *Homo*, in contrast, resembles modern humans, with a ratio of 99% for KNM-ER 1813 and 101% for OH 24.

Since the position of the medial part of the articular eminence is relatively fixed in hominins, as just seen, we can conclude that much of the variation in the lateral pro-

Table 3.26 Absolute Measures of Cranial Base Breadth (mm)

Specimen/Sample		FO-FO (25)	CC-CC (26)	Bi-entoglenoid Br. (24)	Bi-articular Tubercle Br. (23)	PA-PA	Articular Eminence Br. (28)[a]
Australopithecus afarensis	Mean	56	55	79	150	29	37
A.L. 444-2		52	59	80	150	29	37[b]
A.L. 333-45		59	51	78	140	(35)	(37)[c]
Homo sapiens	Mean (n = 10)	49	56	74	116	30	24
	Range	46–56	51–62	70–79	107–127	27–34	21–26
	SD	3	4	3	6	2	2
Gorilla gorilla female	Mean (n = 10)	47	46	64	132	27	37
	Range	40–52	40–51	56–70	125–137	22–33	32–41
	SD	3	3	5	3	3	2
Gorilla gorilla male	Mean (n = 10)	52	53	72	149	30	42
	Range	46–57	44–60	62–80	143–156	23–38	36–47
	SD	3	5	6	3	4	2
Pan troglodytes female	Mean (n = 10)	44	42	59	111	20	28
	Range	40–49	37–47	52–68	105–122	17–25	24–32
	SD	2	2	5	4	3	2
Pan troglodytes male	Mean (n = 10)	45	42	61	112	24	28
	Range	38–49	35–52	52–70	101–124	18–29	23–32
	SD	3	5	6	7	3	2
A. africanus	Mean	47	47	64	118	23	32
Sts. 5		48	48	65	115	25	30
MLD 37/38		45	45	62	120	21	33
Early *Homo*	Mean	48	48	59	113	23	30
Sts. 19		43	48	60	116	25	30
KNM-ER 1813		51	n/a	57	112	20	31
OH 24		49	48	61	110	24	30
A. boisei	Mean	54	59	81	155	30	38
OH 5		59	61	89	155	29	37
KNM-ER 406		59	65	85	155	29	40
KNM-ER 407		41	53	65	n/a	28	34
KNM-ER 13750		n/a	n/a	86	154	n/a	38
KNM-ER 23000		57	58	80	154	32	41
A. aethiopicus							
KNM-WT 17000		57	55	80	144	29	38

[a]Measured direct from entoglenoid process to articular tubercle.
[b]Measured on the right; on the left it measures 35 mm.
[c]Measured on the left; on the right it measures 34 mm.
Br., breadth; FO, foramen ovale; CC, carotid canal; PA, petrous apex.

trusion of the articular tubercle must stem from the variation in the mediolateral breadth of the eminence itself (that is, the width of the mandibular condyle). Indeed, a clear relationship exists between this dimension and the degree of the eminence's lateral protrusion. In modern humans, the eminence is very short mediolaterally, constituting only 20% of the biporial saddle breadth, whereas in A.L. 444-2, the mediolateral length of the eminence occupies 28% of this breadth. In *A. boisei* and *A. aethiopicus*, the length of the eminence constitutes about 30% and 29%, respectively, of this distance. The corresponding values in A.L. 444-2 and other hominins, except *H. sapiens*, resemble those in gorillas. In chimpanzees, the values are lower.

The high value for early *Homo* (27%) must arise from a medially, rather than a laterally, extended articular eminence. Indeed, the distance between the entoglenoid processes in early *Homo* is smaller in relation to the width of the cranial base than in other hominin taxa. Thus, early *Homo* contradicts our generalization about the fixed position of the entoglenoid processes.

In gorillas, the mediolateral length of the articular eminence is also great (occupying 30% of the biporial saddle breadth in females and 29% in males), somewhat greater than in chimpanzees (25% in both females and males). However, the distance between the entoglenoid processes in the gorillas is very small (50% of the cranial

Table 3.27 Shape of the Cranial Base (%)

Specimen/Sample		Biforamen Ovale Br. (25)/ Biorb. Br. (54)	Biforamen Oval Br. (25)/ Biporial Saddle Br. (14)	Bicarotid Canal Br. (26)/ Biorb. Br. (54)	Bicarotid Canal Br. (26)/ Biporial Saddle Br. (14)	Bi-entoglenoid Br. (24)/Biorb. Br. (54)	Bi-entoglenoid Br. (24)/ Biporial Saddle Br. (14)	Biarticular Tubercle Br. (23)/ Biorb. Br. (54)	Biarticular Tubercle Br. (23)/ Biorb. Saddle Br. (14)	Bi-petrous Apex Br./ Biorb. Br. (54)	Bi-petrous Apex Br./ Biporial Saddle Br. (14)	Articular Eminence Br. (28)/ Biorb. Br. (54)	Articular Eminence Br. (28)/ Biporial Saddle Br. (14)
Australopithecus afarensis	Mean	54	42	61	42	86	60	159	111	37	25	40	28
A.L. 444-2		55	39	62	45	84	61	157	114	31	22	39	28
A.L. recon. (A.L. 333-1 + A.L. 333-45)		(52)	45	(59)	39	(89)	(59)	(161)	(107)	(42)	(28)	(40)	(27)
Homo sapiens	Mean (*n* = 10)	52	41	59	47	77	61	122	97	32	25	25	20
	Range	48–56	39–45	55–65	44–50	72–82	58–64	116–127	93–102	29–36	21-28	21–28	16–23
	SD	3	2	4	2	3	2	5	3	3	2	2	2
Gorilla gorilla female	Mean (*n* = 10)	47	39	47	38	64	52	134	107	27	22	37	30
	Range	42-52	33-45	42-52	33-43	59-69	48-57	127-139	101-113	23-32	18-24	32-42	25-34
	SD	3	3	3	3	3	2	4	3	2	2	3	2
Gorilla gorilla male	Mean (*n* = 10)	46	36	48	37	63	50	130	103	26	21	37	29
	Range	41–51	33–39	41–54	31–41	55–69	41–57	120–140	98–112	20–34	16–28	32–42	25–32
	SD	2	2	5	3	5	4	6	3	4	3	3	2
Pan troglodytes female	Mean (*n* = 10)	49	37	47	36	65	51	121	96	22	17	30	24
	Range	42–54	31–45	43–50	31–39	57–73	47–54	114–129	91–101	18–29	13–23	25–36	20–28
	SD	3	5	2	3	6	2	5	3	3	3	2	2
Pan troglodytes male	Mean (*n* = 10)	49	39	46	36	67	54	124	99	26	21	31	25
	Range	45–53	35–41	40–52	31–41	60–73	48–62	116–136	94–106	21–32	17–26	26–35	20–29
	SD	2	2	4	3	4	4	5	2	3	2	2	2
A. africanus	Mean	56	44	56	44	76	59	135	110	29	22	35	29
Sts. 5		56	45	56	45	76	61	135	108	29	24	35	28
MLD 37/38		n/a	42	n/a	42	n/a	57	n/a	111	n/a	19	n/a	30
Early *Homo*	Mean	59	43	56	42	69	53	130	102	26	20	35	27
Sts. 19		n/a	39	n/a	43	n/a	54	n/a	105	n/a	20	n/a	27
KNM-ER 1813		60	45	n/a	n/a	67	50	132	99	24	17	35	27
OH 24		57	45	56	40	71	56	128	101	28	22		28
A. boisei	Mean	58	44	62	46	84	64	151	117	29	22	38	30
OH 5		59	42	61	44	89	64	155	111	29	21	37	27
KNM-ER 406		57	44	63	49	83	64	150	117	28	22	40	30
KNM-ER 13750		n/a	n/a	n/a	n/a	80	67	147	121	n/a	n/a	37	30
KNM-ER 23000		n/a	45	n/a	45	n/a	62	n/a	120	n/a	n/a	n/a	32
A. aethiopicus													
KNM-WT 17000		61	42	59	42	85	61	153	110	31	22	41	29

Br., breadth; n/a, not available.

base breadth), which is consistent with the typically narrow distance between structures in the center of the cranial base, as already described.[24] Thus, in comparison to the hominin cranial base, the lateral extension of the articular eminence is less than would be expected given its length. In male gorillas, the distance between the tubercles constitutes only 103% of the cranial base breadth, as compared to 114% in A.L. 444-2 and 117% in *A. boisei.*

In his monograph on OH 5, Tobias recognized this difference between OH 5 and gorillas. Tobias refers to the mediolateral length of the mandibular fossa rather than of the articular eminence, and he writes:

> If we measure the distance apart of the left and right entoglenoid processes, they are seen to be further apart absolutely and relatively in *Zinjanthropus* than in gorilla. An index may be devised relating the distance between the outer faces of the left and right entoglenoid processes to the biporial distance or to the "*biglenoid distance*" (the distance between the outermost points of the glenoid articular surface). The former index (the *interglenoid–biporial index*) gives values (per cent) of 64.5 for *Zinjanthropus* and ? 62.3 for *Paranthropus* (SK 48), but only 48.8 and 52.6 for two gorilla crania. The latter index (*interglenoid–biglenoid index*) is 56.2 in *Zinjanthropus* and ? 55.0 in SK 48, but 44.2 and 48.4 in two gorillas. These *ad hoc* indices confirm the view that the glenoid fossa approaches nearer the midline in gorilla than in *Zinjanthropus.* (Tobias 1967: 40)

Tobias elaborates and then concludes, "it seems that *Zinjanthropus* has thus achieved an increase in dental, mandibular and condylar size, without narrowing the space between the two halves of the mandible."

It is clear, therefore, that A.L. 444-2 adheres to the pattern described for robust *Australopithecus,* which can be applied to *Australopithecus* generally. The comparable degree of lateral extension of the articular tubercles in A.L. 444-2 and robust species stems from a wide articular eminence combined with a mediolaterally wide central cranial base.

Relationship between Breadth and Length Measurements; Angular Measurements

As demonstrated, the breadth measurements of various structures in the cranial base differ less among hominoid taxa than the length measurements, which differ substantially. Thus, the length measurements are probably the primary source of many, if not most, of the differences observed in angular measurements and ratios between the width and length measurements. By way of illustration, we can cite the ratio of the biforamen ovale distance to the cranial base width, which is the same in A.L. 444-2 and female gorillas though these two taxa differ substantially in the distance measured from the coronal plane of basion to that of the foramen ovale. Hence, it is clearly the latter, the anteroposterior dimension, that affects the angle formed by the lines from basion to each foramen ovale (an obtuse angle in A.L. 444-2 is about 110°, and an acute one in the female gorillas is about 60°).

The quadrilateral of four foramina. In hominins, the distance between the openings of the carotid canals tends to be slightly greater than the distance between the foramina ovale, with the reverse relationship holding true for African apes (Figure 3.57 and Tables 3.26 and 3.28). Modern humans display extreme tapering, with the distance between the openings of the carotid canals constituting 114% of the biforamen ovale distance. In chimpanzees, the values are 97% for females and 96% for males, whereas in gorillas, the index is 98% in females and 103% in males (recall the relatively small biforamen ovale distance in the latter).

Of all the hominins in our sample, *A. afarensis,* with a mean value of 115% (calculated from the A.L. 444-2 value of 113% and the A.L. 333-45 value of 116%), is the closest to modern humans. The mean value for two *A. africanus* specimens is 100%, a perfect rectangle, and for three *A. boisei* specimens, 105%. In KNM-WT 17000, the region appears deformed (note the asymmetry between the left and right sides of this specimen [Figure 3.57]) and yields a value of 96%.

The relatively short anteroposterior distance between the two pairs of foramina in hominins gives the quadrilateral a squat appearance. In modern humans, for example, this distance constitutes only 36% of the biforamen ovale breadth, a proportion that does not differ greatly from the 38% in A.L. 444-2 (when A.L. 333-45 is averaged in with A.L. 444-2, the figure rises slightly, to 39%). In chimpanzees, the mean is 46% for females and 50% for males. At 58% for females and 63% for males, the gorilla values are significantly higher; hence, the quadrilateral is much longer than it is wide. The lowest values are seen in early *Homo* and *A. boisei,* in which the mean is 30%. Note that in *A. africanus* the mean is much higher; at 47%, it is almost the same as in the chimpanzees. Also worth mentioning is the high value (40%) in Sts. 19, which we have attributed to *Homo.* Without this specimen, the early *Homo* mean would drop to the extremely low value of 28%. Finally, because of the differences between hominins and the African apes in the bicarotid canal distance, the discrimination that the index yields between these groups would be more dramatic if we were to base the index on that dimension rather than on the biforamen ovale distance.

The small anteroposterior distance between the two pairs of foramina and the shape of the quadrilateral in hominins are primarily the outcome of variation in the

Table 3.28 Angular Relationships within the Cranial Base

Specimen/Sample		Rectangle Proportions (31)/(26) (%)	CC-CC (26)/ FO-FO (25)* (%)	Articular Eminence Angle[b]	Tympanic Angle[a,b]	Petrous Angle[a,b]	Petrotympanic Angle[b]
Australopithecus afarensis	Mean	40	102	79	104	149	140
A.L. 444-2		38	113	79	105	143	144
A.L. 333-45		41	91	n/a	103	154	135
Homo sapiens	Mean (n=10)	36	114	72	107	136	146
	Range	30–42	106–122	65–78	102–114	121–145	128–180
	SD	4	6	5	3	5	21
Gorilla gorilla female	Mean (*n* = 10)	58	98	76	n/a	n/a	115
	Range	52–63	90–106	72–80	n/a	n/a	110–122
	SD	3	6	2	n/a	n/a	4
Gorilla gorilla male	Mean (*n* = 10)	63	103	79	n/a	n/a	118
	Range	57–73	92–113	74–85	n/a	n/a	103–128
	SD	5	7	3	n/a	n/a	7
Gorilla gorilla mixed sex	Mean (*n* = 30)	n/a	n/a	n/a	95	162	n/a
	Range	n/a	n/a	n/a	88–102	150–171	n/a
	SD	n/a	n/a	n/a	3	3	n/a
Pan troglodytes female	Mean (*n* = 10)	46	97	79	n/a	n/a	117
	Range	41–51	90–106	75–86	n/a	n/a	109–126
	SD	3	5	3	n/a	n/a	5
Pan troglodytes male	Mean (*n* = 10)	50	96	80	n/a	n/a	117
	Range	43–59	87–104	70–86	n/a	n/a	102–131
	SD	5	5	4	n/a	n/a	8
Pan troglodytes mixed sex	Mean (*n* = 30)	n/a	n/a	n/a	96	159	n/a
	Range	n/a	n/a	n/a	86-105	150-168	n/a
	SD	n/a	n/a	n/a	5	5	n/a
A. africanus	Mean	47	100	68	104	156	124
Sts. 5		46	100	69	103	155	128
MLD 37/38		47	100	66	93	150	123
Sts. 25		n/a	n/a	n/a	104	162	122
Early *Homo*	Mean	32	106	74	100	143	139
Sts. 19		40	112	73	98	149	129
KNM-ER 1813		25	109	n/a	95	136	143
OH 24		31	98	75	108	144	144
A. boisei	Mean	30	111	64	105	137	146
OH 5		32	103	61	102	135	147
KNM-ER 406		31	110	70	100	134	146
KNM-ER 407		24	129	n/a	112	139	147
KNM-ER 13750		n/a	n/a	64	n/a	n/a	n/a
KNM-ER 23000		32	102	60	102	140	142
A. aethiopicus							
KNM-WT 17000		35	96	72	105	140	146

[a]For extant humans and apes, as well as fossil hominins, data for tympanic and petrous angles are from Dean and Wood (1982). We recalculated them in reference to the sagittal plane (see Figure 3.58).

[b]See Figure 3.58 for explanation.

*Bicarotid canal distance as a percentage of the biforamen ovale distance.

anteroposterior and bilateral positions of the openings of the carotid canals, as mentioned earlier, and are also closely related, as discussed in the following sections.

Inclination of the tympanic element. The long tympanic element in A.L. 444-2 is delineated by the segment that stretches from the inferior margin of the external auditory meatus ("TP" in Dean and Wood, 1982) to the center of the carotid canal opening ("CC" in Dean and Wood, 1982). In A.L. 444-2, the tympanic element is more obliquely oriented (posterolateral to anteromedial) than in the African apes. The tympanic's inclination in A.L. 444-2 is associated with the carotid canals' more anterior location—closer to the coronal plane of foramen ovale and farther from the bitympanic (or biporial) line. In the African apes, the tympanic element is oriented almost perpendicular to the sagittal plane: this means that the carotid foramen is located farther posteriorly from the external auditory meatus than it is in A.L. 444-2, a configuration that produces the relatively large anteroposterior distance between the carotid foramina and the coronal plane of foramen ovale, as seen in Table 3.25. The mean angle is 95° in gorillas and 96° in chimpanzees (see Dean and Wood, 1982, the source of most of the comparative data discussed below[25]), as compared to 105° in A.L. 444-2 and 103° in A.L. 333-45 (Figure 3.58). *A. afarensis* shares with many other hominins a tympanic element that is diagonally oriented to the sagittal plane (see Table 3.28).

Inclination of the petrous element. The anterior position of the carotid foramen in A.L. 444-2, whose center serves as one terminus for the measurement of petrous pyramid length, results in a petrous element that is both shorter from its apex to the foramen and more diagonally oriented (on the coronal plane) than in the African apes (again, the source of our comparative data is Dean and Wood, 1982). In A.L. 444-2, the petrous element forms an angle of 143° with the midline. The petrous is more sagittally oriented in A.L. 333-45, where the angle is 154°, which accords with its more posterior carotid foramina. Although the petrous element is missing in Hadar specimen A.L. 417-1c (the basioccipital/sphenoid fragment associated with the individual's mandible and maxilla), the gap that it leaves in the preserved part of the cranial base suggests a more diagonal orientation, like that of A.L. 444-2. The mean petrous angle of 149° for the Hadar hominins stands between those of the African apes (159° in chimpanzees and 162° in gorillas) and modern humans (136°). *A. africanus*, with a mean petrous angle of 156°, is closer to the African apes, whereas four *A. boisei* specimens (OH 5, KNM-ER 406, KNM-ER 407, and KNM-ER 23000) provide a mean of 137°, which is essentially identical to that of modern humans. Among hominin taxa, the main discrimination seems to be between *A. boisei*, with humanlike, diagonally oriented petrous elements, and *A. africanus*, with apelike, sagittally oriented petrous elements, as Dean and Wood (1982) concluded. *A. afarensis* occupies the middle ground between these extremes.

Figure 3.58 Craniometric landmarks, lines, and angles in the cranial base of A.L. 444-2. Numbered points refer to the following landmarks: 1, porial saddle; 2, lateral tympanic point; 3, articular tubercle; 4, entoglenoid apex; 5, carotid foramen; 6, foramen ovale; 7, basion; 8, opisthion; 9, petrous apex; 10, inion. See text for description and discussion.

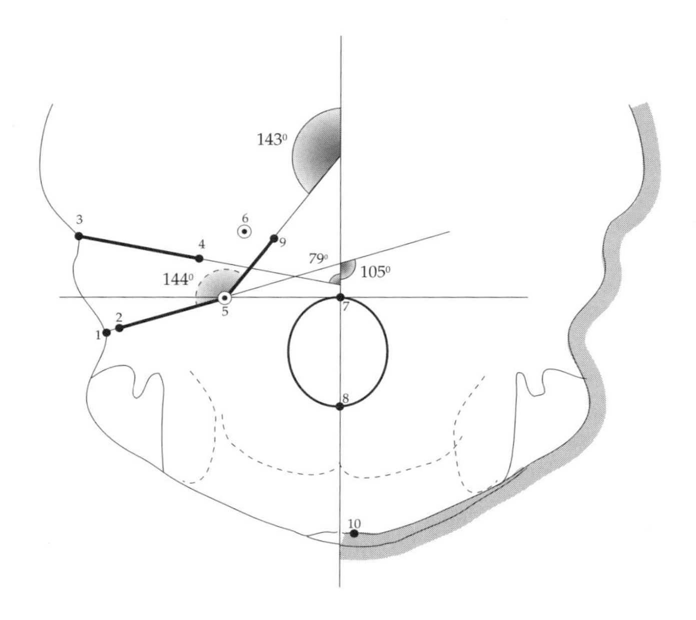

Inclination of the articular eminence. The mediolateral axis of the articular eminence in A.L. 444-2 forms an angle of 79° with the sagittal plane. As such, the angle is not significantly different from that of the extant hominoids in our sample (Table 3.28). Thus, the differences that exist in the topography and morphology of the eminence cannot be attributed to its orientation.

Gorillas, particularly males, exhibit the greatest resemblance to A.L. 444-2 in terms of both the semicylindrical shape of the eminence and its orientation. In females, the angle measures 76°, and in males, it is 79°. A similar orientation is found in chimpanzees (where the angle is 80° in females and 79° in males), although here the morphology is quite different: the eminence is flat and disklike in shape.

The most significantly different angle from that of A.L. 444-2 is seen in *A. boisei*, with 64° calculated as the mean of four specimens (OH 5, KNM-ER 406, KNM-ER 13750, and KNM-ER 23000)—15° less than in A.L. 444-2. As a result of the diagonal orientation of the articular eminence and also of its long mediolateral dimension (particularly in comparison to that of the chimpanzees), the articular tubercle at the lateral end of the eminence in *A. boisei* is carried substantially forward.[26] Thus, in all other hominoids, including A.L. 444-2, the bitubercle line is at the same coronal level as the foramen ovale or posterior to that plane, whereas in *A. boisei*, this line is usually anterior to the coronal plane of foramen ovale. The horizontal distance between the coronal plane of the articular tubercle and that of the entoglenoid process lends the anterior edge of the articular eminence a sigmoid shape in *A. boisei* (see further discussion in Chapter 5). The diagonal orientation of the eminence also contributes to the medial shifting of the most posterior part of the temporal foramen toward the wall of the braincase, as described in the section "The Temporal Foramen and Adjacent Anatomical Structures."

In *A. aethiopicus*, the angle formed by the articular eminence and the sagittal plane is 72°, between the *A. boisei* mean and the A.L. 444-2 value. However, the topographical relief of the *A. aethiopicus* eminence is very flat, and the anatomical detail of its glenoid region differs greatly from that of *A. boisei* and A.L. 444-2 (see Chapter 5). At 68°, the mean inclination of the articular eminence of two *A. africanus* specimens (Sts. 5 and ML 37/38) lies between the Hadar cranium value and the *A. boisei* mean. Although not a single specimen of *A. robustus* permits reliable quantification of the morphology and orientation of the articular eminence, specimens of this species on which the area is sufficiently preserved (TM 1517, SK 46, SK 48, SK 83, and SKW 29) resemble A.L. 444-2 more than they do *A. boisei*.

Inclination of the pterygoid plates (on the sagittal plane). The inclination of the pterygoid plates that bridge the maxillary tubercle behind M^3 and the foramen ovale (adjacent to which lies the root of the plates) is naturally influenced by both the horizontal and vertical distances between them.[27] KNM-WT 17000 is one of the most extreme examples, with its highly angled, triangular pterygoid plates (in a lateral view). In the African apes, the plates are also slanted, whereas in modern humans they are upright, with a more rectangular shape (Clarke, 1977). Reconstruction of the pterygoid plates, based on their original superior and inferior ends, demonstrates that the plates' inclination in A.L. 444-2 falls between the angles exhibited by African apes and modern humans. Because it is correlated with the degree of prognathism, the inclination of the pterygoid plates in *A. africanus* varies according to the degree of prognathism, which itself is rather variable in this species' hypodigm. Hence, the orientation of the pterygoid plates in Sts. 5 is more diagonal than in A.L. 444-2, whereas in the orthognathic Sts. 71, the plates appear to be much steeper than in A.L. 444-2. In Sts. 19, as discussed by Clarke (1977), the orientation of the pterygoid plates is more vertical than is usually encountered in *A. africanus*.

Conclusions regarding the configuration of the cranial base. Comparative examination of the cranial base anatomy of A.L. 444-2 leads to the following conclusions:

- The magnitude of prognathism is not related to a long basioccipital segment, such as seen in the great apes. The co-occurrence of strong prognathism and an anteroposteriorly abbreviated basioccipital, evident in A.L. 444-2, is also evident in the extremely prognathic *A. aethiopicus* cranium, KNM-WT 17000. Although this specimen is by far more prognathic than male gorillas, its basioccipital element is substantially shorter. The contribution of the cranial base to prognathism apparently resides in the anteroposterior length of the segment that lies anterior to the biforamen ovale line and extends from the coronal plane of the spheno-occipital synchondrosis to that of M^3.[28] The major contributor to the short basioccipital segment appears instead to be the anterior migration of the foramen magnum. The crowded appearance of the center of the cranial base in A.L. 444-2 stems primarily from the extremely anterior position of the foramen magnum in relation to the quadrilateral of the four foramina. A similarly anterior position and the resulting crowded effect are observed elsewhere only in some *A. boisei* and *A. robustus* specimens (Figure 3.57).
- Some of the variation in width measurements and in the orientation of certain cranial base elements in the African apes and A.L. 444-2 (along with other hominins) appears to be a direct outcome of the forward crowding of the foramen magnum and the necessary reorganization of the structures that are located anterior and lateral to it. These differences in position and orientation are summarized schematically in Figure 3.59. The anterior shift of the foramen magnum in A.L. 444-2 and other hominins displaces the carotid foramina more anterolaterally than in the African apes. This displacement, in turn, brings about four changes in the petrous and tympanic elements:

    1. A modification of their inclination relative to the sagittal plane
    2. An opening of the angle between them
    3. A decrease in their length
    4. A broadening of the posterior side of the quadrilateral of the four foramina (the distance between the openings of the carotid canals) along with a reduction in its anteroposterior dimension, as the openings of the carotid canals are shifted more anteriorly (particularly in relation to the bitympanic line)

This scheme is supported by an apparent relationship between the position of the foramen magnum along the anteroposterior axis and the juxtaposition of all the cranial base elements described here. Even though they phrase it slightly differently, Dean and Wood (1981) attribute many of the differences between the base of the skull in hominins and in apes to the anterior migration of the foramen magnum:

> However, although this implies that the forward migration of the foramen magnum and rotation of the petrous axis are modifications that have occurred more than once in human evolution, it is by no means certain that the causal mechanisms

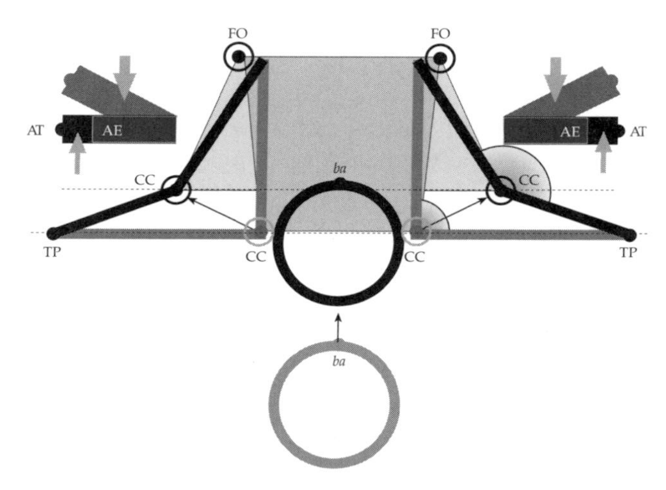

Figure 3.59 Schematic summary of the fundamental differences in the cranial base of hominins and the African apes. Light gray lines: great ape configurations; dark gray and black lines: hominin configurations. The thick gray arrows point to schematic representations of the articular eminence's orientation in *A. boisei*. The thin gray arrow points to a representation (black) of the increase in the length of the articular eminence from chimpanzees to A.L. 444-2. The landmarks are as follows: AT, articular tubercle; AE, articular eminence; TP, lateral tympanic point; CC, carotid foramen; FO, foramen ovale; *ba*: basion. In the hominins the advancement of the foramen magnum to the vicinity of the bicarotid line brings about the lateral displacement of the carotid foramina. This displacement modifies the "quadrilateral" of the four foramina: the tall, narrow, posteriorly tapering shape in apes transforms into the broad, squat, anteriorly tapering shape in the hominids; the right angle between the tympanic and petrous elements in the great apes becomes an obtuse angle in the hominins; and the orientation of the tympanic and petrous elements in the hominins becomes diagonal relative to the sagittal plane.

> are the same in each case. It is quite positive that the rotation of the face beneath the large neurocranium in the *Homo* lineage and the massive jaws and large, but relatively orthognathic, facial structure in the robust australopithecines have been responsible for similar modifications in the basicranium. (Dean and Wood, 1981: 70)

Taking in the whole picture, we may conclude that the relationships and angles in the *center* of the cranial base in A.L. 444-2, a hominin over 3 Myr old, exhibit a virtually modern human pattern. No other element in the skull of A.L. 444-2 comes as close to the anatomy of modern *H. sapiens* in this respect. Noteworthy is the fact that many of these similarities are shared as well with the robust species of *Australopithecus*. Thus, we may not need to invoke homoplasy to explain the similar morphology in the two later taxa. Nevertheless, *A. africanus* exhibits a greater similarity to the African great apes in these relationships, as realized by Dean and Wood (1981, 1982). Again, the angular relationships in the *A. africanus* cranial base conform to the relatively posterior position of its foramen magnum.

"Composite Reconstruction" of 1984/1988 in Light of A.L. 444-2

In the absence of a complete adult skull in the 1970s hypodigm of *A. afarensis*, Kimbel et al. (1984) published a composite skull reconstruction based on 12 adult specimens from Hadar. New insights led to a slightly revised reconstruction a few years later (Kimbel and White, 1988a). The new Hadar skull represents the first opportunity to test the hypothesis about the functional relationships implicit in the composite reconstruction (Kimbel et al., 1984). We provide brief comments here.

The most obvious differences between the composite reconstruction and A.L. 444-2 are in the shape of the cranial vault and the elevation of the face (Figure 3.60). Compared to the composite reconstruction, A.L. 444-2 features a steeper, more convex slope and a higher elevation of the frontal squama. The very flat slope and low elevation of the frontal in the composite reconstruction have been recognized as partly artificial (Kimbel and White, 1988a: 545), which is now confirmed by the new Hadar skull. Rotating the facial skeleton of the composite reconstruction ventrally would lower the snout and introduce a more klinorynch hafting of the face on the brain case, as suggested by A.L. 444-2, and, at the same time, necessitate the steeper frontal squama.

Calvarial height (measured above the FH in the biporion plane) in A.L. 444-2 is 16% greater than in the composite reconstruction, and calvarial length (g–o) is approximately 11% greater. However, calvarial breadth (as gauged by the biauricular breadth) of the two specimens is similar. These comparisons confirm the visual impression (in posterior view) that the calvaria in the composite reconstruction is squatter, with more sloping sides, than that of A.L. 444-2. Indeed, this morphology in the composite reconstruction is represented entirely by the actual morphology of specimen A.L. 333-45 (see Figure 3.49).

Compared to the condition in A.L. 444-2, the mandible corpus depth comprises a much larger percentage of infraorbital facial height in the composite reconstruction, where the index achieves a very high value of 91% (cf. 67% in A.L. 444-2 and 69% in A.L. 417-1; see the section "The Cranium with the Occluded Mandible"). This comparison reinforces the visual impression that A.L. 333-1, constituting the major facial portion of the composite skull, is vertically much shallower than that of A.L. 444-2 (but not unusually so relative to its biorbital breadth; see Table 3.3). As a result of this shallowness, the

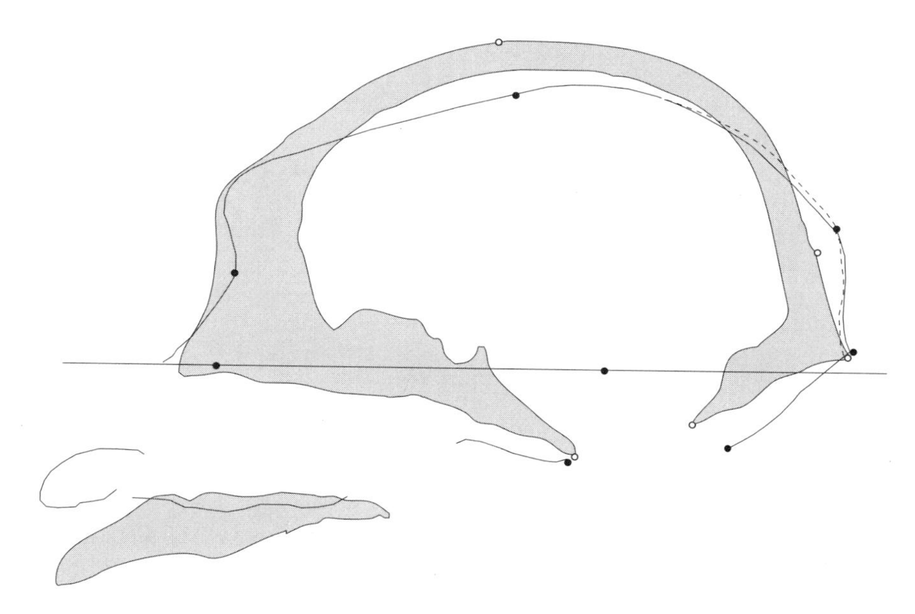

Figure 3.60 Comparison of the midsagittal craniograms of Kimbel and White's 1988 reconstruction of *A. afarensis* and A.L. 444-2 (shaded gray outline).

index reaches its unusually high value. However, we do not believe that this discrepancy is significant, in view of the fact that the reconstruction's mandibular (A.L. 333w-32+60), facial (A.L. 333-1), and palatal (A.L. 200-1a) components come from three different individuals. As a further illustration of the hazards involved with joining skull components of different individuals, we note that the maxillary dental arcade of A.L. 417-1d occludes nearly perfectly with Lucy's mandible, A.L. 288-1i; yet a substantially shallower total facial height is achieved with this artificial combination than when A.L. 417-1d is occluded with its own, much deeper, mandible (A.L. 417-1a)!

The A.L. 444-2 skull shows a less apelike forward projection of the palate (relative to sellion) than does the composite reconstruction. In A.L. 444-2 a little more than half of the palate length projects anterior to sellion (approximately the same amount as in A.L. 417-1d), whereas in the composite reconstruction approximately two-thirds of the palate length does so. The magnitude of this difference would be reduced somewhat if the face of the reconstruction is ventrally rotated, as suggested above. In contrast, relative to the anterior origin of the masseter muscle, the palate projects to the same degree in the composite reconstruction and A.L. 444-2 (approximately 60% in both). This apparent conflict is resolved when it is realized that the anterior masseter origin itself is further forward on the face of the composite reconstruction than it is on A.L. 444-2, as revealed by the masseter position index (108% in the composite reconstruction, compared to 97% in A.L. 444-2). The anterior flare of the zygomatic bone's thick inferior margin in A.L. 333-1—which constitutes the peripheral part of the face in the composite reconstruction—is responsible for this difference; the position of the body of the zygomatic approximates the coronal plane of sellion in both specimens. The functionally significant overlapping index is virtually identical in both the composite reconstruction and A.L. 444-2, with approximately 26% of the distance from the articular eminence to prosthion occupied by the overlap between the origin of the masseter muscle and the palatal length. As discussed with reference to A.L. 444-2, this figure is greater than that observed in the great apes, but it is far less than what is seen in *A. boisei*, in which the overlap occupies as much as 50% of the distance between articular eminence and prosthion.

Pattern of Cranial Cresting

The A.L. 444-2 skull provides an important perspective on the pattern of cranial cresting in *A. afarensis*, as the specimen permits, for the first time, a look at the cresting phenomenon in all parts of a single cranium in this taxon. By necessity, previous analyses of cranial cresting in *A. afarensis* relied on fragmentary specimens that represented chiefly the middle and posterior parts of the calvaria, as almost nothing of the adult frontal bone was known (Kimbel et al., 1984; Kimbel and Rak, 1985; Kimbel, 1988) until the discovery by T. D. White of a small fragment of sagittally crested frontal squama in the 1970s collection from A.L. 333 (A.L. 333-125: Asfaw, 1987; Kimbel and White, 1988a). Some recent discussion has focused on the functional implications of intraspecific variation in hominin cranial cresting patterns (for example, in the species *A boisei*: Brown et al.,1993; Wood et al., 1994). Here, too, the Hadar skull, as well as other recently discovered specimens, make an important contribution.

Description of the Temporal Lines and Crests

The path of the temporal lines is discernible over most of the calvaria of A.L. 444-2. Erosion of the external bone table, in some places severe, makes it difficult to interpret the morphology and exact position of the lines in some locations, but evidence from the whole calvaria facilitates an accurate reconstruction of their course (see Figure 3.31).

The temporal lines are well marked in the supraorbital region, where, as elevated crests, they stand about 1 mm above the superior surface of the supraorbital bars (see the description of the frontal bone, Chapter 5). The temporal crests pass medially and slightly posteriorly along the posterior border of the supraorbital bars, for a preserved distance of 21.8 mm on the right and 26.9 mm on the left. Posteromedial to these points on each side, the path of the temporal lines is interrupted by patches of missing bone. It is evident, however, that the lines had to have followed a tight arc to reach a point 15 mm lateral to the midline in the plane of the postorbital constriction, where they are next observed. In this plane, preservation on the left side permits a brief glimpse of the detailed structure of the temporal line: it is very thin, delicate, double crease, with inferior and superior lines individually discernible less than 1 mm apart. From this point posteriorly on the frontal the paths of the temporal lines are difficult to follow with certainty owing to poor preservation of the external surface; they are detectable intermittently, but often their only signal is the fairly abrupt contour change in coronal section between the interline (temporalis-free) median strip and the inferolaterally sloping facies temporalis of the frontal squama. Nevertheless, it is possible to conclude that the temporal lines are not as strong on the frontal squama as in the supraorbital region, despite their steady convergence on the midsagittal plane as they pass posteriorly, reaching about 8 mm lateral to midline in the plane of maximum frontal breadth (bicoronalia) and 3.5 mm lateral to midline on the coronal suture at bregma. Here the right temporal line suddenly veers laterally and then back toward the midline again just medial to an unusual, marked, localized swelling on the external surface along the parietal bone's coronal margin (see description of parietal bones, Chapter 5). The temporal line's abruptly erratic path and the odd thickening of the external table are presumably correlated with an abnormality in the origin of the right temporalis muscle.

The small biparietal section attached to the frontal immediately posterior to bregma is too eroded to reveal the path of the temporal lines. Behind this segment there is a gap in the bone surface of about 9 mm; then, the temporal lines are again visible, approximately 31 mm posterior to bregma. At this point the temporal lines assume the form of smooth, elevated crests (resembling banister railings), about 4 mm wide and 1 mm high, superimposed on the sagittal margins of the parietals. In fact, the exposed sagittal sutural edges clearly show that the medial surfaces of the temporal crests were in contact in the midline, although they do not appear at this point to have fused completely. It is appropriate to describe the temporal lines at this point as forming a bifid compound sagittal crest. Thus constituted, the compound crest sits in the middle to posterior part of the parietal sagittal arc and occupies about 50% of the arc distance (103 mm). The crest begins to disintegrate into diverging left and right temporal lines beginning about 22.5 mm above lambda. At this point, as for most of the distance along which the compound crest is formed, the lateral edges of the inferior temporal lines are decorated with a string of small tubercles, which represent the termini of a series of light ridges that stream posteromedially across the body of the parietal bones (see description of the left parietal bone, Chapter 5). These ridges presumably signify the attachment of intramuscular septa of the posterior, horizontally oriented fibers of the temporalis muscles, which appear, on this evidence, to have been very well developed.

For much of the temporal lines' bilaterally asymmetric, but steadily diverging course across the posterior part of the parietals and the occipital squama, the inferior and superior lines are easily discernible on each side: the superior lines appear as thin, smooth wrinkles in the external surface, and the inferior lines appear as rugose, irregularly beaded ridges (see Figure 3.61). The right superior line is located about 9 mm lateral to the midline at the level of lambda. Below this level the right pair of lines follows a gently sinusoidal path inferolaterally before sharply swinging laterally, intersecting the nuchal crest 23 mm lateral to the midline to form a posteriorly thickened compound temporal/nuchal (T/N) crest that spans the remaining chord distance to asterion. The compound crest achieves progressively greater salience laterally as it travels in a smooth arc toward asterion. On the left side the temporal lines run a bit crazily below the level of lambda, initially mirroring the inferolateral path of the right lines, then reversing course in a very tight hook along the lambdoidal margin of the parietal immediately inferior to the level of lambda, and finally resuming a normal, diverging path before turning laterally to parallel the vertically undulating profile of the nuchal crest. A compound T/N crest is not formed on the left (and, consequently, the nuchal crest on this side is not thickened as on the right); the closest approach of the superior temporal line and the nuchal crest's inferior edge is 3.8 mm, occurring 39.5 mm lateral to the midline. The superior line traverses the asterionic region immediately above asterion itself and appears to continue as the mastoid crest on the pars mastoidea of the temporal bone. Meanwhile, the inferior temporal line, running on the occipital

Figure 3.61 Detailed anatomy of the asterionic region in A.L. 444-2, showing detail of the nuchal crest and asterionic articulation. (Above) Posterior view shows the low divergence of the temporal lines on the occipital squama and the formation of a compound temporonuchal crest on the right side; and, beneath the blunt-edged supramastoid crest, the narrow, laterally extended impression of the (lost) asterionic ossicle that formed the roof of the asterionic notch. (Center) Posterolateral view shows, above the nuchal crest, the imprint of the now missing posteriorly extended plate of temporal squama. (Below) Posterolateral view shows the broken plate of temporal squama removed to reveal mastoid air cells in the base of the squama above the supramastoid crest.

squama 4.0 mm above its partner and achieving increased prominence laterally, approaches the asterionic region and then abruptly disappears; apparently, the balance of its course was on a posteriorly extended plate of temporal squama that is now lost because of breakage (the broken cross section on the temporal is clearly seen; see description of the temporal bones, Chapter 5).

The Cresting Pattern

Of all currently described early hominin taxa, *A. afarensis* most closely approaches the probable ancestral hominin pattern of cranial cresting, which is the product of differential hypertrophy of the posterior fibers of the temporalis muscle (Rak, 1983; Kimbel et al., 1984; Kimbel and Rak, 1985; Kimbel, 1988). Sagittal crests occur, or are inferred to occur, both in males and, less frequently, in females. The temporal lines are most closely approximated posterior to bregma, and when sagittal crests occur, they are most well developed in the occipitoparietal region. A compound T/N crest is always developed when a sagittal crest is present. The compound T/N crest, strongest laterally, extends across the asterionic region and on to the pars mastoidea of the temporal bone. As a result, the sutural articulation at asterion is frequently in the form of the asterionic notch (Kimbel and Rak, 1985; see further discussion below).

The new Hadar skull further amplifies the primitive aspects of the *A. afarensis* cranial cresting pattern. In the shape of the path taken by the temporal lines on the vault, the posterior location of the most developed portion of the temporal lines or sagittal crest, strong lateral development of the compound T/N crest, the small "bare area" of the occipital bone, and the asterionic notch sutural articulation, A.L. 444-2 reveals a strong emphasis on the posterior (horizontal) fibers of the temporalis muscles (Figure 3.61).

The long, straight but gradually converging course of the temporal lines on the A.L. 444-2 frontal squama, with initial confluence of left and right pairs in the posterior half of the parietal sagittal arc, is a configuration very much allied to that observed in female gorillas and, occasionally, male chimpanzees (Figure 3.62). Although erosion of the external bone table makes interpretation of temporal line morphology uncertain in some locations, there can be little doubt that the temporal lines are most strongly expressed well posterior on the vault, within and posterior to the zone of sagittal crest formation. Even after the sagittal crest disintegrates into individual temporal line pairs just above the lambdoidal suture, the heavily beaded inferior lines are more rugged than they are on the anterior part of the calvaria. Glimpses of the uneroded left temporal line pair in the plane of the postorbital constriction demonstrates the relative weakness of the lines anteriorly.

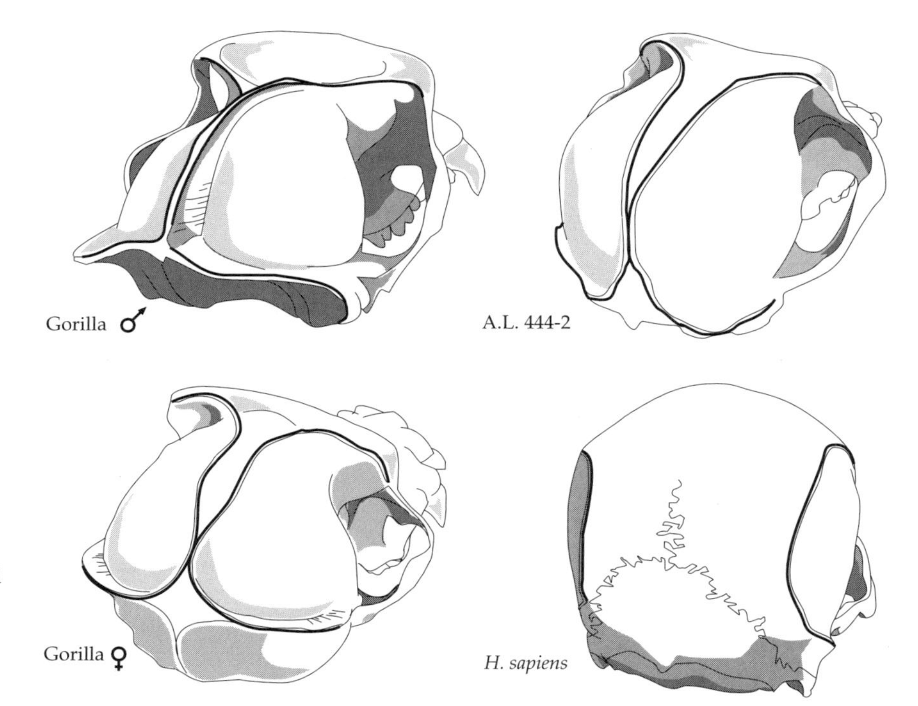

Figure 3.62 Morphology of the sagittal crest in A.L. 444-2 and other hominoids. Specimen A.L. 444-2 exhibits the generalized configuration seen in the apes, in which the temporal lines meet on the posterior slope of the calvaria and posterolaterally participate in a compound temporonuchal crest. In some *A. afarensis* specimens, such as A.L. 333-125, the sagittal crest extends anteriorly onto the frontal squama.

The medial to lateral gradient in the strength of the compound T/N crest in A.L. 444-2 is repeated in a new occipital fragment, A.L. 439-1 (Figure 3.63). In this specimen, the nuchal lines medial to the formation of the compound crests are barely discernible, but once they are joined by the inferior temporal lines, about 20 mm lateral to the midline, they form massive, posteroinferiorly protruding ridges that sweep inferolaterally toward the asteria, causing the entire lateral sections of the calvaria to flare out and dramatically thicken in cross section across the occipitomastoid sutural face. We see this pattern in other Hadar calvariae, such as A.L. 162-28 and A.L. 333-45, too (Figure 3.63).

In comparison to *A. afarensis*, *A. boisei* presents a very different cresting pattern. In OH 5 and KNM-ER 406 the temporal lines are very strongly expressed in the postorbital plane, taking the form of posteriorly protruding crests that overhang the anterior wall of the temporal fossa (see discussion of frontal bone, Chapter 5). Moreover, in every sagittally crested *A. robustus* and *A. boisei* cranium, the crest commences on the frontal squama and then extends across the coronal suture to occupy the anterior half of the parietal sagittal arc. This is also the case in at least some *A. afarensis* individuals, however, as a fused sagittal crest extends across the coronal suture for 33 mm on the Hadar frontal fragment A.L. 333-125. Moreover, as emphasized by Brown et al. (1993), there is variation in the degree to which the sagittal crest extends posteriorly on the vault of *A. boisei*, with specimens such as OH 5 and KNM-CH 304 having crests with greater posterior extension than either KNM-ER 406, KNM-ER 13750, or KNM-ER 23000 has. Thus, the contrast between *A. afarensis* and *A. boisei* may not be absolute, but there is undoubtedly a marked difference in central tendency that most likely relates to species-specific patterns of facial prognathism and the functional role of the different dental arcade segments.

Circumstantial confirmation of this comes from *A. aethiopicus* cranium KNM-WT 17000, which combines extraordinary facial prognathism with a posteriorly hypertrophied sagittal crest that extends nearly to inion. However, in spite of this manifestly generalized hominin morphology in KNM-WT 17000, its cresting pattern departs from the great ape configuration. In the apes the elevated inion coincides with the highest part of the sagittal crest, which lies far posterior on the calvaria (in the posterior part of the occipital sagittal arc), whereas in KNM-WT 17000 inion is low, approximating the FH, and the tallest part of the crest is more anterior, in the middle third of the parietal sagittal arc (see Figures 3.64 and 3.65). Although *A. afarensis* and *A. aethiopicus* approach it most closely, no hominin thus far discovered retains these fundamental elements of the great ape cranial cresting pattern.

Nuchal crest morphology distinguishes *A. boisei* crania from those of *A. afarensis*. This is best appreciated in basal view (Figure 3.53). Typically the most rugose and projecting part of the nuchal crest in *A. boisei* occurs approximately midway between inion and asterion. Here, a prominent protuberance or upwardly directed lip (near

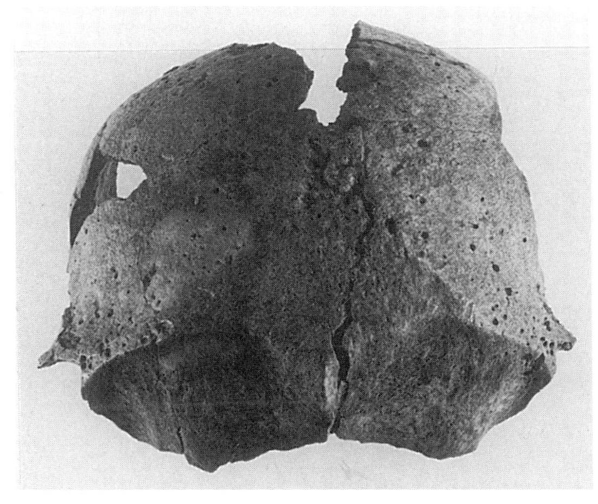

Figure 3.63 Occipital view of *A. afarensis* specimens A.L. 439-1 (*above*) and A.L. 162-28 (*below*). The compound temporonuchal crest increases in prominence laterally; the compound crest must have crossed the asterionic region onto the pars mastoidea of the temporal bone, as is common in the great apes and in *A. afarensis*.

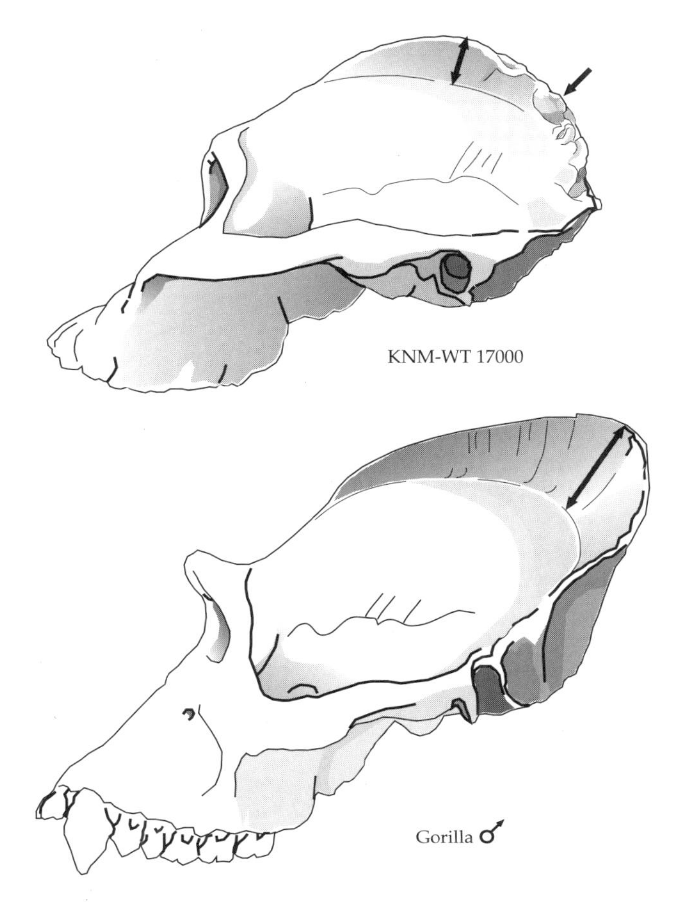

Figure 3.64 Essential differences between the sagittal crest of KNM-WT 17000 and that of a male gorilla. The double-headed arrows indicate the tallest section of the crest. Whereas in KNM-WT 17000 the crest is tallest in its middle section, which lies some distance anterior to a low inion, the tallest part of the crest in the male gorilla is located posteriorly (extending even beyond the back of the braincase) and coincides with an elevated inion. The single-headed arrow indicates the most rugose section of the crest in KNM-WT 17000.

the confluence of the splenius capitis and obliquus capitis superior insertion scars) marks a break in the transverse contour of the nuchal crest, giving the posterior margin of the calvaria a "segmented" appearance. Lateral to this, as seen very well in KNM-ER 406, KNM-ER 13750, OH 5, and probably KGA 10-525 (judging from the posterior and vertical views in Suwa et al., 1997), a saddle is formed between the occipital part of the nuchal crest and the mastoid, the deepest point in the saddle falling at or near asterion (Figure 3.66). In addition to these specimens, Omo 323-1976-896 and KNM-ER 407 show this feature, but with a less pronounced degree of development.

The situation in *A. afarensis* is clearly different. In A.L. 444-2, A.L. 439-1, and A.L. 333-45, the compound T/N crest, gaining prominence laterally, forms a broad, smooth arch that curves onto the mastoid without an appreciable break in its posteriorly convex contour. This morphology is essentially the same as that found in chimpanzees and female gorillas with compound crests developed over the lateral part of the calvaria.

The "Bare Area" of the Occipital Bone

The muscle origin–free "bare area" of the occipital bone (Dart, 1948) is defined by Tobias (1967: 21) as "that part of the bare area of the skull occupying the squama occipitalis and lying between the superior nuchal lines and the lambdoid suture"; the smaller the bare area, calculated as the area of a triangle with lambda at the apex, the more extensive is the temporalis muscle coverage of the occipital squama. The bare area of the occipital bone in A.L. 444-2 is the largest in the *A. afarensis* sample:

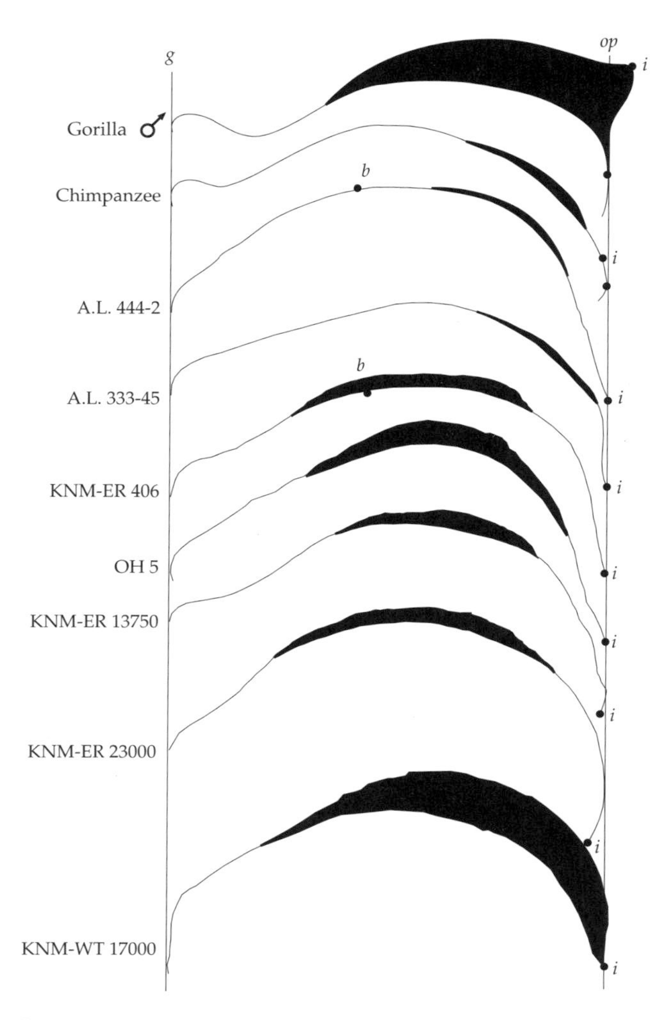

Figure 3.65 The size and position of the sagittal crest in various hominoids. The horizontal distance between the two vertical lines represents the distance between glabella and opisthocranion.

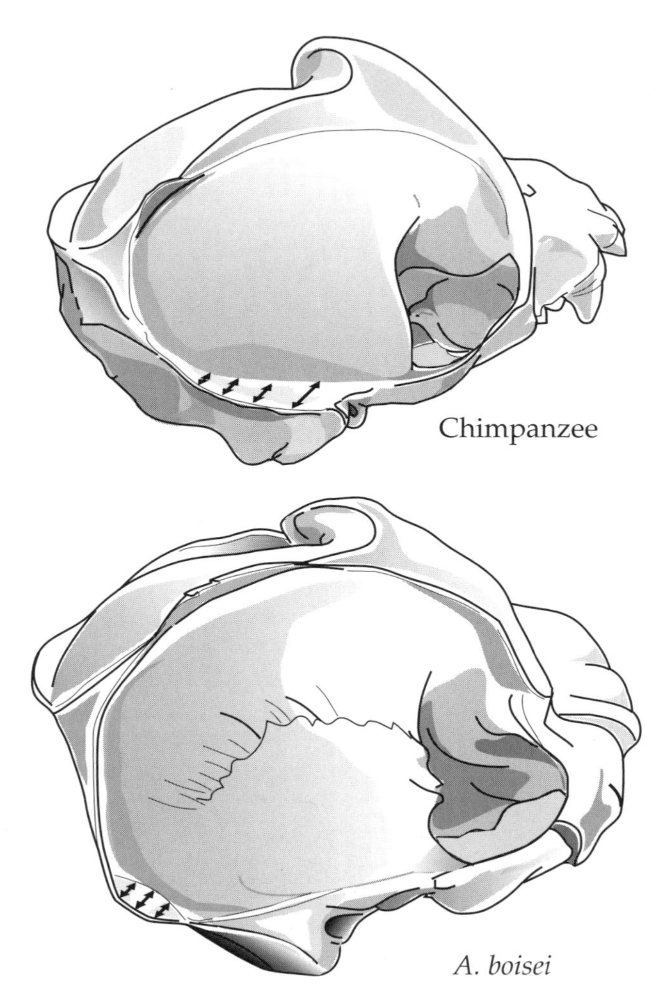

Figure 3.66 Sagittal and nuchal crests in the chimpanzee and *A. boisei*. In the chimpanzee, the sagittal crest is located on the posterior slope of the braincase, and in *A. boisei*, at the summit. The compound temporonuchal crest in *A. boisei* is confined to the middle third of the nuchal line, where its protrusion is greatest, and then diminishes laterally. In chimpanzees, the compound crest forms on the lateral third of the nuchal line and intensifies laterally, crossing onto the pars mastoidea of the temporal bone. The latter pattern is found in *A. afarensis*.

SPECIMEN	BARE AREA	SEX ASSIGNED
A.L. 444-2	551 mm^2	Male
A.L. 162-28	500 mm^2	Female
A.L. 439-1	380 mm^2	Male
A.L. 333-45	361 mm^2	Male
KNM-ER 2602	333 mm^2	Female

Alongside these data we provide our judgment of sex assignment based on the overall level of cranial size and robusticity (see Kimbel et al., 1984; Kimbel and Rak, 1985; Kimbel, 1988); there is no obvious intraspecific patterning in the absolute degree of posterior temporalis coverage in *A. afarensis*. However, as cranial vault size varies rather widely in this sample, it is useful to see the effect of standardizing the bare area by a size proxy, in this case by dividing the bare area by the biasterion chord:

SPECIMEN	ASTERION–ASTERION	RELATIVE BARE AREA SIZE	SEX ASSIGNED
A.L. 162-28	79	6.33	Female
A.L. 444-2	103	5.35	Male
KNM-ER 2602	76 ± 2	4.38	Female
A.L. 333-45	95 ± 2	3.80	Male
A.L. 439-1	100	3.80	Male

When standardized in this fashion, the smaller *A. afarensis* crania, presumptive females, tend to have a relatively large bare area (i.e., relatively less temporalis muscle coverage of the posterior cranial vault), with the notable exception of A.L. 444-2. The small, uncrested (but fragmentary and unmeasurable) occipital A.L. 288-1a, which visually demonstrates a large bare area of the occipital, further substantiates the separation in the *A. afarensis* sample of the male and female specimens. This is in keeping with expectation based on the well-documented pattern of cranial cresting dimorphism in the extant great apes (Ashton and Zuckerman, 1956; Robinson, 1958).

In the two specimens with the largest relative bare areas (A.L. 162-28 and A.L. 444-2), the temporal lines diverge from the midline superior to the lambdoidal suture, that is, on the parietal bones, and the "height" of the bare area triangle is identical to the height of the planum occipitale (la–i chord). This is also the case on all three specimens that comprise the measurable *A. boisei* sample, despite the fact that the planum occipitale tends to be *higher* than in *A. afarensis* (see discussion in Chapter 5). In both absolute and relative terms, a much larger bare area of the occipital characterizes the measurable *A. boisei* sample than the *A. afarensis* sample:

SPECIMEN	BARE AREA	ASTERION–ASTERION	RELATIVE BARE AREA SIZE
KNM-ER 406	1015 mm^2	93	10.91
OH 5	735.5 mm^2	89	8.26
Omo 323-1976-896	782 mm^2	95	8.23

The fact that the measurable *A. boisei* sample is composed solely of male individuals, which are expected to have more extensive temporalis muscle coverage of the vault than females, further highlights the distinction between this species and *A. afarensis*. On the female *A. boisei* cranium KNM-ER 407, for example, the temporal lines do not even cross the lambdoidal suture, and the bare area of the occipital is calculated simply as the area of the planum occipitale itself. (The Chesowanja *A. boisei* occipital fragment KNM-CH 304, described as having simple temporal crests that descend the posterior calvaria close to the midline on the occipital squama and a "small bare area between inion and the point where the parasagittal crests diverge" [Gowlett et al., 1981: 127], may be an exception to this generalization.) It thus seems clear that *A. afarensis* features much greater coverage of the posterior calvaria by the temporalis muscle, irrespective of calvarial size, than *A. boisei*. Undoubtedly this is a manifestation of a more horizontal (posteriorly oriented) temporalis muscle resultant in *A. afarensis* (Kimbel and Rak, 1985). Only *A. aethiopicus*, represented by the single cranium KNM-WT 17000, has a smaller bare area of the occipital (both absolutely and relatively) than *A. afarensis*, which is consistent with other evidence for the highly hypertrophied posterior temporalis fibers that originated from the surface of a small brain case (see above).

Intraspecific Distribution of Cranial Crests

The Hadar skull A.L. 444-2 and other new calvarial specimens from Hadar provide further evidence for a great apelike pattern of intraspecific variation in the distribution of cranial crests in *A. afarensis*. Eleven specimens permit the assessment of cranial crests; on six of these the status of both the sagittal crest and the nuchal crest can be ascertained.

A sagittal crest, or at least the intimate approximation of the temporal lines in the sagittal midline, characterizes seven of eight specimens in which the feature can be detected on some part of the calvaria. Only on A.L. 288-1, a female, is there a clear gap between the left and right temporal lines over the vault's entire midsagittal span. Two other specimens judged to be female on metrical grounds (A.L. 162-28 and KNM-ER 2602; see Kimbel, 1988) do, in fact, possess a sagittal crest. On a fourth presumptive female, the left parietal fragment of A.L. 701-1, the left pair of temporal lines are adjacent to the midsagittal plane at bregma, 20 mm posterior to which they veer toward the sagittal suture, leaving the inferior line immediately lateral and parallel to the midline at the posterior break 47 mm posterior to bregma. No sagittal crest is formed along this distance, but even in this small fragment there is strong evidence of an extensive temporalis origin area on a small, probably female cranium.

Of the five larger specimens judged as male, a sagittal crest is present in three: A.L. 333-125, A.L. 444-2, and A.L. 457-2. A fragment of frontal squama with the coronal margin (Asfaw, 1987), A.L. 333-125 is the sole specimen with a sagittally crested frontal (only A.L. 288-1 and A.L. 444-2 can be compared directly), although present at the bregmatic angle of the A.L. 457-2 parietal fragment is the right half of a low, unfused crest that may certainly have extended onto the frontal bone. We can infer the formation of a sagittal crest in two additional specimens.

The partial calvaria A.L. 333-45, a young adult judging from the open state of the vault sutures, features intimately approximated temporal line pairs that straddle the sagittal margins of the parietal bones and, in parallel, extend onto the planum occipitale of the occipital bone. Even a very conservative projection of growth leads to the conclusion that the temporal lines would have developed into a compound sagittal crest with attainment of full adulthood.

On A.L. 439-1, a fragment of a large occipital bone with an unfused lambdoidal suture, the superior temporal lines merge in a thin median crest ca. 17 mm below lambda on the planum occipitale; the inferior temporal lines, following the suture's path immediately adjacent to the superior lines, are virtually on the midline at the lambdoidal angle of the squama. In the great apes, union of the superior temporal lines is always a developmental prerequisite to the formation of a compound sagittal crest, although in some male chimpanzees and female gorillas the superior lines alone may form a low sagittal crest. This morphology in A.L. 439-1 matches that of the crested female KNM-ER 2602: the superior temporal lines coalesce on the planum occipitale, above which, on the parietals, the inferior lines contribute to the formation of a much stronger compound crest.

Six of eight adult *A. afarensis* specimens that provide evidence of compound T/N crests show this structure at least unilaterally. Whereas all four specimens judged to be male have compound T/N crests (A.L. 444-2, A.L. 439-1, A.L. 333-84, A.L. 333-45; asterionic morphology indicates the presence of a compound crest on the calvaria from which the A.L. 333-84 temporal fragment derived [Kimbel and Rak, 1985: 41–42]), two females show these crests unilaterally (A.L. 162-28 and KNM-ER 2602), and two other females lack them altogether (A.L. 224-9 and A.L. 288-1).

Although compound T/N crests are ubiquitous in both sexes of all great ape species except the bonobo, *A. afarensis* remains the only hominin species to exhibit these crests in crania of both sexes. Moreover, with each instance of a sagittal crest in *A. afarensis*, a compound T/N crest is also present. This is the pattern described for the great apes 40 years ago by Ashton and Zuckerman (1956), which Robinson (1958) showed conclusively did *not* apply to cranially crested *Australopithecus* specimens known to him because, inter alia, although sagittal crests were common in *A. robustus*, compound T/N crests did not occur in this taxon. With respect to the southern African early hominins, Robinson's conclusion remains true (a crushed cranium from Swartkrans, SKW 29—described by Grine and Strait, 1994—shows a sagittal crest in the vicinity of the coronal suture, but a compound T/N crest is demonstrably absent.)

Cranial crest distribution in *A. boisei* underscores the great apelike crest distribution pattern in *A. afarensis.* Six of seven *A. boisei* males possess sagittal crests, but only three of these also have compound T/N crests, whereas neither of the two female specimens has sagittal or compound T/N crests.

Until the discovery of Stw. 505, a large male cranium of *A. africanus*, no *A. africanus* skull preserved a sagittal crest (although its occasional presence in the taxon was predicted by, e.g., Wolpoff, 1974). As documented by Stw. 505, the sagittal crest in *A. africanus* is low and posteriorly placed along the frontal–occipital arc (Lockwood and Tobias, 1999). This condition is similar to *A. afarensis* rather than to *A. robustus* or *A. boisei*, but the absence on the Sterkfontein cranium of a compound T/N crest is a clear departure from the morphology of sagittally crested *A. afarensis* crania, all of which evince confluent sagittal and nuchal crests. No cranium of *A. africanus* yet discovered exhibits a compound T/N crest.

Robinson (1958) ascribed the unique pattern of cresting in *Australopithecus* to the remodeling of the hominin occipital bone with the acquisition of upright posture, but, given the marked differences between *A. afarensis* and other early hominin bipeds, a more appropriate explanation relates it to the specialized aspects of the *Australopithecus* masticatory apparatus (e.g., Tobias, 1967; Jolly, 1970; Wolpoff, 1974; White et al., 1981; Rak, 1983; Kimbel et al., 1984; Kimbel and Rak, 1985).

Sutural Articulation at Asterion

In *A. afarensis*, as in many extant catarrhine primates, the ontogenetic extension of the compound T/N crest across the asterionic region to the pars mastoidea of the temporal bone results in a complex set of growth-mediated spatial adjustments in the articulation of the parietal, occipital, and temporal bones and a characteristic adult morphology called the "asterionic notch" (Kimbel and Rak, 1985). The asterionic notch configuration has been related through ontogenetic analysis to the hypertrophy of the posterior fibers of the temporalis muscle, which develop to a greater extent and more rapidly than the anterior fibers during growth (for details, see Kimbel and Rak, 1985). The A.L. 444-2 skull preserves the asterionic notch in great detail; its association with a posterior sagittal crest and an extensive compound T/N crest further corroborates the causal link to the differential development of the posterior temporalis fibers.

On the right side of A.L. 444-2 a deep, transversely elongate, diamond-shape depression traverses the asterionic region between the pars mastoidea and the squama of the temporal bone (Figure 3.61). The deeper, lateral, part of the imprint, the morphogenetically transformed parietal notch (incisura parietalis), originally lodged what we describe (see Chapter 5) as a separately ossified "parietal notch bone," now lost, corresponding in position and form to the parietal bone's mastoid angle. Comparison of A.L. 444-2 with other *A. afarensis* temporal bones evincing the asterionic notch morphology (A.L. 162-28, A.L. 333-45, A.L. 333-84) confirms the homology of the missing ossicle of the new skull with the laterally flared mastoid angle of the parietal bone. The roof of the triangular asterionic notch is the parietal notch ossicle; its medial wall is the occipitomastoid margin of the occipital bone; together they embrace the pars mastoidea of the temporal bone.

In A.L. 444-2 the shallower, medial, part of the diamond-shape imprint marks the extension of the parietal notch ossicle across the parietomastoid suture to overlap the lateral extremity of the junction between the parietal and the occipital bones (Figure 3.61). The lower margin of this medial part is the superior lip of the compound T/N crest. Above it is a second, somewhat smaller indentation for a posteriorly extended spur of temporal squama, now broken and lost, that apparently overlapped the posterolateral corner of the parietal bone. A break above the supramastoid crest reveals the squamosal suture's pathway on the parietal bone and its continuity with this second feature (Figure 3.61). Similar traces of squamous temporal overlap of the parietal bone, widely observed in the crania of extant great apes, is seen on the laterally flared mastoid angle of the left parietal bone in Hadar calvaria A.L. 162-28 (Kimbel and Rak, 1985).

Whereas most hominin species retain the edge-to-edge articulation of the temporal, parietal, and occipital bones at asterion, which is common to all hominoids at birth, adult males of *A. boisei* develop a unique pattern of temporal bone overlap of the parietal bone and parietal bone overlap of the pars mastoidea and the occipital squama, which is hypothesized to reinforce a very thin vault against the powerful contractions of the masticatory muscles (Kimbel and Rak, 1985; see the recently discovered exception in the thick-walled calvaria of KNM-ER 23000 [Brown et al., 1993: 155]). However, two cases unrelated to *A. afarensis*—the KNM-WT 17000 cranium of *A. aethiopicus* and the *H. habilis* specimen KNM-ER 1805—both evince a well-developed asterionic notch sutural articulation (Kimbel and Rak, 1985; R. Leakey and Walker, 1988). In these specimens the asterionic notch is consistent with other evidence for hypertrophy of the posterior fibers of the temporalis muscles, including posterior sagittal crests and massive compound T/N crests that traverse the asterionic region. They independently verify the functional influence of the masticatory apparatus on the morphology of the asterionic region.

Pattern of the Venous Sinuses

Evidence from the endocranial aspect of the occipital bone and the endocranial cast demonstrates that in A.L. 444-2 the occipital-marginal venous sinus system is dominant but is asymmetrically developed. As midline bone on the endocranial aspect of the parietal bones is destroyed, the status of the sulcus for the superior sagittal sinus cannot be assessed. On the occipital, however, the sulcus is barely detectable; often in *A. afarensis* this sulcus is imperceptible (A.L. 162-28, A.L. 288-1, A.L. 333-45, and KNM-ER 2602) or negligibly developed (A.L. 224-9 and A.L. 439-1). There is a very weak, but palpable, depression immediately above the internal occipital protuberance, and this can be traced over the protuberance into a broad, shallow sulcus for the occipital–left marginal sinus system. Along the thickened left inner rim of the foramen magnum, a well-defined marginal sulcus achieves a width of 4 mm before breakage terminates its pathway to the jugular notch. The right transverse limb of the cruciate eminence bears a relatively weak, 2.5-mm wide groove for the transverse sinus that begins abruptly in the middle of the limb, runs toward the internal occipital protuberance, then, shy of the center of the protuberance, hooks sharply downward to flow into a relatively ill-defined, 7.5-mm wide sulcus for the right marginal sinus. The left transverse limb of the cruciate eminence lacks markings for venous sinuses.

Neither sigmoid sulcus is well developed. Apparently the sigmoid sinuses received no contributions from the transverse sinuses; instead, a deep groove for the right petrosquamous sinus and negligible ones for both superior petrosal sinuses suggest that, compared to the marginals, the sigmoid sinuses carried relatively little venous blood. The fact that the right transverse sinus makes its first appearance in the middle of the transverse limb of the cruciate eminence suggests that its more lateral part ran within the fold of the tentorium cerebelli rather than on the endocranial bone surface.

The pattern of venous sinus drainage described for A.L. 444-2—the dominant occipital-marginal venous tract—is essentially that previously documented for six of seven *A. afarensis* individuals (A.L. 162-28, A.L. 288-1, A.L. 333-45, A.L. 333-105, A.L. 333-114, A.L. 333-116, but not LH 21; Kimbel, 1984). Three other new Hadar specimens shed additional light on the state of this anatomical system in *A. afarensis*. The most complete of these is the adult occipital squama fragment, A.L. 439-1. Here, a mild, 5.0-mm wide sulcus for the superior sagittal sinus crosses the internal occipital protuberance to continue on the depressed, fan-shaped inferior limb of the cruciate eminence as the sulcus for the occipital sinus. The sulcus for the occipital sinus coexists with a 3.3-mm wide, sharply delimited sulcus for the right transverse sinus, and a softer, narrower (2.5 mm) sulcus for the left transverse sinus. The right sulcus commences, as in A.L. 444-2, in the middle of the transverse limb of the cruciate eminence (the specimen is broken short on the left side), and both travel toward the internal occipital protuberance, where the right transverse sinus apparently merged with the confluence of sinuses while the left swept inferiorly and flowed into the occipital sinus. Although the foramen magnum's margins are not preserved, the marginal sinuses appear to have been paired, judging by the faint suggestion of a low ridge dividing the depression that lodged the occipital sinus into left and right compartments. Nevertheless, there is no doubt that A.L. 439-1 possessed a fully developed occipital-marginal drainage system.

Two small occipital fragments, A.L. 224-9 and A.L. 427-1b, show well-developed transverse sinus sulci. On

A.L. 427-1b, all that can be said is that the right transverse sinus (6.5 mm wide medially, narrowing laterally to about 4.0), extended across the squama to the occipitomastoid margin, presumably to flow into the sigmoid sinus lodged on the temporal bone. A well-developed, 3.5-mm wide transverse sinus sulcus travels across the entire left half of the squama on A.L. 224-9, which, in addition, preserves the superior and inferior limbs of the cruciate eminence. The inferior limb of the cruciate eminence, which is intact from the internal occipital protuberance inferiorly to opisthion, lacks any trace of a sulcus for the occipital sinus, and the chipped right margin of the foramen magnum shows no evidence of a marginal sinus sulcus. The superior limb of the eminence presents a weak sulcus for the superior sagittal sinus that deviates to the right of the internal occipital protuberance to continue either as the sulcus for the right transverse sinus or that for the occipital sinus. Unfortunately, preservation of the right side of the squama does not permit a decision on the destination of the superior sagittal sinus, but if it was into the occipital sinus, the sulcus for the latter would be placed well to the right of the cruciate eminence's inferior limb. The evidence does not discount the presence of an occipital-marginal sinus system on specimens A.L. 224-9 and A.L. 427-1b, but both of them demonstrate that if such a venous system was present, it coexisted with a fully developed transverse sinus system, at least unilaterally.

In sum, the new sample of Hadar crania reinforces previous studies that show a very high frequency of occipital-marginal venous sinus drainage systems in *A. afarensis* (Falk and Conroy, 1983; Kimbel, 1984). Taking into account the total sample now available, the frequency of these systems in the species is eight of nine, or 89% (see below). Of the four specimens in which the status of the sinus system can be studied bilaterally (A.L. 162-28, A.L. 333-45, A.L. 439-1, A.L. 444-2), three exhibit evidence of paired marginal sinuses, whereas one (A.L. 333-45) has a marginal sinus on one side only. The new Hadar sample includes two specimens documenting well-developed transverse-sigmoid venous drainage routes; whether these coexisted with occipital-marginal drainage systems cannot, on present evidence, be determined.

It is well established that a high frequency of dominant occipital-marginal venous drainage systems is shared by *A. afarensis*, *A. robustus*, and *A. boisei*, although functional and phylogenetic interpretations of this similarity have differed (Falk and Conroy, 1983; Kimbel, 1984; Falk, 1990; Tobias and Symons, 1992). In *A. africanus* the usual, though not universal, pattern is for the transverse-sigmoid sinus system to carry the largest volume of blood from the endocranial cavity—which is the predominant pattern in humans and great apes. However, at least two of seven (ca. 29%) individuals attributable to *A. africanus* exhibit evidence of enlarged occipital-marginal sinus systems: the Taung juvenile (Tobias and Falk, 1988) and Stw. 187a, a basioccipital fragment that has taxonomically diagnostic features on its external surface (see Dean, 1985) and bears the distal segment of a distinct right marginal sinus sulcus endocranially (relevant bone surface is not preserved on the left side). This percentage, although based on a small sample, approximates the highest frequency encountered in Kimbel's (1984) survey of venous drainage patterns in modern human skeletal populations. In any case, it remains true that *A. africanus* evinces the occipital-marginal sinus system less commonly than does either *A. afarensis* or the robust *Australopithecus* species, where it is present almost without exception (although a transverse-sigmoid drainage route has been inferred for the Konso *A. boisei* cranium KGA 10-525; Suwa et al., 1997). The one known exception is LH 21, the child's partial skeleton from Laetoli, which is part of the *A. afarensis* hypodigm, yet whose occipital fragments shows sulci for a well-developed transverse-sigmoid drainage system bilaterally with no evidence of the alternate occipital-marginal route (Kimbel, 1984). To Falk et al. (1995) this implies that different hominin species were present at Hadar and Laetoli, a subject to which we will return in the last chapter of this monograph.

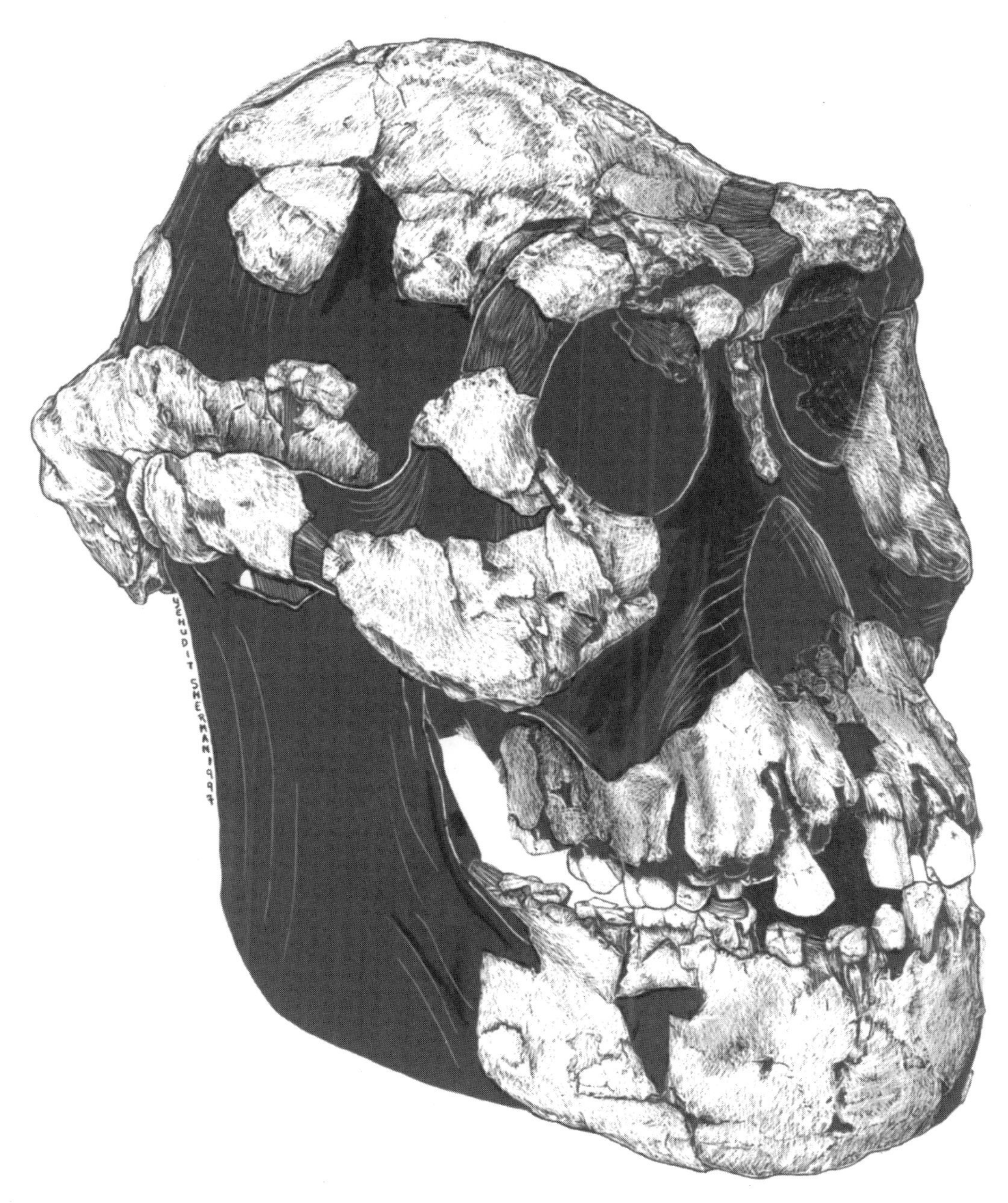

Figure 3.1 Artist's rendering of the reconstructed skull of A.L. 444–2, anterolateral view (75% natural size).

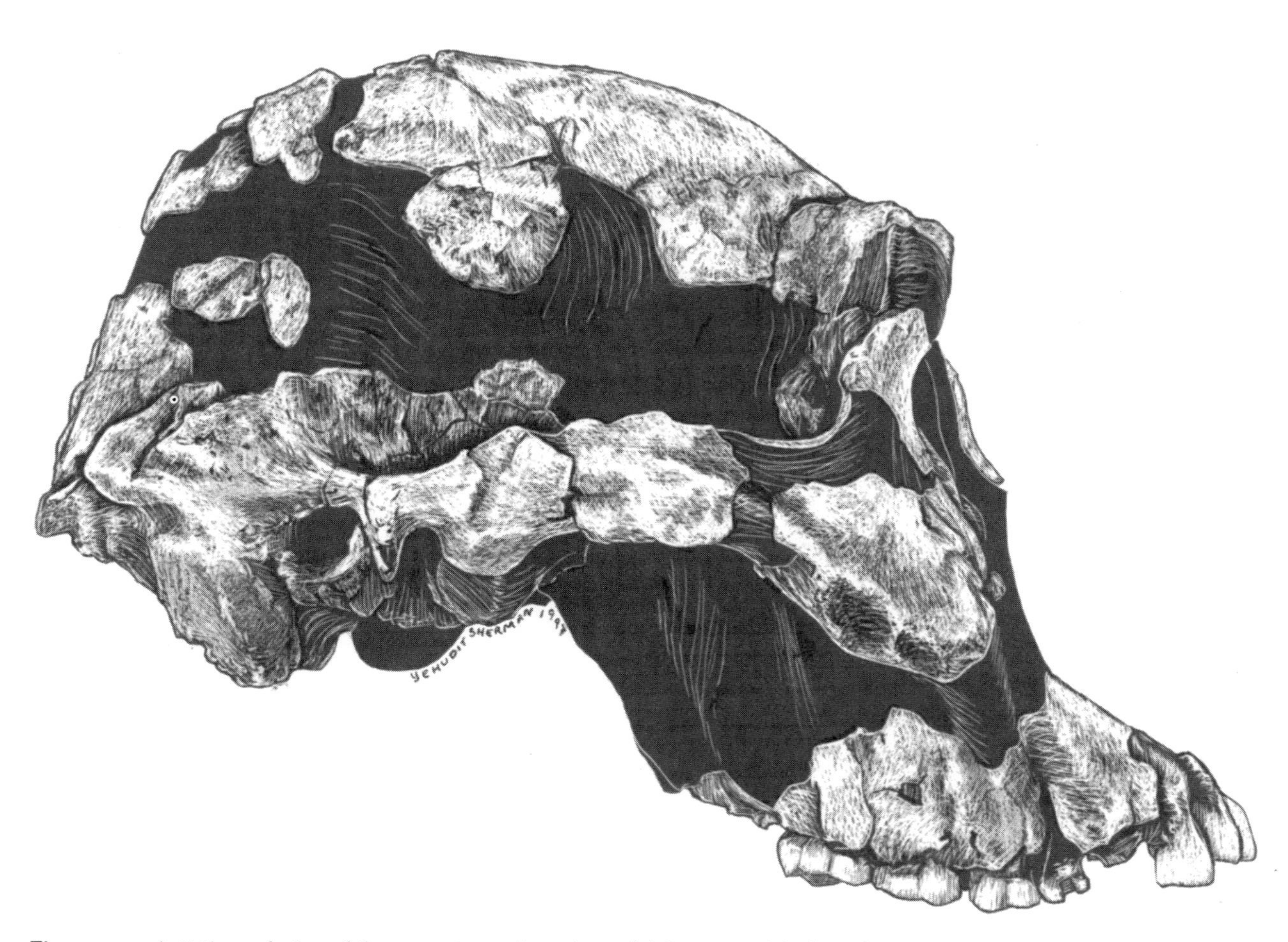

Figure 3.5 Artist's rendering of the reconstructed cranium of A.L. 444–2, right lateral view (75% natural size).

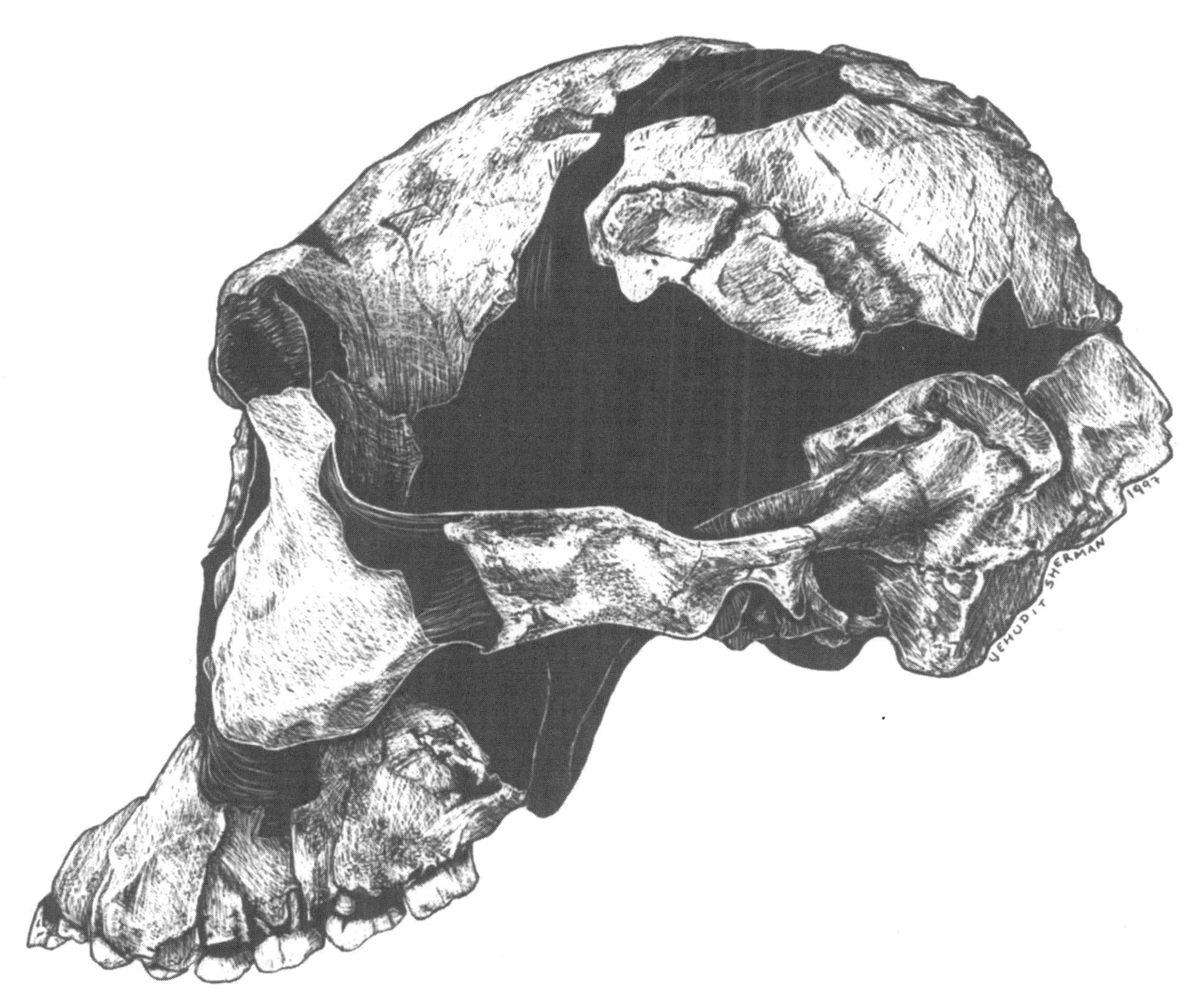

Figure 3.6 Artist's rendering of reconstructed cranium of A.L. 444–2, left lateral view (75% natural size).

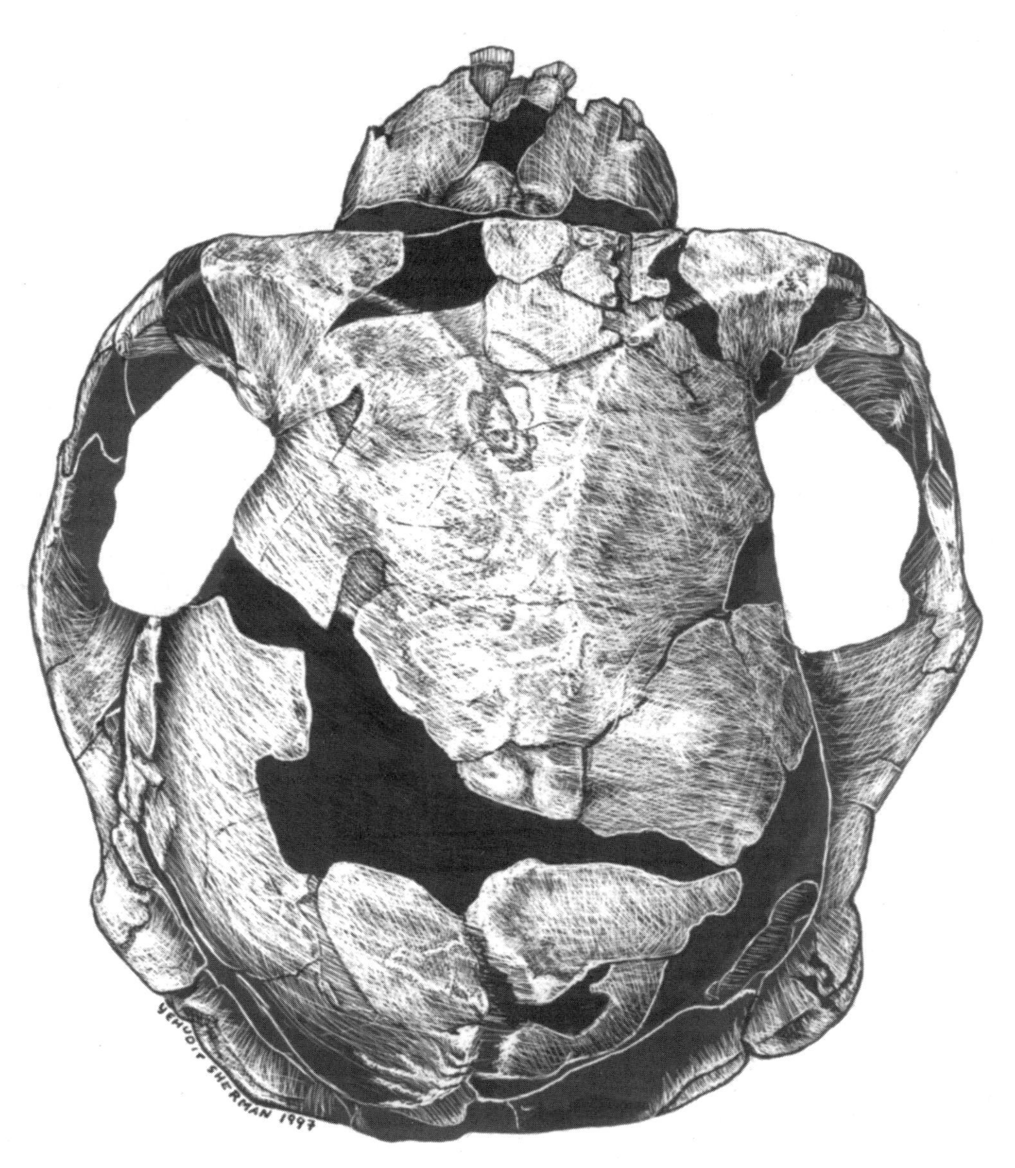

Figure 3.31 Artist's rendering of the superior view of the A.L. 444–2 reconstruction (75% natural size).

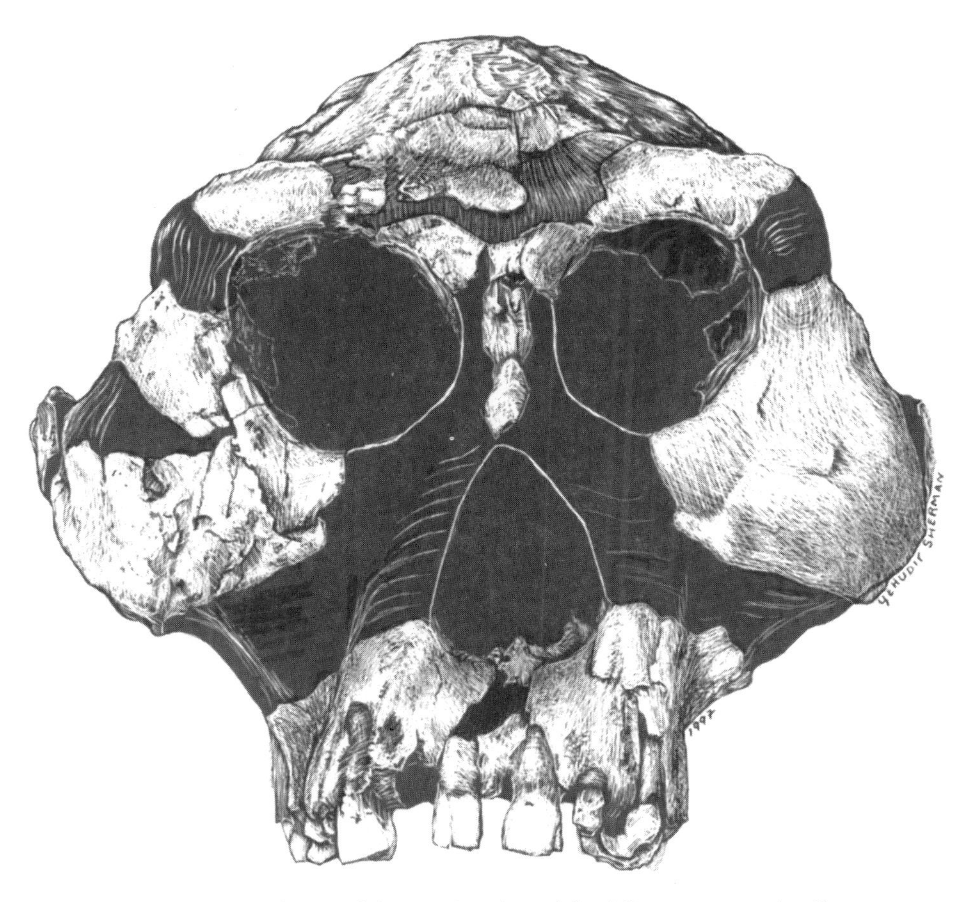

Figure 3.36 Artist's rendering of the anterior view of the A.L. 444–2 reconstruction (75% natural size).

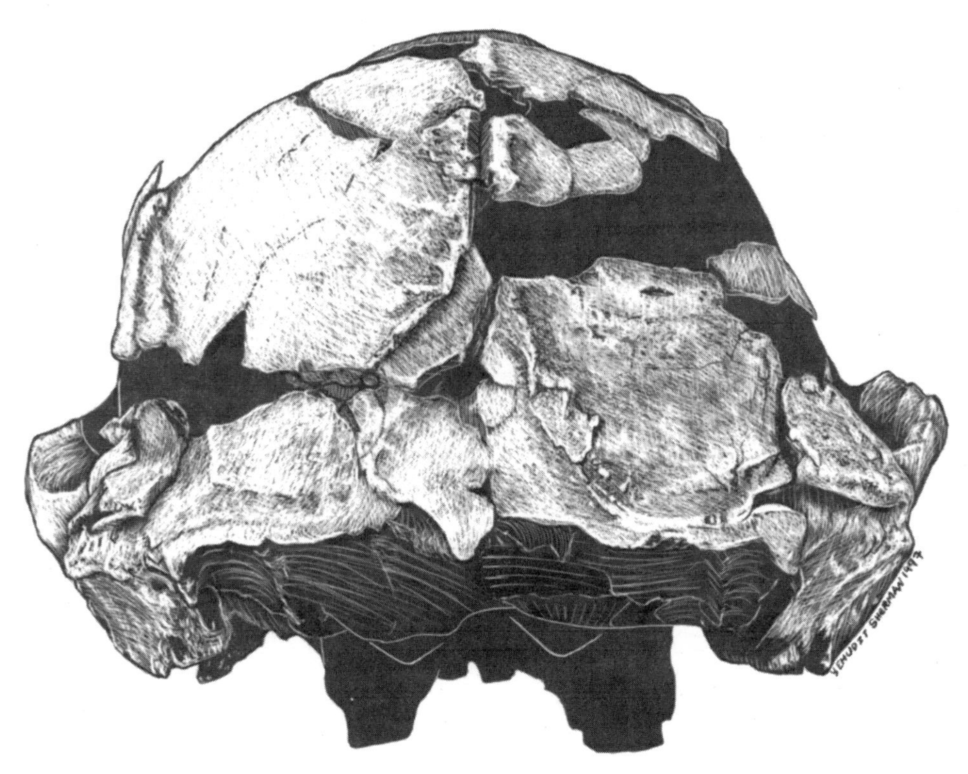

Figure 3.47 Artist's rendering of the posterior view of the A.L. 444–2 reconstruction (75% natural size).

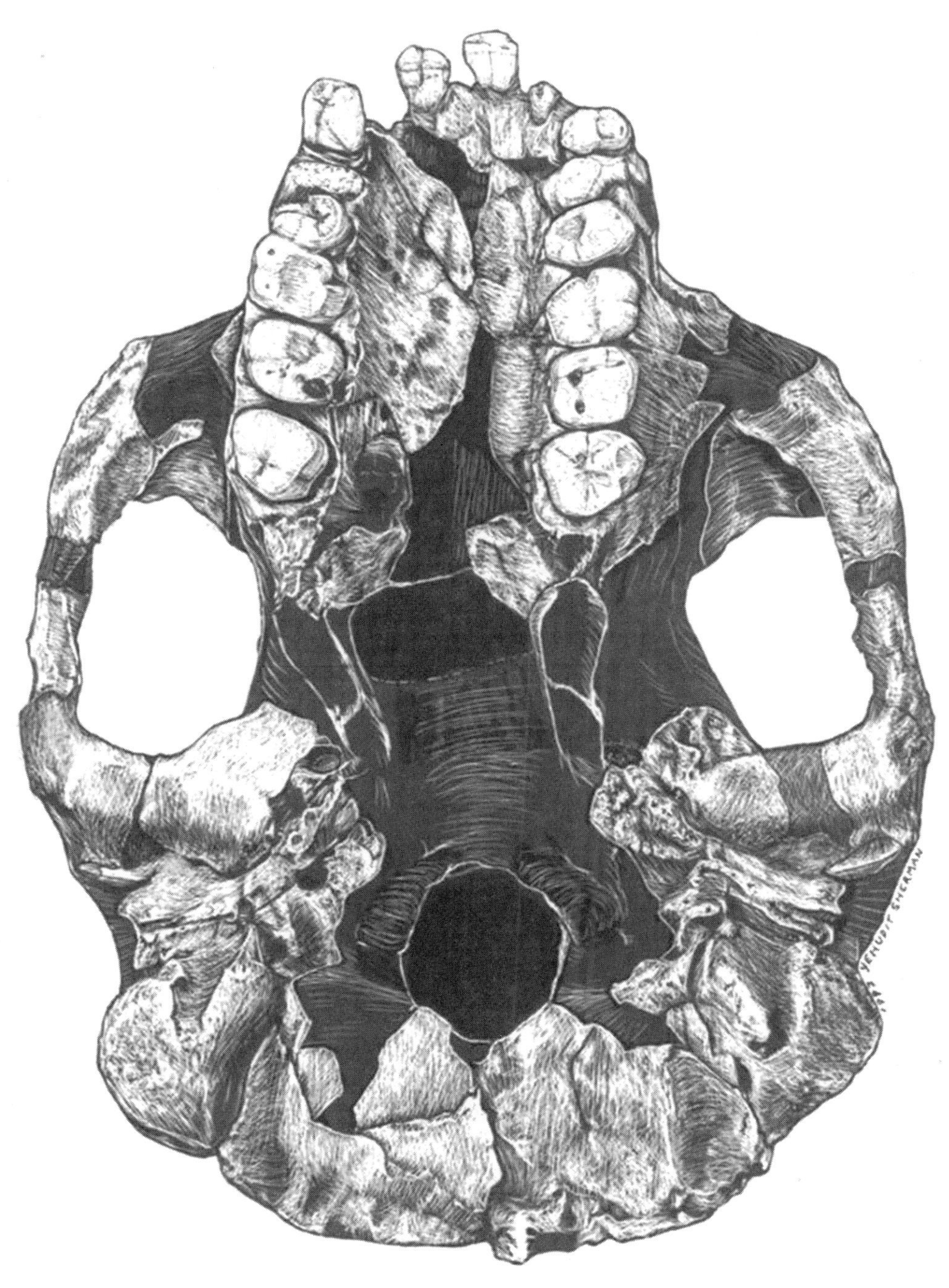

Figure 3.50 Artist's rendering of the basal view of A.L. 444–2 reconstruction (75% natural size).

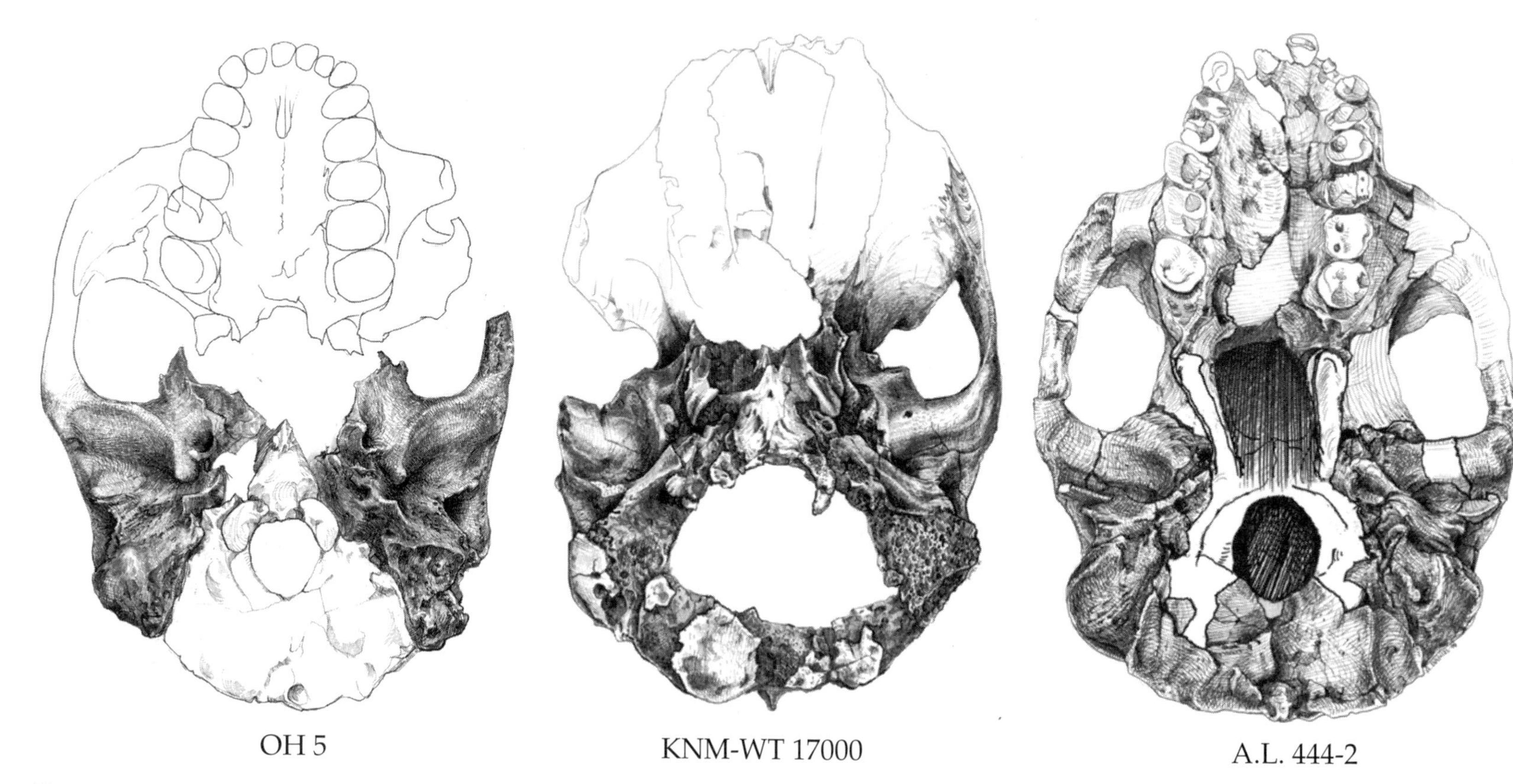

Figure 5.15 The anatomy of the glenoid region in *A. boisei* (OH 5), *A. aethiopicus* (KNM-WT 17000) and *A. afarensis* (A.L. 444–2) (45% natural size).

4

Endocranial Morphology of A.L. 444-2

Ralph L. Holloway and Michael S. Yuan

The original endocast of A.L. 444-2 consisted of a single plastic cast, colored to show the original fragments (light brown) and the reconstructed missing parts (black) (Figure 4.1). This we label the Rak-Kimbel endocast, which was based on the reconstruction of cranial and facial fragments. Because distortion was severe enough to interfere with morphological description and measurements, and especially the assessment of endocranial capacity, a plaster endocast was received from Yoel Rak in 1998 for purposes of modification. This newer plaster endocast formed the basis for the original endocast reconstruction done by R.L.H., who based the reconstruction on the less distorted side (left) and then doubled its water-displaced volume to achieve the final endocranial volume. As will become clear in our descriptions, this first method required several additions and subtractions to compensate for missing portions, for flash lines left from the casting process, and for distortion remaining in the reconstruction.

We concluded that a more accurate reconstruction would result if the portions of the original endocast were separated from reconstructed elements and approximated on a plasticene "core" so that distortion could be effectively eliminated. The second method, which was accomplished mostly by M.S.Y. with minimal guidance from R.L.H., permitted a range of possible reconstructions of the actual brain endocast pieces and provided a range of endocast volumes. This reconstruction methodology, referred to as the "dissection method," eliminated most of the distortion and obviated the need to correct for flash lines.

Although both methods provide a final endocranial capacity very close to what must have been the actual living brain volume of A.L. 444-2, we consider the dissection method to be the more accurate one.

Distortion of the Endocast

Distortion of the endocranial cast mirrors that of the cranium. While the right parietotemporal area appears to be depressed, the left parietotemporal area shows signs of bulging in compensation. In addition, due to a gap that runs anteroposteriorly along the left temporal lobe (see Figure 4.1), there is an artificial increase in the distance between the base of the endocast and its apex of about 3–8 mm on the left side. That is, relative to the right side, the left temporal portion of the endocast is depressed inferiorly by as much as 8 mm, and the left cerebellar lobe by about 3 mm, as described in Chapter 2. When viewed from posterior aspect and aligned properly on the midsagittal plane, the left temporal and cerebellar lobes project inferiorly more than the right ones do.

Posterior cerebral/cerebellar relationships are distorted on the right side. Due to a displacement crack that runs across the occipital squama, the bone table appears "folded." We noted that these relationships are anatomically more accurate on the left side, despite the aforementioned descent on this side.

The right cerebellar lobe is distorted posteriorly about 5 mm beyond the left side, which appears undistorted. The right occipital pole (OP) also appears displaced posteriorly relative to the left OP; this asymmetry does not represent a true petalial pattern. The left temporal lobe is also displaced inferiorly relative to the right side, by approximately 5 mm.

The right cerebrum required some addition of plasticene in the posterior frontal lobe, near the temporal and sphenoid juncture, and the greater wings required removal of plaster from the Rak-Kimbel reconstruction.

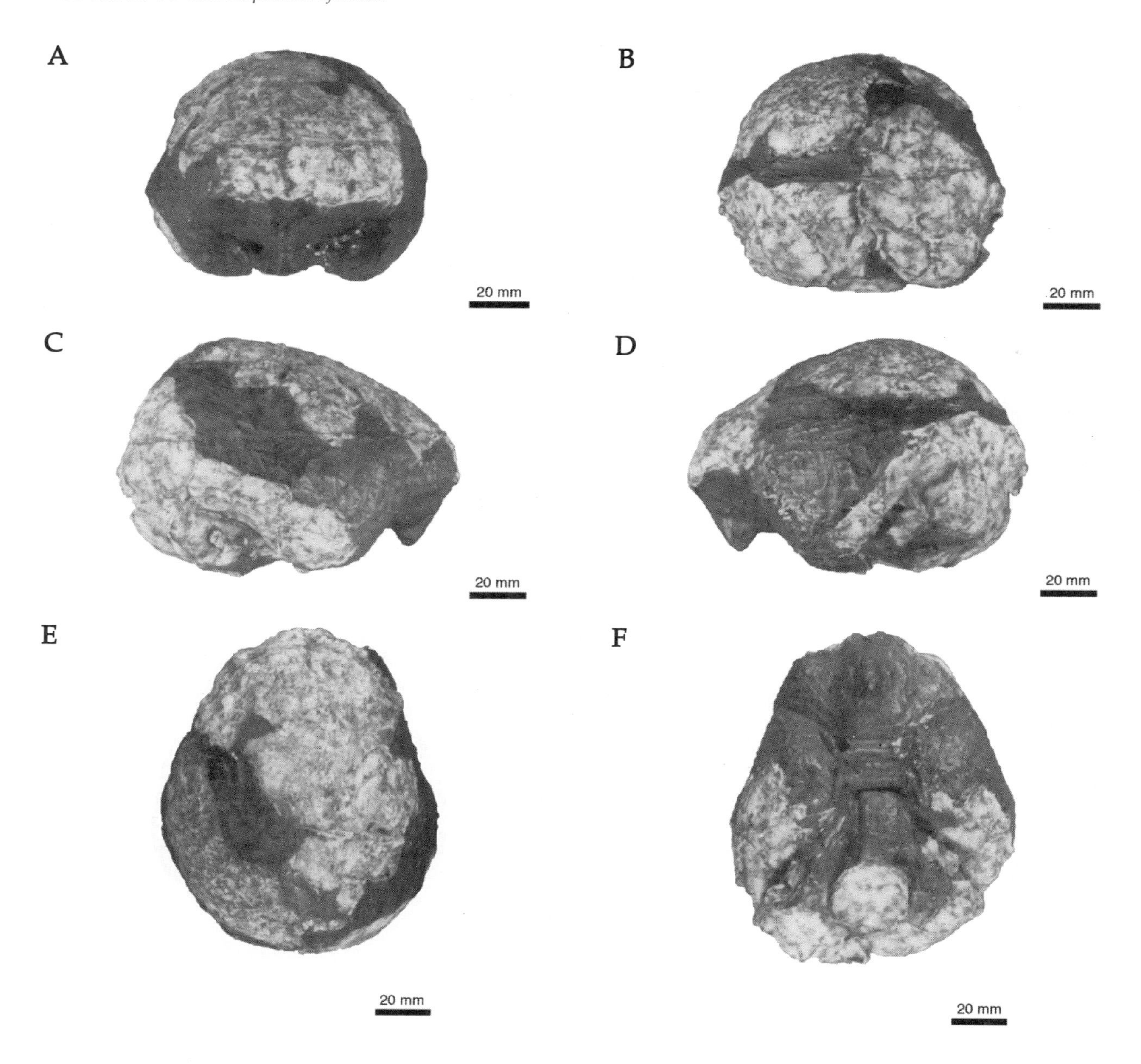

Figure 4.1 The A.L. 444-2 Rak-Kimbel plastic reconstructed endocast. (A) Anterior view. (B) Posterior view. (C) Right lateral view. (D) Left lateral view. (E) Superior view. (F) Inferior view.

The left side, however, appears to have been distorted laterally and required carving down.

The distortion of the cranium gives a dorsal height of the cerebrum above the cerebellar lobes that appears too high, and the occipital planum appears steeper in comparison to any of the other *Australopithecus* endocasts. On the left side of the cast the posterior temporal lobe is displaced laterally relative to the most posterior part of the petrous cleft.

Assessment of Endocranial Volume

Hemisection Method

As measured by water displacement, the capacity of the original plastic endocast made by Yoel Rak and Bill Kimbel is 572 ml. Subsequent modification by them to correct for displacement in the left temporal region (see above) yielded a capacity of 534 ml.

We independently measured endocranial capacity on an *uncorrected* plaster endocast received from Yoel Rak (Figure 4.2) at 615 ml, based on five measurements using weight in air minus weight in water. This endocast was sawed in half in the midsagittal plane, and the less-distorted left half was reconstructed for a more accurate volume displacement. (We are grateful to Chet C. Sherwood for undertaking this task.)

After sculpting out the sphenoid wing region, which in the original reconstruction did not accurately reflect the greater wing's indentation between frontal and temporal lobes, the left side of the endocast gave a volume of 320 ml by water displacement, which, assuming perfect symmetry, yields a cranial capacity of 640 ml. This value was based on the average of five water displacements. Estimating the flashline (approximately 1 mm in thickness) area to be approximately 51 mm^2, the corrected volume of the left side would be 320 – 51 = 269 ml. Doubled, the value is 538 ml, which is virtually the same as the corrected volume measured by Rak and Kimbel (534 ml).

The right half of the plaster endocast was also measured, giving a volume of 253 ml, which when doubled yields a volume of 506 ml. While this might appear to be a major discrepancy between left and right sides, it must be remembered that the saw cut itself measured 1.0 to 1.5 mm in thickness, and the cut was purposely made slightly to the right of the midsagittal plane.

Thus a correction was made to the right side as follows. The left half of the endocast was placed in an Agfa Studiostar scanner, and the resulting outline of the medial surface was analyzed with Sigmascan Pro 4.0 to ascertain the area of this piece, which was 75 mm^2. Assuming that the missing portion of the right half adjacent to the midsagittal plane was approximately 1 mm thick, the "lost" volume would be 75 ml. Adding both halves and the "lost" portions together, gives 320 + 253 + 75 – 70 = 578 ml.

The right side was not reconstructed because it is the more distorted side, and it is likely that the discrepancy between 572 ml and 578 ml is accounted for by this distortion. This discrepancy is ca. 1%, and can surely be treated as a combination of measurement error and minor distortion. Using a volume of 538 ml (based on the left side), and allowing for an error within 5%, the volume would vary between 511 ml and 565 ml. We believe it would be reasonable to estimate the endocranial volume as 540 ml.

Placing the left half of A.L. 444-2 endocast against the right halves of either OH 5 (530 ml) or SK 1585 (530 ml) indicates that the Hadar endocast is larger, so that while 540 ml appears to be the most accurate of our determinations, the true capacity may have been closer to 550 ml.

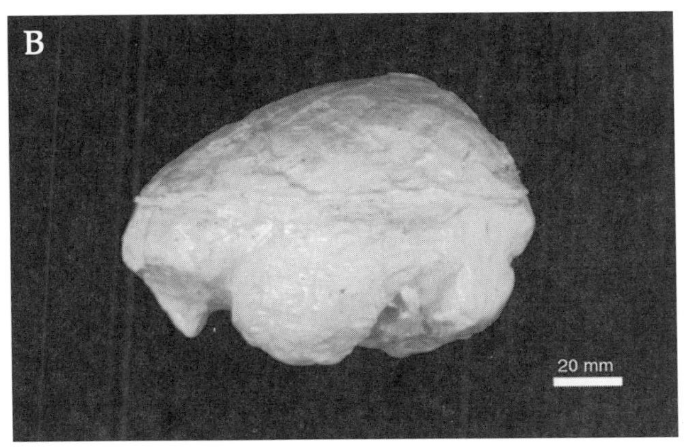

Figure 4.2 The A.L. 444-2 Rak-Kimbel plaster reconstructed endocast. (A) Right lateral view. (B) Left lateral view.

Complete and Hemi-Endocast Reconstruction: Dissection Method

Basic Structure

The Rak-Kimbel reconstructed A.L. 444-2 endocast was molded using alginate dental compound. The plaster cast from this mold was dissected into three major sections that bear the original endocranial bone surface. These three parts of the endocast are (1) the right and left frontal–right parietal segment, (2) the left parietal segment, and (3) the right and left temporal–right and left occipital–foramen magnum (FM) segment. The detached fragments were then reassembled, based on their anatomical contours to rebuild a continuous and complete brain endocast. This dissection-reassembly reconstruction procedure is referred to as the "Dissection Method" (Yuan and Holloway, 2000).

Three endocast reconstructions were made using different approaches to provide a reasonable range for the

A.L. 444-2 endocranial morphology and capacity. These approaches were (1) a maximum-capacity complete endocast reconstruction, (2) a minimum-capacity complete endocast reconstruction, and (3) a left hemi-endocast reconstruction. Reconstruction procedures are described in the following sections.

Complete Endocast Reconstruction: Maximum-Capacity Reconstruction

Our purpose in the first approach was to reconstruct the A.L. 444-2 endocast with the largest possible endocranial capacity commensurate with the existing morphology in order to find a reasonable upper boundary for endocranial capacity. Efforts were made to make the reconstruction consistent with the external morphology of the A.L. 444-2 cranial remains. These include the anterior–posterior distance, the bilateral width, and the biforamen ovale distance on the cranial base. Bicerebellar distance (width measured across the right and left lateral cerebellar lobes) was also used during this reconstruction process. Adjustments were made on the endocranial height, as this distance appeared to be unusually large when compared to undistorted *Australopithecus* endocasts.

PROCEDURES

a. The right and left frontal–right parietal segment was dissected into one frontal fragment and three small right parietal fragments according to the crack lines. They were then reassembled.
b. The left parietal segment was set to align with the reassembled right and left frontal–right-parietal segment. These two segments formed the upper part of the endocast.
c. The lower part of the endocast, the right and left temporal–right and left occipital–FM segment, was dissected into two subunits: the left temporal–left occipital fragment with FM and the right temporal–right occipital fragment without FM.
d. The left temporal–left occipital fragment was used as a reference to reset the right one. The right temporal–right occipital fragment was repositioned both anteroposteriorly and superoinferiorly. The right temporal–right occipital fragment was moved forward approximately 5 mm and upward 2–3 mm in order to adjust cerebellar level for symmetry and alignment.
e. The right occipital region (planum occipitalis) appeared to be flatter than the left, a feature of compressive distortion. The left occipital region was used as the template. Plasticene was added to compensate for the artificial flatness of the right occipital pole.
f. The midsagittal plane was determined to realign the upper and lower parts. The missing morphology was filled in with plasticene to complete the endocast reconstruction.

This maximum reconstruction yielded a volume of 587 ml by water displacement (Figures 4.3–4.7).

Complete Endocast Reconstruction: Minimum-Capacity Reconstruction

The purpose of the second approach was to reconstruct the endocast of A.L. 444-2 with the smallest possible en-

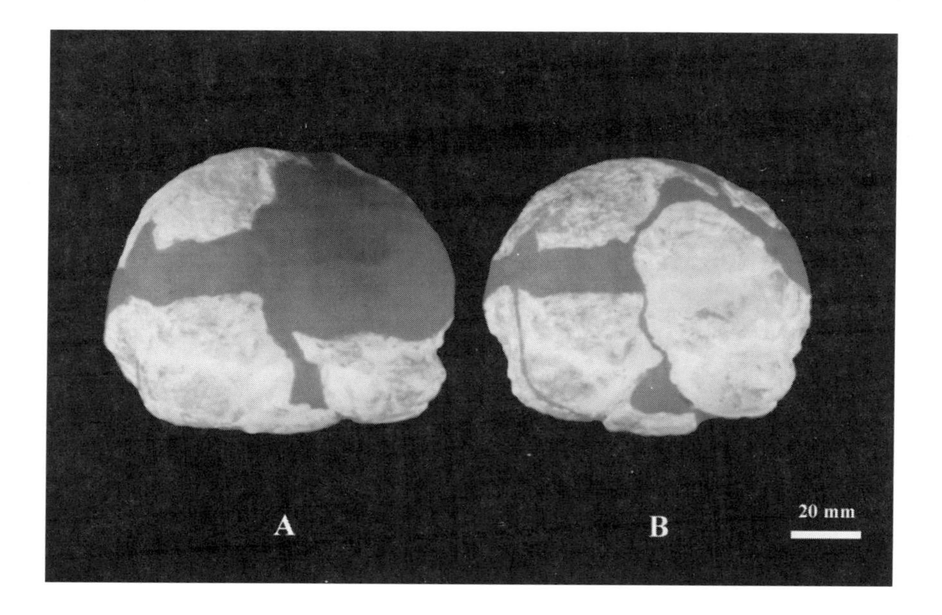

Figure 4.3 The A.L. 444-2 Holloway-Yuan reconstructed endocasts, posterior view. (A) Maximum capacity reconstruction. (B) Minimum capacity reconstruction.

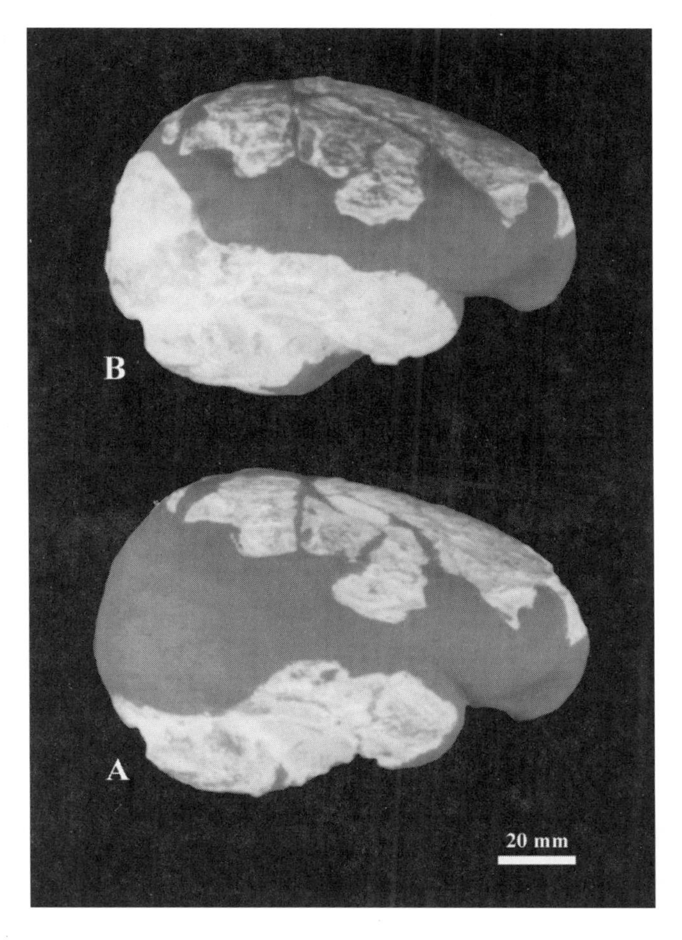

Figure 4.4 The A.L. 444-2 Holloway-Yuan reconstructed endocasts, right lateral view. (A) Maximum capacity reconstruction. (B) Minimum capacity reconstruction.

docranial volume to provide a realistic lower limit to endocranial capacity. The dissection method was employed to reconstruct the endocast as described above. Efforts were made to achieve minimum anterior–posterior distance, bilateral width, endocranial height, and bicerebellar distances.

PROCEDURES

The procedures followed were the same as for the maximum model, except

a. The left temporal–left occipital fragment (with FM) was further dissected into a left temporal piece and a left occipital piece (with FM) based on the crack line going through the middle of the left cerebellum. This was used only as a guide to the anatomical relationships, not as an exact template. All of the dissected left pieces were reassembled by using the right temporal-occipital fragment as the template. The left temporal piece was repositioned superiorly by rotating it upward approximately 2–3 mm.
b. The right temporal–right occipital and left temporal–left occipital fragments were then realigned by adjusting their mutual spatial relations in three dimensions in order to reduce asymmetry and to achieve a minimum breadth. If the right fragment was utilized as a reference, the left fragment was brought backward about 5 mm.
c. The right occipital region (planum occipitalis) appeared to be only slightly flattened after the adjustments. The right occipital region was noted to show compressive distortion on the cranial fragment. No compensation was made to adjust the right occipital pole regions.
d. The midsagittal plane was determined to reassemble the upper and lower parts. The missing

Figure 4.5 The A.L. 444-2 Holloway-Yuan reconstructed endocasts, left lateral view. (A) Maximum capacity reconstruction. (B) Minimum capacity reconstruction.

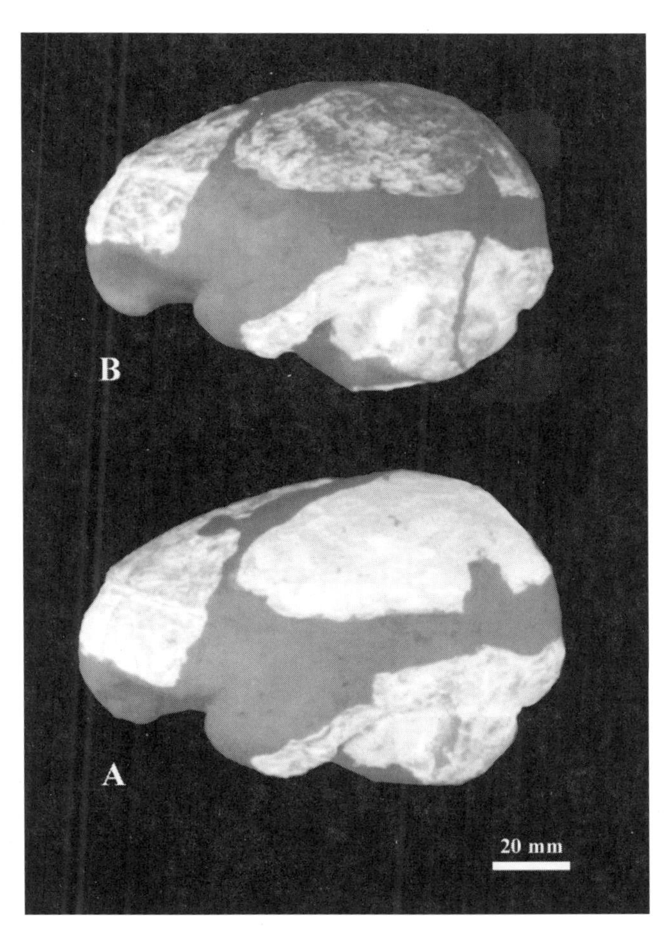

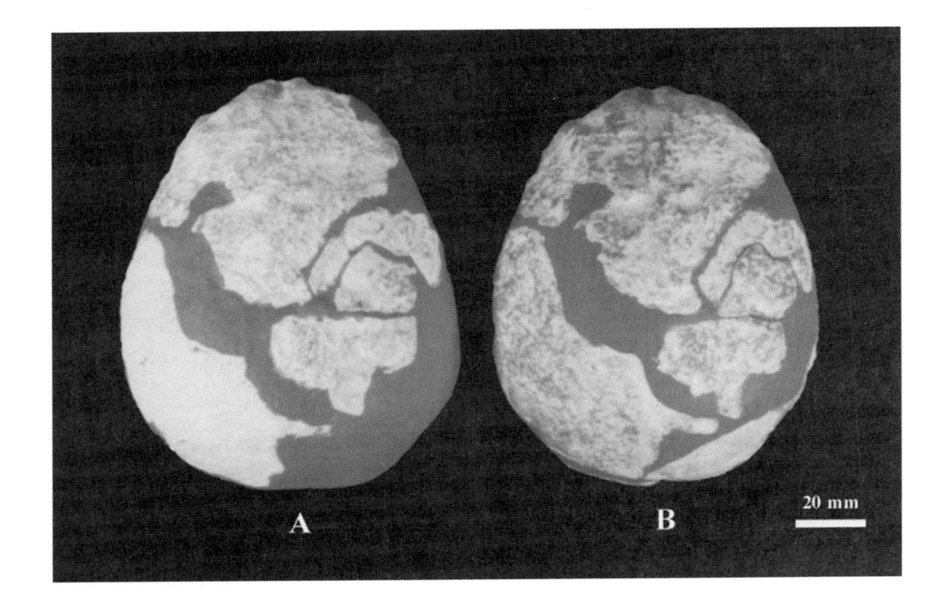

Figure 4.6 A.L. 444-2 Holloway-Yuan reconstructed endocasts, superior view. (A) Maximum capacity reconstruction. (B) Minimum capacity reconstruction.

morphology was filled in with plasticene to complete the endocast reconstruction.

This minimum endocast reconstruction yielded a volume by water displacement of 513.5 ml (Figures 4.3–4.7).

Hemi-Endocast Reconstruction: Left Hemisphere Reconstruction

The hemi-endocast reconstruction with dissection method was performed on the less-distorted left half as follows:

PROCEDURES

a. The right and left frontal–right-parietal segment was dissected into right and left portions based on the landmarks of the anterior median groove between the two frontal lobes and the groove's extension line. The right portion was discarded.
b. The left parietal segment demonstrated traces of sagittal sulcus at the medial posterior border. After trimming the portion of the right hemisphere that remained in the segment, the left parietal segment was set to align with the left frontal segments.

Figure 4.7 The A.L. 444-2 Holloway-Yuan reconstructed endocasts, basal view. (A) Maximum capacity reconstruction. (B) Minimum capacity reconstruction.

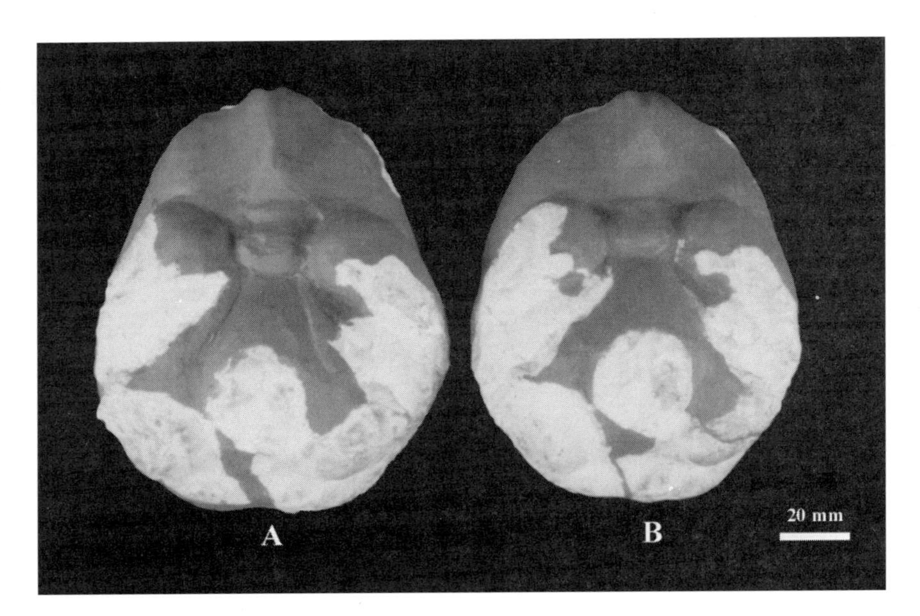

c. After dissecting the right and left occipital lobes based on the preexisting crack line, the left temporal and occipital segment were used for reassembly. The midsagittal plane was used to divide the FM into two halves. The right portions were discarded.
d. All of the left segments were reassembled to complete the hemi-endocast.

Using the water displacement method, the hemi-endocast yielded a volume of 285.7 ml, which doubled provides a volume for the total endocast of 571.4 ml.

Conclusion on A.L. 444-2 Endocranial Capacity

Results of assessment of A.L. 444-2 endocranial volume are summarized in Table 4.1. We suggest that the most reasonable estimate for the Hadar A.L. 444-2 brain endocast would be a mean of the minimum and maximum values, which is 550 ± 10 ml. While the hemi-endocast reconstruction is larger than this by some 20 ml, or about 5%, we believe the closest approximation to the plastic/plaster reconstruction provides the best overall estimate. This is very close to the 540 ml suggested for the plaster reconstructed left side, assuming symmetry.

We provide some linear chord and arc measurements for the sake of comparison with other endocasts, particularly those of *A. robustus* and *A. boisei* endocasts (Tables 4.2 and 4.3).

Morphological Description

The endocast is practically devoid of convolutional detail in either the frontal or occipital regions, making it impossible to comment on the possibility of a reduced primary visual striate cortex or a posteriorly placed lunate sulcus, which could be taken as evidence for reorganization of the cerebral cortex. Given the distortion of the endocast, particularly in the right occipital region, and the missing right lateral frontal portion, the petalial pattern of asymmetries cannot be realistically assessed, and thus nothing can be suggested regarding possible hemispheric specialization or handedness.

Table 4.1 A.L. 444-2 Endocranial Capacity Measurement Summary

Endocast subject	Estimate (ml)	Comments
Rak-Kimbel plastic endocast, first version	572	Distortion corrected, 534 ml
Rak-Kimbel plastic endocast, second version	558	
Rak-Kimbel plaster endocast	615	Non-corrected, with excessive flashline By weight measurement and water displacement
Rak-Kimbel plaster endocast: hemi-section method		The saw-cut was purposely made to the right of the midsagittal plane, as the left half appeared to be less distorted.
Left: Left half (large half after sawing)	320	Non-corrected.
Corrected	269	With deduction of the estimated flashline 51 ml.
Estimated complete volume from the left half	538	±5% range, 511.1 ml–564.9 ml
Right: Right half (small half after sawing)	253	Non-corrected.
Estimated complete volume from right half:	506	±5% range, 480.7 ml–531.3 ml
Combined and corrected:		
Right half: 253 ml		
Left half: 320 ml		
Estimated loss in sawing to halves: 75 ml		
Estimated error in flashline: 70 ml		
Corrected	578	Right 253 ml + left 320 ml + lost by sawing 75 ml–flashline 70 ml
Best estimate	540	
Holloway-Yuan endocasts: dissection method		Reconstructed from the Rak-Kimbel plastic endocast replicas.
Complete reconstruction with maximum capacity:	587	
Complete reconstruction with minimum capacity:	514	
Average [(maximum + minimum) / 2]	550	
Hemi-endocast reconstruction on the left half:	285.7	
Total [285.7 x 2]	571	

ml, milliliters

Table 4.2 Linear Measurements on the Reconstructed Endocasts (mm)

Measurement	R-K	H-Y max.	H-Y min.	H-Y hemi.[b]
Depth temporal poles to vertex	92	83	81	82
Left OP–FP chord	130	130	128	128
Right OP–FP chord	131	130	128	n/a
Left OP–FP, dorsal arc	180	182	178	178
Right OP–FP, dorsal arc	185	183	177	n/a
Left OP–FP, lateral arc	167	170	165	165
Right OP–FP, lateral arc	168	170	165	n/a
Lateral circumference through OP–FP's	378	379	363	185 (370)
Max. bi-cerebellar breadth, chord	85	91	84	44 (88)
Max. breadth, chord (temporals)	105	106	99	57 (114)
Max. breadth, arc	173	173	154	80 (160)
Bregma–lowest cerebellar point, chord	104	101	97	94
Bregma–lambda, chord	86	98	90	86
Bregma–lambda, arc	102	102	100	95
Bregma–lambda, chord/arc	0.84	0.96	0.9	0.91
Volume, ml	572[a] (558)	587	513	571

[a]The R-K plastic endocast was first measured as 572 ml. The most recent reconstruction of the endocast by Rak and Kimbel revised the volume to 558 ml.

[b]Values in paranthesis in the H-Y hemisphere column are estimates of the true dimensions obtained by doubling the measurements taken on the left hemisphere.

OP, occipital pole; FP, frontal pole; R-K, Rak-Kimbel endocast; H-Y min., Holloway-Yuan endocast of the minimum capacity reconstruction; H-Y max., Holloway-Yuan endocast of the maximum capacity reconstruction; H-Y hemi., Holloway-Yuan hemi-endocast of the left hemisphere reconstruction; n/a, not accessible.

Gross dimensions and indices derived from them are provided for the A.L. 444-2 endocast in Tables 4.2–4.4. The cerebral height/length index of 70.5% falls within the range of other early hominins, as does the biparietal width/cerebral length index value of 80.5%. None of the metrics or indices shows unambiguous evidence of systematic patterning among early hominin species.

Several features of endocast size, shape, and morphology in A.L. 444-2 are informative in a comparative context. As can be seen on the left side, the occipital poles of the cerebrum strongly overhang (by ca. 8 mm) the cerebellar lobes, which are situated relatively far forward on the underside of the endocast. As noted by several authors (e.g., Tobias, 1967; Holloway, 1972), this morphology is often, but not always, encountered in endocasts of *A. robustus* (i.e., SK 1585) and *A. boisei* (e.g., OH 5 and KNM-ER 407; but not KNM-ER 23000; Brown et al., 1993). In *A. africanus*, the cerebellar lobes and occipital poles usually project backward subequally (as in Sts. 5 or Sts. 60 [also called TM 1511]), similar to what is seen in chimpanzees. This is the same pattern seen in the endocast of the KNM-WT 17000 cranium of *A. aethiopicus* (Holloway, 1988). Marked cerebral overhang over the cerebellum, which is often regarded as a hallmark of modern human endocast morphology, characterizes the *H. habilis* endocast OH 24 (Tobias, 1991), but not that of KNM-ER 1805 or KNM-ER 1813. In fact, this feature is quite variable, both in extant great apes and in modern *Homo sapiens*. In two other *A. afarensis* endocasts, A.L. 162-28 and A.L. 333-45, the cerebellar lobes are tucked under the occipital poles but not to the extent observed in A.L. 444-2 (see Holloway, 1983; Holloway and Kimbel, 1986).

The cerebellar lobes of A.L. 444-2 appear fairly small in size not only relative to the whole endocast but also in comparison to other adult *Australopithecus* cerebellar lobes. Tables 4.5 and 4.6 provide some cerebellar measurements and indices on the A.L. 444-2, adult African apes, *Australopithecus*, and modern human endocasts (see Figure 4.8). In addition, we estimated the absolute and relative cerebellar volume for comparison. The approximate volume is calculated for heuristic purposes and represents a crude estimate (see Table 4.7 regarding the data comparisons to Stephan et al., 1981). Although additional data may improve the accuracy of estimating cerebellar volumes, this information does reveal the extraordinarily small size of A.L. 444-2 cerebellar lobes in contrast to those of other *Australopithecus* and living African apes.

The shape of the cerebellar lobe on the A.L. 444-2 endocast, best appreciated in left lateral view, is distinctive (Figure 4.2). The cerebellar lobe bears a dominant, vertically and transversely flattened, posterolaterally directed surface that is truncated anterolaterally by the sigmoid sinus. Posteromedially, just lateral to the confluence of sinuses, this flattened surface quickly

Table 4.3 Linear Measurements on *Australopithecus* Endocasts (mm)

Measurement	OH 5[a]	SK 1585[a]	KNM-ER 23000	H-Y hemi.[c]
Depth temp. poles to vertex	85	86	ca. 81	82
Left OP–FP, chord	126	n/a	124	128
Right OP–FP, chord	129	129	124	n/a
Left OP–FP, dorsal arc	171	n/a	170	178
Right OP–FP, dorsal arc	170	180[b]	172	n/a
Left OP–FP, lateral arc	167	n/a	164	165
Right OP–FP, lateral arc	170	165	168	n/a
Max. bi-cerebellar breadth, chord	85	ca. 82	90	44 (88)
Max. breadth, chord	100	ca. 98	102	57 (114)
Max. breadth, arc	155	ca. 160	158	80 (160)
Bregma–lowest cerebellar point	94	95	ca. 102	94
Bregma–lambda, chord	84	81	80	86
Bregma–lambda, arc	95	96	85	95
Bregma–lambda, chord/arc	0.88	0.84	0.94	0.91
Endocranial capacity, ml	530	530	497	571

[a]From Holloway 1972: 179, Table 1.
[b]In Holloway (1972: Table 1) the value is given as 190 mm, but 180 mm is more accurate. The maximum breadth arc is from this study and not Holloway (1972).
[c]Values in parentheses in the H-Y hemisphere column are estimates of the true dimensions obtained by doubling the measurements taken on the left hemisphere.
OP, occipital pole; FP, frontal pole; H-Y hemi., Holloway-Yuan hemi-endocast of the left hemisphere reconstruction; n/a, not accessible.

grades into a smoothly convex prominence corresponding to the most posteriorly protruding part of the cerebellum. In some ways, cerebellar shape in A.L. 444-2 recalls that of robust *Australopithecus* endocasts, such as SK 1585 and OH 5 (Tobias, 1967; Holloway, 1972). In these specimens, the cerebellar lobe is similarly dominated by a flattened posterolateral surface, but unlike in A.L. 444-2 this surface is more extensive anterolaterally, lending the robust cerebellar lobe an elongated triangular form in lateral view (Holloway, 1972). In addition to the SK 1585 and OH 5 endocasts, this morphology can be reconstructed from endocranial evidence of the occipital bone in Omo 323-1976-896 and KNM-ER 23000. In contrast to this pattern, the endocast of *A. africanus* (as well as that of *H. habilis* cranium KNM-ER 1813) appears more apelike. The surface of the cerebellar lobe is evenly convex in its entirety, and it lacks a flattened, dominant face. In these endocasts, more of the total surface of the cerebellar lobe is visible in posterior view than in the A.L. 444-2 and robust endocasts, where the

Table 4.4 Indices Based on Linear Measurements of Hominid Endocasts (%)

Cast/Index	W/L	H/L	B/L	H/W	B/W	H/B
A.L. 444-2						
R-Y	0.805	0.705	0.797	0.876	0.99	0.885
H-Y max.	0.815	0.638	0.777	0.783	0.953	0.822
H-Y min.	0.773	0.633	0.758	0.818	0.98	0.835
H-Y hemi.	0.891	0.64	0.734	0.719	0.825	0.872
SK 1585	0.759	0.666	0.738	0.877	0.969	0.905
OH 5	0.781	0.664	0.734	0.85	0.94	0.904
KNM-ER 23000	0.823	0.653	0.823	0.794	1.000	0.794
Taung	0.728	0.703	0.779	0.965	1.069	0.902
Sts. 60	0.834	0.704	0.817	0.843	0.979	0.861
Sts. 5	0.737	0.746	0.86	1.011	1.116	0.866
Chimpanzee	0.871	0.734	0.798	0.842	0.915	0.919
Gorilla	0.731	0.671	0.753	0.918	1.300	0.891

Data adapted from Holloway 1972, p. 180, Table 2.
W, maximum width; L, maximum length, frontal to occipital poles; B, bregma to lowest cerebellar point; H, vertex to deepest temporal lobe portion.

Table 4.5 Cerebellar Measurements on A.L. 444-2, African Ape, *Australopithecus*, and Modern Human Endocasts

Cast/Measurement[a]	Volume Total (ml)	Bi-cerebellar Width-A (mm)	Bi-cerebellar Width-B (mm)	Diagonal P-A (mm)	Diagonal M-L (mm)	Depth P-A (mm)	Est. Volume Cerebellum (ml)[d]
A.L. 444-2							
R-K recon.	572	85	80	37.5	29	39	42.4
H-Y max.	587	91	85	37.5	29	39	42.4
H-Y min.	513	85	79.5	37.5	29	39	42.4
H-Y hemi.	571	87	81	37	27	39	39.0
African apes[b]							
Gorilla gorilla							
Male (*n* = 23)	555	91	87	48.7	33.2	50.7	82.6
Female (*n* = 14)	453	85.1	80.4	43.9	31	44.9	61.6
Mean (*n* = 37)	516	88.8	84.5	46.9	32.4	48.5	74.7
Pan troglodytes							
Male (*n* = 7)	408	80.9	77	44.9	28.4	44.4	56.9
Female(*n* = 8)	373	78.9	74.3	42.3	27.3	43.6	50.2
Mean (*n* = 15)	389	79.8	75.5	43.5	27.8	44	53.3
Pan paniscus							
Male (*n* = 18)	359	78.7	74.6	43.4	27	43.9	51.7
Female (*n* = 19)	340	77.3	73.1	42.6	27.9	42.8	51.1
Mean (*n* = 37)	349	78	73.8	43	27.4	43.4	51.4
Hominins[c]							
A. afarensis							
A.L. 444-2	550	85	80	37.5	29	39	42.4
A.L. 333-45	485	83	78	42	35	41	60.3
Mean (*n* = 2)	518	84	79	39.8	32	40	51.3
Robust							
SK 1585	530	83	78	41	29	41	48.8
OH 5	530	91	86	46	32	44	64.8
KGA 10-525	540	94	88	46	34.5	39	61.9
KNM-WT 17000	410	84	78	37	31	36	41.3
KNM-ER 23000	491	89	84	43	34	38	55.6
Mean (*n* = 5)	500	88.2	82.8	42.6	32.1	39.6	54.5
A. africanus							
Sts. 60	436	88	85	46	31	42	59.9
Sts. 5	485	80	76	39	34	39	51.7
Mean (*n* = 2)	461	84	80.5	42.5	32.5	40.5	55.8
Modern human							
Mean (*n* = 14)	1452	111.9	104.8	66.3	43.9	65.9	192.5

Volume total: total endocast or endocranial capacity

Bi-cerebellar width-A: maximum bi-cerebellar width including sigmoid sinus.

Bi-cerebellar width-B: maximum bi-cerebellar width excluding sigmoid sinus

Diagonal P-A: maximum diagonal posteroanterior width of cerebellum

Diagonal M-L: maximum diagonal mediolateral width of cerebellum; measured from the line segment perpendicular to the Diagonal P-A line segment.

Depth P-A: the posteroanterior distance from the aperture of internal acoustic meatus to the posterior margin of cerebellum along the parasagittal plane.

Est. Volume Cerebellum: estimated cerebellar volume of the endocast.

Equation: $2 \times [(.5) \times (\text{Diagonal P-A}) \times (\text{Diagonal M-L}) \times (\text{Depth P-A})]$

[a]See Figure 4.8 for measurement illustration.

[b]All the African apes and modern human endocasts belong to the Ralph L. Holloway endocast collection.

[c]Measurements of the fossil specimens are taken either from the more complete side, when both are present, or the only existing side. If both sides are intact and undistorted, the average of both sides is recorded.

[d]The estimated cerebellar volume of an endocast is a crude approximation for comparative purposes. It should not be taken as the true or absolute cerebellar volume.

Table 4.6 Indices Based on the Cerebellar Measurements of A.L. 444-2, African Ape, *Australopithecus*, and Modern Human Endocasts

Cast/Index	Volume Total (ml)	Est. Volume Cerebellum (ml)	Est. Vol. Cerebellum/ Volume Total (%)	Diag. (P-A)/ Diag. (M-L) (%)	Diag. (P-A)/ Depth (P-A) (%)
A.L. 444-2					
R-K recon.	572	42.4	7.4	1.29	0.96
H-Y max.	587	42.4	7.2	1.29	0.96
H-Y min.	513	42.4	8.3	1.29	0.96
H-Y hemi.	571	39.0	6.8	1.37	0.95
African apes					
Gorilla gorilla					
Male (n = 23)	555	82.6	14.9	1.47	0.96
Female (n = 14)	453	61.6	13.6	1.42	0.98
Mean (n = 37)	516	74.7	14.4	1.45	0.97
Pan troglodytes					
Male (n = 7)	408	56.9	14.0	1.59	1.01
Female (n = 8)	373	50.2	13.5	1.57	0.97
Mean (n = 15)	389	53.3	13.7	1.58	0.99
Pan paniscus					
Male (n = 18)	359	51.7	14.4	1.61	0.99
Female (n = 19)	340	51.1	15.0	1.54	1.00
Mean (n = 37)	349	53.3	14.7	1.58	0.99
Australopithecus					
A. afarensis					
A.L. 444-2	550	42.4	7.7	1.29	0.96
A.L. 333-45	485	60.3	12.4	1.20	1.02
Mean (n = 2)	518	51.3	10.1	1.25	0.99
Robust					
SK 1585	530	48.8	9.2	1.41	1.00
OH 5	530	64.8	12.2	1.44	1.05
KGA 10-525	540	61.9	11.5	1.33	1.18
KNM-WT 17000	410	41.3	8.7	1.19	1.03
KNM-ER 23000	491	55.6	11.3	1.26	1.13
Mean (n = 5)	500	54.5	10.6	1.33	1.08
A. africanus					
Sts. 60	436	59.9	13.7	1.48	1.10
Sts. 5	485	51.7	10.7	1.15	1.00
Mean (n = 2)	461	55.8	12.2	1.32	1.05
Modern human					
Mean (n = 14)	1452	192.5	13.3	1.51	1.01

Notes: As for Table 4.5.

majority of the surface is visible in lateral view. Endocasts of *A. africanus* demonstrating this morphology include Taung, Sts. 5, probably Sts. 60 and, based on the morphology of the exposed cerebellar fossae of the occipital bone, MLD 1 as well.

The anterior poles of the frontal lobes combine to give the front region of the A.L. 444-2 endocast a full, rounded contour that contrasts with the sharply tapering anterior contour of some of the robust *Australopithecus* endocasts (see Falk et al., 2000). In this feature, the A.L. 444-2 endocast is similar to those of *A. africanus* and *Homo*. Another adult Hadar specimen adds information on this feature. The frontal bone fragment A.L. 438-1b preserves the fossae for the left and right frontal poles, which reflect a relatively broad and even anterior margin for the frontal lobes, as seen in A.L. 444-2.

Meningeal patterns on the A.L. 444-2 endocast are largely obscured due to the poor surface quality. It is only possible to conclude that the middle meningeal artery issued from the foramen spinosum and formed posterior and anterior branches, the latter lost with the missing temporal lobe portions.

The venous sinus pattern in A.L. 444-2 and its comparative context are discussed in Chapter 3.

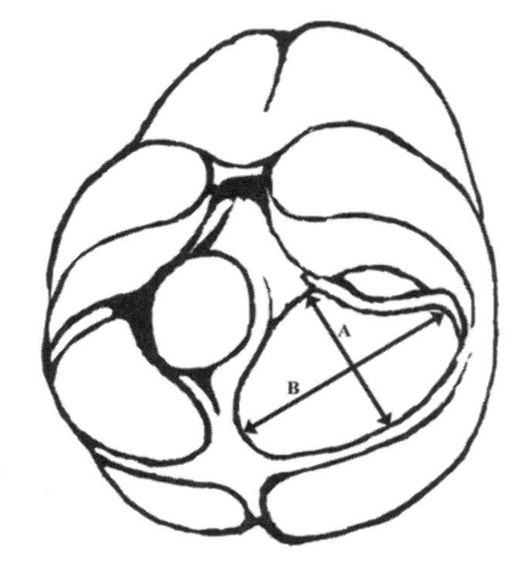

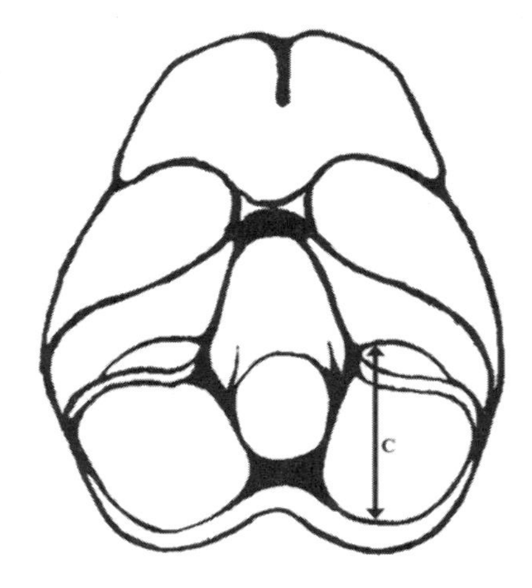

Figure 4.8 Schematic drawing of the basal view showing the three measurements of the cerebellum used in this study. A, diagonal P-A; B, diagonal M-L; C, depth P-A.

Discussion

Endocast Reconstructions

The Hadar A.L. 444-2 specimen presented one of the most challenging reconstructions in the senior author's (R.L.H.) long history of working on fossil hominin endocasts. Although most of the endocast is present, severe distortion and displacement of the two hemispheres made reconstruction difficult, particularly with regard to the true positions and orientations of the temporal lobes.

Endocranial Capacity

With an estimated endocranial capacity of ca. 550 ± 10 ml, the *A. afarensis* specimen A.L. 444-2 possessed a larger brain than either of the other adult individuals of this species for which capacity estimates are available (see Table 4.8) (A.L. 162-28: ca. 400 ml; A.L. 333-45: ca. 485 ml [Holloway, 1996]). The mean endocranial volume for these three adult specimens is 478 ml (SD of 75.2 ml), which is 8.4% larger than the mean for *A. africanus* (mean = 441 ml, SD = 19.6 ml, *n* = 6; see Tobias, 1991: 708, Table 181 [excluding Stw. 505; see below]; see also Holloway, 1970: 199, Table 1), but 6.8% smaller than the mean for *A. boisei* (mean = 513 ml, SD = 11.5, *n* = 4; see Tobias, 1991: 708, Table 181).

When the Stw. 505 cranium is added to the *A. africanus* sample, assuming an endocranial capacity of ca. 515 ml for this specimen (Conroy et al., 1998), the mean value for this taxon rises to 452 ml (also see Lockwood and Kimbel, 1999, and Hawks and Wolpoff, 1999, for questions about the Stw. 505 capacity estimated by Conroy et al., 1998). The average *A. afarensis* endocranial capacity is still 5.8 % larger than the average for *A. africanus* with Stw. 505 included.

The resulting coefficient of variation for endocranial capacity in *A. afarensis* is 15.7%, although with n = 3, little

Table 4.7 The Brain Volume Data of Stephan et al. (1981)

Specimens	Total Brain Volume (ml)	Cerebellum Volume (ml)	Cerebellar Vol./ Total Brain Vol. (%)	Est. Cerebellar Vol./ Total Endocast Vol. (%) (this study)
Gorilla gorilla	470	69.2	14.72 (*n* = 1)	14.36 (*n* = 37, mixed)
Pan troglodytes	382	43.7	11.43 (*n* =1)	13.72 (*n* = 15, mixed)
Pan paniscus[a]	n/a	n/a	n/a	14.74 (*n* = 37, mixed)
Modern *Homo*	1452	137.4	10.97 (*n* = 1)	13.26 (*n* = 14, mixed)
A.L. 444-2 (this study)				7.71
R-K recon.				7.41
H-Y max.				7.22
H-Y min.				8.27
H-Y hemi.				6.82
Australopithecus (this study; includes a, b, and c below)				10.83 (*n* = 9)
a. *A. africanus* (Sts. 5 and Sts. 60)				12.20 (*n* = 2)
b. *A. afarensis* (A.L. 444-2 and A.L. 333-45)				10.07 (*n* = 2)
c. Robust (SK1585, OH 5, KGA 10-525, KNM-WT17000, and KNM-ER 23000)				10.57 (*n* = 5)

[a]Data for *P. paniscus* were not presented by Stephan et al., 1981.

Table 4.8 Endocranial Capacities of *Australopithecus afarensis* (ml)

Specimen	Volume	Reference
A.L. 162-28	350–400	Falk (1985)
	ca. 400	Holloway (1983)
	375–400	Holloway (1996, 2000)
A.L. 333-45	485–500	Holloway (1983, 2000)
	485	Holloway (1996)
A.L. 333-105	352 (adult)	Falk (1985)
(juvenile)	343 (adult)	Falk (1988)
	310–320 (juvenile)	Holloway (1983, 1996, 2000)
	ca. 400 (adult)	Holloway (1983)
A.L. 444-2	ca. 550	This study

confidence can be placed in the reliability of this figure. Nevertheless, based on the data in Holloway (2000: 143, Table 1), the CV for *A. afarensis* is greater than that for *A. africanus* (7.7%, *not* including Taung and Stw. 505) or *A. boisei* (4.3%, *not* including KNM-WT 17000, KNM-WT 17400 and KNM-ER 23000). Using extant hominoid coefficients of variation (ca. 8%-15%) as a guide (see Tobias, 1991: 716, Table 183), the CV for *A. afarensis* is high yet not excessive, while those of *A. africanus* and *A. bosiei* appear low.

In the case of *A. afarensis,* the high CV is clearly affected by the large capacity of the A.L. 444–2 endocast, which may stem from the very large overall size of this individual's skull (as documented in Chapter 3), which in turn is apparently due in part to the late temporal position of the specimen in the Hadar stratigraphic sequence (Lockwood et al., 2000). The low CV in the *A. africanus* and *A. boisei* samples, in contrast, likely illustrates artificially depressed variation in endocranial capacity (Holloway, 1970; Tobias, 1991; Lockwood and Kimbel, 1999). The large endocranial volumes for A.L. 444–2 and Stw. 505 demonstrate the need for caution in accepting the currently documented ranges of endocranial capacity as accurate estimates of population parameters.

Reorganization

The large dorsal height of the cerebrum and the pronounced degree of posterior cortical overhang relative to the cerebellum suggest affinities between the A.L. 444-2 endocast and those of robust *Australopithecus*, such as SK 1585 and OH 5. This morphological pattern is not matched in chimpanzee or *A. africanus* endocasts, which retain the plesiomorphically low, squat endocast shape and posterior projecting cerebellar lobes. Although two of three early *Homo* endocasts lack cerebellar tucking, the presence of this character is standard in later *Homo*. The fact that this pattern is also lacking in the endocast of *A. aethiopicus* specimen KNM-WT 17000 would suggest that the geologically older *A. afarensis* might well be apomorphic relative to putative descendant conditions in these taxa.

The A.L. 444-2 endocast shows features more frequently found in hominins than in the great apes and thus suggests some advancement of brain morphology, possibly involving some cortical reorganization. The cerebral overlap of the cerebellar lobes plus posterior cerebral height suggests that the posterior region of the brain underwent some expansion by 3.0 Ma, most probably with a concomitant reduction of primary visual striate cortex, or Area 17 of Brodmann. This finding has already been documented for *A. afarensis* specimen A.L. 162-28 (Holloway, 1983).

Since surface preservation is so poor, little can be said about the cerebral gyal and sulcal patterns. The issue of the placement of the lunate sulcus remains unresolved. The issues of Broca's cap or language association areas cannot be investigated either, since most of the involved external morphology is missing or lacks convolutional details.

Cerebellar Size and Shape

The A.L. 444-2 cerebellar lobes appear smaller than those found on either SK 1585, OH 5, or KNM-ER 23000 (Brown et al., 1993). Confirmation of this observation and the preliminary quantitative data in Tables 4.5–4.7 will require volumetric comparative data that are not available at this time. Although the cerebellar lobes in A.L. 444-2 appear more similar in shape to those of robust *Australopithecus* than to those of *A. africanus*, such qualitative shape comparisons also require morphometric data for verification.

Prefrontal Lobe Morphology

In contrast to the posterior morphology of the endocast, the morphology of the prefrontal region points to affinities with *A. africanus* and *Homo*. The prefrontal region of *A. afarensis* in A.L. 444-2, A.L. 438-1b, and A.L. 333-105 has a pattern more similar to that of *A. africanus* than to that of *A. boisei*.

Mosaic Brain Evolution

Mosaic evolution has been invoked to explain the coexistence of primitive and derived traits and conditions within the same individual or population (Holloway, 1973; McHenry, 1975; White, 1979; Holloway and Post, 1982; Rae, 1997). Most recently, Barton and Harvey (2000) argued that mosaic changes throughout the brain may have played an important part in mammalian brain evolution. In this context we are intrigued that the posterior region of the A.L. 444-2 endocast appears to resemble robust *Australopithecus* endocasts, while the anterior region appears to be similar to those of *A. africanus* and *Homo*. This evidence raises the possibility that a mosaic pattern of hominin brain evolution may have already been in place before 3 Ma.

5

Elements of the Disarticulated Skull

The Frontal Bone

The 1970s collection of hominin cranial remains from Hadar is notoriously weak in its representation of the frontal bone. Besides the complete but distorted frontal of the A.L. 333-105 juvenile (Kimbel et al., 1982), only two very incomplete adult specimens provided glimpses of frontal morphology: A.L. 288-1 (Johanson et al., 1982b) and A.L. 333-125 (Asfaw, 1987). With the recovery of the almost complete frontal bone of A.L. 444-2, we are able to fill one of the last remaining gaps in our knowledge of the Hadar hominin adult skull. Another frontal specimen, A.L. 438-1b, contributes important information on the glabellar and supraglabellar regions, which are missing or poorly preserved in A.L. 444-2.

Superior View

The A.L. 444-2 frontal bone features prominent, laterally projecting supraorbital bars, strongly convergent temporal lines, and a transversely broad squama with only moderate postorbital constriction (Figures 3.31 and 5.1). The minimum distance between the temporal lines (30 mm) in the plane of the postorbital constriction is much smaller than the postorbital constriction itself (77 mm), creating on each side an extensive, almost horizontally inclined facies temporalis that, in coronal section, slopes gradually from the inferior temporal lines to the medial wall of the temporal fossa. In between the temporal lines, the supraglabellar region bears a mild hollow that grades smoothly onto the superior surface of the supraorbital bars. Neither a supratoral sulcus nor a trigonum frontale is present.

The supraorbital bars are wide anteroposteriorly, measuring 16 mm at the right lateral break, about 42 mm lateral to the midline. The preserved portions of the anterior supraorbital margins are aligned coronally, forming right angles with the midsagittal line. At the lateral break on each side, the margin actually occupies a slightly more anterior plane than the middle part of the margin, suggesting an anteriorly prominent superolateral corner of the orbit. At the medial break through the left supraorbital, about 22 mm lateral to the midline, the anterior margin begins to swing out toward glabella (this area is damaged on the right side). The extent of anterior glabellar protrusion is suggested by the preserved supraglabellar plate, whose superior surface projects in the midline about 5 mm beyond the anterior supraorbital margins. As this supraglabellar surface shows no evidence of dropping over on to the anterior face of the glabellar mass, the actual position of glabella was at least a few mm anterior to this point.

Strong temporal crests parallel the anterior supraorbital margins, demarcating and truncating the supraorbital bars posteriorly. On the right, the temporal crest stands about 1 mm above the superior surface of the bar, rather than forming a sharp-edged, posteriorly extended shelf that overhangs the anterior wall of the temporal fossa (Figure 5.1). Indeed, on the right side, the anterior wall of the fossa actually encroaches on the upper surface of the supraorbital bar just at the point where the temporal crest begins to curve on to the squama.

The temporal crests travel medially to a point 26 mm lateral to the midline (slightly medial to midorbit), then arc broadly onto the squama and steadily converge pos-

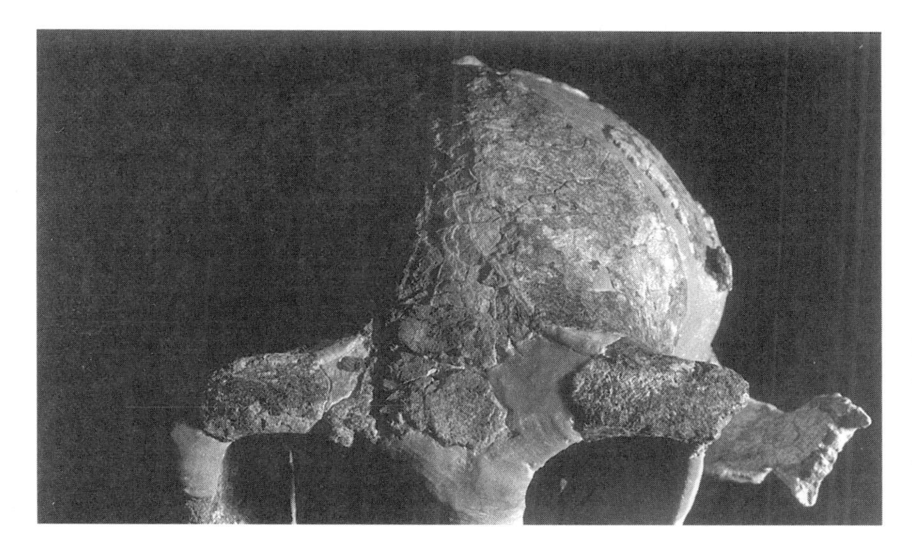

Figure 5.1 Anterosuperior view of the supraorbital elements of A.L. 444-2. Note the elevated position of the temporal lines relative to the supraorbital surface, their extreme medial encroachment relative to the breadth of the frontal squama, and the laterally thickening supraorbital bars.

teriorly toward the coronal suture. A full account of the cresting pattern of A.L. 444-2 is provided in Chapter 3.

Anterior View

On the left, undisturbed side, the supraorbital arch is broad and flat. The maximum superior projection of the arch, which occurs above the middle of the orbit, is about 3 mm higher than the floor of the supraglabellar hollow. The curve of the supraorbital arch is more pronounced medially than laterally; at the lateral break—above the lateral orbital wall—the superior surface of the supraorbital bar remains nearly horizontal: it has not yet begun to bend downward toward the frontozygomatic suture. It is therefore possible to infer that the supraorbital corner was "squared off" rather than inferolaterally sloping.

The supraorbital element has distinct superior and anterior surfaces. Laterally, the intersection of these surfaces is blunt, and, here, above the lateral wall of the orbit, the bar measures 11 mm inferosuperiorly. Medially, the intersection of the surfaces is sharper and the bar is thinner, measuring 8.5 mm at the medial break. There are no traces of a supraorbital notch or foramen, probably due to the loss of the medial segments of the supraorbitals. The margo supraorbitalis itself is blunt but denotes an abrupt transition between the anterior supraorbital surface and the orbital roof.

Anteriorly, the temporal lines mark a strong medial to lateral change in the frontal squama's coronal contour. Between the temporal lines and behind the supraglabellar hollow, the surface of the squama is flat, whereas lateral to the temporal lines the barely convex facies temporales slope gradually to the steep, convex medial walls of the temporal fossae. This topography is reflected in the bregma–pterion chord/arc index value of 88%, which indicates a strongly curved squama in coronal section. As the temporal lines converge posteriorly, this curvature becomes progressively flatter.

Lateral View (Left)

The chord length from the most anterior preserved median point on the supraglabellar plate to bregma is 86 mm; chord length from the deepest point in the supraglabellar hollow to bregma is 73 mm; the reconstructed glabella–bregma chord is 93 mm. The squama's midsagittal contour describes a very low, even arc that commences anteriorly in the coronal plane of the supraorbital bars, dipping slightly in a mild supraglabellar hollow, and then gradually curves smoothly and convexly to bregma. The superior surface of the supraorbital bar slopes upward and backward in the same attitude as the sagittal rise of the squama, with the posteriorly bounding temporal crest forming a small superior step on the most elevated point on the bar.

Internal Aspect

The endocranial surface of the frontal squama exhibits strong relief—reflecting frontal lobe convolutions—especially in comparison to the surface of the adhering parietal segments, which is nearly smooth. The blunt, thick posterosuperior end of the frontal crest is preserved, but there is no trace of a sulcus for the superior sagittal sinus that emanates posteriorly from it (this absence does not appear to be due to loss of surface bone).

Breakage of the glabellar mass has exposed pockets of the frontal sinus, which invade bilaterally the medial third of the orbital roofs and the squama as far posteriorly as the rear of the supraglabellar area (Figure 5.2). Preserved curvature of the orbital roofs permits the in-

Figure 5.2 Frontal sinus cavities in the frontal bone of A.L. 444-2. Lateral view of the natural break through the right frontal squama exposing the posteriormost extension of the frontal sinus cavity.

ference of a fairly narrow interorbital region, estimated at 19 mm. The preserved segment of the left orbit's lateral wall passes backward at a slight angle to the sagittal plane, indicating a fairly deep orbital cavity (>35 mm).

Coronal Suture

In the left temporal fossa the parietal margin presents an unfused coronal sutural surface that can be followed from the postorbital constriction superiorly for a chord distance of 48 mm; along this length the suture is accompanied by a distinct liplike thickening of the external table. The external aspect of the suture is completely obliterated for the remaining 37 mm to bregma; portions of the parietals remain attached to the frontal along the fused midsagittal segment of the suture. Internally, faint, fissure-like traces of the coronal suture can be traced from the midline for about 25 mm laterally on the left side and 23 mm on the right. A similar-appearing segment of sagittal suture intersects the coronal suture remnants and thus enables one to pinpoint bregma. At bregma, the limbs of the coronal suture meet at a sharp angle of about 110°–112°.

Thickness

In general, the frontal squama is thick. Along the left limb of the coronal suture, the squama measures between 2.5 and 5.5 mm (ca. 3.0–6.5 mm, if one compensates for differential bone loss); it thins out toward the temporal fossa. At pterion it measures 4.5 mm (+0.5) thick, and at bregma, 7.5 mm (+0.5).

Comparative Morphology

The frontal bone of A.L. 444-2 presents a morphological pattern that, in many respects, is unique among early hominins and extant great apes. Its distinctive aspects derive chiefly from a transversely broad squama coupled with modest postorbital constriction and strong medial incursion of the temporal lines, along with heavy supraorbital bars with prominent, "squared off" superolateral corners.

The Frontal Squama

In A.L. 444-2, the sagittal length of the frontal squama, measured from the supraglabellar hollow to bregma (86 mm), is large, in part due to the posteriorly angled limbs of the coronal suture that place bregma itself in a posterior position. This dimension of the Hadar cranium exceeds the length of the KNM-ER 406 and KNM-ER 13750 frontals of *A. boisei* by nearly 1 cm and is identical to the mean value of the frontal's *entire* midsagittal chord (glabella–bregma) for two specimens of *A. africanus* (Sts. 5 and Sts. 71). However, when the reconstructed position of glabella is used as the anterior terminus of the squama, the length of the frontal in A.L. 444-2 is nearly the same as in the two (male) *A. boisei* specimens (Table 5.1), which have strongly protruding glabellar blocks compared to the slightly built glabellar area reconstructed for A.L. 444-2. (As noted in Chapter 3, our reconstruction of the glabellar region in A.L. 444-2 is based on the morphology of A.L. 438-1b, an undoubtedly male individual, which demonstrates very little anterior protrusion of the glabellar block. See also below.) The glabella–bregma chord distance in A.L. 444-2 is 25% longer than that of our sample of chimpanzee frontals, which are not sexually dimorphic in this respect.

The Belohdelie hominin frontal bone fragment (BEL-VP 1/1) is much shorter sagittally than that of A.L. 444-2, although neither glabella nor the supraglabellar region is preserved on the former specimen. The midsagittal length of the BEL-VP 1/1 frontal from the plane of the postorbital constriction to bregma is 50 mm (Asfaw, 1987); in A.L. 444-2 this distance is 61 mm.

The width of the frontal squama in the coronal plane of the postorbital constriction is large in A.L. 444-2, falling outside the ranges for chimpanzees, gorillas, and *Australopithecus*, which have small postorbital dimensions, with the notable exception of the Belohdelie specimen (see Table 5.1 and Figure 5.3). The postorbital breadths of A.L. 444-2 and BEL-VP 1/1 fall, respectively, 21% and 25% above the mean value for 11 *Australopithecus* crania, among which there is virtually no taxonomically patterned variation, and well within the range of values for early *Homo* specimens (*H. habilis* and *H. rudolfensis*; see Table 5.1).

The great ape data demonstrate a virtual absence of sexual dimorphism in this feature, and this pattern apparently is borne out by the *A. boisei* sample, which includes two specimens widely considered to be female (KNM-ER

Table 5.1 Measurements of the Frontal Bone

Specimen/Sample		Min. Frontal Breadth (mm)	Glabella–Bregma Chord (mm)	Inner Biorbital Breadth (mm)	Frontobiorbital Index[e] (%)
A. afarensis					
A.L. 444-2		77	93	95	81
A. africanus					
Sts. 5		64	77	84	76
Sts. 71		56	68	80	70
Stw. 505[a]		71	n/a	97	73
TM 1511[b]		69	n/a	n/a	n/a
	Mean	65	73	87	73
	SD	7			
A. robustus					
SK 48		67	n/a	93	72
A. boisei					
OH 5		69	n/a	97	71
KNM-ER 406		62	92	100	62
KNM-ER 407		67	n/a	n/a	n/a
KNM-ER 732		60	80	86	70
KNM-ER 23000		66	94	90	73
KNM-ER 13750[c]		64	90	105	61
	Mean	65	89	96	67
	SD	3	6	8	6
A. aethiopicus					
KNM-WT 17000		60	n/a	94	64
Australopithecus sp.					
BEL-VP 1/1		80	n/a	n/a	n/a
Early *Homo*					
KNM-ER 1470		80	93	99	81
KNM-ER 1813		70	80	85	82
OH 24		75	90	90	83
OH 16		72	n/a	n/a	n/a
Great apes[d]					
Pan troglodytes males	Mean	71	70	92	77
	Range	62-77	64-77	83-105	70-86
	SD	4	4	6	5
Pan troglodytes females	Mean	70	71	88	80
	Range	65-76	59-79	75-100	76-88
	SD	3	6	6	3
Gorilla gorilla males	Mean	70	91[f]	111	63
	Range	61-77	90-92	105-118	57-74
	SD	5	n/a	4	5
Gorilla gorilla females	Mean	69	76	99	69
	Range	66-74	66-83	94-109	66-72
	SD	2	5	4	2

n/a = not available

[a]Data for Stw. 505 are from Lockwood and Tobias, 1999.

[b]Minimum frontal breadth of TM 1511 was measured on a cast made from a matrix impression of the frontal squama housed in the Transvaal Museum, Pretoria.

[c]The outer bone table of KNM-ER 13750 is severely abraded; measurements are ±2.0 mm.

[d]Great ape samples are composed of 10 male and 10 female crania for each species; data from Kimbel et al., 1984.

[e]Frontobiorbital index is (minimum frontal breadth/inner biorbital breadth) × 100.

[f]$n = 2$

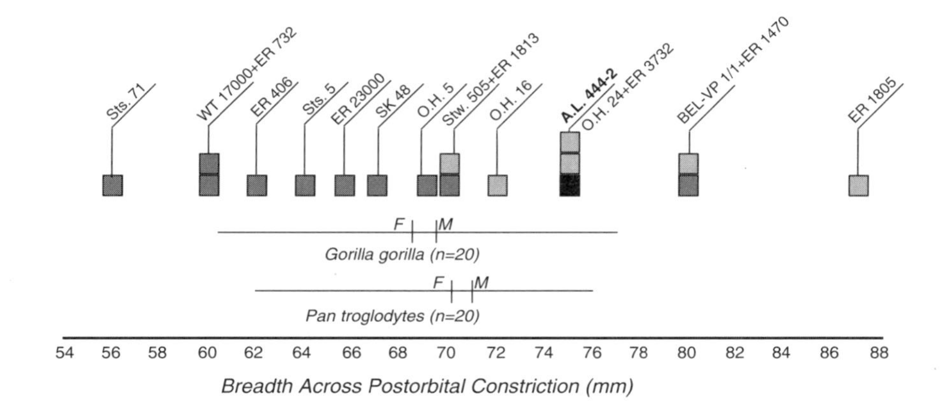

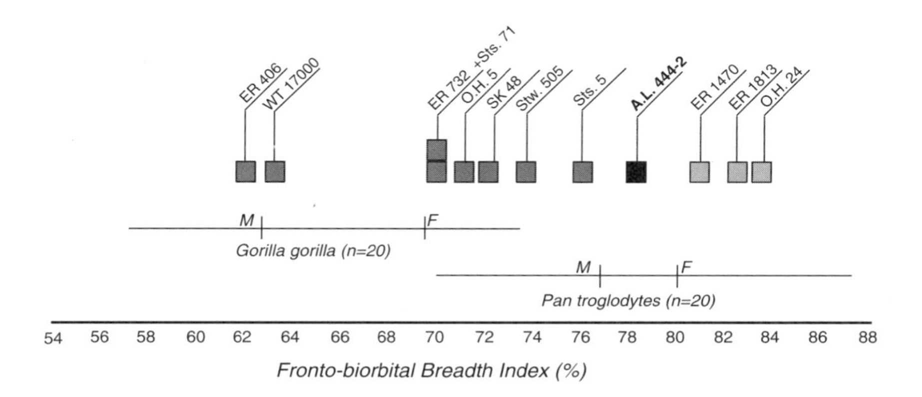

Figure 5.3 Absolute and relative postorbital breadths in A.L. 444-2 and other hominoids. The frontobiorbital breadth index expresses the distance across the postorbital constriction as a percentage of the inner biorbital breadth. Value for BEL-VP 1/1 from Asfaw (1987) and measured by the authors on a cast.

407 and KNM-ER 732) that do not as a group differ in this respect from the presumed male specimens (Table 5.1). It is therefore unlikely that the broad postorbital region of A.L. 444-2 (and BEL-VP 1/1) is particularly related to its male status; the region's breadth is more likely a taxon-specific character that is common to both sexes.

In absolute terms, the *A. boisei* frontal is not any narrower postorbitally than that of *A. africanus*. Only in relation to the width of the face (as seen in the frontobiorbital breadth index; Table 5.1) and zygomatic arches (as discussed in Chapter 3) does the robust australopith cranium manifest an extreme degree of postorbital constriction.

Consistent with the large postorbital width in A.L. 444-2, the preserved section of the lateral wall of the left orbit is set at a very small angle to the sagittal plane as far posteriorly as the coronal plane of the postorbital constriction. In contrast, where a narrow postorbital constriction is combined with a broad upper face, as in *A. boisei*, the lateral wall of the orbit becomes much more coronally oriented, effectively truncating the anteroposterior depth of the orbit. This highly unusual morphology is well seen in *A. boisei* specimens KNM-ER 406, KNM-ER 732, KNM-ER 13750, and OH 5.

This relationship can also be explored metrically (Figure 5.3). When the postorbital breadth is expressed as a percentage of inner biorbital breadth (frontobiorbital breadth index: Kimbel et al., 1984), the index value of 81% for A.L. 444-2 also exceeds the *Australopithecus* range (62%–76%), and it groups with the values for chimpanzees and the small sample of early *Homo* specimens (80.8%–83.3%) (the preservation of the Belohdelie frontal does not allow calculation of this index). As compared to the chimpanzee, the large-bodied gorilla maintains a small relative postorbital width. Crania of robust *Australopithecus*, with an absolutely narrow postorbital constriction and especially wide faces, fall at the low end of the hominin range; the index value for *A. africanus* specimen Sts. 5 (76%) is closest to that of A.L. 444-2, and this similarity is due to the narrow upper face of the Sterkfontein specimen (seen also in Sts. 71). For the Hadar specimen, it can be concluded that the postorbital width of the frontal squama is absolutely and relatively large compared to other crania of *Australopithecus*.

The great discrepancy between the breadth of the frontal squama and the minimum distance between the temporal lines on the frontal of A.L. 444-2 has profound morphological consequences. The large postorbital width places the medial wall of the temporal fossa so far from the midsagittal plane that the facies temporalis is nearly horizontal as it begins its descent from the temporal line into the temporal fossa. In the large, sagittally crested robust *Australopithecus* crania, such as OH 5, KNM-ER 406, and KNM-WT 17000, a narrow postorbital width places the medial wall of the temporal fossa much closer to the midsagittal plane so that the facies temporalis descends directly and steeply into the fossa. This difference in shape may be expressed by the bregma–pterion chord/arc index. In *A. boisei* cranium KNM-ER 406, for example, the coronal margin of the squama shows little curvature (index = 97%; Wood, 1991a), as compared to A.L. 444-2 (88%). In presumptive female *A. boisei* crania (KNM-ER 407 and KNM-ER 732), as well as in *A. africanus* (Sts. 5 and Sts. 71), the postorbital width is small, but the temporal lines usually maintain a relatively lateral position on the frontal such that, compared to A.L. 444-2, a greater proportion of the facies temporalis is vertically oriented. Again, BEL-VP 1/1, alone among known *Australopithecus* frontals, is fundamentally similar to A.L. 444-2 in terms of the large postorbital width, the strong temporal line incursion on the frontal squama, and the horizontally inclined facies temporalis. The small median fragment of sagittally crested frontal squama from Hadar (A.L. 333-125) also suggests this morphology.

In contrast to robust *Australopithecus* frontals, in which the depth of the basinlike trigonum frontale is correlated with the degree of medial encroachment of the temporal lines (Kimbel et al., 1984), in A.L. 444-2 the mild supraglabellar hollow appears unrelated to the degree of temporal line incurvation. The frontal trigone of robust *Australopithecus* crania passes backward into a long, sagittally flat or mildly concave median strip that gives rise to the low, convex midsagittal arc of the squama *posterior* to the coronal plane of the postorbital constriction. In A.L. 444-2, in contrast, the convex curve of the squama rises directly out of the supraglabellar plateau, *anterior* to the plane of the postorbital constriction. In this respect, the frontal squama of *A. africanus* is more similar to that of A.L. 444-2 than to those of *A. boisei*, *A. robustus*, or *A. aethiopicus*, although frequently there is barely any supraglabellar depression at all in *A. africanus* and the squama is generally steeper than in A.L. 444-2, rising right off of the superior surface of the supraorbital elements. The right lateral view of BEL-VP 1/1 indicates a morphology exactly like that seen in A.L. 444-2, with the sagittal contour of the squama medial to the temporal line already convex in the coronal plane of the postorbital constriction. In the other new Hadar adult frontal fragment A.L. 438-1b, which preserves the entire glabellar region, the supraglabellar area is completely flat, both sagittally and transversely (see below).

The Supraorbital Elements

In no other hominin cranium so far known is there evidence of an anteriorly prominent and "squared off" superolateral corner of the orbit as in A.L. 444-2. The near-coronal orientation of the anterior supraorbital margins (in superior view) differs from the condition in *A. boisei* (KNM-ER 406, KNM-ER 732, KNM-ER 13750, OH 5, and Omo 323-1976-896) and especially *A. robustus* (SK 46, SK 48, and SK 52), in which the anterior margins retreat diagonally toward the lateral corner of the orbit from glabella—so strongly in the latter species that it results in a triangular appearance to the supraorbital elements in superior view. Frontals of *A. africanus* are somewhat variable in this feature, but overall they are more similar to *A. robustus* than to any other taxon (note the strong similarity of Stw. 505 to SK 46 in this regard).

The "squared off" superolateral orbital corner in A.L. 444-2 (as seen in frontal view; see Figure 3.36), inferred from the horizontal orientation of the superior supraorbital surface above the lateral wall of the orbit, differs markedly from the homologous morphology in other *Australopithecus* crania. In the comparable sagittal plane of other *Australopithecus* frontals, the superior surface of supraorbital element already faces more laterally than superiorly, having begun its outwardly convex descent toward the frontozygomatic suture and thereby creating the sloping superolateral orbital corner that is so common in *Australopithecus* (Figure 5.4). Superolaterally squared-off supraorbital tori are, of course, a very prominent feature of the majority of chimpanzee and gorilla crania, whose stereotypical "torus and sulcus" supraorbital morphology is otherwise remarkably distinct from that of any known *Australopithecus* specimen, including A.L. 444-2.

The medial to lateral increase in superoinferior supraorbital thickness is another indication of a prominent superolateral supraorbital corner in A.L. 444-2. *Australopithecus* frontals usually show the opposite gradient in supraorbital thickness (medial thickness being greater, as in SK 46, SK 48, SK 52; Stw. 13, Stw. 252, Stw. 505; Sts. 5, Sts. 17, Sts. 71; OH 5; and KNM-ER 732), or are of uniform thickness (as in KNM-ER 406, KNM-ER 13750, and KNM-ER 23000) (Figure 5.4). Although there is moderate variation in this character in chimpanzees and gorillas (it is not, however, sexually dimorphic), by far the most common morphology in the African great apes is the medial to lateral increase in thickness observed in A.L. 444-2.

In lateral view, the superior supraorbital surface in *Australopithecus* is an anteroinferior to posterosuperior

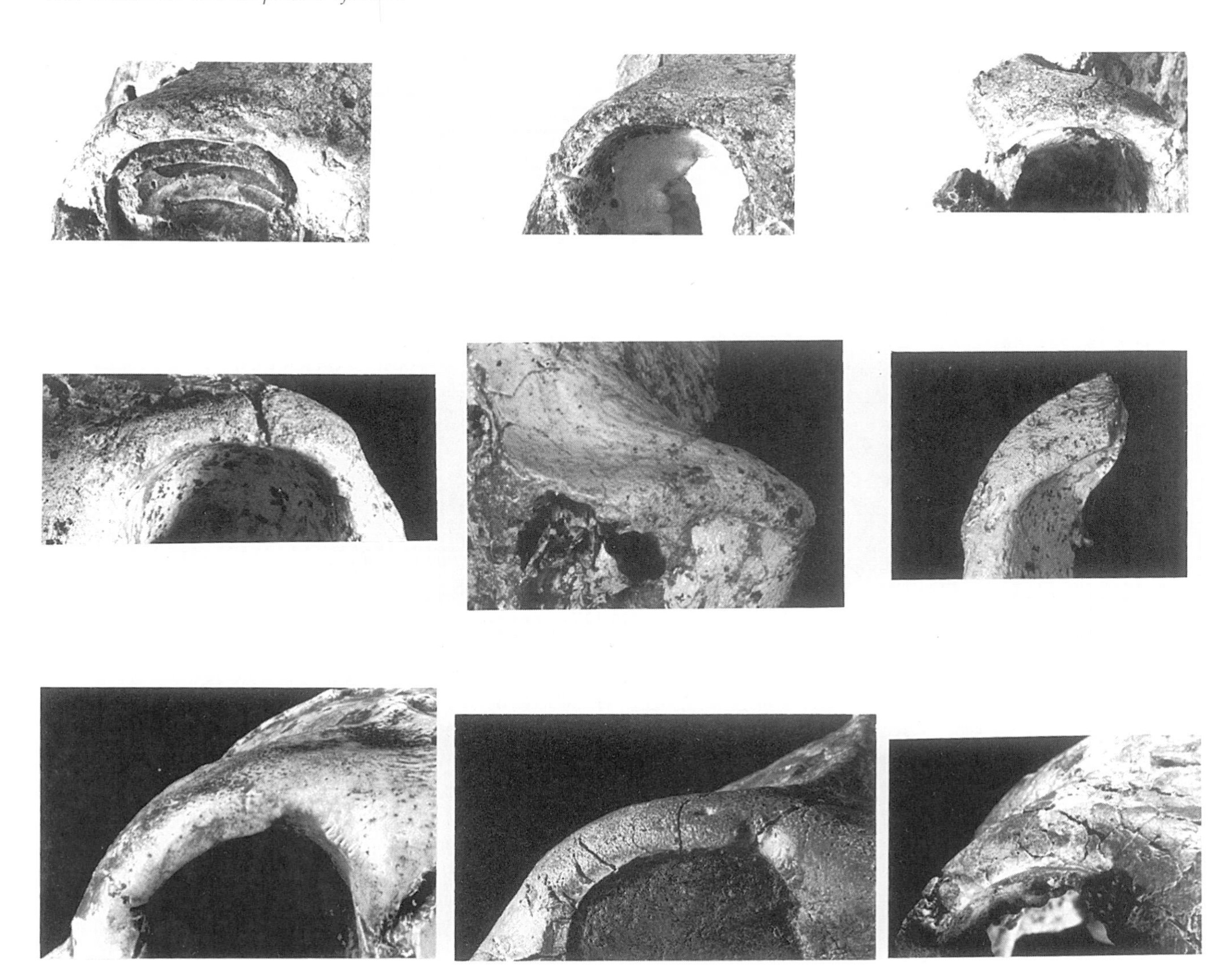

Figure 5.4 Supraorbital morphology in various *Australopithecus* crania. Note the tendency of the supraorbital element to vertically thin and round off laterally. In A.L. 444-2 the supraorbital thickens laterally to a sqaured off superolateral corner. (Top row, *left to right*) *A. africanus* specimens Sts. 5, Sts. 71, and Sts. 17. (Center row) *A. robustus* specimens SK 48, SK 46 (partially crushed), and SK 52. (Bottom row) *A. boisei* specimens OH 5 (cast), KNM-ER 406, and KNM-ER 732.

inclined plane, at the summit of which sits the temporal line, which is a sharp posterior prolongation of the superior surface that usually overhangs the anterior wall of the temporal fossa (Rak, 1983). This is well seen in every adult frontal bone of *A. africanus*, *A. aethiopicus*, *A. robustus*, and *A. boisei*. A different configuration distinguishes the African great apes, as well as *H. habilis* and *H. erectus*, from *Australopithecus*: the supraorbital torus has a vertically protruding, sagittally convex superior surface behind which the temporal line resides at a more *inferior* position than the torus itself. The Hadar specimen resembles *Australopithecus* in the inclination of its flattened superior supraorbital surface, but the steplike projection of the temporal crest above the summit of this surface sets A.L. 444-2 apart from the usual *Australopithecus* condition. In the Belohdelie frontal, the lateral part of the superior supraorbital surface is gently convex sagittally—like that of many chimpanzees—and the temporal line rides the posterosuperior summit of the supraorbital element but does not peak above it, as in the *Australopithecus* pattern. The stepped topography of the inclined superior supraorbital surface in A.L. 444-2 finds a close parallel in the morphology of several Late Miocene hominoid taxa, such as *Dryopithecus laietanus* and *Ouranopithecus macedoniensis* (see, e.g., de Bonis and Koufos, 1993; Begun, 1994; Moyà Solà and Köhler, 1995), which, at least in this respect, diverge from the African great ape supraorbital anatomy as characterized above.

In A.L. 444-2 the supraorbital element retains distinct superior and anterior surfaces along its entire length. In all *A. boisei* frontals, the supraorbital bar twists about its long axis such that above the lateral two-thirds of the orbit a single, anteroinferiorly sloping surface merges with the margo supraorbitalis. Some *A. robustus* specimens (SK 48 and SK 52) recall *A. boisei* in this respect, while others (SK 46) retain distinct surfaces. Specimens that have distinct superior and anterior supraorbital surfaces do not exhibit the twisted long axis of the supraorbital element that is so prominent in *A. boisei*. This is the case, for example, in the African great apes, in all *A. africanus*, and in the sole frontal of *A. aethiopicus* (KNM-WT 17000). The Hadar specimen and BEL-VP 1/1 share the more common—and apparently more generalized—morphology.

Thickness of the Frontal Bone

The A.L. 444-2 frontal squama is thicker than most *Australopithecus* and African great ape frontals. Frontal thickness at bregma (ca. 8 mm) is more than double the mean figure for chimpanzees, which are not sexually dimorphic in this respect. In a sample of five *A. boisei* crania, thickness in the region of bregma ranges from 4 to 8 mm (not including the sagittal crest), with a mean of 5.9 mm. Interestingly, crania generally accepted as female (KNM-ER 407 and KNM-ER 732) have thinner frontals than the male specimens (KNM-ER 733 and KNM-ER 23000), although the heavily built OH 5 is a notable exception to this tendency, with parietal thickness near bregma of only 5.5 mm (Tobias, 1967). Two *A. africanus* frontals have thickness recordings near bregma of 4.5 and 5.5 mm (Sts. 5 and Sts. 71, respectively).

In addition to A.L. 444-2, four Hadar specimens provide information on frontal thickness at or near bregma, and this newly enlarged sample confirms that *A. afarensis* possesses, on average, a relatively thick frontal squama: A.L. 288-1 (5.0 mm), A.L. 333-125 (8.0 mm), A.L. 444-2 (8.0 mm), A.L. 457-2 (9.1 mm), A.L. 701-1 (7.2 mm). The mean thickness at bregma for these five specimens is 7.5 mm. Thick vault bone in the region of bregma is inferred for KNM-ER 2602, a fragment of a small *A. afarensis* calvaria (Kimbel, 1988). With a reading at bregma of 8.0 mm (Asfaw, 1987), the Belohdelie specimen shares a thick frontal squama with the *A. afarensis* sample.

The A.L. 438-1b Frontal Fragment

This frontal bone fragment is the only calvarial part associated with the A.L. 438-1 adult mandible and partial upper limb skeleton (Kimbel et al., 1994). Its importance is enhanced due to the fact that it perfectly preserves aspects of frontal morphology that are lost or damaged in the A.L. 444-2 skull (Figure 5.5).

Medial portions of the supraorbital tori and orbital roofs are preserved, along with adjacent parts of the glabellar block and frontal squama. Frontal sinus cavities are exposed, and the fossae receiving the anterior poles of the frontal lobes of the cerebrum are present endocranially.

The medial portions of the supraorbital elements are vertically thin, measuring about 8 mm inferosuperiorly immediately lateral to the supraorbital notch on the left side. Breaks on both sides give the strong impression that the thickness of the supraorbital elements increased laterally, as in A.L. 444-2. A superior view reveals that the supraorbital elements are clearly divided into distinct superciliary and supraorbital components. The superciliary eminences, whose surfaces have a vermiculate texture, are comma-shaped mounds that protrude weakly but palpably above the superior surface of the squama. Shallow, posterolaterally directed sulci separate them from the supraorbital components; the latter show a smooth rather than vermiculate surface texture. In its separate superciliary component, A.L. 438-1b is similar to most *A. africanus* frontals (Lockwood and Tobias, 1999); in contrast, robust *Australopithecus* supraorbitals typically comprise topographically undivided bars stretching from the glabellar block to the superolateral corners of the orbits (Clarke, 1977).

In superior view, the anterior margins of the supraorbital elements are oriented coronally, but medially they swing suddenly forward to meet the glabellar block, whose maximum anterior projection is about 9 mm beyond the coronal plane of the anterior margins. At the lateral breaks, some 25 mm lateral to midline, the anterior supraorbital contour begins to migrate forward, indicating that, as in A.L. 444-2, the anteriorly most prominent portion of the supraorbital elements occurred laterally rather than medially. As already noted, medial anterior prominence of the supraorbitals is the typical condition of the robust *Australopithecus* frontal bone, and it is also a frequent feature of *A. africanus* morphology.

Using the orbital roofs to orient A.L. 438-1b, it is evident that the frontal squama rises at an angle of about 30° to the horizontal. The superior surface of the squama is completely flat on both the sagittal and coronal planes. Not a trace of a supraglabellar depression or a trigonum frontale is present. Anteriorly the squamal surface dips slightly between the superciliary eminences before it abruptly joins the anterior surface of the glabellar block. On the sagittal plane that corresponds to the position of the supraorbital notches, approximately 20 mm lateral to the midline, the squama is still essentially flat, giving only the slightest hint (on the right side) of curving inferiorly to the region of postorbital constriction. Consequently, there is no trace of the temporal line bordering the su-

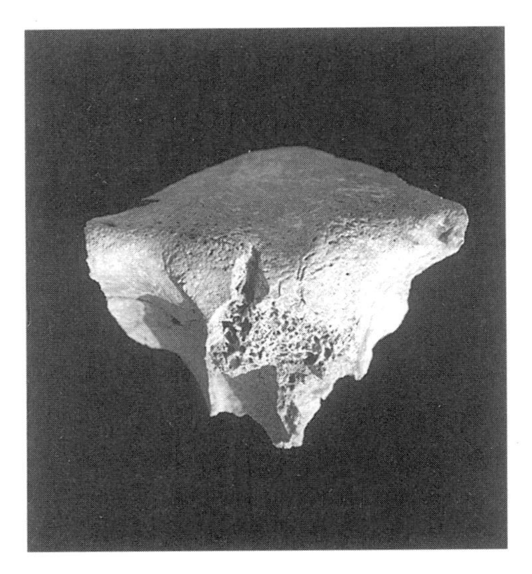

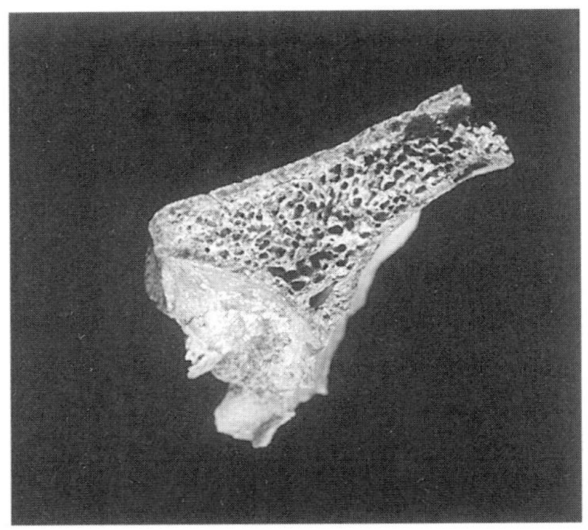

Figure 5.5 *A. afarensis* frontal bone fragment A.L. 438-1b. Anterior view (*above*), lateral view (*center*), superior view (*below*). Note the low, flat squama; thick squama cross section; the coronally oriented supraorbital element; and the flat facet adjacent to the superomedial corner of each orbit (see text for discussion).

praorbital element at this position. The homologous sagittal plane on every robust *Australopithecus* frontal known to us passes lateral to the temporal line and thus finds a coronally convex squamal surface already falling into the tightly constricted postorbital region.

Maximum anterior glabellar projection occurs at a vertical level approximating one-quarter to one-third the estimated height of the orbit (measured from the superior margin). The glabellar region itself appears as a parallel-sided block; estimated interorbital width is ca. 21 mm. On each side of the glabellar mass, adjacent to the superomedial corner of the orbit, is located a roughly oval, 8 × 10 mm, weakly concave facet. Each facet is accentuated by a distinct peripheral rim, strongest inferiorly, that demarcates it from the surrounding supraorbital and glabellar surfaces. The location of these facets, which we have not encountered elsewhere in the hominin fossil record, corresponds to the site of origin of the corrugator muscle of facial expression (see Figure 5.5).

The frontal sinuses are poorly developed. Small chambers, one on either side of midline, are confined to the lower part of the interorbital mass and do not extend into either the orbital roofs or the glabellar block itself. The superoanterior wall of each chamber lies about 16 mm from the external surface of the glabellar region, implying that the glabellar mass is composed of solid bone. Crania of *A. boisei* demonstrate a considerable degree of variation in the size and location of the frontal sinuses (compare KNM-ER 733 with its tiny sinus cavities, to KNM-ER 23000, with its enormous sinus cavities). Such variation appears also to characterize *A. afarensis*, as in A.L. 444-2 the frontal sinus extends laterally into the orbital roofs and frontal squama, as described above.

Although bregma is not preserved on the A.L. 438-1b frontal fragment, this specimen exhibits great squamal thickness, measuring 10.8 mm (just to the right of midline, avoiding the influence of the frontal crest), approximately 35 mm posterior to the coronal plane of the superior orbital margins. This strongly contrasts with the squamal thickness of the frontal fragment associated with the robust *Australopithecus* specimen Omo 323-1976-896, which is only half as thick (5.5 mm) at a point roughly the same distance behind the orbital margins.

The Parietal Bones

External Aspect

The parietals are rectangular, with a longer sagittal dimension: the reconstructed sagittal margin chord length is 93 mm, and the right lambdoidal margin chord is 60 mm, approximately 65% of the sagittal dimension. Overall, the external surface is smoothly domed, with only a hint of bossing at the parietal eminence, which is located posterior to the bone's center. Coronal curvature generally decreases along an anterior to posterior gradient, the lambdoidal margins being fairly flat despite the strong lateral flare of the mastoid angles.

Although most of the coronal margins are destroyed, the posterosuperior course of the coronal suture's limbs on the frontal permits the reconstruction of a very obtuse bregmatic angle for each parietal. A 30-mm long trace of squamosal suture is present on the left parietal, representing the upper edge of the posterior part of the suture's arch as it descends toward asterion. It is marked by a series of very fine, irregularly spaced corrugations for the corresponding inner face of the overlapping temporal squama; in fact, a very thin plate of temporal squama remains tightly adherent to part of this surface. Above the sutural face there are a few short striae parietalis, but abrasion of the external surface has probably removed the majority of them. Emanating from the superior edge of the sutural surface, a set of four or five mild, horizontal ridges sweeps across the parietal posteriorly, traversing the parietal eminence. (These ridges are difficult to see under other than unidirectional light, but are palpable.) Apparently reflecting the origin of horizontally inclined posterior temporalis muscle fibers, these ridges become more prominent posteriorly and terminate in rugose tubercles along the lateral edge of the inferior temporal line. The temporal lines, in the form of blunt, raised crests (or a bifid compound crest), approximate and parallel the sagittal margins of the parietals all along the preserved midline (see description of cresting pattern in Chapter 3). On the right anterior segment just posterior to the coronal suture, there is a prominent, mediolaterally elongate bulge, 16 mm long, centered about 20 mm lateral to the midline. This area is lost on the left side. It appears that the bulge may be related to an irregularity in the course of the temporal line, but the temporal line is not well preserved here. (Brown et al. [1993] describe a very similar irregularity on the external table of both parietal bones in KNM-ER 23000.)

Endocranial Aspect

Details of the endocranial surface are reasonably well preserved only on the left side. The meningeal vessel pattern appears to be simple. A single groove runs posterosuperiorly across the parietal, fading out before reaching the midline. This is apparently for the posterior ramus of the middle meningeal artery, which probably originated from a common trunk with the anterior ramus. It does not appear to give off subsidiary twigs, and the posterior surface of the parietal is otherwise devoid of meningeal vessel grooves. Evidence for the anterior ramus is lacking, possibly because of a loss of endocranial surface bone along the parietal's coronal margin.

Parietal Thickness

The parietal is relatively thick. Parietal thickness near bregma is 8.6 mm (includes 0.5 mm to compensate for eroded external table), and at the left parietal eminence is 7.5 mm. Generally, the parietal becomes thinner anteriorly and inferiorly. On the left side, a maximum thickness of 8.5–9.0 mm is achieved posteroinferiorly, along the lambdoidal margin. Destruction of the endocranial surface negates measurements on the right anterior segment, but the parietal is visibly quite thick here, too.

Comparative Morphology

The sagittal chord of the A.L. 444-2 parietal is absolutely very long compared to those of other *Australopithecus* crania. It is almost exactly a centimeter (12%) longer than the mean value for four *A. boisei* parietal bones; and it is longer still—15 mm (19%)—than the mean for four *A. africanus* parietals (see Table 5.2). The parietal sagittal length is even shorter in chimpanzees than in *A. africanus* (Rak and Howell, 1978), which appears to have the shortest parietal bones of any hominin (this does not take into account Stw. 505, whose cranium is significantly larger than that of any previously known *A. africanus* specimen but whose parietal sagittal dimensions cannot be determined). The Hadar specimen's sagittal length is well within the range of parietal sagittal dimensions for early *Homo*, however, falling closest to KNM-ER 1470 and the subadult OH 7. Neither the coronal margin nor the temporal margin dimensions can be obtained on A.L. 444-2; the shape of the parietal can only be judged (using conventional metrics) on the basis of the lambdoidal margin/sagittal margin chord ratio. As the A.L. 444-2 lambdoidal margin chord value of 60 mm is not unusual for *Australopithecus* parietal bones (Table 5.2), the low index of 65% must signal a relatively long sagittal dimension rather than a short lambdoidal dimension for the Hadar specimen. In fact, it is relatively very long compared to the condition in other *Australopithecus* species for which there are reasonable samples: the mean ratio for three *A. africanus* parietals is 71%, and for four *A. boisei* parietals it is 74%. Modern humans show a very similar ratio (72%), whereas chimpanzees have a much higher index (79%), consistent with their squarer parietal bones (see Table 5.2). The only specimen in the early hominin sample that shows the relative sagittal elongation of the parietal like that seen in A.L. 444-2 is the immature *H. habilis* specimen OH 7 (Table 5.2), in which the low index of 63% is achieved through the same combination of an extremely long sagittal dimension and a "typical" early hominin lambdoidal dimension. Interestingly, the shape of OH 7's parietal strongly contrasts with that of KNM-ER 1470

Table 5.2 Measurements of the Parietal Bone

Specimen/Sample		SMC (mm)	SMA (mm)[e]	SMC/SMA × 100	LMC (mm)	LMC/SMC × 100
Australopithecus afarensis						
A.L. 444-2		93	106	88	60	65
A. africanus	Mean	78	86	91	56	72
Sts. 5		85	92	92	60	71
Sts. 25		76	84	90	54	71
Sts. 71		74	78	95	n/a	n/a
MLD 37/38		77	88	88	55	71
A. boisei	Mean	83	86	92	60	74
KNM-ER 406		81	86	94	62	77
KNM-ER 407		78	85	92	55	71
KNM-ER 13750[a]		88	n/a	n/a	61	69
KNM-ER 23000[b]		86	n/a	n/a	66	77
OH 5[c]		n/a	n/a	n/a	57	n/a
H. habilis	Mean	86	92	94	62	72
KNM-ER 1813		79	82	96	64	81
OH 7		99	105	94	62	63
OH 13		81	87	93	60	74
OH 16		n/a	n/a	n/a	63	n/a
H. rudolfensis						
KNM-ER 1470		95	101	94	72	76
P. troglodytes (n = 10)[d]		63	66	95	50	79
H. sapiens (n = 15)[d]		120	134	90	86	72

SMC, sagittal margin chord; SMA, sagittal margin arc; LMC, lambdoidal margin chord.
[a]Data from Walker and Leakey, 1988.
[b]Data from Brown et al., 1993.
[c]Data from Tobias, 1967.
[d]Data from Rak and Howell, 1978.
[e]Slight discrepancies relative to values in Table 3.7 result from different measurement techniques.

(type specimen of *H. rudolfensis*), in which a long sagittal margin is coupled with a long lambdoidal margin, yielding a much higher index of 76% (Table 5.2). It is possible, but not certain, that this difference is due in part to the immaturity of OH 7, as relatively late growth changes involving the lateral expansion of the asterionic region may produce strong lateral flare of the mastoid angle of the parietal bone (Kimbel and Rak, 1985). Although such lateral prominence is not usually a feature of the *Homo* calvaria, it is well developed in KNM-ER 1805, a probable male specimen of *H. habilis.*

In the posterior position of the parietal eminence, the A.L. 444-2 parietal bone differs from those of extant great apes, in which the parietal eminence is located anterior to the bone's center. In most australopiths the eminence is located centrally (as in modern humans) or, less commonly, just posterior to the bone's center (as in the Hadar specimen).

As with the frontal bone, the parietal bone is thick in A.L. 444-2 compared to that in other early hominins, both in bregmatic region (8.6 mm) and at the parietal eminence (7.5 mm). Among the *Australopithecus* specimens available for comparative measurement (n = 7 individuals; Table 5.3), parietals tend to be quite thin in the bregmatic region. Only KNM-ER 23000, an unusually thick-boned, young adult *A. boisei* calvaria (Brown et al., 1993), approaches the Hadar specimen's thickness near bregma. As emphasized by Tobias (1991), thin parietal bones are also prevalent in *H. habilis*; our measurements show a mean thickness at bregma identical to that calculated for two individuals of *A. africanus*. The three parietal bones of *H. rudolfensis* tend to be thicker than those of *H. habilis*, a millimeter thicker at bregma on average (Table 5.3).

A similar picture emerges when parietal thickness at the parietal eminence is considered, although in this case we have several *A. afarensis* specimens available for comparison with A.L. 444-2 (Table 5.3). However, a comparison of sample means reveals only minor differences among the early hominin taxa under consideration. In the sample from Hadar there are both thick and thin parietals, the latter associated with small specimens we consider as female (A.L. 162-28 and A.L. 288-1). An almost identical level of sexual dimorphism characterizes the *A. boisei* sample, with big males (OH 5 and KNM-ER 23000) having much thicker bone than females (KNM-ER 407 and KNM-ER 732) at the parietal eminence.

Table 5.3 Measurements of Parietal Bone Thickness (mm)

Specimen/Sample		At Bregma	At Parietal Eminence
A. afarensis	Mean	n/a	5.7
A.L. 444-2		8.6	7.5
A.L. 333-45		n/a	7.0
A.L. 288-1		n/a	4.6
A.L. 162-28		n/a	3.5
A. boisei	Mean	5.7	5.6
KNM-ER 407		5.0	4.0
KNM-ER 732		4.0	4.0
KNM-ER 23000[a]		8.2	7.5
OH 5[b]		5.5	7.0
A. africanus	Mean	5.1	6.2
Sts. 5		4.1	5.9
Sts. 71		6.0	6.0
MLD 1		n/a	6.8
H. habilis	Mean	5.1	5.5
OH 7[c]		4.2	5.0
OH 13		5.5	4.0
OH 24		4.0	5.0
KNM-ER 1813		6.0	6.5
KNM-ER 1805		6.0	7.0
H. rudolfensis	Mean	6.2	n/a
KNM-ER 1470		6.0	7.5
KNM-ER 1590[c]		6.7	n/a
KNM-ER 3732		6.0	n/a

[a]Data from Brown et al., 1993.
[b]Data from Tobias, 1967.
[c]OH 7 and KNM-ER 1590 are subadults.

The Temporal Bones

Description is based on the better-preserved right temporal bone unless otherwise noted.

Lateral Aspect

Maximum anteroposterior length of the temporal bone, from asterion to the sphenosquamosal suture in the temporal fossa, is 85 mm. Superior and anterior to the external auditory meatus (EAM) the temporal squama's facies temporalis is vertical, but progressively reclines posteriorly so that in the asterionic region it is a wide, nearly horizontal, inferolaterally sloping shelf (Figures 3.47 and 5.6). On each side there are two "squamosal" sutures descending into the asterionic region instead of one; it appears that facies temporalis is actually composed of two separately ossified bones, the squama proper and superior to this a "squame–parietal ossicle" (Brothwell, 1981: 94), that together extend inferiorly from the summit of the squamosal suture to the deeply invaginated incisura parietalis (i.e., the junction of the squamosal and parietomastoid sutural margins of the temporal bone) in the asterionic region (Figure 3.6). Within the incisura on the left side can be seen what would normally be the laterally flared mastoid angle of the parietal bone but in A.L. 444-2 is a separately ossified "parietal notch bone." On the right side the incisura is empty, leaving a transversely elongated (20.5 mm), diamond-shaped imprint of the separate parietal notch bone that extends from the incisura on to the fused parietal and occipital bones just above asterion (Figure 5.6). Although convoluted by the extra ossified elements, the morphology of the A.L. 444-2 asterionic region bears all of the hallmarks of the asterionic notch sutural pattern (see description of cresting pattern in Chapter 3).

The superior margin of the incisura parietalis is coincident with the supramastoid crest, a blunt, linear ridge that angles anteroinferiorly from the asterionic region, across the pars mastoidea to the EAM, where it forms the posteroinferior limit of a large suprameatal triangle. A shallow, 4.5-mm wide supramastoid sulcus separates the supramastoid crest from the prominent, roughened mastoid crest below it. Laterally, owing to the inferior angulation of the supramastoid crest, the crests begin to converge and, just behind the EAM, they unite, obliterating the sulcus. The lateral projection of the mastoid crest very slightly exceeds that of the supramastoid crest, thus occupying the most lateral position on the posterior part of the calvaria.

The massive, pyramidal pars mastoidea features a highly rugose mastoid crest that marks the insertion of the sternocleidomastoid and splenius capitis muscles and an extensive, flattened posterolaterally directed face, which terminates in a well-developed, crestlike tip that is strongly inflected anteromedially under the cranial base. The right mastoid tip projects vertically 29 mm below porion (this distance is slightly smaller on the left) and is very close to the coronal plane of porion.

The right EAM is circular, with inferosuperior and anteroposterior dimensions of 11.8 mm. The tympanic forms the inferior half of the EAM's anterior wall but none of its posterior wall, which is thus formed entirely by the anterior face of the pars mastoidea. Thickness of the tympanic's lateral margin varies between 3.0 and 3.5 mm, with maximum thickness achieved inferiorly. Immediately posterosuperior to the EAM is a strong suprameatal spine, which defines the inferior limit of a triangular suprameatal fossa whose major anteroposterior axis runs from the supramastoid crest forward for 17.5 mm to auriculare. In the coronal plane of the articular eminence of the mandibular fossa the zygomatic process has a humped superior margin (see discussion in Chapter 3). Maximum vertical depth of the right zygomatic process is 21 mm (24 mm on the left) and it is 8 mm thick mediolaterally.

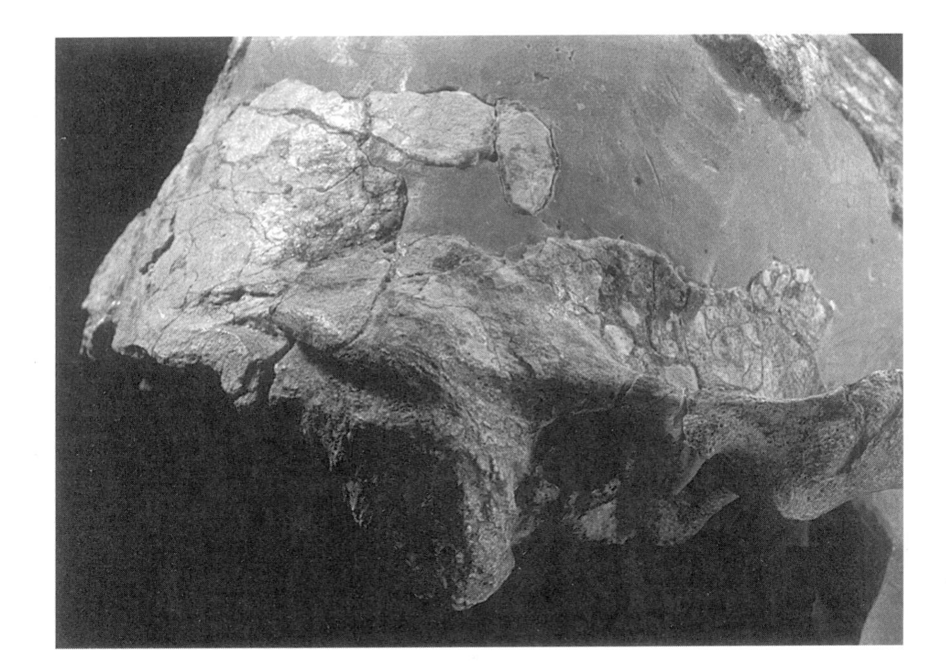

Figure 5.6 Lateral view of the right temporal bone of A.L. 444-2. Note the proximity of the mastoid process tip to the coronal plane of porion and the anterior disintegration of the compound temporonuchal crest on the pars mastoidea. Note also the empty slot for the (lost) asterionic ossicle.

Superior Aspect

The root of the zygomatic process diverges at an angle of about 35° to the sagittal plane, strongly overhanging the external auditory meatus to define the base of a ~20 mm broad, mediolaterally concave supraglenoid gutter. At the anterior edge of the gutter, the zygomatic process deflects medially into line with the sagittal plane. From the root of the zygomatic process at auriculare to the anterior edge, the sagittal length of the gutter is 34.5 mm.

The base of the temporal squama is very thick: maximum thickness above the EAM is 10.5 mm, increasing anteriorly to 13 mm at the level of the front edge of the supraglenoid gutter. On the left side, the broken section immediately above the EAM reveals several large mastoid air cells, indicating aggressive proliferation of mastoid pneumatization anteriorly into the squama itself. Thickness of the sphenosquamosal suture face ranges from 4.5 mm at the base of the facies temporalis to 7 mm immediately anterior to the foramen ovale. Across the right occipitomastoid suture, maximum thickness is 21 mm.

Basal Aspect

Measured according to the method of Wood (1991a), the mandibular fossa is wide mediolaterally (35 mm), but relatively short anteroposteriorly (19 mm), producing a length/breadth index of 54.3%. The articular eminence, with a maximum direct width of 37.5 mm, weakly projects inferiorly, shows little sagittal curvature medially but moderate convexity laterally, and is delimited from the ceiling of the mandibular fossa posteriorly by a distinct lip. The lowest point on the sagittal cross section through the middle of the eminence is in a coronal plane posterior to the edge of the temporal foramen. An expansive preglenoid plane fronts the medial half of the eminence. The line denoting the anterior attachment of the articular synovial capsule is clear and does not extend onto the preglenoid plane. The anteroposterior breadth of the circumscribed articular surface narrows from 15.5 mm laterally to 13.5 mm medially.

Beyond the sharp lateral lip of the eminence, the superolaterally inclined external face of the zygomatic process root is roughened and pitted by the attachment of the joint capsule and the temporomandibular (lateral) ligament. The roof of the mandibular fossa has an abruptly defined lateral termination as the result of the distinct angle it forms with the lateral face of the zygomatic process. Medially the eminence gives rise to a stout entoglenoid process whose prominent, inferiorly projecting apex—composed entirely of temporal squama—forms the medial wall of the articular fossa. None of the sphenoid remains, exposing the temporal's sphenosquamosal sutural face. The lateral margin of the foramen ovale is present, and it is evident that the foramen straddles the suture. The foramen spinosum, located on the entoglenoid process apex, appears also to be split by the suture.

The lateral margins of the articular eminence and postglenoid process co-occupy the most lateral position on the cranial base. Nearly all of the postglenoid process is situated lateral to the plane of the EAM. The process, about 8 mm deep inferosuperiorly and with a sharp inferior margin, lies in a coronal plane 2.5 mm anterior to that on which the tympanic element is situated.

The tympanic element is fairly long (lateral margin to center of carotid foramen = 31 mm), but its lateral edge (the plane of the EAM) is recessed under the base. It is tubular rather than platelike in form, with an undifferentiated crista petrosa that denotes a smooth contour change between the minor inferior and major anteroinferior faces. From the lateral margin 10 mm medially, the tympanic is deeply incised and convex inferosuperiorly. Medial to the incisions, the tympanic axis bends anteriorly and the major face flattens both inferosuperiorly and mediolaterally. Midway along its length, at the level of the stylomastoid foramen, the vertical depth of the anteroinferior face is 14.5 mm, but it shallows laterally as the tympanic curves up behind the postglenoid process. Medially the tympanic gives rise to a small, pneumatized tubercle just anterior to the carotid foramen, from whose anterior margin the tubercle is separated by a strong 2.5-mm wide groove. The foramen, measuring 6.8 mm mediolaterally by 4.8 mm anteroposteriorly, opens inferiorly and slightly medially. On the left side, at the foot of the occipitomastoid suture, 20.5 mm medial to the tympanic's lateral margin, a small vagina of the styloid opens inferomedially into a narrow groove bounded posteriorly by a prominent paramastoid process that projects inferiorly and laterally from the pars jugularis of the occipital bone. Immediately posterolateral to the vagina is a round 2.0-mm wide stylomastoid foramen. There is no vaginal process of the styloid and no evidence of an ossified styloid process.

Except for part of the right jugular fossa, little basal petrous anatomy remains owing to breakage on the right and crushing on the left. The mastoid tips lie on a slightly more medial plane to that of the EAM. Medial to the left tip is a deep, triangular digastric fossa (rather than a groove) that fans out anteriorly to an 11 mm width and which impinges slightly on the posterior face of the mastoid. The fossa is bounded medially by a very robust (6-mm wide), pneumatized occipitomastoid crest whose inferior projection is so pronounced that it equals that of the mastoid tip itself. On the right side the occipitomastoid crest is thicker (9 mm) and the digastric fossa is irregularly shaped due to its indenting of the mastoid tip's medial face. The occipitomastoid crest is actually the inferior prolongation of the mastoid's extensive occipitomastoid sutural face; that is, the crest and suture overlap. The sutural surface is exposed on both sides of the specimen, owing to the breakage of the thin, downturned lateral margins of the occipital's planum nuchale that originally had overlapped it.

A distinct, 1.3-mm wide groove for the occipital artery traverses the anterior end of the left occipitomastoid crest parallel and 1.5 mm lateral to the occipitomastoid suture adjacent to the jugular process of the occipital bone. This groove is not visible on the right, possibly due to breakage along the suture. On each side a small, but distinct mastoid foramen pierces the mastoid's occipitomastoid sutural face. Reconstructing the broken occipital indicates that the mastoid emissary vein penetrated the external surface through the sutural crevice itself.

Endocranial Aspect

The posterior face of the petrous pyramid is nearly vertical and 16.5 mm deep laterally. The internal acoustic meatus, located 3.3 mm below the superior petrous margin, measures 5.5 × 5.0 mm and has a sharp lateral margin that overhangs the meatal aperture. Just posterior to the meatus, the area of the subarcuate fossa is depressed and slightly overhung by the superior margin. Below the depressed area is a strong bulge over the posterior semicircular canal, immediately lateral to which the posterior surface bears a strong oval hollow that dramatically undercuts the superior margin. On the left side the hollowing of the posterior surface is much less marked, and here the petrous preserves a very large aperture for the vestibular aqueduct in the medial part of the hollow, probably indicating a large endolymphatic sac.

The superior margin of the right petrous is blunt medially and sharp laterally where it overhangs the posterior face. On the left, the superior margin is thick and rounded medially and dull laterally; consequently it does not overhang the posterior face except along its most lateral segment. The anterior surface of the right petrous drops gently but steadily into the middle cranial fossa. The arcuate eminence is indistinct, and, aside from a shallow circular depression laterally, the preserved anterior surface is featureless. Apical anatomy is not preserved on either side.

A very faint but palpable sulcus for the superior petrosal sinus rides along the superior margin of the petrous on the left, entering the sulcus for the sigmoid sinus just medial to the junction of the petrous with the inner calvarial wall. A superior petrosal sulcus is not discernible on the right side. Much deeper is the 2.4- to 3.0-mm wide groove for the petrosquamous sinus, which can be followed along the posterior half of the eponymous junction on the right temporal, notching the superior margin as it descends to join the sigmoid sulcus, which is ill defined within the depth of the lateral hollow on the posterior face. Only faint, anastomosing vascular grooves and two tiny foramina (marking the outlet of the mastoid emissary veins) mark the floor of the hollow.

The left temporal bone bears no trace of a sulcus for the petrosquamous sinus. However, the middle meningeal vascular trunk carves an unusual path on this side. From the foramen spinosum, the meningeal sulcus travels along the petrosquamous junction laterally as far as the calvarial wall before splitting into three deeply impressed branches, two of which climb up onto the anterior surface of the petrous before resuming their ascent toward the parietal bone. The left sigmoid sulcus is slightly more developed than the right, perhaps because the hollowing of the posterior face is weaker on this side. On neither side does the sulcus for the transverse sinus appear to connect to the sigmoid sulcus; this connection is a signpost of a dominant occipital-marginal sinus system (see description of the venous sinus pattern in Chapter 3).

Comparative Morphology

Previous comparative studies of the *A. afarensis* cranium highlighted the strongly plesiomorphic (apelike) temporal bone anatomy relative to those of other *Australopithecus* species (White et al., 1981; Kimbel and Rak, 1985; Kimbel et al., 1985; Picq, 1985). In many ways, the temporal bones of A.L. 444-2 reiterate this primitive pattern. However, the new Hadar skull adds information on variation in temporal bone anatomy that may have phylogenetic significance vis-à-vis even more ancient hominin species such as *A. anamensis* (M. Leakey et al., 1995) and *Ard. ramidus* (White et al., 1994, 1995).

Temporal Squama

Two features of the temporal squama have commanded attention in previous discussions of *A. afarensis* cranial morphology: (1) the posterior flattening of the facies temporalis in the asterionic region, and (2) the extensive proliferation of mastoid pneumatization into the base of the squama (Kimbel et al., 1984; Kimbel and Rak, 1985).

The progressive flattening of the facies temporalis as one moves posteriorly along the base of the squama from auriculare to the asterionic region is seen in all three adult *A. afarensis* temporal bones in which the region is preserved (A.L. 333-45, A.L. 333-84, and A.L. 444-2). These are all male specimens that exhibit the asterionic notch configuration in association with a compound temporal/nuchal crest that traverses the asterionic region. Kimbel and Rak (1985) demonstrated in the gorilla how, with growth, the initially vertical posterior part of the facies temporalis "folds" back on itself and flattens out into a horizontal shelf as the temporalis muscle migrates posteriorly into the asterionic region. In *A. afarensis*, as in the great apes, this is part of the suite of ontogenetic modifications comprising the asterionic notch sutural pattern as a morphogenetic concomitant of the cresting process.

Shelving of the facies temporalis in the asterionic region occurs in very few crania outside of *A. afarensis*. The squama stays quite vertical, or at most slightly inclines medially, above the supramastoid crest in almost all *A. boisei* calvariae—even the most heavily crested ones (KNM-ER 406, KNM-ER 407, KNM-ER 732, KNM-ER 13750, KNM-ER 23000, and Omo 323-1976-896). Only in OH 5 is there a horizontal supramastoid shelf—a posterior extension of the zygomatic process root—but even here the facies temporalis is nearly vertical where it inclines on the weakly flared mastoid angle of the parietal bone above asterion. It is this vertical supramastoid profile, combined with the remarkable lateral projection of the pars mastoidea below the supramastoid crest, that gives the *A. boisei* calvaria its characteristic profile in occipital view (see Figure 3.48). A single exception to the typical *A. boisei* description is encountered among the crania of *A. robustus*: TM 1517, whose morphology recalls that seen in OH 5.

Two non–*A. afarensis* hominins do appear to have the Hadar type of squamous shelf in the asterionic region: KNM-WT 17000 (*A. aethiopicus*) and KNM-ER 1805 (*H. habilis*). That both of them also share with *A. afarensis* evidence for strong posterior temporalis muscle components, as well as massive compound temporal/nuchal crests and the asterionic notch articulation (Kimbel and Rak, 1985; R. Leakey and Walker, 1988), nicely illustrates the developmental link among this suite of characters.

In *A. afarensis* temporal bones, as in those of the great apes, mastoid pneumatization proliferates anteriorly as far as the sphenosquamosal suture, resulting in great thickness of the temporal squama at the base of the calvaria's lateral wall (Kimbel et al., 1984). Extreme pneumatization of the squama places A.L. 444-2 at the upper end of the range of squama thickness in the Hadar sample (measured just above the floor of the supraglenoid shelf; see Table 5.4). The mean thickness for five *A. afarensis* temporals (9.9 mm) is substantially greater than for any other hominin species, including the most craniodentally robust taxon, *A. boisei*, even though one male cranium of the latter species, KNM-ER 23000 (Brown et al., 1993), has a squamal thickness equal to that of A.L. 444-2. This *A. boisei* specimen has notably thick bones throughout the membranous calvaria (Brown et al., 1993), as does *A. afarensis*. However, relatively thin frontal, parietal, and occipital bones, combined with an extremely thick temporal squama in great ape crania, show that thickening due to diploic expansion and that resulting from mastoid air cell penetration of the temporal squama are not necessarily correlated. Notwithstanding KNM-ER 23000, the fact that no other undoubted male *A. boisei* cranium shows the degree of squamosal inflation of any specimen in the Hadar temporal bone sample (composed chiefly if not entirely of male individuals) underscores the value of this character for distinguishing *A. afarensis* from other early hominin taxa.

Zygomatic Process

The A.L. 444-2 cranium is the first adult *A. afarensis* specimen to preserve the zygomatic process of the temporal

Table 5.4 Thickness of the Temporal Squama[a] (mm)

A. afarensis		Robust *Australopithecus*		*A. africanus*		Other Hominins		African Apes	
A.L. 444-2	10.5	OH 5	6.0	Sts. 5	5.5	KNM-BC 1	6.0	Chimpanzee m	10.5
A.L 58-22	10.5	KNM-ER 407	5.5	Sts. 71	5.5	Sts. 19	8.5	Chimpanzee m	12.5
A.L. 166-9	9.5	KNM-ER 732	5.0			KNM-ER 1470	6.0	Chimpanzee m	11.0
A.L. 333-84	9.1	KNM-ER 23000	10.5			O.H. 13	5.2	Chimpanzee f	7.0
A.L. 333-45	10.0	Omo 323-1976-898	6.0			L. 894-1	4.5	Chimpanzee f	7.0
		TM 1517	4.5			SK 847	5.0	Chimpanzee f	11.0
Mean	9.9	Mean	6.3	Mean	5.5	Mean	5.9	Gorilla m	14.0
SD	0.62	SD	2.16	SD		SD	1.42	Gorilla f	11.0

[a]Mediolateral thickness of the base of the temporal squama measured in the plane of the anterior edge of the supraglenoid gutter.

bone (this structure is intact on the young juvenile cranium A.L. 333-105; see Kimbel et al., 1982). Two features are worthy of comparative note. The first, as already described in Chapter 3, is the abrupt upward convexity in the superior margin located just anterior to the coronal plane on which the tubercles of the articular eminences reside (Figures 5.6 and 5.7). This upward convexity is frequently encountered in orangutan and gorilla crania, but it is rare in chimpanzees and modern humans. Among *Australopithecus* species, the convexity is observed in *A. robustus* (e.g., SK 46, SK 48, and SK 52) and *A. aethiopicus* (KNM-WT 17000), but not in *A. boisei* or *A. africanus*. However, unlike the "humped" superior margin of the zygomatic process in gorillas, orangutans, *A. robustus* (and probably *A. aethiopicus*), in which the superior margin returns to a lower position anteriorly, in A.L. 444-2 the superior margin remains elevated, taking a straight path to the zygomaticotemporal suture.

The second noteworthy feature of the zygomatic process is found at its root, on the lateral and basal surfaces immediately lateral to the articular tubercle. This is the site of the deep masseter muscle origin and lateral ligament attachment, which in A.L. 444-2 is (in basal view) a superolaterally sloping plane marked by heavy pitting all along the lateral side of the articular tubercle, giving the latter an abrupt, well-defined lateral edge. In *A. boisei*, as first noticed by Clarke (1977), the surface of the zygomatic process root lateral to the articular tubercle is not a single plane but, instead, is divided by a very strong ridge that extends upward from the lateral edge of the articular tubercle on the lateral surface of the root of the process, dividing it into two surfaces: a smooth, concave posterior surface that is topographically continuous with the roof of the mandibular fossa; and a strongly excavated, rugose, anterior surface that bridges the front of the articular tubercle to the masseteric surface along the wide inferior margin of the zygomatic process (Figure 5.7). In fact, the homolog of the dividing ridge is seen in the Hadar skull but is manifested merely as a faint, vertical crest that originates in the posterolateral corner of the articular tubercle rather than, as in *A. boisei*, in the middle or anterior part of the tubercle's circumference. In A.L. 444-2 the ridge is insufficiently expressed to divide the lateral face of the zygomatic process root into discrete surfaces.

Every adult *A. boisei* temporal bone manifests a distinct division to one degree or another. It is most extreme in KNM-ER 406 and KNM-ER 13750, but is also clearly visible in KNM-ER 732, KNM-ER 23000, and OH 5. Adult crania of *A. robustus* likewise show the bipartite division of the zygomatic process root's lateral surface (i.e., TM 1517, SK 46, SK 48, and SK 83), although the surface posterior to the dividing ridge is oriented at a right angle to the roof of the fossa, whereas in *A. boisei* these surfaces usually flow into one another (but not always; compare OH 5 or Omo 323-1976-896 with SK 48). Interestingly, the single cranium of *A. aethiopicus* (KNM-WT 17000) is very similar to A.L. 444-2 in lacking the division. In the great apes, even in male gorillas, this lateral surface is smooth and undivided.

Mastoid Process

In A.L. 444-2, and in all other adult *A. afarensis* crania so far known, the maximum lateral projection of the mastoid process is situated high up on the pars mastoidea, coincident with the conjoined supramastoid/mastoid crests, and is vertically aligned with the center of the external auditory meatus. From the point of maximum projection to the tip, the mastoid's lateral contour (seen in occipital view) describes a long, straight, inferomedially angled line (Figures 3.47 and 5.8). This is essentially the anatomy of the great ape mastoid process, except that in these species the point of maximum lateral projection is even higher on the temporal bone, superior to the level of porion, owing to the very high position of the nuchal crest from which the supramastoid and mastoid crests descend toward the mastoid process. Most other early hominins retain this aspect of *A. afarensis* morphology as a fundamental component of their mastoid anatomies,

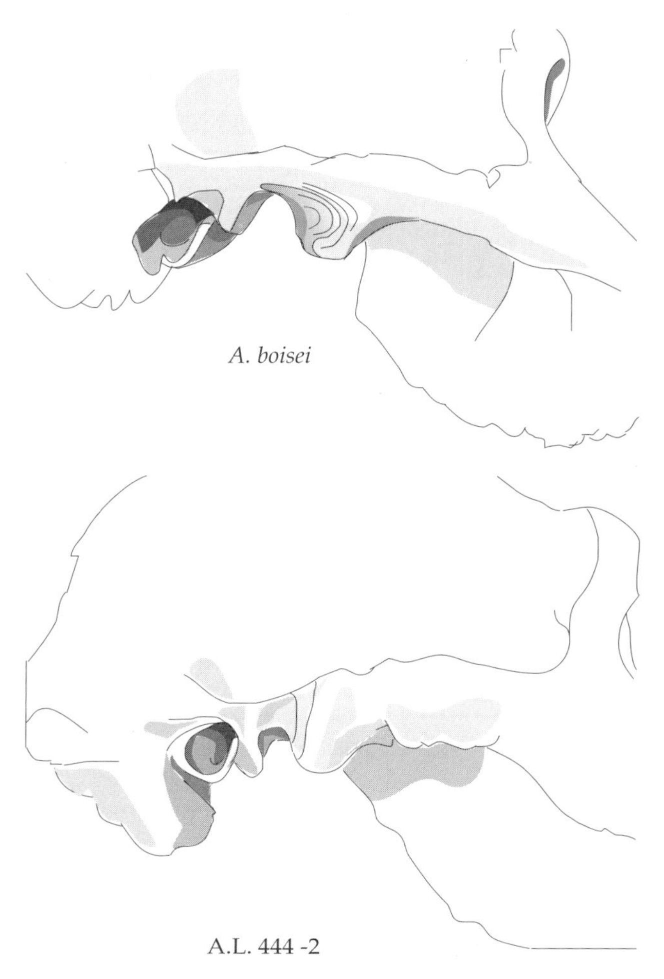

Figure 5.7 The lateral surface of the articular tubercle and the posterior root of the zygomatic arch. Note the concave bone surface of the tubercle in *A. boisei* and the flat, convex surface in A.L. 444-2.

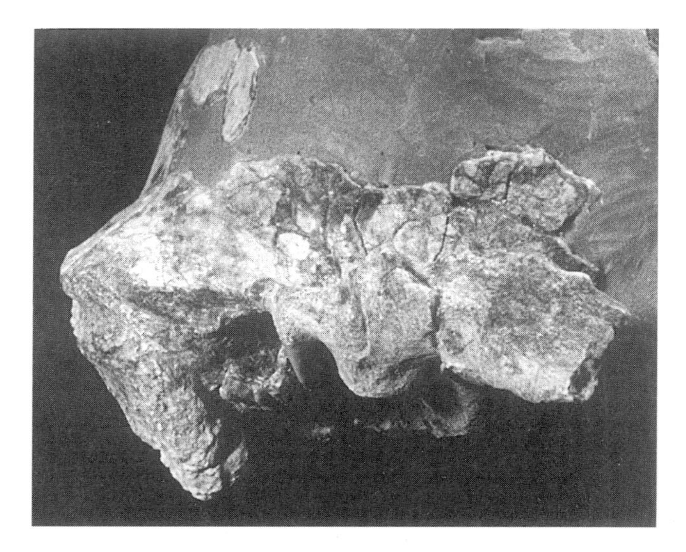

Figure 5.8 Anterolateral view of right temporal bone of A.L. 444-2. The lateral contour of the mastoid describes a long, straight line that angles anteromedially beneath the cranial base.

even though in almost all of them the supramastoid and mastoid crests are widely separated vertically: *A. africanus* (MLD 37/38, Sts. 5, and Sts. 71), *A. robustus* (SK 46, SK 52, and SK 83; but not TM 1517) and *Homo habilis* (KNM-ER 1813, Stw. 53, and Stw. 98; but not KNM-ER 1805).

The mastoid process of *A. boisei* contrasts with this common hominin morphology. The maximum lateral projection of the mastoid process occurs much more inferiorly, actually *below* the level of the external auditory meatus in most cases (Figure 3.48). Related to the low position of maximum projection is the fact that the lateral margin of the process, rather than slanting inferomedially under the cranial base, is a laterally convex bulge (in occipital view), with only a minor inflection of the mastoid tip under the base of the calvaria. This morphology is observed in every known *A. boisei* temporal bone, both male (KNM-ER 406, KNM-ER 23000, OH 5, and Omo 323-1976-896) and female (KNM-ER 407 and KNM-ER 732).

The inflection of the mastoid process under the cranial base (in lateral aspect) has a marked anterior component in A.L. 444-2, bringing the mastoid tip close to the coronal plane of porion (Figure 5.9). A good way to quantify this character is to calculate the horizontal distance between the coronal plane of porion and that of the mastoid process tip as a percentage of the projected horizontal distance between porion and asterion ("mastoid tip position index"; Table 5.5). The smaller the calculated value, the closer the mastoid tip is to porion. For A.L. 444-2 this figure is ca. 29%, which is very close to the values for two other adult *A. afarensis* temporal bones (A.L. 333-45 and A.L. 333-84). The difference between the means for *A. afarensis* (ca. 28%) and *A. boisei* (ca. 42%) is striking, and examination of specimens such as OH 5, KNM-ER 732, and KNM-ER 23000 demonstrates its morphological basis (Figure 5.9). In these specimens the mastoid process appears as an equilateral pyramid, with an extensive, swollen lateral face that is sharply delimited from the posterior face, which is mostly hidden in lateral view. Inferiorly converging anterior and posterior margins of approximately equal length terminate in the mastoid tip much nearer to the midpoint of the distance between porion and asterion.

No known *A. afarensis* mastoid process has this appearance; instead, it is dominated by a single planar face, posterolaterally directed, that occupies the great majority of the mastoid's lateral aspect, all but obliterating the lateral face proper (see Figures 5.6, 5.8, and 5.9). Seen in

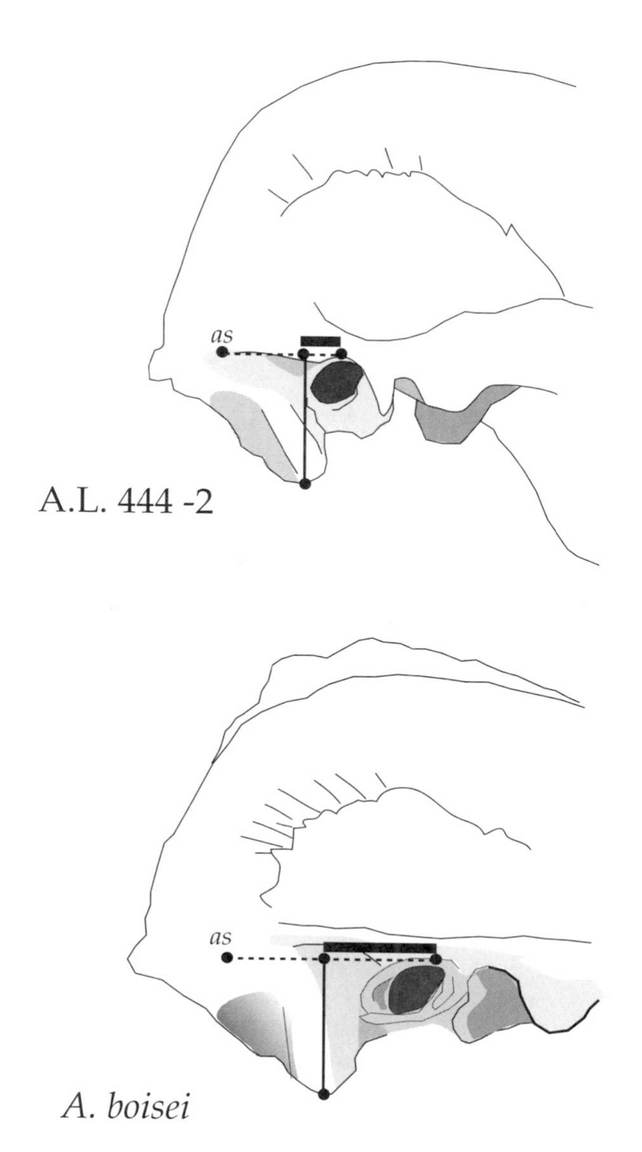

Figure 5.9 Relationship between the tip of the mastoid process and the external auditory meatus in A.L. 444-2 and *A. boisei*. Note the horizontal proximity of the tip of the mastoid to the external auditory meatus in A.L. 444-2, compared with the much greater distance between these structures in *A. boisei*. The dashed line represents the horizontal distance between asterion and porion; the heavy solid line represents the horizontal distance between the mastoid tip and porion.

lateral aspect, the long, straight posterior margin angles strongly anteroinferiorly to converge with the shorter, vertical anterior margin in the mastoid tip, which is located a little more than 1 cm behind porion. All of these elements are present in each of the adult mastoids in the Hadar sample (A.L. 333-45, A.L. 333-84, A.L. 333-112, and A.L. 444-2), and they contribute to the characteristic mastoid morphology in chimpanzees and gorillas as well. Related to this feature, in both groups the compound temporal/nuchal crest traverses the asterionic region, running far anterolaterally on the pars mastoidea, essentially pulling the posterior face of the mastoid almost entirely into open view (in lateral aspect). Figure 5.6 shows that in A.L. 444-2 the mastoid crest, which bounds the posterior face of the mastoid laterally, travels nearly half the distance to porion before arching anteroinferiorly toward the mastoid tip.

The most likely explanation for the profound difference between the autapomorphic mastoid process morphology of *A. boisei* and the plesiomorphic mastoid of *A. afarensis* is found in a basal view of the temporal bone. In *A. boisei* the tympanic element is both flattened toward the roof of the mandibular fossa and horizontally inclined toward the anterior wall of the pars mastoidea, with which the lateral part of its inferior margin is often blended into a single unit (e.g., OH 5 and KNM-ER 406), effectively extending the area of the glenoid region posteriorly. Consequently, a large anteroposterior distance lies between the coronal planes on which the roof of the external auditory meatus (porion) and the inferior margin of the tympanic plate are situated (and which is why several *A. boisei* temporal bones have a diagonally oriented meatal axis). This distance is purchased at the expense of the close proximity of the mastoid process tip to porion. Temporal bones of *A. boisei* that have this tympanic plate morphology in its most exaggerated state include OH 5 and KNM-ER 23000; they have, as well, the highest mastoid tip position index values (Table 5.5). Less highly modified tympanic plates in Omo 323-1976-896 and KNM-ER 406 are associated with mastoids that retain more of the plesiomorphic anatomy, including inferred lower mastoid tip position index values.

For the most part, the mastoid processes of *A. robustus* are more similar to those of *A. afarensis* than to those of *A. boisei*. This is unsurprising given the more generalized tympanic element in *A. robustus* relative to that of *A. boisei* (see description of the glenoid region, below). Although no *A. robustus* specimen permits calculation of the mastoid tip position index, mastoid processes are usually inflected anteriorly, with mastoid tips located only a short distance behind porion (SK 46, SK 83, and SKW 2581). One specimen from Swartkrans, SKW 11, differs from this pattern, with a wide distance (18.5 mm) separating the noninflected mastoid tip from the coronal plane of porion. Some specimens show a single, posterolaterally directed face (TM 1517 and SK 83), while others resemble *A. boisei*, with discrete lateral and posterior faces (SK 46 and SKW 11). The mastoid morphology of *A. aethiopicus* cranium KNM-WT 17000 resembles that of *A. robustus* (especially SK 46) and less-modified *A. boisei* specimens (Omo 323-1976-896).

Measurable specimens of *A. africanus* also suggest a closer morphological affinity with the Hadar sample in

Table 5.5 Mastoid Tip Position Index[a]

Specimen/Sample		Porion-Mastoid tip (mm)	Porion–Asterion (mm)	Index (%)
A. afarensis				
A.L. 444-2		12	42	29
A.L. 333-45		11	43	26
A.L. 333-84		11	38	29
	Mean	11	41	28
A. boisei				
OH 5		18	41	44
KNM-ER 406[b]		14	39	36
KNM-ER 732		15	35	43
KNM-ER 23000		17	38	45
	Mean	16	38	42
A. aethiopicus				
KNM-WT 17000		15	42	36
A. africanus				
MLD 37/38		14	39	36
Sts. 71		10	35	29
	Mean	12	37	32
Homo				
KNM-ER 1813		11	39	28
SK 847		13	ca. 40	33
Stw. 53		13	ca. 41	32

[a]"Mastoid Tip Position Index" expresses the horizontal (projected) distance between porion on the mastoid tip as a percentage of the horizontal distance between asterion and porion.
[b]For KNM-ER 406, the mastoid tip position is estimated.

terms of the relative position of the mastoid tip (mean position index = 32%; see Table 5.5). The mastoid processes in specimens such as MLD 37/38, Sts. 5, and Sts. 71 are in this sense "miniature" versions of what is seen in *A. afarensis*, with relatively long posterior margins and partially visible posterior surfaces in lateral aspect. In *H. habilis*, despite the close proximity of the mastoid tip to porion (mean index = 30%, $n = 2$), some specimens show a tendency toward a confined lateral surface (KNM-ER 1813), as in *H. erectus* (e.g., KNM-ER 3733 and the Zhoukoudian crania) and modern humans, while others more closely resemble *A. africanus* (Stw. 53).

Despite its large size and rugosity, the mastoid process of *A. afarensis* remains highly plesiomorphic, whereas that of *A. boisei* exemplifies a much more distinctive—indeed, apomorphic—configuration. As hinted at above, the *A. boisei* mastoid morphology is based in part on a radical alteration in the configuration of the glenoid region. These latter changes are discussed in detail in subsequent sections.

Basal Anatomy of the Pars Mastoidea

Olson (1981, 1985) used basal mastoid anatomy variation to diagnose multiple hominin clades in the East African Middle Pliocene. Based on allegedly specialized topographic relationships among the digastric fossa, the occipitomastoid crest, and the groove for the occipital artery in calvaria A.L. 333-45, Olson suspected the presence of a species of robust *Australopithecus* in the Hadar sample. This conclusion was challenged by Kimbel et al. (1985), who characterized the basal mastoid anatomy of *A. afarensis* as fundamentally apelike and essentially primitive for the hominin clade. The new Hadar skull, which beautifully preserves basal mastoid anatomy on both temporal bones (Figure 5.10), reiterates in detail the apelike anatomy of the mastoid seen already in A.L. 333-45 and A.L. 333-84 (Kimbel et al., 1985). Moreover, our comparative studies reveal, on the one hand, similarities between A.L. 444-2 and the other Hadar specimens and, on the other, marked differences between the Hadar specimens and those few crania of *A. boisei* in which the basal mastoid anatomy is intact.

The bounding of a triangular digastric fossa by a discrete, crestlike mastoid tip and a highly pneumatized, inferiorly projecting occipitomastoid crest coinciding with the occipitomastoid sutural margin is a pattern common to A.L. 333-45, A.L. 333-84, A.L. 333-112, and A.L. 444-2. In addition, in all of the Hadar specimens, the apex of the digastric fossa extends only a short distance posteriorly to impinge on the lowermost part of the mastoid's posterolateral face (compare the preceding description

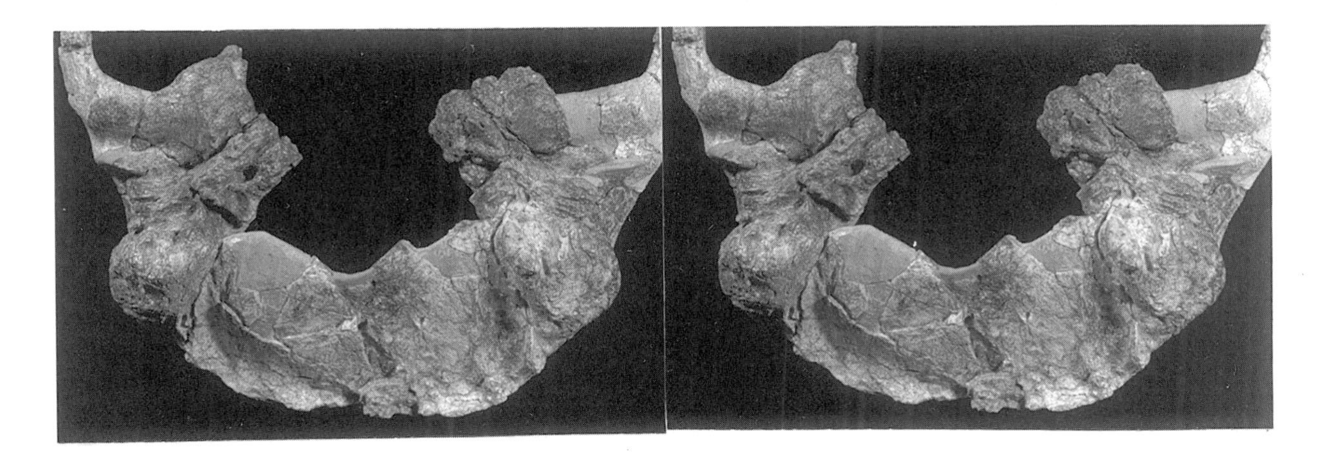

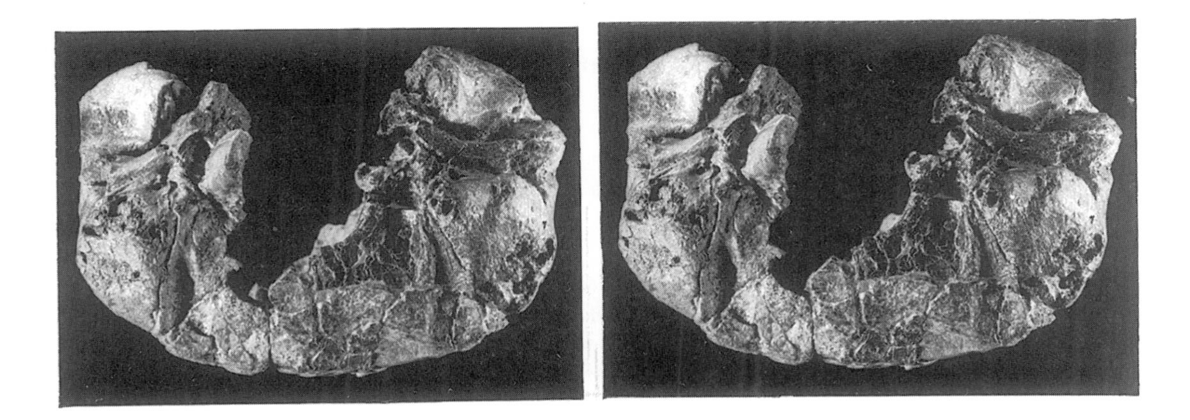

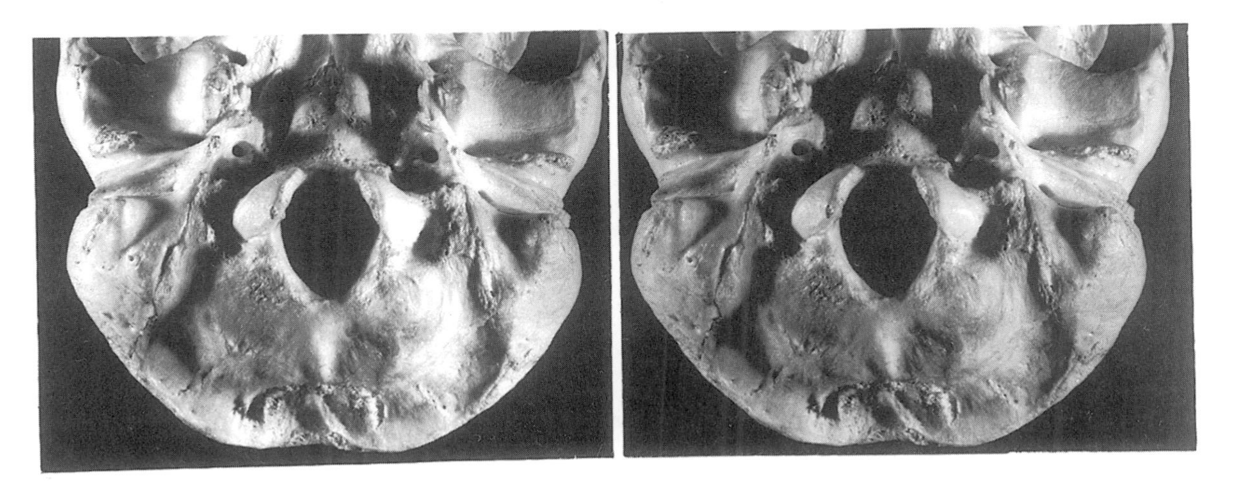

Figure 5.10 Stereophotographs of the cranial bases of A.L. 444-2 (*above*), A.L. 333–45 (*center*) and female gorilla (*below*). All approximately 40% natural size. As in the great apes, the tympanic element of the temporal bone in *A. afarensis* is tubular in sagittal cross section and lies entirely behind the large postglenoid process. The digastric groove in all three specimens is an anteroposteriorly short triangle that barely indents the posterior face of the pars mastoidea.

with that found in Kimbel et al., 1985: 129–130). Of the four, the new skull has by far the deepest and largest digastric fossa, followed by A.L. 333-45, then A.L. 333-112 (which is badly eroded), and then A.L. 333-84. In A.L. 333-45 and A.L. 333-84 the fossa is a shallow depression whose floor is a paper-thin sheet that covers the mastoid air cell mass. The mastoid tip is barely distinguishable topographically from the floor of the digastric fossa in A.L. 333-84, as in many great ape crania. The prominence of the mastoid tip and the occipitomastoid crest on A.L. 444-2 is greater than what is usually seen in the great apes (especially chimpanzees and orangutans), though it is not uncommonly encountered among gorillas.

As discussed in the previous section, the mastoid processes in *A. boisei* have an appearance (in lateral view) that is unique among *Australopithecus* species. This characterization can be extended to the basal anatomy as represented in OH 5 and KNM-ER 23000, the only two crania of *A. boisei* in which the digastric fossa is well preserved (Figure 5.11). In both specimens the fossa is an elongate, roughly triangular excavation that extends backward almost all the way to the posterior contour of the temporal bone in basal view. This has the effect of completely cleaving the posterior face of the pars mastoidea, confining the mastoid process itself to the lateral part of the temporal bone's basal aspect. In contrast, in *A. afarensis*, as in the African great apes, the very weak posterior indentation by the digastric fossa leaves the mastoid process firmly "connected" to the mass of the pars mastoidea via the circumferentially continuous posterolateral face. Two adult crania of *A. robustus* preserve this region in good condition—SK 83 (left side) and SKW 11—and they appear to follow the *A. boisei* pattern. Elsewhere among the hominins only in *Homo* is a complete isolation of the mastoid process achieved by a posterior extension of the digastric fossa all the way to the rear margin of the temporal bone (see, for example, KNM-ER 1805; Figure 5.11).

The basal mastoid morphology of A.L 444-2 is a highly exaggerated version of what is seen in *A. africanus*, at least as represented in Sts. 5 and MLD 37/38, the only crania of the species to preserve this region in detail (Figure 5.12). Kimbel et al. (1985: 128–129) argued that the Makapansgat specimen in particular has African apelike basal mastoid anatomy, and it would thus seem that both *A. afarensis* and *A. africanus* retain the primitive hominin state in this regard (which is concordant with the evaluation of the mastoid in lateral view, per discussion in the previous section). The only significant difference between these species in terms of mastoid anatomy is the generally greater size and inflation of this structure in the entirely male *A. afarensis* sample.

Size and Shape of the Mandibular Fossa

The mandibular fossa of A.L. 444-2 is very broad mediolaterally, but it is short anteroposteriorly, producing a low length/breadth index of 54.3% that falls within the cluster of index values for other *A. afarensis* temporal bones (measurements discussed in this section follow the method of Wood, 1991a: 291) (Table 5.6). Intersample differences

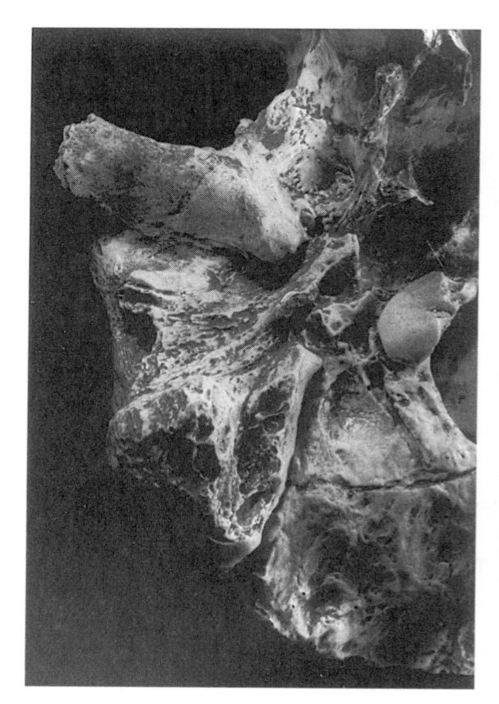

Figure 5.11 Basal view of OH 5 (*A. boisei*) (*left*) and KNM-ER 1805 (*H. habilis*) (*right*). Natural size. In both specimens the anteroposteriorly elongate digastric groove nearly reaches the posterior edge of the calvaria.

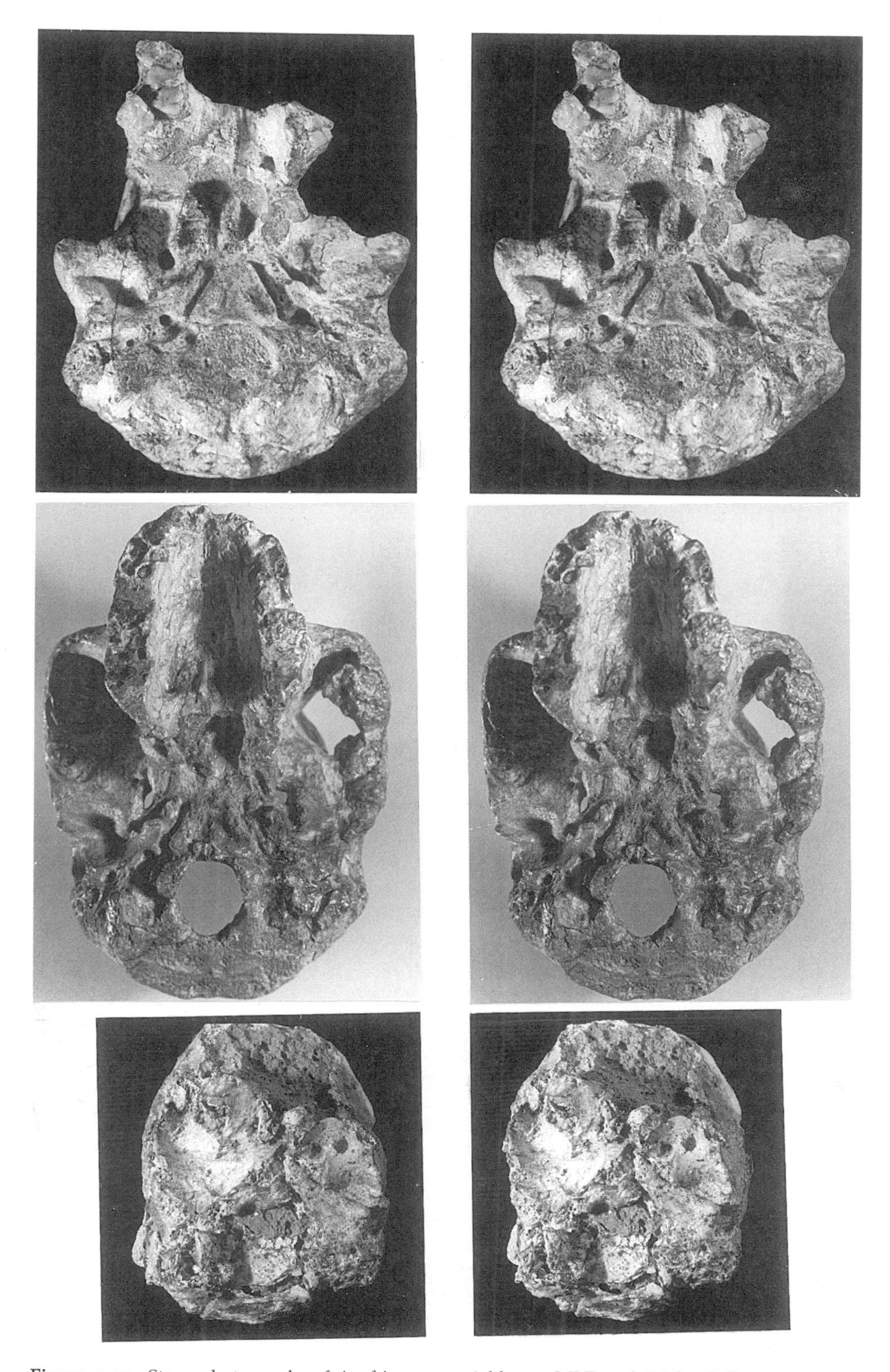

Figure 5.12 Stereophotographs of *A. africanus* cranial bases: MLD 37/38 (*above*), Sts. 5 (*center*), and Sts. 25 (*below*). All approximately 40% natural size. This figure illustrates the wide range of morphological variation in the *A. africanus* glenoid region, especially in the shape of the tympanic element and its topographic relationship to the postglenoid process anteriorly and the mastoid process posteriorly.

Table 5.6 Measurements of Hadar Temporal Bones (mm)

Measurement	A.L. 444-2	A.L. 166-9	A.L. 58-22	A.L. 333-45	A.L. 333-84	Hadar Mean	Hadar SD
Articular eminence width (direct)	37.5	30.5	31.0	33.0	n/a	33.0	3.2
Articular eminence breadth[a]	35.0	28.0	29.0	32.0	n/a	31.3	3.2
Articular eminence length[a]	19.0	14.5	15.0	17.5	n/a	16.4	2.1
Length/Breadth Index[a]	54.3	51.8	51.7	54.7	n/a	54.7	1.6
Mandibular fossa depth[a]	7.5	4.5	4.2	3.5	n/a	4.9	1.8
Petrous length (CC-PA)	21.0	n/a	n/a	20.6	n/a	20.8	n/a
Tympanic length (LT-CC)	35.0	n/a	n/a	28.5	29.2	30.9	n/a
Tympanic depth	14.5	12.3	n/a	11.0	10.0	12.0	1.9
External auditory meatus area (mm^2)[b]	109.3	69.2	n/a	92.6[c]	47.7	79.7	26.9
Mastoid height (vertical below porion)	29.0	n/a	n/a	27.0	27.0	27.7	n/a

CC, carotid canal center, after Dean and Wood, 1982; PA, petrous apex, after Dean and Wood, 1982; LT, lateralmost point on inferior tympanic margin, after Dean and Wood, 1982.
[a]Measurement after Wood, 1991a.
[b]Calculated as the area of an ellipse.
[c]Value is average of 94.2 (L), 91.0 (R) mm.

in mandibular fossa proportion are evident in *Australopithecus* and are due largely to the greater anteroposterior length in *A. robustus* and *A. boisei* than in *A. afarensis* (Table 5.7). For example, the mediolateral breadth of the mandibular fossa is virtually the same in A.L. 444-2 and OH 5, but the *A. boisei* specimen's fossa is 32% longer anteroposteriorly in absolute terms (Table 5.7; Figure 5.13). Among the reasons for the absolutely and relatively great anteroposterior length of the robust *Australopithecus* mandibular fossa is the tendency for the topographically lowest point on the articular eminence (the anterior terminus of the length measurement) to lie far anteriorly, and this essentially coincides with the posterior margin of the temporal foramen; in *A. afarensis*, in contrast, as in all other hominoids, the lowest point on the eminence is located behind the anterior limit of the mandibular fossa. Another factor appears to relate to the strong posterior inclination of the tympanic plate in many robust *Australopithecus* temporals. These morphological alterations in the temporal bones of *A. robustus* and especially *A. boisei* (discussed in detail in subsequent sections) help to create an enormous, anteroposteriorly enlarged glenoid "basin" that is unique among the hominoids.

In the *A. africanus* sample the mandibular fossa is, on average, slightly narrower mediolaterally than in the Hadar sample, while fossa length is marginally greater. In some specimens, such as Sts. 71, it is clearly an abbreviated fossa breadth that drives the index value upward into the robust range, while in others, such as MLD 37/38, there is an undoubted anteroposterior lengthening of the fossa compared to *A. afarensis* specimens with similar fossa breadths (e.g., A.L. 333-45). Thus, *A. afarensis* has an anteroposteriorly short mandibular fossa (mean = 53.1%) compared to the other *Australopithecus* species (range of sample means = 61%–65.6%; Table 5.7). This is conveyed graphically in Figure 5.13.

Articular Eminence, Preglenoid Plane, and Entoglenoid Process

Although much of the morphology of the glenoid region has been previously available in several adult *A. afarensis* temporal bones (A.L. 58-22, A.L. 166-9, A.L. 333-45), A.L. 444-2 is the first specimen to present the entire region in well-preserved detail. Earlier studies (e.g., Kimbel, 1986) emphasized the plesiomorphic qualities of the *A. afarensis* glenoid region—in particular, the horizontally oriented preglenoid plane; the very weak articular eminence resulting in a topographically "open" mandibular fossa; the massive, inferiorly pointing entoglenoid process; and the vertically shallow, "tubular" tympanic element. This morphological set gives the glenoid region a topographical map with very low relief in the sagittal plane, and this is quite similar to what is observed in chimpanzees (e.g., Biegert, 1957; Picq, 1985).

The glenoid region anatomy of the A.L. 444-2 skull (Figure 5.10), with its more convex, diagonally set articular eminence, steeper preglenoid plane, and more upright (though still "tubular") tympanic element, represents a shift away from the chimpanzee-like condition toward one that may be characterized as slightly more "humanlike." However, several of these humanlike features are manifested in the glenoid region of the gorilla, as compared to the chimpanzee: that is, a more convex articular eminence and a steeper preglenoid plane (e.g., Biegert, 1957).

If we envision the human and great ape glenoid regions as situated at opposite ends of a morphocline, then

Table 5.7 Mandibular Fossa Dimensions[a]

Specimen/Sample		Fossa Breadth (mm)	Fossa Length (mm)	Index (%)
A. afarensis				
A.L. 444-2		35.0	19.0	54.3
A.L. 58-22		29.0	15.0	51.7
A.L. 166-9		28.0	14.5	51.8
A.L. 333-45		32.0	17.5	54.7
	Mean	31.0	16.5	53.1
	SD	3.2	2.1	1.6
A. africanus				
Sts. 5		27.0	15.0	55.6
Sts. 71		26.0	16.0	61.5
MLD 37/38		31.0	20.0	64.5
TM 1511		32.0	20.0	62.5
	Mean	29.0	17.8	61.0
	SD	2.9	2.6	3.9
A. robustus				
TM 1517a		30.0	20.5	68.3
SK 48		33.0	21.0	63.6
SK 83		34.0	22.0	64.7
	Mean	32.3	21.2	65.6
	SD	2.1	0.8	2.5
A. boisei				
OH 5 (cast)		36.0	25.0	69.4
KNM-ER 406		37.0	23.0	62.2
KNM-ER 407		33.0	21.0	63.6
KNM-ER 732		33.0	20.0	60.6
KNM-ER 23000		39.0	23.0	59.0
	Mean	35.6	22.4	63.0
	SD	2.6	1.9	4.0
Other hominins				
KNM-WT 17000		32.0	20.0	62.5
KNM-BC 1		33.0	19.5	59.1
Sts. 19		28.0	15.0	53.6

[a]Mandibular fossa mediolateral breadth and anteroposterior length measured according to the method of Wood (1991a: 291). Shape index expresses length as a percentage of breadth × 100.

the *A. afarensis* glenoid morphology occupies the most generalized position on the morphocline of any *Australopithecus* species (with the possible exception of *A. anamensis*; M. Leakey et al., 1995; Ward et al., 1999, 2001), with A.L. 444-2 being the *least* generalized of the four *A. afarensis* specimens now known. Size differences aside, in terms of the morphology of the mandibular fossa, articular eminence, and preglenoid plane, the new Hadar skull resembles some *A. africanus* specimens such as Sts. 5 and TM 1511. No *A. africanus* specimen shows the flat, open mandibular fossa morphology of *A. afarensis* crania as in A.L. 166-9 and A.L. 333-45, and no *A. afarensis* specimen shares the suite of derived glenoid region characters as evidenced in cranium MLD 37/38, which pulls *A. africanus* toward an even more derived position on the morphocline of glenoid region structure. There remains a marked difference in the mandibular fossa morphologies of the two species' hypodigms.

Two aspects of mandibular fossa morphology in MLD 37/38 represent noteworthy departures from that of other *A. africanus* crania. In the Sterkfontein representatives of this species (i.e., TM 1511, Sts. 5, Sts. 25, Sts. 71, and Stw. 13), the maximum convexity, or summit, of the articular eminence is situated behind the plane of the temporal foramen's posterior edge, as in *A. afarensis* and the great apes. In the Makapansgat specimen, in contrast, the articular surface slopes continuously, with its summit coinciding with the posterior edge of the foramen. As a result, in lateral view, the articular eminence has the appearance of having been flexed downward around the transverse axis through the fossa.

The second major departure from the usual *A. africanus* morphology in MLD 37/38 is the orientation of the articular eminence as seen in basal view (Figure 5.12). Here, the eminence is observed to twist about its transverse axis: laterally, it faces predominantly inferiorly, but medially it gradually twists to face more posteriorly. Associated with this medial twisting of the eminence is the fact that the entoglenoid process is tilted backward, overlapping the upper edge of the tympanic element and partially obscuring the continuity between the petrosquamous and petrosphenoid fissures. The anterior surface of the entoglenoid process, rather than rising into the infratemporal fossa, is a flat, nearly horizontal plane sandwiched between the entoglenoid apex and the posterolateral margin of the foramen ovale. It truncates the articular eminence summit medially and forms a sharp angle with the relatively steep, narrow preglenoid plane.

The details of the MLD 37/38 mandibular fossa highlight the more derived morphoclinal position of the *A. africanus* glenoid region relative to that of A.L. 444-2 and, in several respects, it foreshadows later developments in the *Australopithecus* glenoid region.

Other than in its absolutely greater size and depth, the mandibular fossa morphology of *A. robustus* (SK 48, SK 83, and TM 1517) differs little from that seen in *A. africanus* specimens Sts. 5, Stw. 13, and TM 1511. The greater depth of the fossa in *A. robustus* stems from the combined effect of stronger inclination of the posterior slope of the articular eminence and the elevation of the fossa ceiling to approximate the plane of the FH (in SK 48 and SK 83, but *not* in TM 1517 [contra Picq, 1985], which is more generalized in this respect). The preglenoid plane covers approximately the same area as in *A. africanus*. The articular eminence maintains the same orientation over its entire mediolateral extent (there is no twisting, such as just described for MLD 37/38), and its most inferior point is just posterior to the edge of the temporal foramen (in the relatively posterior location of the eminence summit,

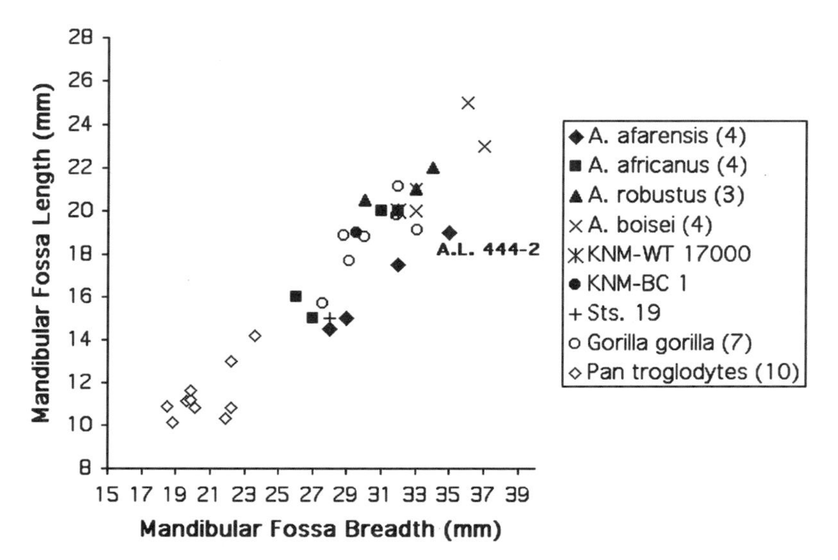

Figure 5.13 Bivariate scattergram of early hominin mandibular fossa breadth and length. The *A. afarensis* mandibular fossa is anteroposteriorly short relative to breadth. Note especially the marked narrowness of the *A. afarensis* fossa in comparison to *A. robustus* and *A. boisei* fossae of equivalent breadth.

TM 1517 is most similar to *A. africanus*). In morphology and orientation, the entoglenoid process in *A. robustus* is also similar to the more generalized *A. africanus* condition (excluding MLD 37/38), although the process arises from a more extensive base in crania of the former taxon.

It is in *A. boisei* that we encounter a truly metamorphosed glenoid region, one that diverges sharply from the relatively plesiomorphic glenoid region of all other *Australopithecus* species. In addition to extraordinary dimensions, especially anteroposterior length (Table 5.7; Figure 5.13), the mandibular fossa of *A. boisei* presents a number of highly unusual features:

- The articular fossa is very deep—in fact, so deep that its roof's highest point is located well above the FH—and is fronted by a very long, steep articular eminence, whose slope ends abruptly at the edge of the temporal foramen.
- The preglenoid plane, likewise steeply inclined superiorly, is extremely restricted mediolaterally.
- The articular eminence twists about its transverse axis so that medially it faces almost completely posteriorly.
- Related to the twisting of the articular eminence, the entoglenoid process is "rocked" backward on its base so that the apex points posteriorly and overlaps the tympanic element, thereby creating a flat, inferiorly directed platform between the medial end of the articular eminence and the lateral margin of foramen ovale that Du Brul (1977) termed the "medial glenoid plane."

This suite of characters is observed in the following crania of *A. boisei*: OH 5, KNM-ER 406, KNM-ER 407, KNM-ER 732, KNM-ER 13750, KNM-ER 23000, and KNM-WT 17400. Its distribution is so pervasive that it may be considered one of the most characteristic aspects of this species' cranial morphology (Figures 5.11 and 5.14).

However, two geologically early specimens often considered to belong to *A. boisei* do *not* share this character constellation: Omo 323-1976-896 and KNM-WT 17000. The latter cranium is alternatively attributed to a distinct taxon, *A. aethiopicus*, based in part on its relatively primitive glenoid region (Walker et al., 1986; Kimbel et al., 1988; R. Leakey and Walker, 1988). Despite their large dimensions and their differences in mandibular fossa depth (the Omo fossa is substantially the deeper of the two), the Omo and West Turkana temporal bones share several key attributes of the common hominin glenoid region pattern not found in the *A. boisei* specimens listed here, including (Figure 5.15):

- Inferiorly facing articular eminence (no transverse "twisting"), with the lowest point on the convexity of the eminence located posterior to the edge of the temporal foramen
- Expansive, more horizontally oriented preglenoid plane
- Triangular, inferiorly pointing entoglenoid process (no "medial glenoid plane")

Inasmuch as these shared character states are plesiomorphic, they do not imply that the Omo specimen is attributable to *A. aethiopicus*; in fact, Alemseged et al. (2002) have reaffirmed the assignment of Omo 323-1976-896 to *A. boisei*.[1] The suggestion by Walker and Leakey (1988) that these two specimens represent early stages in the differentiation of an anagenically changing *A. boisei* lineage needs to be tested by additional fossil discoveries in the 2.3 to 2.7 Myr time period in eastern Africa.

Figure 5.14 Basal view of the glenoid region of *A. boisei* specimens KNM-ER 406, male (*left*) and KNM-ER 407, female (*right*). Natural size. In the male cranium, especially, autapomorphic modifications convert the glenoid region into an enormous triangular basin bounded by the posterior edge of the temporal foramen anteriorly and the anterior face of the mastoid process posteriorly.

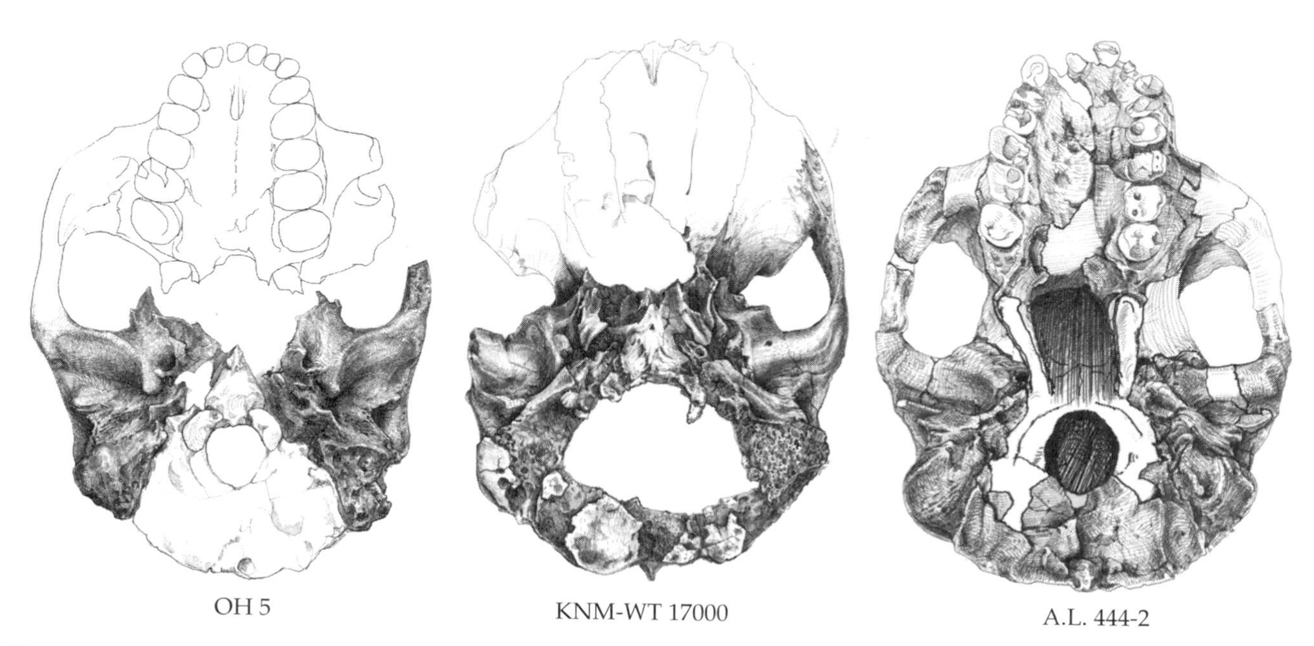

Figure 5.15 The anatomy of the glenoid region in *A. boisei* (OH 5), *A. aethiopicus* (KNM-WT 17000) and *A. afarensis* (A.L. 444-2). The *A. aethiopicus* cranium lacks many of the specializations of the *A. boisei* glenoid region, and in this respect is morphologically intermediate between *A. boisei* and the more plesiomorphic *A. afarensis*. See the picture gallery following page 122 for an enlarged view.

Postglenoid Process

In A.L. 444-2 the large, heavily pneumatized postglenoid process is situated far laterally, with its lateral end lying on the same sagittal plane as the articular tubercle. The process extends so far laterally that it appears as a hump on the outline of the cranial base in a basal view (Figure 3.50). Of all the hominin and African ape specimens that we examined (including A.L. 166-9 and A.L. 333-45, the other *A. afarensis* specimens in which the structures are preserved), only Sts. 5 exhibits such a configuration. In fact, in the latter, the lateral projection of the postglenoid process slightly exceeds that of the articular tubercle.

The difference in the degree of lateral extension of the postglenoid process and the external auditory meatus (a larger distance between these two structures is seen on the left than on the right) results in a unique relationship between the process and the external auditory meatus in A.L. 444-2. The process extends so far laterally that its *medial* end lies on the same sagittal plane as the meatus. The disparity in the degree of lateral extension of the structures underscores the process as an isolated unit. Such a relationship characterizes A.L. 333-45 as well.

The posterior wall of the mandibular fossa is composed entirely of the postglenoid process in A.L. 444-2, as in all known *A. afarensis* temporal bones. A distinct, but variable, transverse groove separates the process from the tympanic element to the rear; the groove is very prominent in A.L. 444-2 and least developed in A.L. 166-9. In other words, in *A. afarensis* the tympanic element and the postglenoid process are situated on two different coronal planes. This condition is ubiquitous in the great apes and other catarrhines and is therefore judged to be primitive for hominins. In modern humans the postglenoid process, which is greatly reduced in both inferior and lateral projection, occupies the same coronal plane as the vertically oriented tympanic plate; in basal view, the lower margin of the process is often completely hidden by the tympanic.

Despite the significant modification of tympanic form in *A. africanus* (see the following discussion), the postglenoid process is large, laterally projecting, and situated on a coronal plane anterior to that on which the tympanic is situated. This relationship is well seen in MLD 37/38, Sts. 5, probably Sts. 25, and undoubtedly Stw. 266, the latter a small fragment of temporal bone from Sterkfontein. In Sts. 19, however, the anteroposterior distance between the coronal planes on which the postglenoid process and the tympanic plate reside is eliminated (that is, these structures appear to merge with one another in basal view). This is one reason we consider Sts. 19 to belong to the genus *Homo* (Kimbel and Rak, 1993), a subject to which we return in the following discussion.

In both *A. robustus* and *A. boisei*, however, the offset between these structures is all but completely eliminated. The postglenoid process in these species tends to be surprisingly underdeveloped, given the overall size and rugosity of the cranium (see, for example, KNM-ER 13750, KNM-ER 23000, and TM 1517) and the superior margin of the platelike tympanic element merges with the process in many specimens: TM 1517, SK 48, SK 83, KNM-ER 406, KNM-ER 407, KNM-ER 23000, OH 5, and probably KNM-WT 17400.

The rear wall of the mandibular fossa in *A. robustus* and *A. boisei* is actually the anterior face of the tympanic element, and it would seem that, as in humans, the replacement of the postglenoid process in this role led to its reduction. This characteristic morphology does appear to be present in the Omo 323-1976-896 temporal, but not in KNM-WT 17000 (*A. aethiopicus*), which retains the primitive gap between the large postglenoid process and the tympanic element.

Size of the External Auditory Meatus

With equal vertical and horizontal axes of 11.8 mm, the EAM of the A.L. 444-2 cranium is quite large, exceeding the meatal dimensions of three other Hadar adult temporal bones, which, however, are very variable in this respect (Table 5.6). By comparison with the great apes, hominins tend to have large EAM dimensions (e.g., Weidenreich, 1943). M. Leakey et al. (1995) listed a small, elliptical EAM as diagnostic of *A. anamensis*, based on temporal fragment KNM-KP 29281. The size of this specimen's EAM is 39.8 mm^2 (calculated as the area of an ellipse from measurements in Leakey et al., 1995), only a little less than that of the smallest EAM in the Hadar adult sample (A.L. 333-84: 47.7 mm^2). The mean area value for four Hadar adult individuals is 79.7 mm^2, with an upper extreme of 109.3 mm^2 (A.L. 444-2).

Length of the Tympanic Element

The tympanic plate in A.L. 444-2 is longer mediolaterally (35.0 mm, measured according to the method of Dean and Wood, 1982) than in two other *A. afarensis* temporal bones (mean = 30.9 mm, n = 3; see Table 5.6). Great ape tympanics are much longer than this, whereas in modern humans the tympanic is greatly reduced in mediolateral length. Among the other species of *Australopithecus*, only *A. robustus* and *A. boisei* exhibit long tympanics (Clarke, 1977) with dimensions of some specimens falling into the lower part of the gorilla range (Dean and Wood, 1982). In contrast, the *A. africanus* tympanic is short, with many specimens overlapping the human spectrum of variation (Dean and Wood, 1982). It is in those taxa with the greatest degree of lateral expansion of the cranial base—associated with heavy cresting and pneumatization of the temporal root of the zygomatic arch—namely, the great apes, robust *Australopithecus* species, and *A. afarensis*—that we find the longest tympanic elements.

Notwithstanding this similarity, there is a consistent difference among these taxa regarding the extent to which the tympanic extends laterally relative to the roof of the external auditory meatus. Thus, in *A. robustus* and *A. boisei* the inferior margin of the tympanic extends laterally as far as, or occasionally even farther than, the roof of the meatus (e.g., OH 5, KNM-ER 406, KNM-ER 23000, TM 1517, SK 47, SK 83, and SK 821), as is usually the case in chimpanzees and gorillas. As Clarke (1977) has observed, the lateral prolongation of the tympanic element in robust *Australopithecus* crania results in the tip of the mastoid process occupying a sagittal plane well medial to that on which the tympanic's lateral margin resides (as seen in basal view)—even though the mastoid process itself is not strongly inflected under the cranial base. This is not the case in A.L. 444-2 and the other adult *A. afarensis* temporal bones (A.L. 333-45 and A.L. 333-84); here, the most lateral extent of the tympanic is medial to the roof of the meatus at porion, and the tip of the mastoid process and the lateral edge of the tympanic are aligned on more or less the same sagittal plane.

Form of the Tympanic Element

Hominoid temporal bones exhibit taxonomic variation in the morphology of the tympanic element, as recognized many years ago by Weidenreich (1943). In humans the tympanic element assumes the form of a plate, vertically oriented, with an upper and sharp lower margin (the crista petrosa) that bounds a single anteriorly directed face, which is concave both mediolaterally and inferosuperiorly. Medially, the crista petrosa is drawn out as the sharp vaginal process of the styloid that ensheaths the ossified styloid process anterolaterally. Laterally, the tympanic plate is usually intimately applied to the anterior wall of the pars mastoidea, closing off the tympanomastoid groove in which, anterior to the digastric fossa, sits the stylomastoid foramen. A second process emanates from the tympanic plate medially. This bony spur, termed the "eustachian process" (Dean, 1985), denotes the site of origin of the levator palati and the tensor palati muscles and arises from the upper margin of the tympanic (obscured in basal view) and the body of its anterior face to overlie the lateral aspect of the petrous pyramid's basal surface. The eustachian process is inconstant in humans and, when present, is usually quite small: the levator palati muscle usually arises from the cartilage of the eustachian tube, whereas the tensor palati muscle originates from the scaphoid fossa between the medial and lateral pterygoid plates.

In all great apes the tympanic element is recessed in a transverse furrow between the prominent postglenoid process anteriorly and the anterior wall of the pars mastoidea posteriorly, but it is in intimate contact with neither: the tympanic is situated on a coronal plane completely behind that of the postglenoid process, and the tympanomastoid groove is open laterally. The chimpanzee tympanic element is a horizontally oriented tube, with anterior and posterior margins bounding a smooth, sagittally convex inferior surface. Gorillas and orangutans often have taller, more vertically oriented tympanics, but even here the tympanic is sagittally rounded, broader anteroposteriorly, and more posteriorly inclined than in humans; they are not platelike. The crista petrosa is usually undifferentiated in chimpanzees, there being no bony fronting of the vagina of the styloid. However, the crista petrosa is often present in gorillas and orangutans in the form of a low, blunt ridge that divides the inferior from the anterior face of the tympanic element. The eustachian process is well developed in the chimpanzee and gorilla (Dean, 1985), where it appears as a prominent flange or spike arising from the anteromedial corner of the flattened medial section of the tympanic that overlies the sagittally oriented petrous element. It lies below the bony opening of the eustachian tube, rather than on the petrous itself, as in humans.

In *A. afarensis* the tympanic element occupies a transverse recess between the postglenoid process and the mastoid process, very much like the great ape condition (Figure 5.10). The tympanomastoid groove is open laterally, and the tympanic sits completely behind the postglenoid process. The two surfaces of the A.L. 444-2 tympanic—the narrow inferior surface and the more extensive anteroinferior surface—are only barely demarcated from one another due to an undifferentiated crista petrosa, so it is appropriate to describe the tympanic element in this specimen as "tubular." It is relatively upright, and thus has anterosuperior and posteroinferior margins, similar to Hadar temporal bone A.L. 333-84. Together these specimens present a picture that is more reminiscent of the gorilla than of any other hominoid. However, another specimen from locality 333, A.L. 333-45, differs from these two in that, very much as in the chimpanzee, the tympanic is horizontally oriented, with a single, anteroposteriorly broad, convex inferior face bounded by anterior and posterior margins, and no crista petrosa at all. Specimens A.L. 333-45 and A.L. 444-2 show that the eustachian process is a moderately developed bony spur and is separated from the rim of the carotid foramen by a distinct groove or depression aligned with the long axis of the tympanic element. The process arises from the superior margin and the face of the tympanic itself and overlaps the basal aspect of the petrous, as in modern humans.

The apelike form of the *A. afarensis* tympanic plate is not encountered in other species of *Australopithecus*, with the exception of a single specimen of *A. africanus* (Sts. 25; see below) and the fragmentary *A. anamensis* temporal bone KNM-KP 29281 from Kanapoi (M. Leakey et al., 1995; Ward et al., 2001). Some of the modifications typical of modern human tympanic plate morphology are seen already in the crania of *A. africanus*, which is thus easily distinguished from *A. afarensis*. Derived, platelike tympanic morphology in this taxon includes greater uprightness of the plate; a dominant, biconcave, anteriorly or anteroinferiorly directed face; and a sharp, freely projecting crista petrosa. This platelike tympanic morphology is observed in Sts. 5, Sts. 71, MLD 37/38, Stw. 266, and Stw. 505 (but not in Sts. 25, which has a more tubular, plesiomorphic form of tympanic; see Figure 5.12). These specimens remain primitive, however, in the absence of an inferiorly projecting vaginal process of the styloid and because the postglenoid process, the lateral end of the tympanic plate, and the anterior wall of the pars mastoidea are all situated on different coronal planes, separated by distinct gaps. This is not the case in Sts. 19, however, which we have elsewhere (Kimbel and Rak, 1993) affiliated with *Homo* on the basis of its derived glenoid region (we were not the first to notice how closely the morphology of this specimen approaches that of later *Homo*; Broom et al., 1950; Clarke, 1977; Dean and Wood, 1982; but see Ahern, 1998, for a contrasting opinion). The Sts. 19 cranial base exhibits an apomorphic compression of the coronal planes on which the postglenoid process, tympanic element, and anterior face of the pars mastoidea reside: that is, a merging of the superior margin of tympanic and the postglenoid process in basal view, and lateral obliteration of the tympanomastoid groove (Figure 5.16). This cranium also departs from the *A. africanus* condition in its advanced differentiation of the crista petrosa, which is prolonged into a well-developed vaginal process of the styloid.

Dean (1985) drew attention to the extremely well developed eustachian process at the medial end of the *A. africanus* tympanic plate. The fashion in which this tubercle is formed transforms the entire tympanic, giving *A. africanus* a highly characteristic morphology in addition to its humanlike modifications (see Figure 5.12). Seen from anterior aspect, the tympanic plate of *A. africanus* is paddle-shaped, vertically deepest laterally. Depth gradually decreases medially as the free inferior margin (the crista petrosa) and the superior margin converge, the inferior margin forming the anterolateral rim of the carotid foramen before merging with the superior margin in the eustachian process. The inferomedial side of the process arises directly from the rim of the carotid foramen, as can be seen in Sts. 5, Stw. 505, Stw. 266, and MLD 37/38. Again, Sts. 19 presents a different pattern of morphology. Here the absence of the apparently autapomorphic derivation of the eustachian process from the crista petrosa and the carotid foramen margin seen in *A. africanus* (i.e., there is a gap separating the process from the crista petrosa) renders this Sterkfontein specimen more similar to other hominins (Figure 5.16).

The *A. africanus* morphology differs considerably from that characterizing *A. afarensis*. First, in the Hadar specimens the tympanic element is actually shallower laterally than medially (A.L. 333-45, A.L. 333-84, A.L. 444-2).

Figure 5.16 Stereophotograph of cranial base of Sts. 19. Approximately 40% natural size. The morphological pattern in the glenoid region is consistent with an attribution of Sts. 19 to *Homo*.

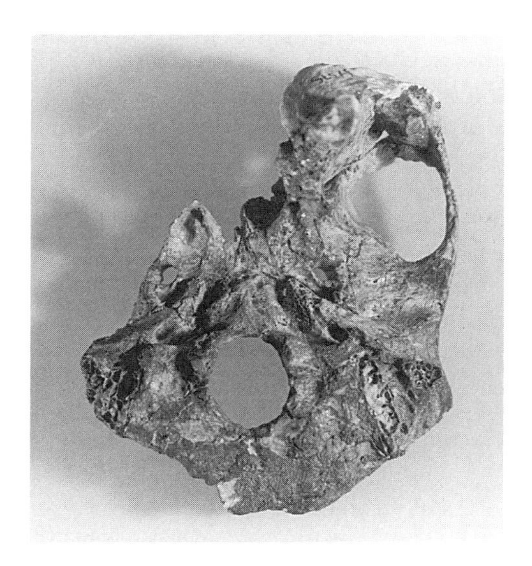

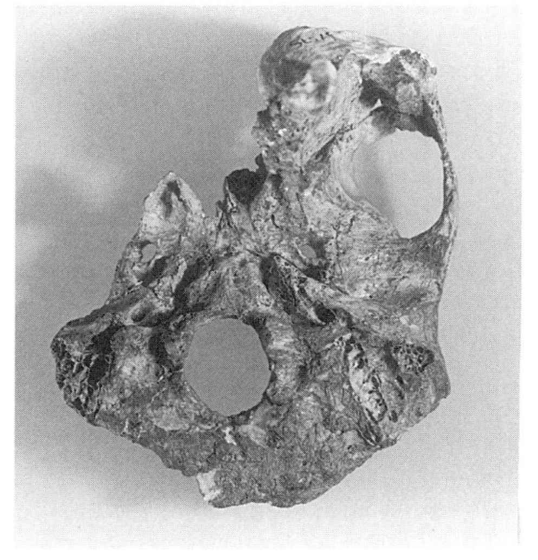

This difference is due, in part, to a portion of the vertical depth of the tympanic's face being hidden (in anterior view) by the large postglenoid process in *A. afarensis*, but also because of a real increase in depth laterally associated with the assumption of an upright, platelike form in *A. africanus*. Second, the relatively small eustachian process in *A. afarensis* arises topographically from deep within the squamotympanic fissure and the face of the tympanic but not from the crista petrosa. In the Hadar specimens the process is separated from the rim of the carotid foramen by a distinct groove or depression that is aligned with the long axis of the tympanic. That these differences are not simply a function of the large size of the eustachian process in *A. africanus* is seen in a comparison between Stw. 266 (which, despite its relatively small, spatulate process, is endowed with the characteristic *A. africanus* tympanic form) and A.L. 444-2 (which has a laterally shallow tympanic associated with a fairly thick eustachian process that is separated from the carotid foramen's rim by a deep groove). Furthermore, the basicranial anatomy adhering to the natural endocast of the juvenile specimen from Taung preserves the medial end of the tympanic element, whose inferior and superior margins contribute to the formation of a eustachian process in exactly the same fashion as in adult *A. africanus* counterparts. Taung's morphology contrasts with the tympanic configuration on Hadar juvenile A.L. 333-105, where the inferior margin of the tympanic and the eustachian process are separated by a wide groove.

Unexpectedly, crania attributed to robust *Australopithecus* species have tympanic element morphologies that in many ways approximate the modern human condition, providing a strong contrast with the primitive morphology of *A. afarensis*. This is seen principally in:

- Greatly increased inferosuperior depth and uprightness of the tympanic, converting it into a single-faced plate
- A markedly oblique orientation (posterolateral to anteromedial) of the tympanic axis in basal view (Dean and Wood, 1982), which, combined with the posteromedial to anterolateral orientation of the articular eminence, gives the entire glenoid region the outline of a medially pointing triangle (see especially *A. boisei* specimens OH 5, KNM-ER 406, KNM-ER 23000, and, despite damage, KNM-ER 13750; this feature is highlighted with reference to the basicranium as a whole in Chapter 3).
- Further reduction, if not obliteration, of the gaps between the coronal planes on which the basal glenoid region structures are located (i.e., superior tympanic margin and postglenoid process; crista petrosa and the anterior face of the pars mastoidea)
- A sharp, projecting crista petrosa, with the frequent occurrence of a prominent flange-like vaginal process of the styloid
- Reduction in size of the postglenoid and eustachian processes

All of these characteristics are also associated with the modern human glenoid region morphology.

Yet there are fairly consistent differences between the tympanic morphologies of *A. robustus* and *A. boisei*, and in these respects *A. robustus* remains relatively plesiomorphic:

- In *A. robustus* (TM 1517, SK 46, SK 47, SK 48, SK 83, SKW 11, and SKW 2581) there remains a deep though narrow tympanomastoid groove separating the crista petrosa from the anterior face of the pars mastoidea, whereas in several *A. boisei* specimens this gap is eliminated completely (i.e., OH 5 and KNM-ER 406; but not KNM-ER 23000 or KNM-ER 407). In Omo 323-1976-896 and KNM-WT 17000 the groove is reduced to a very narrow crevice laterally, much narrower than is usually the case in *A. robustus*.
- In *A. robustus*, although the crista petrosa is sharp medially, it "fans out" laterally (in basal view) because of its division into anterior and posterior ridges that run to the lateral margin of the tympanic (the posterior ridge is the one juxtaposed to the anterior face of the mastoid). This lends the tympanic a "trumpet" shape (Rak and Clarke, 1979) in basal view, a configuration that among East African robust crania is approached only by KNM-ER 23000.
- The tympanic plate in *A. boisei* is often flattened superiorly and inclined posteriorly, giving the external auditory meatus (EAM) an oval shape with a diagonal long axis (in lateral view; see OH 5, KNM-ER 406, KNM-ER 407, KNM-ER 23000, and probably KNM-ER 732). In contrast, the EAM in *A. robustus* (and in Omo 323-1976-896 and the KNM-WT 17000 cranium of *A. aethiopicus*) is usually vertically elongate, the plesiomorphic condition. In some *A. boisei* specimens (OH 5 and KNM-ER 406) this inclination has proceeded to such an extent that the major face of the tympanic plate actually constitutes the *ceiling* rather than the posterior wall of the glenoid cavity, and the crista petrosa is "smeared" on the anterior face of the pars mastoidea, no longer discernible as a separate entity lateral to the vaginal process of the styloid (i.e., resulting in a nearly complete obliteration of the tympanomastoid groove) (Figures 5.11, 5.14, and 5.15). In these specimens, in particular, it is necessary to distinguish topographically between

the mandibular fossa proper (that is, the true articular component of the joint) and the larger, triangular glenoid "cavity" that encapsulates it, formed as the result of the unique deployment of the huge tympanic plate. This latter, more extensive topographic cavity is bounded anteriorly by the summit of the articular eminence and posteriorly by the anterior wall of the pars mastoidea. Such a morphological pattern is unknown among the crania of other *Australopithecus* species and can be considered one of the supreme autapomorphies of the *A. boisei* skull.

In *H. habilis* (*sensu stricto*) there is a marked reduction or elimination of the gaps that separate the postglenoid process from the superior margin of the tympanic plate and the crista petrosa from the anterior face of the pars mastoidea (i.e., lateral closure of the tympanomastoid groove), as we have also seen in *A. robustus* and especially *A. boisei*. Thus, Stw. 53, KNM-ER 1813, KNM-ER 3735, KNM-ER 3891, and OH 24 show near complete overlap of the coronal planes on which the postglenoid process and the superior margin of the tympanic plate reside, and in Stw. 53, OH 24, KNM-ER 3891, and Stw. 98 and Stw. 329 (a juvenile), two other temporal bones from Sterkfontein that we assign to *H. habilis*, the tympanomastoid groove is sealed laterally, as the crista petrosa is here appressed to the anterior wall of the pars mastoidea (the tympanics are broken off laterally in KNM-ER 1813). Another early *Homo* specimen showing the elimination of the spaces between these basal temporal structures is SK 847, whose tympanic plate form is very similar to that of modern humans, especially in the development of a vaginal process of the styloid and the near complete elimination of the tympanomastoid groove (see also Clarke, 1977).

Regarding the derivation of the eustachian process, it is with *A. afarensis* that these early *Homo* specimens compare most favorably. Although few *H. habilis* temporal bones preserve the medial part of the tympanic plate in decent shape, on those that do, the thickened crista petrosa does not appear to give rise to a well-developed vaginal process of the styloid. Nor is it continuous with the small eustachian process, from which the crista is separated by a transverse depression in the flattened, inferiorly directed medial part of the main tympanic face (see Stw. 53, Stw. 98, and OH 24). The absence of an appreciable vaginal process of the styloid and the separation of the eustachian process from the crista petrosa are primitive retentions of the great ape condition that are also prevalent in *A. afarensis*.

Summary

From a purely structural point of view, the crania of *A. boisei* occupy the most derived position on the morphocline of the *Australopithecus* glenoid region (Figure 5.17). At the opposite end, much closer to the African great ape condition, sits *A. afarensis*, with A.L. 444-2 occupying the least primitive part of this species' morphological spectrum. This is seen chiefly in the greater degree of vertical relief in the sagittal plane engendered by the more prominent articular eminence and more upright (but still tubular) tympanic element, compared to A.L. 166-9 and A.L. 333-45. Although it is not yet well known, it is possible that the glenoid region of *A. anamensis* is even closer than *A. afarensis* to the great apes' position on the morphocline (M. Leakey et al., 1995; Ward et al., 1999, 2001). With a more prominent articular eminence (on average) and a relatively upright, platelike tympanic element, further—and uniquely—transformed by modifications to the eustachian process, the *A. africanus* glenoid region is substantially more derived than that of *A. afarensis*. Makapansgat cranium MLD 37/38 differs from the Sterkfontein representatives of *A. africanus* in the coincidence of the coronal planes occupied by the summit of the articular eminence and the posterior edge of the temporal foramen. In this, MLD 37/38 recalls *A. boisei*. Although *A. aethiopicus* (KNM-WT 17000) features a flatter articular eminence and shallower mandibular fossa than does *A. africanus*, which would place it in a more primitive position on the morphocline, sealing of the tympanomastoid gap laterally and more coronally oriented petrous bones in KNM-WT 17000 are more derived states than in crania of the latter taxon, unless Sts. 19 is included in its hypodigm. Reduction in height of the postglenoid process; the tendency toward elimination of the gap between the coronal planes on which it and the vertically deep tympanic plate reside; lateral extension of the tympanic to or beyond the sagittal plane of porion; prolongation of the petrous crest into a bony vaginal process of the styloid; deepening of the mandibular fossa through increased height of the articular eminence; and extension of the digastric fossa posteriorly to near the rear calvarial border—all are derived states that position *A. robustus* and *A. boisei* at similar points on the glenoid region morphocline. With the exception of the lateral extension of the tympanic, all of these states, plus the coronal alignment of the petrous bones and reduction or elimination of the tympanomastoid gap—states shared by these two robust species and *A. aethiopicus*—are features of the temporal bone in *Homo*. Finally, as already noted, *A. boisei* can be located at the extreme derived end of the morphocline by virtue of its uniquely transformed glenoid region: coincidence of the summit of the articular eminence and posterior edge of the temporal foramen; lateral to medial twisting of the articular eminence such that the entoglenoid process "points" posteriorly; further deepening of the mandibular fossa, with the deepest point positioned above the FH; and flattening of the tympanic plate to form

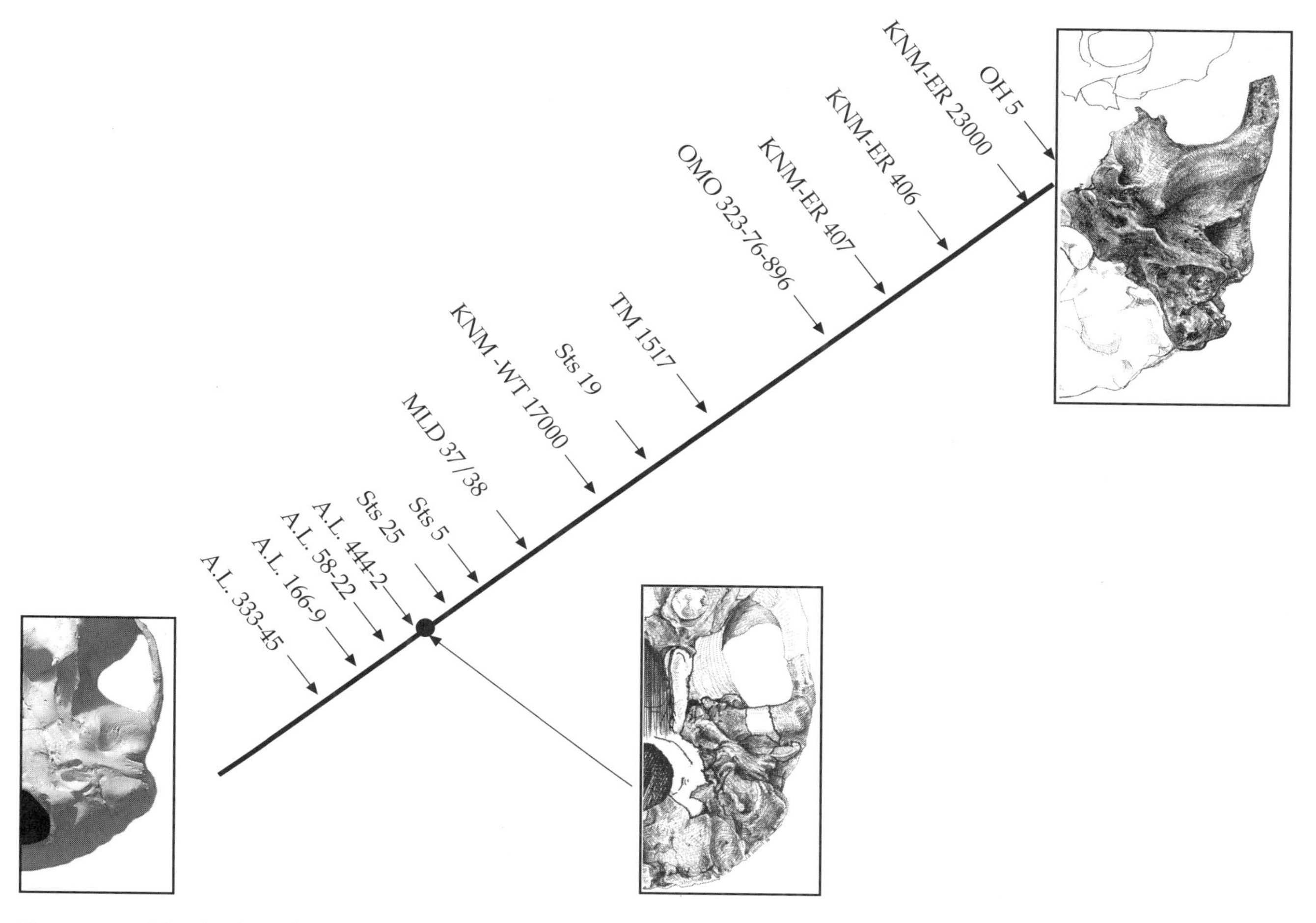

Figure 5.17 Morphocline of the early hominin glenoid region. The plesiomorphic end of the morphocline is at lower left (represented by the chimpanzee) and the apomorphic end is at upper right (represented by *A. boisei*). The position of individual fossils on the morphocline is based on their relative approximation to these two poles. Cranium Sts. 19 represents the ancestral configuration for *Homo*.

part of the ceiling, as well as the posterior wall of the mandibular fossa, its inferior margin blending with the anterior face of the pars mastoidea. When combined with its uniquely modified mastoid morphology, the temporal bone of *A. boisei* ranks with its face as one of the most autapomorphic cranial regions in the genus *Australopithecus*.

The Occipital Bone

Posterior Aspect

The occipital squama of A.L. 444-2 is very broad transversely (biasterionic chord breadth = 103 mm), but short sagittally (occipital sagittal chord = 57 mm), yielding a very low occipital length/breadth index of 55%. A remarkably short planum occipitale (la–i chord = 31 mm) in such a large skull dictates the small occipital sagittal dimension and produces a high ratio of occipital scale chords of 142% (see Table 5.8). As a result of the large biasterionic breadth and the short planum occipitale, the lambdoidal margin of the squama is broad and low, describing a wide angle at lambda. In spite of the complete fusion of the lambdoidal suture externally, crevice-like traces of the suture are seen endocranially and in the broken cross section at the midline. Between the inferiorly diverging sets of temporal lines the planum occipitale is flat sagittally (chord/arc index = 100%), as well as mediolaterally. Lateral to the temporal lines, as it begins to curve anterolaterally, the squama weakly flares posteriorly to meet the nuchal crest, formed by an inferior extension of the planum occipitale, whose free lower margin strongly overhangs the nuchal plane. It is this inferior extension

Table 5.8 Measurements of the Occipital Bone

Specimen/Sample		Biasterion Chord	OSC (mm)	OSA (mm)	OSC/Biasterion Ch. (%)	OSC/OSA (%)	la–ast Ch. (mm)	la–i Ch. (mm)	la–i Arc (mm)	la–i Arc/ OSA (%)	i–o Ch. (mm)	Occipital Scale Index (%)
A. afarensis												
A.L. 444-2		103.0	57.0	70.0	55.3	81.4	60.0	31.0	29.0	41.4	44.0	142
A.L. 439-1		100.0	>63	>70.5	n/a	n/a	64.0	35.5	n/a	n/a	>34	n/a
A.L. 333-45		94.5	60.5	72.0	64.0	84.0	57.5	31.6	33.3	46.3	38.0	120
A.L. 162-28		79.2	n/a	n/a	n/a	n/a	50.1	25.1	n/a	n/a	n/a	n/a
A.L. 288-1		82.0	51.2	58.8	71.7	87.1	n/a	30.7	31.8	54.1	27.2	89
KNM-ER 2602		76.0	n/a	n/a	n/a	n/a	52.0	31.0	n/a	n/a	n/a	n/a
	Mean	89.1	56.2	66.9	63.7	84.2	56.7	30.8	31.4	47.3	36.4	117
	SD	11.5	4.7	7.1	8.2	2.9	5.7	3.3	2.2	6.4	8.5	27
A. africanus												
Sts. 5		75.0	58.0	72.5	77.3	80.0	52.0	38.4	40.2	55.4	31.7	83
Sts. 71		74.0	64.0	75.0	86.5	85.3	50.5	32.0	35.0	46.7	38.5	120
MLD 1		86.0	62.1	75.0	72.2	83.9	60.0	36.0	39.0	52.0	37.6	104
MLD 37/38		79.5	56.7	70.0	71.3	81.0	54.3	35.6	37.2	53.1	31.6	89
	Mean	78.6	60.2	73.1	76.8	82.6	54.2	35.5	37.8	51.8	34.9	99
	SD	5.5	3.4	2.4	7.0	2.5	4.2	2.6	2.2	3.7	3.7	17
A. bosiei												
OH 5[a]		89.2	57.8	82.0	64.8	70.5	57.2	36.2	36.5	44.5	43.7	121
KNM-ER 406		93.0	61.0	70.0	65.6	87.1	61.8	34.5	34.5	49.3	36.0	104
KNM-ER 407		81.0	62.0	68.0	76.5	91.2	53.0	39.0	39.0	57.4	31.5	80
KNM-ER 13750		96.5	n/a	n/a	n/a	n/a	59.4	36.0	n/a	n/a	n/a	n/a
KNM-CH 304		n/a	n/a	n/a	n/a	n/a	n/a	34.3	n/a	n/a	n/a	n/a
KNM-ER 23000		96.5	65.5	n/a	67.9	n/a	68.5	n/a	38.0	n/a	34.0	91
	Mean	91.2	61.6	73.3	68.7	82.9	60.0	36.0	36.8	50.4	36.3	99
	SD	6.5	3.2	7.6	5.4	11.0	5.8	1.9	1.9	6.5	5.3	18

OSC, occipital sagittal chord (la–o); OSA, occipital sagittal arc (la–o); occipital scale index, i–o ch./la–i ch.; ch., chord.

[a]Data from Tobias, 1967.

of the occipital plane that is in part responsible for the sharply flexed midsagittal contour of the squama, measured as a right angle at inion (see Chapter 3).

The external occipital protuberance (and, hence, inion) is naturally sited 4 mm to the left of the calvaria's midline. It is a massive, inferiorly pointing triangle from whose apex the limbs of the nuchal crest diverge upward to form deep notches on each side before running laterally on a somewhat irregular, sinusoidal path toward the asteria. The lowest points on the nuchal crest limbs, on the left side occurring about 28 mm lateral to the midline (the crest is chipped on the right), fall on approximately the same transverse line as inion.

The occipital squama is impressively thick, measuring 10.2 mm at lambda, and 19.5 mm at asterion.

Inferior Aspect

The nuchal crest is evenly arched between the asteria. With its knife-edge sharp lower margin, the crest forms a nearly vertical skirt projecting maximally about 6 mm below the nuchal plane surface on the left side. The planum nuchale is fairly long (i–o chord = ca. 44 mm) and flat midsagittally; its left half is flat from side to side as well, whereas the right half has a mild, central swelling that renders the surface mediolaterally convex. This asymmetry appears to be related to the differential development of nuchal muscle insertion sites on the two sides.

The morphology of the planum nuchale bears witness to extremely well developed nuchal muscles (Figure 5.10). Dominant among the nuchal muscle attachment scars are concave, exclamation-point-shaped impressions for the superior oblique muscles, deeper and more medially encroaching on the left side, whose insertions abutted the nuchal crest itself. The lateral boundaries of the superior oblique impressions coincide with the thin, down-turned occipitomastoid sutural margins, and their medial margins are the sharp lateral arms of the inferior nuchal line. Medial to the superior oblique impression and anterior to the relatively weak transverse arms of the inferior nuchal line, the nuchal plane presents a fairly flat, rectangular area (ca. 68 mm wide at the level of the inferior nuchal line) bearing light, but palpable, thumb-shaped imprints of the suboccipital muscle (rectus capitis posterior major and minor) insertions. On each side, the insertion scars of the major and minor recti lie side by side, extending posteriorly to approximately the same point and rendering the coronally oriented transverse arm of the inferior nuchal line as a double arch. The transverse arms of the inferior nuchal line divide the nuchal plane into anterior and posterior halves and meet the lateral arms at a right angle. Between the inferior nuchal line and the nuchal crest, prominent, irregular markings for the insertions of the semispinalis capitis muscles lie on either side of the external occipital protuberance, abutting the exposed anterior face of the nuchal crest. On the left side the semispinalis insertion is divided from the superior oblique scar laterally by a very rugose, linear spur that runs from the nuchal crest to the lateral arm of the inferior nuchal line. On the right side the homologous feature is much weaker—it is merely a roughened line—and the superior oblique insertion is more confined laterally.

The bone of the nuchal surface around the base of the external occipital protuberance is crushed, and the path and condition of the external occipital crest are obscured by bone flaking along the sagittally oriented displacement crack through the nuchal plane. Insufficient of the foramen magnum's margins remain to be able to reconstruct the shape of the foramen with confidence. The right pars lateralis bears a prominent, rectangular (5 mm anteroposterior by 9 mm mediolateral) "paramastoid" process, which is an inferior extension of the occipital portion of the jugular notch. The inferior surface of the process is too abraided to tell whether or not it bares a facet signifying contact with the transverse process of the atlas vertebra (as it does in some other *A. afarensis* occipitals, such as A.L. 333-45). Posterior to the paramastoid process, across a small area of missing bone, the nuchal surface of the pars lateralis bears a light, curved line, presumably a trace of the rectus capitis lateralis muscle insertion. Medially this surface is broken as it begins to dive into the (lost) postcondylar fossa.

Endocranial Aspect

Traces of the lambdoidal suture are superimposed on the transverse limbs of cruciate eminence near the internal asteria, where they form V-shaped notches for the mastoid angle of the parietals. On the right, the suture can be traced toward midline as it parallels a mild "step" on the endocranial surface corresponding to a change in character between parietal and occipital lobe topography.

The cruciate eminence is very strong, forming a prominent mound (the internal occipital protuberance) in the middle of the occipital's endocranial surface. Its superior limb is very weak, however, barely rising above the endocranial surface until just superior to the center of the eminence, and lacking any sign of the sulcus for the superior sagittal sinus. The inferior limb, despite suffering cracking and missing bone, is broad, flat, and shorter than the superior limb. The transverse limbs, arising from the eminence at the same vertical level, sharply divide the cerebral from the cerebellar fossae medially; laterally they are much weaker, gradually fading as they approach the asteria. The medial third of the right transverse limb bears a shallow sulcus for the transverse sinus, but the left limb is devoid of venous markings. The endinion and inion are located at approximately the same vertical level, in part because of the strong inferior projection of the external occipital protuberance.

Despite obvious distortion of the posterior endocranial cavity, it is evident that the oval cerebellar fossae are larger and deeper than the cerebral fossae, which are themselves well excavated, especially medially. The deepest parts of the cerebellar fossae form the floor of the posterior cranial fossa, and thus, in the FH, the fossae are positioned under and anterior to the cerebral fossae. Bordering the cerebellar fossae along the posterolateral rim of the foramen magnum bilaterally is a sulcus for the marginal sinus, which is more clearly delimited on the left side. On the right side the marginal sinus sulcus can be followed superiorly to the cruciate eminence, where it flows into the sulcus for the ipsilateral transverse sinus. (See Chapter 4 for further details of cerebellar and posterior cerebral morphology, and Chapter 3 for a description of the venous drainage pattern.)

Comparative Morphology

The occipital bone is the best-represented component of the *A. afarensis* cranium and the locus of many of the distinctly primitive aspects of this species' calvarial morphology. The A.L. 444-2 specimen reiterates much of this primitive pattern while adding information on variation in *A. afarensis* occipital form. Four other new adult specimens have been added to the Hadar hominin occipital sample since 1990, boosting the total number of occipitals attributed to this taxon to nine (A.L. 162-28, A.L. 224-9, A.L. 288-1a, A.L. 333-45, A.L. 427-1b, A.L. 439-1, A.L. 444-1, A.L. 444-2, and KNM-ER 2602).

Size and Shape of the Occipital Squama

Besides the well-developed compound temporal/nuchal crest (see Chapter 3), the most striking aspect of the A.L. 444-2 occipital bone is its great breadth. The biasterionic chord (103 mm) stands well above those of other *A. afarensis* occipitals except for A.L. 439-1, which is only ca. 3 mm narrower than A.L. 444-2. Within the *A. afarensis* sample ($n = 6$) there are two nonoverlapping clusters of biasterionic breadth values, separated by a gap of almost 1.25 cm (Table 5.8), whose composition mirrors the division of the cranial sample into sex categories based on various size and shape attributes. However, here again we must bear in mind that the two largest Hadar occipitals, A.L. 439-1 and A.L. 444-2, originate from the younger part of the Hadar stratigraphic sequence that yields the largest individuals in the *A. afarensis* hypodigm (Lockwood et al., 2000).

The shape index of the squama (occipital sagittal chord/biasterionic breadth) registers A.L. 444-2 with the shortest and broadest occipital in the Hadar sample (55.3%), although there are only two other specimens complete enough for direct comparison. Overall, *A. afarensis* has a short, broad occipital squama relative to the other hominin species samples (Table 5.8). As is the case in the equally small *A. boisei* sample, the female specimen of *A. afarensis* (A.L. 288-1a; 71.7%) has a long, narrow squama relative to the male condition (compare OH 5 and KNM-ER 406 with KNM-ER 407). With a relatively narrow squama and no apparent discrimination by sex, *A. africanus* stands apart from the other species in Table 5.8. However, the range of variation in our small sample of chimpanzees encompasses the occipital shape index for every hominin specimen except A.L. 444-2, with its very broad squama, and Sts. 71, with its strikingly narrow squama, so the importance (if any) of the differences among the hominin samples may only be revealed with an increase in sample size. The average narrowness of the *A. africanus* occipital bone is nevertheless unsurprising in light of its unusually narrow cranial base, as discussed in Chapter 3.

The A.L. 444-2 occipital strongly exemplifies the tendency in larger *Australopithecus* crania (presumed males) for the midsagittal length of the planum nuchale to dominate that of the planum occipitale (producing an index of occipital scales of $>100\%$). In *A. afarensis* the high mean index of 117% is produced not so much by an unusually long nuchal plane as by a very short occipital plane: *A. afarensis* has the smallest occipital plane dimension of the hominin species samples in Table 5.8. The question arises as to whether it is the high position of inion or the low position of lambda that is responsible for occipital plane dominance in the Hadar sample (Kimbel et al., 1984). As shown in Chapter 3, the vertical position of inion approximates the FH in A.L. 444-2, as in virtually all early hominins, and unlike the great ape condition in which inion migrates during growth to a very high position above the FH. Thus, in *A. afarensis*, following the general plan in other *Australopithecus* species, it must be the low position of lambda that dictates the small occipital plane dimension. In A.L. 444-2, this fact is consistent with a very long sagittal margin of the parietal bones, measured as the bregma–lambda distance (see description of the parietal bones).

However, alone among the species of *Australopithecus*, *A. afarensis* includes some female crania that are *inferred* to have nuchal plane dominance in the ratio of occipital scales (i.e., A.L. 162-28 and KNM-ER 2602: Kimbel et al., 1984; Kimbel, 1988). Neither the occipital scale ratio nor its individual components are at all sexually dimorphic in the chimpanzee (early sutural obliteration combined with massive crest formation prevent assessment of this relationship in most adult gorillas and orangutans), but in view of the marked departure of hominins from the chimpanzee in the ontogenetic transformation of occipital bone morphology, this observation can have little relevance to the interpretation of intraspecific variation in *Australopithecus*.

Angulation and Orientation of the Occipital Squama

The occipital bone of *A. afarensis* is notable for its steep, bilaterally convex nuchal plane, with a relatively smooth, even rounded, midsagittal transition to the occipital plane (Kimbel et al., 1984; Kimbel, 1988). These characteristics are developed foremost in smaller specimens interpreted as females (KNM-ER 2602, A.L. 162-28, and A.L. 288-1). The A.L. 444-2 skull and other new discoveries at Hadar introduce further variation into the analysis of the *A. afarensis* occipital bone.

The strongly flexed occipital squama of A.L. 444-2 is reflected not only in the right angle the occipital and nuchal planes make at inion, but also in the occipital sagittal chord/arc index, which is low in cases of strong flexion. This index value of 79% for A.L. 444-2 is smaller than those for three other Hadar occipitals, which range from ca. 84% to 89% (one of these, A.L. 439-1, is broken just short of opisthion but its chord/arc ratio must be very close to actual; see Table 5.8), all of which feature visibly steeper nuchal planes and smoother transitions across the superior nuchal line than does A.L. 444-2 (see also discussion in Chapter 3).

Another recently recovered occipital specimen, A.L. 224-9, a small fragment of squama from an evidently small (female?) cranium (approximately the size of A.L. 288-1a), preserves the external occipital protuberance and portions of the nuchal and occipital planes, permitting the observation of a steep nuchal plane with a smooth, relatively even transition across the superior nuchal line. However, the tendency for only the smaller, presumably female, *A. afarensis* occipitals to have a less sharply angled squama and a steeper nuchal plane (Kimbel, 1988) is countered by still another new specimen, A.L. 439-1, a very large, undoubtedly male individual, which features probably the steepest nuchal plane (and highest occipital sagittal chord/arc index) in the Hadar sample (Figure 3.63).

Flexure of the occipital squama apparently influences the morphology of the nuchal crest in *A. afarensis*. In those specimens in which the nuchal plane is steep, the nuchal crest is formed at the junction between the upward and backward extension of the planum nuchale and the downward and backward extension of the planum occipitale (i.e., A.L. 162-28, A.L. 333-45, A.L. 439-1, A.L. 444-1, and KNM-ER 2602). In all of these specimens the nuchal crest forms part of a continuous rim that runs around the calvaria between the poria. Rather than shelving out posteroinferiorly as in these specimens, the planum occipitale in A.L. 444-2 is quite flat vertically, and the nuchal crest is its direct inferior extension.

Variation introduced by A.L. 444-2 and other new specimens bridges the morphological gap between male and female occipitals of *A. afarensis*. Substantial variation in the morphology of the occipital squama also characterizes *A. boisei*, as emphasized by Brown et al. (1993). As in A.L. 444-2, the squama of type specimen OH 5 is strongly flexed, with a relatively horizontal planum nuchale, and its nuchal crest is an inferior prolongation of the planum occipitale that projects below the nuchal plane surface. In contrast, KNM-ER 406 and KNM-ER 23000, both males, and KNM-ER 407, a female, have a less flexed squama with a steeper planum nuchale. Similarly, in KNM-ER 406 and KNM-ER 13750 the shelflike nuchal crest is an upward and backward extension of the planum nuchale. This is also the pattern in KNM-WT 17000, the cranium of *A. aethiopicus*.

Form of the Nuchal Plane

Among the features of the *A. afarensis* occipital bone considered to be most distinctive and similar to the great ape condition is the bilateral convexity of the planum nuchale, characterized as "broadly arched . . . bilateral, posterolaterally directed plates merging at a midline peak" (Kimbel, 1988: 649). Good examples of this morphology are found in females, both with (A.L. 162-28 and KNM-ER 2602) and without (A.L. 288-1a) well-developed nuchal crests, but the heavily crested male specimen A.L. 333-45 was described as having a transversely flatter nuchal plane.

New Hadar specimens shed light on the variation in this morphological pattern in *A. afarensis*. In a pair of these, A.L. 439-1 and A.L. 444-2, both large, heavily crested males, the two halves of the nuchal area are essentially coplanar, and thus the nuchal plane as a whole is flatter than in the smaller individuals. Two others, A.L. 224-9 and another small occipital fragment, A.L. 427-1b, which has the lateral part of the nuchal plane and the occipitomastoid margin, evince the more convex morphology. The association of A.L. 427-1b with a large (male) maxilla (A.L. 427-1a) may provide an exception to the tendency for only the smaller, female occipitals of *A. afarensis* to display a bilaterally convex nuchal plane. Despite the additional variation in this feature, which cuts across the size spectrum, it remains the case that among the species of *Australopithecus* only *A. afarensis* contains specimens with this apelike occipital morphology. Recently discovered specimens of *A. boisei*, *A. aethiopicus* (R. Leakey and Walker, 1988; Brown et al., 1993), and *A. robustus* (Grine and Strait, 1994) have not altered this assessment.

Internal Anatomy

The anteroinferior position of the cerebellar fossae relative to the cerebral fossae on the internal aspect of the occipital bone of A.L. 444-2 means that the cerebellum was tucked under the occipital poles of the cerebrum, similar to what is seen the OH 5 cranium of *A. boisei* (see further discussion of this feature in Chapter 4). In A.L. 439-1, in contrast, as in A.L. 162-28, A.L. 333-45, KNM-WT 17000 (*A. aethiopicus*), and KNM-ER 23000 (*A. boisei*), occipital bone morphology indicates that the cerebellum was less completely pushed under the occipital poles, but not projecting behind them (Holloway, 1983, 1988; Brown et al., 1993). Interestingly, this split in spatial relationships mirrors differences in external occipital morphology in these hominins, with the latter group characterized by less flexed squamae and relatively steep nuchal planes.

In no Hadar hominin is it the case that the cerebellum protrudes posterior to the occipital poles of the cerebrum. Falk's (1985) conclusion that this was the case for A.L. 162-28 is based on a misorientation of the partial endocast (Holloway and Kimbel, 1986).

Bone Thickness

The occipital squama is extremely thick in A.L. 444-2 but is matched by A.L. 439-1 in thickness measurements recorded at lambda and asterion. These are by far the thickest occipitals of *A. afarensis* (see Table 5.9). Comparative data in Table 5.9 (updated from Kimbel, 1988) show that despite its reputation for thin cranial vault bones (Rak, 1978), *A. boisei* does not, on average, have a particularly thin occipital squama. Although *A. afarensis* tops *A. boisei* in mean asterionic thickness, individual values in the latter species match the high end of the range for the *A. afarensis* sample (i.e., KNM-ER 23000: Brown et al., 1993). The difference is due to the relatively great thickness at asterion in specimens attributed to females of *A. afarensis* (A.L. 162-28 and KNM-ER 2602; compare to female *A. boisei* specimens KNM-ER 407 and KNM-ER 732), which clearly reflects the high frequency of well-developed crests in the asterionic region in female *A. afarensis* crania. *Australopithecus africanus* has the thinnest occipital bones overall, with low values at asterion undoubtedly due to the absence of compound temporal/nuchal crests in this part of the calvaria.

Table 5.9 Thickness of the Occipital Bone (mm)

Specimen/Sample		At Lambda	At Asterion
A. afarensis			
A.L. 444-2		10.2	19.5
A.L. 439-1		10.0	20.0
A.L. 333-45		6.0	15.0
A.L. 333-84[a]		n/a	14.5
A.L. 162-28		6.0	11.6
A.L. 288-1a		6.5	n/a
KNM-ER 2602		9.0	10.0
	Mean	8.0	15.1
	SD	2.0	4.0
A. africanus			
MLD 1		6.8	8.6
MLD 10		8.0	n/a
Sts. 5		7.2	7.0
Stw. 505		n/a	10.0
	Mean	6.0	7.4
	SD	2.7	2.6
A. boisei			
KNM-ER 407		8.7	12.9
KNM-ER 732[a]		n/a	8.5
KNM-ER 733[a]		n/a	15.0
KNM-ER 23000		11.0	20.0
KNM-CH 304[b]		8.0	14.5
OH 5[c]		9.0	13.0
	Mean	9.2	14.0
	SD	1.3	3.7

[a]Thickness at asterion measured across the occipitomastoid sutural surface of the temporal bone.
[b]Thickness at lambda measured lateral to midline sagittal crest.
[c]From Tobias, 1967.

The Maxilla and the Palatine Bone

Deformation affects the left maxilla more than the right; hence, the description of the maxilla is based primarily on the right side.

Superior View

The palatal element of the maxilla is long and narrow. Its width from the outer wall of the right maxillary sinus (on the alveolar process) to the midline is 41 mm, and its length from the coronal plane of prosthion to that of the posterior part of the maxillary tuberosity is 87 mm. The maxillary tuberosity, whose great posterior projection is characteristic of the *A. afarensis* hypodigm, contributes substantially to the overall length of the maxilla in A.L. 444-2 as well.

An extensive maxillary sinus cavity penetrates as far anteriorly as the plane of the nasal aperture and laterally into the root of the zygomatic process of the maxilla (Figure 5.18). The elongated, kidneylike shape of the sinus chamber results from the concavity of its medial wall. Measuring 4.0–5.5 mm, the lateral maxillary sinus walls are quite thick. A prominent blunt ridge that traverses the maxillary sinus cavity floor between the M^2/M^3 junction and the medial wall of the cavity divides the cavity into a large anterior chamber and a small posterior chamber. A lower, thinner crest with a very sharp superior margin further subdivides the rear chamber. This morphology seems to be consistent throughout the *A. afarensis* hypodigm (Kimbel et al., 1997).

The spinal crest (crista spinalis of Gower, 1923), which forms the inferior margin of the nasal opening, gradually gains prominence as it passes laterally to merge

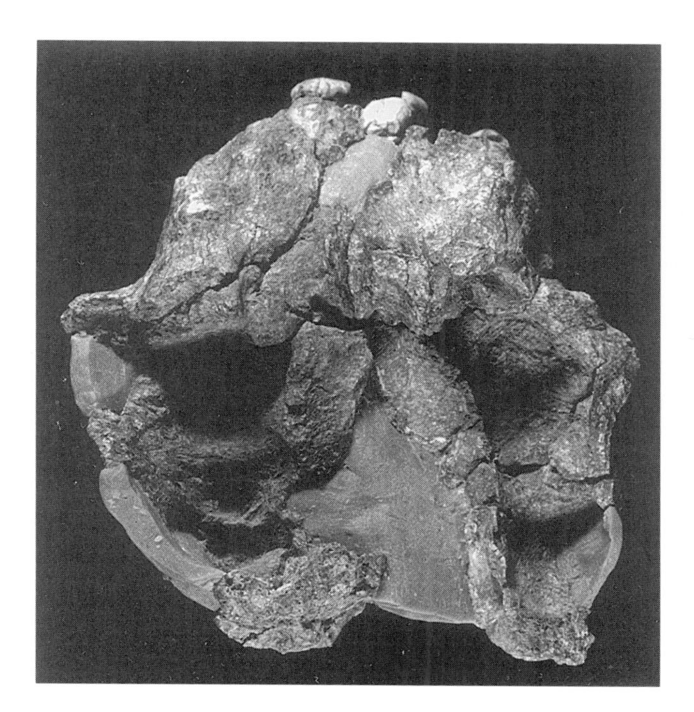

Figure 5.18 Superior view of the maxilla of A.L. 444-2. Natural size.

Figure 5.19 Anterosuperior view of the nasal aperture of A.L. 444-2. On the anterior part of the nasal cavity floor note the extensive intranasal platform surmounted by the bifid midline crest. This morphology is distinctive of *A. afarensis* maxillae. Natural size.

with the thin lateral margin of the nasal aperture (formed by the crista lateralis of Gower, 1923) and create a relatively sharp inferolateral corner. A continuous ridge formed from the merging of the cristae spinalis and lateralis defines the opening of the nasal cavity as a single plane. The spinal crest is not a sharp crest; rather, it is a tight crease in the bone surface, abruptly dividing the subnasal portion from the intranasal portion of the nasoalveolar element. At the midline, a tall, bifid incisive crest measuring 14 mm in length stands above an extensive intranasal platform situated between the spinal crest and the incisive fossa (Figure 5.19). This platform's prominence is enhanced by the enormous vertical drop (15 mm) to the floor of the nasal cavity at the broad, deep incisive fossa. In spite of damage, the intranasal platform's concave floor and the anterior and posterior mediolaterally oriented ridges that demarcate it are clearly visible. This concave surface extends laterally as far as the lateral walls of the nasal cavity, where it appears as a slightly hollowed fossa. Numerous faint mediolateral ridges roughen the floor of this fossa. In this fashion, the concavity accentuates the sharpness of both the crista spinalis and the crista lateralis, which jointly define the nasal aperture's margins.

Although it is partially obscured from view by displaced bone plates of the nasal cavity floor, the incisive fossa appears large. Lateral to the fossa, along the base of the maxillary sinus's medial wall, the floor of the nasal cavity drops abruptly in elevation behind the intranasal platform. Although distortion makes precise evaluation impossible, it appears that the nasal cavity floor maintains a constant elevation as it passes posteriorly toward the (now missing) posterior nasal opening.

Median View

The midline of the hard palate is preserved in the incisor region posteriorly to the P^4/M^1 level. The nasoalveolar component of the palate's cross section is oval in shape and nearly horizontal in orientation. Its length measures 42 mm, and it is 13 mm thick. Overlap of the nasoalveolar component on the palatal component is extensive, measuring 10 mm.

The palate is shallow anteriorly, showing only a very minor inferior deviation from the flat contour of its posterior segment. Thus, almost all of the preserved palatal cross section lies at the horizontal level of prosthion. The palatal elevation drops slightly behind the incisive canal (which is destroyed), creating a mild but palpable hump in the cross section of the palatal roof. Palate depth measured at the P^3/P^4 level is approximately 10 mm, and at the M^1/M^2 level is approximately 12 mm.

Palatal View

Despite remnant distortion, the shape of the A.L. 444-2 palate can be reconstructed as a long and narrow parabolic structure. Estimates of palate length (orale–staphylion = 75 ± 2 mm) and breadth (across the inner alveolar margins at M^2 = 41 ± 1 mm) yield a palatal shape index of 55%. It should be noted that these dimensions of the palate are not those devised in Chapter 3 to express the size of the upper jaw in the context of overall facial size; there the length of the palate is 76 mm, from the coronal level of prosthion to the coronal level of the distal end of M3. The maximum width, the distance between the most lateral inferior edges of the alveolar processes is 82 mm. Using these dimensions, the palatal arch has a slightly greater width than length.

The specimen features a strongly and evenly arched incisor row that, in palatal view, lies mostly anterior to the bicanine line, which is consistent with the presence of bilateral I^2/C diastemata. Distortion makes reconstructing the shape of the postcanine tooth rows tricky, but these appear to have slightly diverged from the midline posteriorly (Figure 5.20).

Lateral View

See the discussion of prognathism, the shape of the facial slope, and the morphology of both the zygomatico-alveolar crest and the root of the zygomatic process of the maxilla in Chapter 3.

Comparative Morphology

Margins of the Nasal Aperture

The confluence of the marginal and lateral crests creating sharp inferior corners and forming a single-planed nasal opening in A.L. 444-2 is matched in every specimen of *A. afarensis* in which the region is preserved (Figure 5.21), except one: the Garusi I maxillary fragment from Laetoli, in which the marginal crests are virtually nonexistent and the corner of the nasal aperture is rounded and indistinct. Indeed, in the Garusi specimen the subnasal and intranasal parts of the nasoalveolar clivus blend indistinguishably. This morphology is very similar to that encountered in chimpanzees, where strong nasoalveolar prognathism coupled with the ill-defined spinal and lateral crests blurs the threshold of the nasal cavity. We return to the Garusi specimen in Chapter 6.

In gorillas the anatomical configuration of the anterior nasal opening is quite different. The intranasal platform seems to "spill out" of the nasal cavity onto the slope of the nasoalveolar clivus (Figure 5.21). The lateral crest, which constitutes the lateral margin of the nasal opening, curves inferomedially along the surface of the clivus toward the bulging jugum of the central incisor. The posterior crest, which demarcates the rear of the intranasal platform, is clearly defined. It descends anteroinferiorly from the medial wall of the nasal cavity toward the midline, as far as the posterior pole of the incisive crest, which is more of a tubercle in gorillas. Since this posterior crest is more medial and positioned higher than the anterior one, which extends downward along the clivus slope as the spinal crest, the surface between the two crests, especially the portion that climbs the medial wall of the nasal opening, faces more anteriorly than medially. Therefore, the surface looks as if it lies outside the nasal cavity proper.

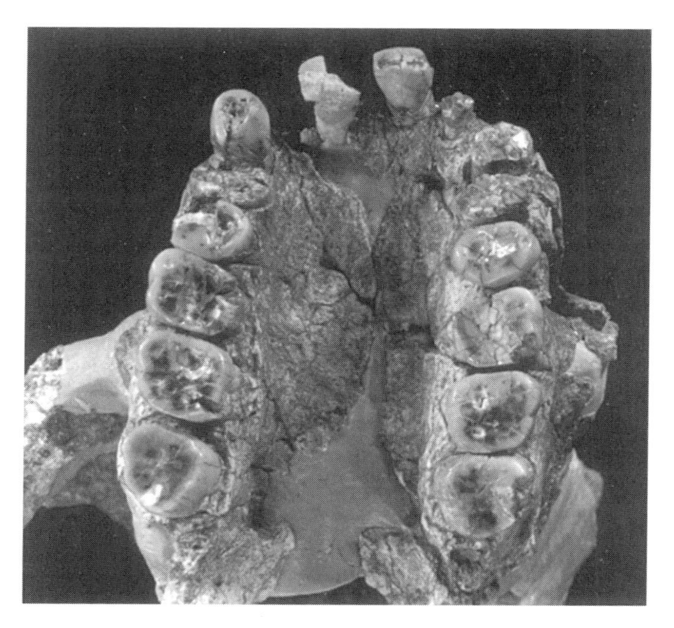

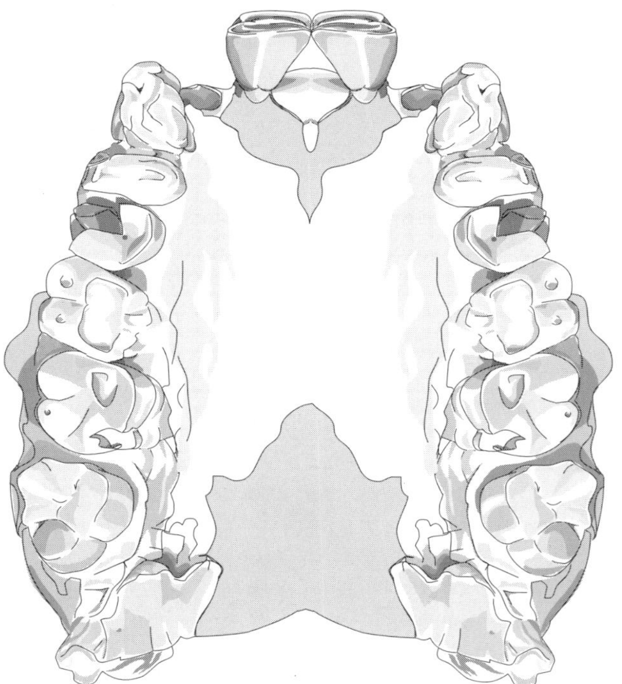

Figure 5.20 Palatal view of the maxilla of A.L. 444-2. The photograph, ca. 80% natural size, shows the palate in its deformed state. The drawing is based on the less deformed right side, of which the left side is a mirror image reconstruction.

Although superficially similar to *A. afarensis*, the nasal aperture morphology in *A. africanus* includes distinctive details. Characteristically, the lateral margin of the nasal

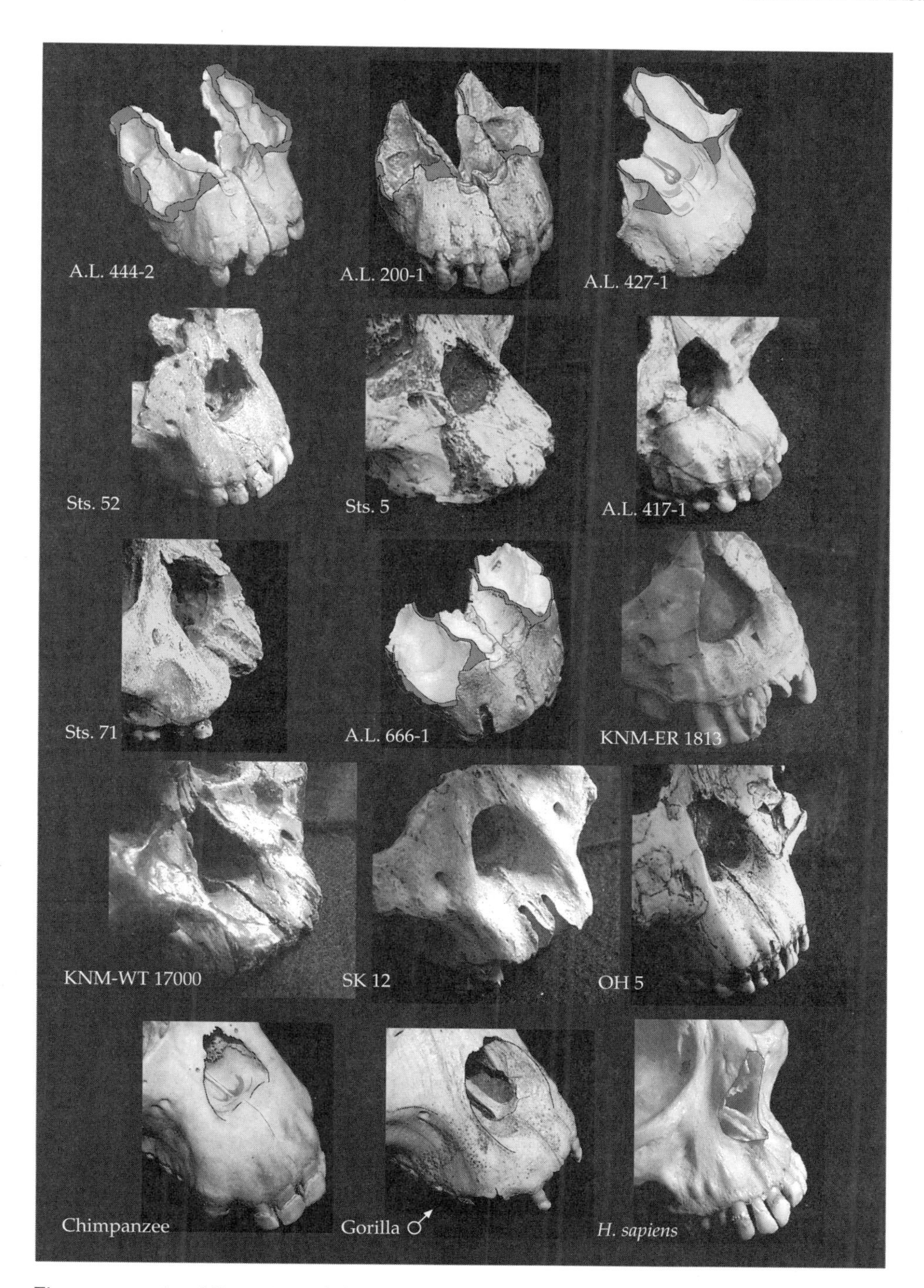

Figure 5.21 An oblique view of the nasal aperture in A.L. 444-2 and other hominoids. The sharp lateral margins in *A. afarensis* (A.L. 444-2, A.L. 200–1, A.L. 427-1, A.L. 417-1) define the nasal aperture and merge with the nasal sill. This arrangement is essentially present in the *Homo* specimens (A.L. 666–1, KNM-ER 1813) also. In *A. robustus* (SK 12) and *A. boisei* (OH 5), a corridor of sagittally oriented walls leads to the nasal opening, and the clivus, between these walls, is transformed into a gutterlike structure. This is not the case in *A. aethiopicus* (KNM-WT 17000). The gorilla specimen differs substantially from all the other hominoids, with its sharp lateral nasal margins descending for some distance on the surface of the nasoalveolar clivus. Not to scale. In some cases morphology has been accentuated by line art.

aperture is blunt rather than thin, resulting in the inferolateral corner having a rounded horizontal cross section. Thus, a distinct, sharp inferolateral corner representing the junction of the spinal and lateral crests is not usually present in this species, in contrast to *A. afarensis.* As explained by Rak (1983), the morphology in this region of the *A. africanus* midface is due to the pervasive influence of the anterior pillar. The anterior pillar, though it encapsulates the canine's jugum in its inferior (alveolar) segment, is not the same entity as the jugum, which is often an independently expressed structure on the face of *A. africanus* and *A. robustus.*

In their description of *A. africanus* cranium Stw. 505, Lockwood and Tobias (1999: 670) note that "strong ridges run diagonally from the inferolateral corners of the nasal aperture to the conjoint central and lateral incisor juga." As suggested by these authors, such crests, although variable in expression, appear to be a unique feature of *A. africanus* among early hominins (although they do recall the gorilla morphology described above). McCollum et al. (1993) also noted the frequent presence of these crests in *A. africanus,* describing them as inferomedial extensions of the lateral crests of the nasal aperture. When they are well developed, as in Stw. 391, Stw. 505, or Sts. 52a, they also delineate an inverted triangular depression centered on the nasoalveolar clivus between the inferior nasal margin and prosthion. At the upper corners of this inverted triangle (i.e., just inferomedial to the corners of the nasal aperture) the gap between the descending lateral crest and the inferior margin of the aperture is occupied by a variably developed fossa prenasalis, which is very similar to modern human anatomy (especially in Stw. 391). This morphology occurs in *A. africanus* specimens with and without anterior pillars (as in Stw. 391; Lockwood and Tobias, 1999), but is not present in A.L. 444-2 or any other *A. afarensis* maxilla, in which anterior pillars are not known to occur.

Nasal Cavity Relationships

Previous research has shown that *A. afarensis* maxillae have a distinctive morphological pattern in the relationships among the bony components of the nasal cavity (Ward and Kimbel, 1983; McCollum et al., 1993; McCollum, 2000). This pattern includes (1) a "stepped" nasal cavity floor in which there is a vertical drop to the incisive fossa; (2) an elevated intranasal platform intervening between the inferior nasal margin and the incisive fossa; (3) a high position of the anterior tip of the vomer, which inserts at the posterior end of the incisive crest. The configuration of the intranasal platform, clearly seen in A.L.444-2, is essentially identical to that observed in A.L. 200-1a, A.L. 333-1, and A.L. 427-1a (Figures 5.21 and 5.22). Smaller Hadar maxillae, such as A.L. 199-1, A.L. 417-1d, and A.L. 442-1, show little development of the incisive crest and intranasal platform, and less topographic relief at the entrance to the nasal cavity, but they are otherwise morphologically comparable to the larger specimens (Figure 5.22).

The morphology of the intranasal platform in *A. afarensis* is similar to that seen in some specimens of early *Homo* and *A. africanus* (see Ward and Kimbel, 1983; McCollum et al., 1993; Kimbel et al., 1997), but differs considerably from the distinctive morphology encountered in robust *Australopithecus.* In contrast to the latter species, both *A. afarensis* and *A. africanus* have a "stepped" nasal cavity entrance and a broad, basinlike incisive fossa. However, most *A. africanus* maxillae show very little, if any, extension of the nasoalveolar clivus into the nasal cavity (the intranasal platform); specimens such as TM 1512, MLD 45, Sts. 5, Sts. 17, Sts. 52a, Sts. 53, and Stw. 73 conform to this description. Other maxillae, including MLD 9, Sts. 71, Stw. 183, and Stw. 391, possess a more distinctly elongated platform between the entrance to the nasal cavity and the incisive fossa.

As recognized initially by Robinson (1953), nasoalveolar morphology is among the most distinctive aspects of the robust *Australopithecus* cranium (Figures 5.21 and 5.22). There is no confusing the morphology described for *A. afarensis* (or *A. africanus*) with the smooth, ill-defined ("unstepped") nasal cavity entrance; the elevated nasal cavity floor; and the coincident anterior nasal tubercle, anterior vomeral insertion, and incisive fossa that define the apomorphic nasal cavity pattern in *A. robustus*, *A. boisei*, and *A. aethiopicus* (Ward and Kimbel, 1983; McCollum et al., 1993).

Size and Shape of the Palate

Internal outline of the palate. Although the palate of A.L. 444-2 approximates the absolute size of the largest *A. boisei* palates, its internal shape index (55%) does not depart from the relatively long, narrow palates common to all *Australopithecus* species, irrespective of overall size (Table 5.10). The sample of four measurable *A. afarensis* palates (A.L. 199-1, A.L. 200-1a, A.L. 417-1d, and A.L. 444-2) yields a mean palatal index value of 54%. A long, narrow palate is a primitive character common to all great apes, contrasting with the relatively short, broad palate of modern humans. The origin of human palate shape is already expressed in the earliest representatives of the genus *Homo*, such as Hadar maxilla A.L. 666-1 and palates of *H. habilis* (Kimbel et al., 1997) (see Table 5.10).

If shallowness of the *A. afarensis* palate is also taken into account (see below), the palate can be easily separated not only from that of *Homo* but also from the *A. boisei* palate, as can be seen in the scattergram in Figure 5.23. The graph clearly indicates that it is palate depth that separates *A. boisei* from *A. afarensis*, whereas the shape

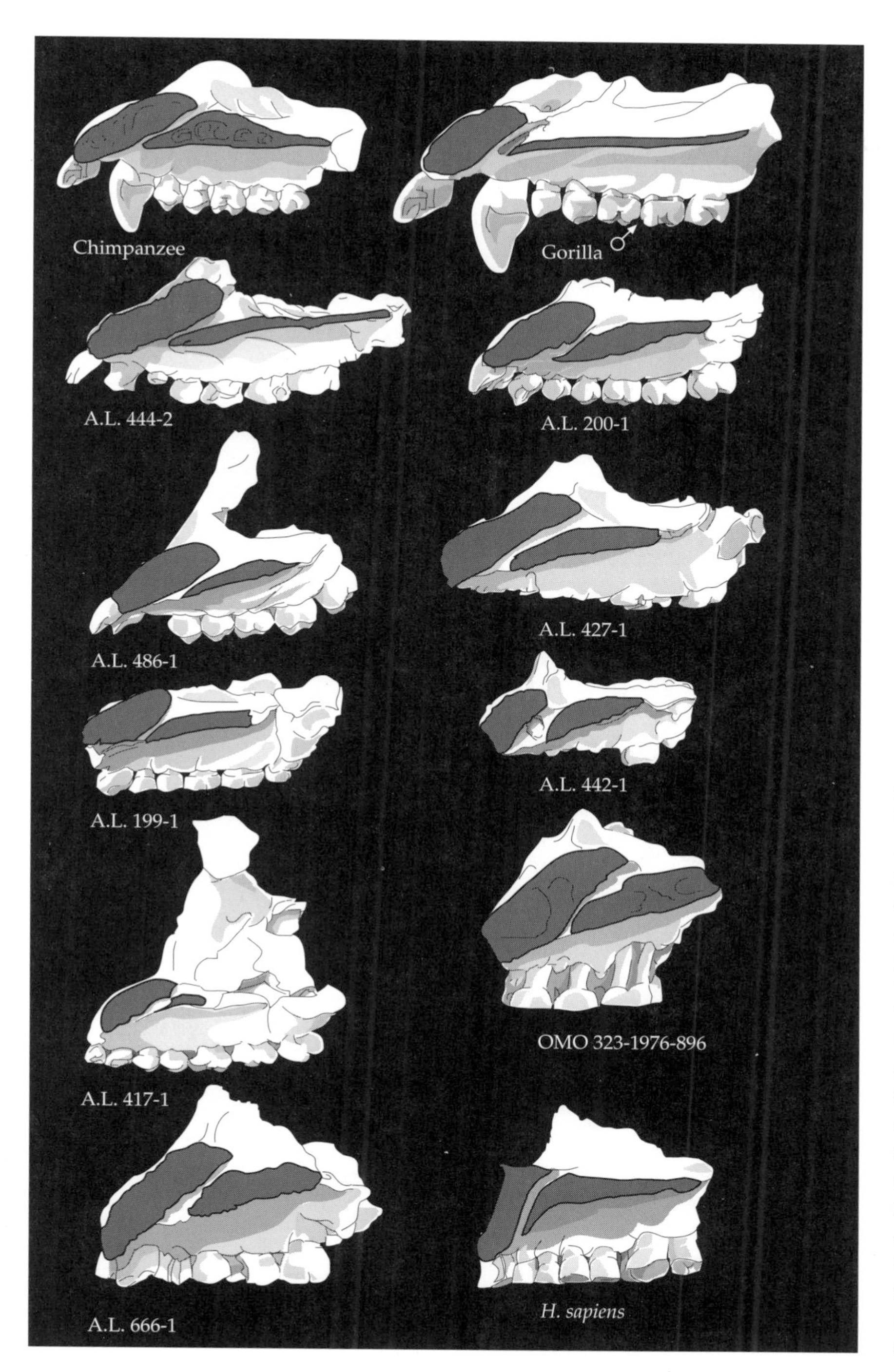

Figure 5.22 Midsagittal cross section of the hard palate and nasoalveolar clivus in A.L. 444-2 and other hominoids. Note the similarity between *A. afarensis* and the chimpanzee, which both exhibit an almost horizontal nasoalveolar clivus that juts forward and substantially overlaps the more posterior section of the palate. This configuration is unlike that of *Homo*, the robust *Australopithecus* specimen (Omo 323–1976–896), or the gorilla specimen, each of which differs in its own way. In the gorilla, the palate is extremely thin, and the nasal cavity floor is substantially lower than the nasal sill. The cross section in the robust *Australopithecus* specimen is extremely thick; the roof of the palate gradually ascends posteriorly. In *Homo*, the nasoalveolar clivus is relatively steep, and the palate is deep.

index separates *Australopithecus* from *Homo* (Kimbel et al., 1997).

The degree to which the alveolar arches diverge or converge posteriorly can be expressed by the ratio between the internal palatal breadth at M^2 (endomolare–endomolare) and the breadth between the lingual aspects of the canine alveoli. A ratio of 100% denotes internally parallel alveolar arches. In A.L. 444-2 the value for this index is 117% (the palatal width increases by 17% between the canines and the M^2s), a little above the mean of 114% for the *A. afarensis* sample as a whole (Table 5.10). Within the measurable sample ($n = 7$), palates range from nearly parallel (A.L. 442-1 = 104%) to moderately divergent (A.L. 417-1d = 123%). Relative to other hominins—with the notable exception of *A. anamensis*, whose index for two specimens is 99.5% (data courtesy of C. V. Ward; see also

Table 5.10 Measurements and Indices of the Palate

Specimen/Sample		Palate Depth (M^1/M^2) (mm)	Palate Length (or-sta) (mm)	Internal Palate Br. at M^2(mm)	Internal Palate Br. at C (mm)	Palate Breadth Index (Br. M^2/Br. C)	Palate Shape Index (Br./lg.)*100	Comments[a]
A. afarensis								
A.L. 199-1		11.0	54.0	32.0	26.5	120.8	59.3	L, CBr est.
A.L. 200-1a		8.5	65.0	33.5	30.1	111.3	51.5	L est.
A.L. 417-1d		14.0	58.0	28.5	23.1	123.4	49.1	L est.
A.L. 427-1		11.0	n/a	32.0	29.5	108.5	n/a	CBr, MBr est.
A.L. 442-1		n/a	n/a	25.0	24.0	104.2	n/a	CBr, MBr est.
A.L. 444-2		12.0	75.0	41.0	35.0	117.1	54.7	CBr, MBr est.
A.L. 486-1		11.2	n/a	33.0	28.9	114.2	n/a	
	Mean	11.3	63.0	32.1	28.2	114.2	53.7	
	SD	1.8	9.2	4.9	4.1	6.8	4.4	
A. africanus								
Sts. 5		18.0	65.3	35.7	27.0	132.2	54.7	CBr est.
Sts. 53		n/a	54.0	32.0	28.0	114.3	59.3	L, CBr, MBr est.
Stw. 73		14.5	58.0	30.0	28.0	107.1	51.7	L, MBr est.
	Mean	16.3	46.6	25.7	21.8	117.9	55.2	
	SD	n/a	5.7	2.9	0.6	12.9	3.8	
A. robustus								
SK 12		12.8	n/a	32.0	26.8	119.4	n/a	MBr est.
SK 46		12.2	n/a	35.0	28.0	125.0	n/a	CBr est.
SKW 11		15.0	60.0	34.6	27.0	128.1	57.7	L, CBr est.
SK 48		15.5	n/a	n/a	n/a	n/a	n/a	
SK 79		13.5	n/a	n/a	n/a	n/a	n/a	
	Mean	13.8	n/a	33.9	27.3	124.2	n/a	
	SD	1.4	n/a	n/a	0.7	2.2	n/a	
A. boisei								
KNM-ER 405		22.0	75.0	38.0	n/a	n/a	50.7	L, MBr est.
KNM-ER 406		20.0	70.0	37.4	28.5	131.2	53.4	CBr est.
OH 5		21.0	79.1	38.2	31.5	121.3	48.3	cast
KNM-CH 1		n/a	72.0	40.8	n/a	n/a	56.7	Tobias, 1991
	Mean	21.0	74.0	38.6	30.0	126.2	52.3	
	SD	1.0	4.0	1.5	2.1	7.0	9.2	
Homo								
A.L. 666-1		16.5	62.5	39.3	30.3	129.7	62.9	L est.
KNM-ER 1813		16.0	53.5	32.7	29.0	112.8	61.1	CBr est.
OH 24		15.0	50.7	35.5	27.0	131.5	70.0	

[a]L, length; CBr, palate breadth at canine; MBr, palate breadth at M2; est, estimate (+/− 1.0–2.0 mm).

Ward et al., 1999, 2001)—*A. afarensis* has, on average, a more parallel-sided palate. In *A. africanus* the mean ratio (118%) is slightly above that of *A. afarensis*, although the three measurable specimens in our *A. africanus* sample span an even greater range of variation than in the latter species. Palates of *A. robustus* and *A. boisei* exhibit substantially greater alveolar divergence posteriorly (124% and 126%, respectively), a characteristic shared with some early specimens of the genus *Homo* (see Table 5.10).

External outline of the palate. As already mentioned, the external palate dimensions of A.L. 444-2 are similar to those of large *A. boisei* specimens, OH 5 and KNM-ER 406 (Table 5.11). However, the similarity in the resulting shape index masks two distinct morphologies. Whereas the palate of A.L. 444-2 is long and parabolic, that of OH 5 has a truncated appearance in the front, which, along with a straight postcanine alveolar element that diverges from the midline as we follow it posteriorly, provides the palate with the shape of an elongated trapezoid. How such great morphological differences can be explained given the similarities in the length and width dimensions is illustrated in Figure 5.24. In *A. boisei*, the elongation of the postcanine dental row and the concomitant reduction of the anterior teeth transform the plesiomorphic dental arcade (and its support-

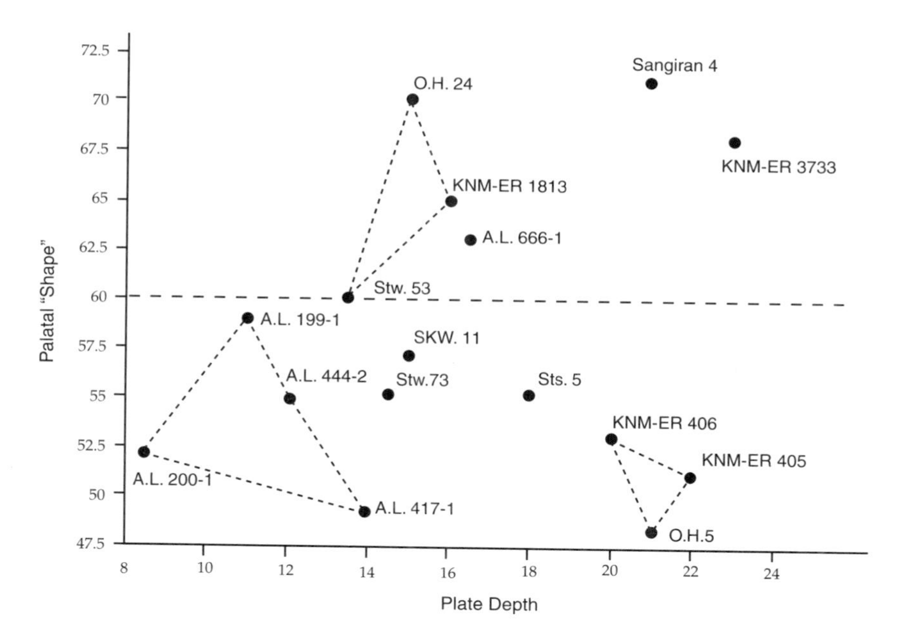

Figure 5.23 Bivariate scattergram of palate depth and the palate shape index. *A. afarensis* has the narrowest, shallowest palate of any *Australopithecus* species, except, perhaps, *A. anamensis* (not shown). The dashed line divides *Australopithecus* (below the line, relatively narrow) and *Homo* (above the line, relatively broad) based on palate shape.

ing bony structure) such as that found in *A. afarensis* into the shape seen in *A. boisei*.

Specimens of *A. robustus* (SK 12, SK 13, SK 46, SK 48, SK 52, SK 79, SK 83, and probably also TM 1517) seem to portray the trapezoidal palate morphology seen in OH 5, as does KNM-WT 17000—and apparently for the same reasons: an increase in the size of the postcanine tooth row and a reduction in the size of the anterior teeth (Robinson, 1956; Tobias, 1967). The presence of the anterior pillars in the corner between the two sets of teeth accentuates the trapezoid shape in *A. robustus*.

In the hypodigm of *A. africanus*, the palate tends to look more generalized—parabolic—than in the robust australopiths (Robinson, 1956). Some of the *A. africanus* specimens (Sts. 17, Sts. 71, MLD 6, and MLD 9) exhibit a palate that is truncated in the front and thus show a greater resemblance to the robust species, while others (Sts. 52a, Sts. 53, and TM 1512) portray the more generalized configuration; Sts. 5 falls in between these groups.

The parabolic shape of the palate in *A. afarensis* is, indeed, plesiomorphic. It is similar to the shape in chimpanzees, especially the females. In male chimpanzees, particularly those that have large canines, the palate takes a more rectangular shape, with the canines forming and defining the anterior corners. Male gorillas exhibit an even more extreme configuration. The enormous canines, wide diastemata, and large incisors join with a posteriorly narrow palate to render the external outline of the palate a very long, narrow trapezoid. Here, however, the shortest side of the trapezoid falls in the back, between the narrowly separated M^3s.

Even though large differences in size can be found in the palatal sample of *A. afarensis* (compare, for example, A.L. 442-1 and A.L. 444-2 in Table 5.11), the palates are all similar vis-à-vis the shape index and are almost identical in actual shape.

The palatal shape index in the taxa to which we have compared A.L. 444-2 in Table 5.11 decreases gradually from modern humans, at one end of the distribution (where the width of the palate by far exceeds its length—the width is 127% of the length), to male gorillas,[2] at the other end, where the index value is only 75%.

In the various taxa under comparison, palate breadth and length play independent roles in determining the index value. In Figure 5.25, the relative width of the palate is plotted against its relative length (relative to biorbital breadth, as previously discussed in Chapter 3). It is evident that the isolated position of the male gorilla mean is the result of both the extreme narrowness and length of the palate. The position of modern *H. sapiens*, in contrast, is a function of the extreme shortness and narrowness of its palate; it is the shortest and narrowest palate in relative terms of all the species examined. Note that the relative width of the palate in male gorillas is almost the same as in modern humans. KNM-WT 17000 occupies yet another extreme position in the graph: here the great width and length of the palate determine its specific position. *A. afarensis* groups with *A. robustus* and *A. boisei*. In its relative palate length, *A. afarensis* resembles chimpanzees and *A. africanus*; however, the great width of the palate in *A. africanus* and the narrowness of the chimpanzee palate pull these taxa in opposite directions, away from the *Australopithecus* group on the graph. The position of *H. habilis* (KNM-ER 1813) is worth pointing out, as its palate width is similar to that of *A. afarensis* and the robust australopithecines but its length sets it apart from that group and brings it closer to *H. sapiens*.

Table 5.11 External Palate Size and Shape

Specimen/ Sample		External Palatal Length (78) (mm)[c]	External Palatal Br. at M^2 (77) (mm)	Ext. Palatal Shape Index (77)/(78) (%)
Australopithecus afarensis				
A.L. 444-2		76	82	109
A.L. 200-1a		71	68	96
A.L. 199-1		59	68	115
A.L. 417-1d		61	61	100
A.L. 486-1		70	64	91
A.L. 427-1a		71	74	104
A.L. 442-1		57	56	98
	Mean (n = 7)	66	68	102
	Range	57-76	56-82	91-115
	SD	7	9	8
Homo sapiens	Mean (n = 10)	52	66	127
	Range	46-58	63-71	122-139
	SD	4	3	6
Gorilla gorilla female	Mean (n = 10)	86	68[a]	79
	Range	82-89	62-73	73-88
	SD	3	4	5
Gorilla gorilla male	Mean (n = 10)	99	74[a]	75
	Range	96-102	70-77	70-79
	SD	2	2	3
Pan troglodytes female	Mean (n = 10)	65	58	91
	Range	58-70	56-60	85-98
	SD	4	2	5
Pan troglodytes male	Mean (n = 10)	71	61	86
	Range	64-84	56-72	81-92
	SD	6	5	5
A. africanus	Mean	67	68	102
Sts. 5		66	68	103
Sts. 70		68	64	94
Sts. 53		64	64	100
Sts. 71		68	74	109
Early *Homo*	Mean	62	66	106
KNM-ER 1813		60	63	105
A.L. 666-1		65	71	109
OH 24		(60)	66	110
SK 847		(64)	64	100
Stw 53		62	65	105
A.robustus	Mean	66	70	107
SK 46		61	72	118
SK 13		66	70	106
TM 1517		69	68	99
A. boisei	Mean	76	80	107
OH 5		79	81	103
KNM-ER 406		73	79	108
KNM-ER 732[b]		(68)	(72)	(94)
A. aethiopicus				
KNM-WT 17000		78	80	103

[a]Measured at M^2 although this is not the point of maximum palate breadth.
[b]Values for KNM-ER 732 not figured in *A. boisei* mean.
[c]Values in parentheses are estimates.

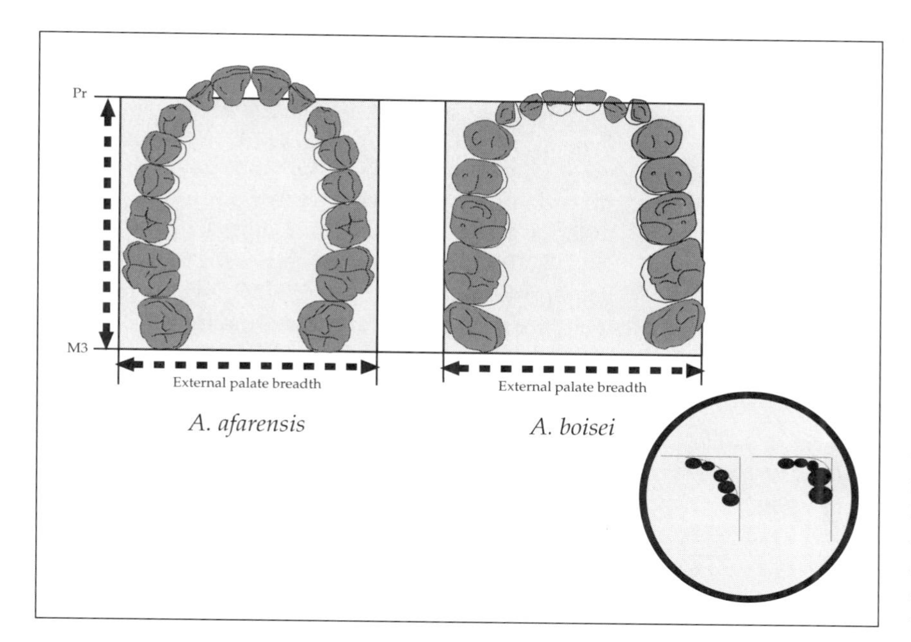

Figure 5.24 Comparison of the palate outlines of *A. afarensis* and *A. boisei.* Because the length and breadth of the palate is the same in both taxa, the shape index is the same; however, their actual shapes differ. The circle in the lower right corner of the figure is a schematic portrayal of the arrangement of the teeth in the anterolateral corner of the palate. In *A. boisei,* truncation and realignment of the anterior dental arch, together with relative expansion of the premolars, create a tighter corner in the palate.

The Palatal Midsagittal Cross Section and Palate Depth

In A.L. 444-2 the uniformly shallow depth of the palate between the incisor and the postcanine regions is typical of *A. afarensis* (see A.L. 199-1, A.L. 200-1a, A.L. 333-1, A.L. 427-1a, A.L. 442-1, A.L. 486-1, and Garusi I), as in the great apes. In the *A. afarensis* specimens, there is little, if any, inferior flexion of the palate between the incisive foramen and prosthion, in contrast to *A. africanus*, early *Homo*, and many *A. boisei* palates. The palatal cross section describes a single plane, from the incisors to the posterior molars (Figure 5.22). In *A. robustus*, however, the palate is thick in cross section and flat anteriorly; it ascends gradually as a single plane and becomes very deep posteriorly. One *A. africanus* specimen, MLD 9, resembles *A. robustus* in this respect but the uniformly flat, shallow palate of KNM-WT 17000 is more like that of *A. afarensis.*

In *A. afarensis*, the posterosuperior slope of the postcanine palatal plane is gentle, deviating relatively little from the horizontal (Figure 5.22), as is also the case in the great apes, *A. africanus*, and *Homo.* The palates of *A. robustus* and *A. boisei* present a very unusual morphology, in which the entire palatine process of the maxilla is strongly tilted upward and backward, marking a substantial posterior increase in palate depth (Tobias, 1967; Ward and Kimbel, 1983). However, because the anterior (premaxillary) palatal plane is also steeply angled in many specimens of these species, the anterior and postcanine palatal planes are still aligned with one another in midsagittal section. This morphology is especially common in *A. robustus* but is seen in some *A. boisei* specimens as well (see cross section of Omo 323-1976-896 in Figure 5.22). More often in *A. boisei*, the premaxillary component tends to be tilted inferiorly relative to the postcanine palatal plane; this angle is strong in OH 5 but apparently less so in KNM-ER 405, KNM-ER 406 (both of which are damaged), and KNM-WT 17400 (an immature individual). This strong angulation of the anterior part of the palate is common in *A. africanus* and *Homo* and is a derived condition among the hominoids.

Average maximum palatal depth (measured at M^1/M^2) is 11.3 mm in *A. afarensis*, which is considerably less than in *Homo* or other species of *Australopithecus*, in which sample means range from 13.8 mm in *A. robustus* to 21.0 mm in *A. boisei* (the value for *A. africanus* is based on too few specimens to be taken at face value) (Table 5.10). However, the exceedingly shallow palate depth measured in A.L. 200-1a (8.5 mm) appears to be exceptional in the context of the expanded *A. afarensis* sample (although the unmeasurable A.L. 333-1 is observably similar to it in this respect).

One new Hadar specimen, A.L. 417-1d, deviates from the pattern described for *A. afarensis* in two fundamental ways: the palate is deeper than in the other Hadar specimens (Tables 5.10 and 5.11), and the anterior palatal surface ascends steeply to the incisive foramen immediately behind prosthion. This specimen's palatal cross section is thus divisible into two planes—a steep anterior one and a more horizontal posterior one—that intersect in an obtuse angle at the incisive foramen. In these respects A.L. 417-1d resembles some specimens of *A. africanus* and early *Homo* (see, for example, TM 1512, Sts. 52a, and A.L. 666-1), although the Hadar specimen is narrower relative to its length and depth than are most of the specimens attributable to either of these groups (see Figures 5.22 and 5.23).

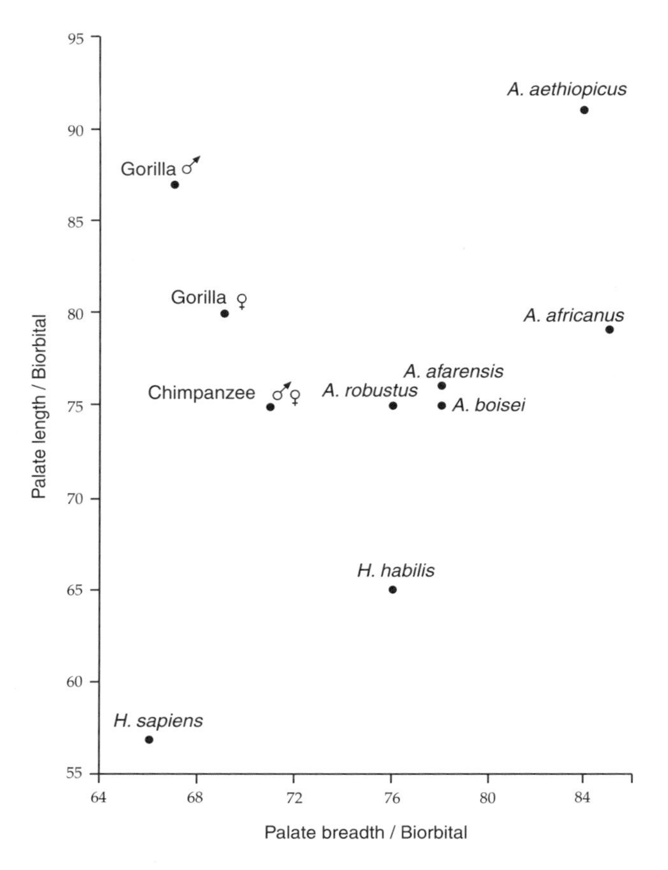

Figure 5.25 Bivariate scattergram of relative external palate breadth and relative external palate length (relative to biorbital breadth). The position of *Homo sapiens* is determined by its relatively short, externally narrow palate, whereas a relatively long, externally narrow palate is what determines the position of the male gorilla. The palate of KNM-WT 17000 (*A. aethiopicus*) is externally very long and wide in relative terms.

The Maxillary Sinus Cavity

The partitioning of the maxillary sinus floor in A.L. 444-2 is similar to that observed in other *A. afarensis* maxillae. Only recently, while comparing another maxilla from the Hadar Formation—the *Homo* specimen A.L. 666-1—with that of the *A. afarensis* hypodigm, did we notice how consistent the partitioning pattern is in *A. afarensis* and how different this arrangement is from that of *Homo* (Figure 5.26). We described these morphological differences as follows:

> In *A. afarensis* the sinus floor is divided into a large anterior chamber and a much smaller posterior chamber by a prominent transverse septum whose position relative to the tooth row varies between M^2/M^3 to distal M^3. Other less significant septa may subdivide the anterior chamber, but the chief division is always located far posteriorly in the sinus cavity. This pattern is clearly expressed in every Hadar *A. afarensis* maxilla in which the character can be judged: A.L. 199-1, A.L. 200-1a, A.L. 413-1, A.L. 427-1a, A.L. 442-1, A.L. 444-2, A.L. 486-1, and A.L. 651-1. On the other hand, the division of the A.L. 666-1 sinus floor into a small anterior and a large posterior compartment by a strong crest running diagonally from the M^1/M^2 level posterolaterally to the P^4/M^1 level anteromedially is a configuration unknown in the Hadar *A. afarensis* sample. . . . Specimens attributed to early species of *Homo* have precisely the same morphology as the new Hadar fossil. Thus, for example, in O.H. 62 and KNM-ER-1470 the maxillary sinus floor is prominently divided anteriorly. Breakage prevents assessment of the state of the posterior sinus floor in both of these specimens, but the presence of an anterior division is sufficient to distinguish them from *A. afarensis* and ally them with A.L. 666-1. The few casts of *A. africanus* maxillae in which the floor of the sinus cavity is exposed (MLD 9, MLD 45[?], Stw. 73, Stw. 183 [a subadult]) definitely lack an anterior division, pointing to an *A. afarensis*-type pattern in this species. It is interesting to note, however, that several maxillae of *A. robustus* (SK 12) and *A. boisei* (KNM-ER 405, KNM-ER 733, KNM-ER 732[?]) seem to show the same anterior segmentation of the sinus floor as seen in the early *Homo* specimens. (Kimbel et al., 1997: 251)

The arrangement seen in *A. afarensis* is almost certainly plesiomorphic, as it appears in the Miocene Moroto palate from Uganda and the *Sivapithecus* specimen GSP 15000 from the Siwaliks of Pakistan. A single chimpanzee specimen with an exposed maxillary sinus was available to us and revealed the *A. afarensis* configuration as well.

The Palatine Bone

In the posterior part of the palate, the horizontal plate of the palatine bone survives. The posterior end of this plate is large and smooth and takes the form of a bony scale that overlaps the more anterior part of the plate. Between these two elements of the horizontal plate is the outlet for the greater palatine nerve. Therefore, the canal of the nerve faces anteriorly. This arrangement is found in every specimen of *A. afarensis,* as well as in *A. africanus* (specimens MLD 37/38 and Sts. 5) and chimpanzees. Although the state of preservation does not enable us to see all the details, the same arrangement appears to be present in *A. robustus* (e.g., SK 47 and SK 48), *A. boisei* (e.g., KNM-CH 1, OH 5, and KNM-ER 406), and *A. aethiopicus*

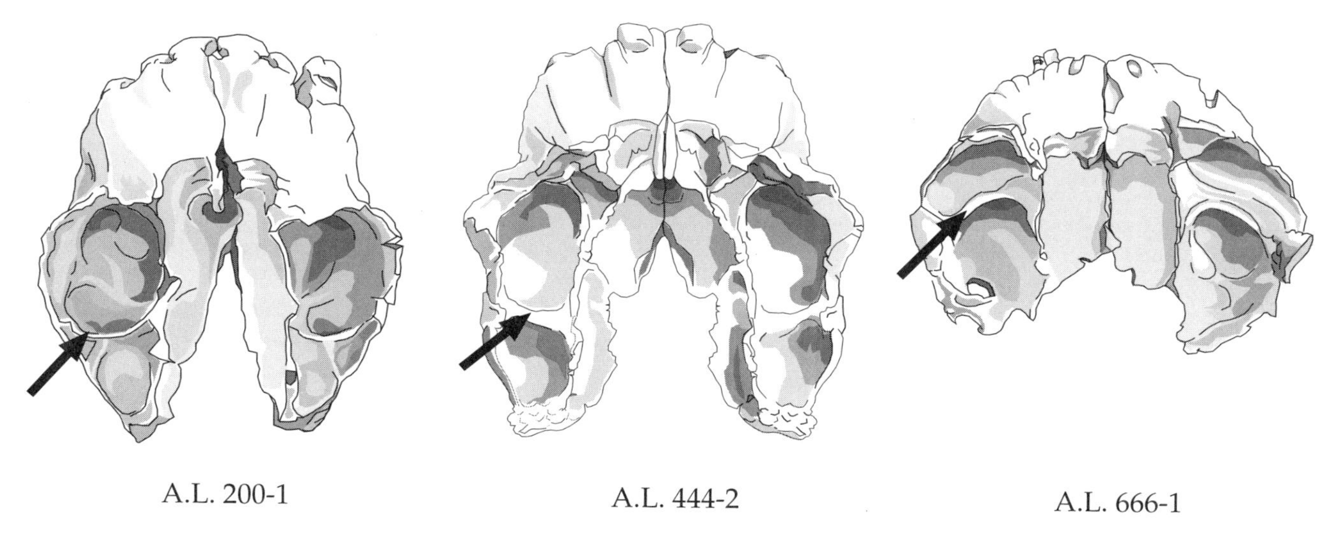

Figure 5.26 The topography of the maxillary sinus floor in *A. afarensis* (A.L. 200-1, A.L. 444-2) and early *Homo* (A.L. 666-1). Note that the transversely oriented septum divides the chamber into two parts (arrow). In *Homo*, the septum creates a small anterior area and a large posterior area, whereas in *A. afarensis*, it produces a large anterior area and a small posterior area.

(KNM-WT 17000). Only in gorillas does the topography of the horizontal plate differ from that of the hominoids just described; the plate has a single smooth, unbroken surface, which seems to be continuous with the palatal element of the maxilla. Situated between the maxilla and the palatine, the foramen of the greater palatine nerve faces mostly inferiorly.

The Nasal Bones

Although the majority of both nasal bones is preserved as a fused unit, neither their superior (nasion) nor inferior (rhinion) termini are present. Thus, total length of the nasal bones can be estimated as 30.0 mm; as preserved, the maximum length is 28.2 mm. The anterior surface of the nasal bones is gently concave sagittally, with the inferior end appearing to swing out anteriorly more than the superior end. Their upper half bears a strong median keel. This keel gradually fades out inferiorly, so that the nasal bones are transversely flat inferiorly but peaked in the midline superiorly. Chunks of the nasal bones' lateral margins are broken away, so it is difficult to be certain about their shape as a unit. However, they do not appear to have the inferiorly tapered shape that is common (though not universal) in robust *Australopithecus* species (Olson, 1985). A shape more like that of the A.L. 333-105 juvenile cranium, which shows superior and inferior expansion and a constricted middle section (Kimbel et al., 1982), is likely.

Two additional specimens of the *A. afarensis* face support these conclusions about the nasal bones of A.L. 444-2. In the small (presumptively female) individual A.L. 417-1d, a segment of the nasal bridge—the right frontal process of the maxilla and part of the nasal bones themselves—is preserved. Here, the nasal bones show the same sagittal concavity, and superior median keeling with inferior flattening, as do those of A.L. 444-2. However, the superior part of the frontal process of the maxilla (which is lost in A.L. 444-2) is everted anteromedially in A.L. 417-1d, implying that the (now missing) inferiormost part of the nasals was supported in a more anterior position than the superior part. The same forward disposition of the inferior end of the nasal bones can be reconstructed for A.L. 486-1, in which the medial margin of the maxilla's frontal process is likewise everted. The A.L. 417-1d maxilla also permits assessment of the nasal bones' shape: they are mediolaterally constricted and expanded inferiorly and especially superiorly, forming the shape of a vertically asymmetric hourglass.

The Zygomatic Bone

The zygomatic bone is immense and massively built. Because of the loss of the bone's original corners, conventional measurements cannot be taken. The chord stretching diagonally between the lowest point on the inferior orbital margin to the masseteric scar measures 43–45 mm. This measurement is influenced by the displacement of a fragment on the inferior orbital margin (see Chapter 3). Measured in this way, the size of the zygomatic resembles that of both OH 5 (40 mm) and male gorillas (43 mm) (Table 5.12). The distance between the zygomaticomaxillary suture and the intersection of the frontal and temporal processes of the bone is immense (48 mm), substantially exceeding the distance in the same segment of OH 5 and of the large male gorillas in our sample.

The topography of the anteriorly facing surface of the zygomatic bone, the facial surface, is nearly flat vertically but moderately convex transversely across the body. Two barely noticeable, blunt protuberances are palpable on the bone surface—one located at the root of the temporal process and the second, slightly more medially and inferiorly to the first. The former accentuates the sulcus running above the lip of the inferior margin. The latter demarcates the ill-defined corner formed by the anteriorly facing infraorbital region and the laterally facing part of the bone. (For a more detailed discussion and comparisons, see Chapter 3.)

The inferior margin of the bone is extraordinarily thick, measuring about 17 mm (Figures 5.27 and 5.28). Surmounting the inferior margin is a well-defined platform that bears the rugose scar denoting the origin of the superficial part of the masseter muscle. This platform is set apart from the surrounding bone surface by a very prominent liplike rim that delineates the masseteric scar and broadens it. So prominent is this surrounding rim that a shallow sulcus separates the facial surface from the inferior margin of the bone. The imprint of the sulcus fades anteriorly but deepens posteriorly. (Taphonomically, this sulcus is of great significance, as it continues onto the lateral aspect of the temporal bone's zygomatic process and in this way helps align the zygomatic and the temporal bones properly, despite the lack of secure contact between them.)

The masseteric scar is an elongate oval, 30 mm long and 10 mm wide at the point of its maximum width. It is divided into at least three subequal sections by two rugose, strongly developed ridges that run diagonally across the bone's inferior margin (Figure 5.27). Although the scar's size attests to a powerful masseter muscle, the bone surface of the scar itself is surprisingly smooth, unlike that of OH 5, which exhibits a much coarser topography full of deep pits and spikes. In A.L. 444-2, the scar faces inferiorly and slightly laterally, whereas in OH 5, it faces mostly laterally and is therefore exposed in an anterior view. The new Hadar zygomatic is similar in this regard to A.L. 333-1, the only other adult *A. afarensis* specimen to preserve the zygomatic bone.

Posterior to the scar of the superficial masseter origin, part of the bone's surface extends posteriorly to the root of the zygomatic bone's temporal process. Two relatively weak ridges paralleling the long axis of the inferior margin demarcate this portion of the bone surface (Figure 5.27). At its anterior end, this stretch of bone measures 11 mm wide and extends 15 mm to the broken, posterior edge of the temporal process. However, examination of the right temporal bone reveals that this topography extends across the zygomaticotemporal suture, yielding a total length of 37 mm. This region probably represents the site of origin of the deep masseter. If this interpretation of the soft anatomy is correct, then the transition from the scar representing the site of the superficial masseter to that representing the site of the deep masseter is the sharpest and most conspicuous of all the hominins and other primates we examined.

On the posterior surface (facies temporalis) of the zygomatic bone, the topography is more convoluted than on the anterior surface. As it descends, the posterior bone surface swells to form a lip so thick that it takes the form of a bony shelf, the medial edge of the thick inferior margin of the zygomatic bone (Figure 5.28). The center of the temporal surface is depressed between this lip and the thick superior (orbital) margin of the bone. In this central region, the zygomatic bone is hollowed and extremely thin (3.5 mm). As seen in the broken posterior cross section of the body, the posterolateral part of the bone was even thinner. This topography is also seen in the other *A. afarensis* specimens in which the region is preserved (A.L. 333-1 and the juvenile A.L. 333-105) but is not found in any other hominins or the African great apes. In these taxa, the temporal surface of the zygomatic bone is concave primarily in the transverse plane. Therefore, the temporal foramina do not display a deeply excavated anterior wall as in *A. afarensis* but, rather, a superoinferiorly oriented gutter.

A substantially different topography is found in *A. boisei*. Because of the anteriorly extended inferior margin of the zygomatic bone (forming the posterior side of the visorlike zygomatic region), the whole temporal surface is exposed in basal view. This posteroinferiorly facing surface is mediolaterally broad and exhibits a flat, even appearance. The *A. aethiopicus* cranium KNM-WT 17000 also displays this topography, since the great degree of prognathism is accompanied by an immense anterior extension of the inferior margin of the infraorbital region. In *A. robustus* this pattern is absent; instead, the generalized topography is seen (as in SK 48, SK 52, TM 1517, and others).

The frontal process of the A.L. 444-2 zygomatic bone, as mentioned in Chapter 3, is extremely thick and wide. As in gorillas, chimpanzees, and most specimens of *Australopithecus*, but unlike the configuration in modern humans, the bone surface faces primarily anteriorly. In A.L. 444-2, the frontal process is slightly concave in a sagittal plane, and on its posterolateral margin lies a distinct marginal process such as that found in modern hu-

Table 5.12 Measurements of the Zygomatic Bone

Specimen/Sample		Zygomatic Height (73) (mm)	Zygomatic Breadth (72) (mm)	Zygomatic Br.(72)/Ht.(73) (%)	Frontal Process Breadth (76) (mm)	Zygomatic Height (73)/ Biorbital Breadth (54) (%)	Zygomatic Breadth (72)/ Biorbital Breadth (54) (%)	Frontal Process Breadth (76)/ Biorbital Breadth (54) (%)
Australopithecus afarensis	Mean	37.5	41.5	111	19	40.5	45.5	21
A.L. 444-2		44	48	109	22	46	51	23
A.L. 333-1[a]		31	35	113	(16)	35	40	(18)
Homo sapiens	Mean (*n* = 10)	24	25	103	13	26	26	14
	Range	20-27	22-28	85-133	11-16	20-29	24-29	12-17
	SD	2	2	17	2	3	2	2
Gorilla gorilla female	Mean (*n* = 10)	40	25	65	11	42	28	12
	Range	38-43	23-27	62-69	9-13	37-45	26-29	10-13
	SD	1	1	2	1	3	1	1
Gorilla gorilla male	Mean (*n* = 10)	43	30	71	16	38	27	14
	Range	40-46	25-34	62-81	13-20	34-43	22-32	11-17
	SD	2	3	7	2	3	3	2
Pan troglodytes female	Mean (*n* = 10)	28	20	71	11	30	22	12
	Range	23-33	15-25	61-78	9-12	26-36	17-26	10-13
	SD	3	3	7	1	3	3	1
Pan troglodytes male	Mean (*n* = 10)	29	20	69	10	31	22	11
	Range	23-34	15-25	61-78	7-13	27-35	19-26	9-13
	SD	3	3	5	2	2	2	1
A. africanus	Mean	25	24	89	11	33	29	13
Sts. 5		27	24	89	12	32	28	14
Sts. 71		27	24	89	9	34	30	11
Sts. 52a		21	n/a	n/a	n/a	n/a	n/a	n/a
A. robustus	Mean	30	31	102	14	33	33	15
SK 48		30	33	110	13	30	33	13
TM 1517a		30	28	93	14	36	n/a	17
A. boisei	Mean	37	37	101	15	36	36	15
OH 5		40	36	90	14	40	36	14
KNM-ER 406		37	43	116	16	36	42	16
KNM-ER 13750		33	(32)	97	14	32	31	14
KNM-ER 732[b]		29	n/a	n/a	(10)	33	n/a	11
A. aethiopicus								
KNM-WT 17000		36	38	106	(16)	38	40	17

[a]Based on the reconstruction in Kimbel et al., 1984.

[b]Values for KNM-ER 732 not figured in the *A. boisei* mean.

n/a, not available; values in parentheses are estimates.

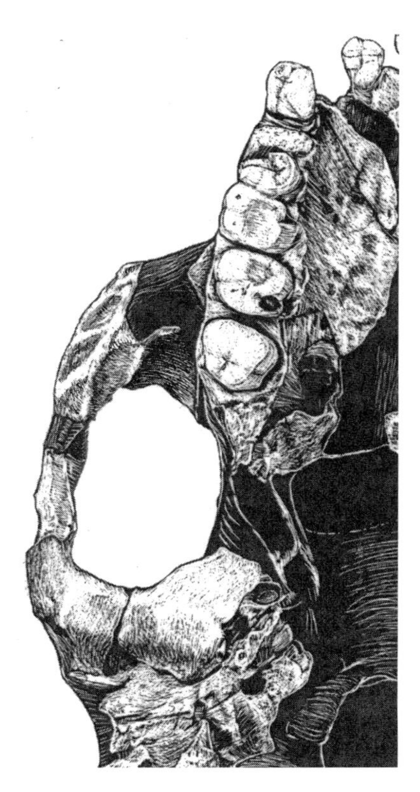

Figure 5.27 Anatomical detail of the masseter muscle origin on the zygomatic arch of A.L. 444-2.

mans. The presence of this process indicates that the frontozygomatic suture was not far superior to the edge of the now broken frontal process. The width of the frontal process at the level of the marginal process is huge, measuring 22 mm. That such a wide frontal process is absent in both the largest male gorilla crania and the most robust *Australopithecus* crania can be attributed primarily to the absence of the marginal process in these species.

Because the meeting point of the frontal and temporal processes is also not preserved, we can confidently infer only that the angle formed by the two processes in A.L. 444-2 was situated at about the level of the inferior orbital margin. This configuration resembles that of modern humans but differs from that seen in the African apes. In the apes, the superior margin of the temporal process of the zygomatic bone (the horizontal arm of the angle) is found substantially below the level of the orbital floor, as discussed in Chapter 3 (see Figure 3.27).

The posterior part of the frontal process—composed of the lateral wall of the orbit and the medial wall of the temporal fossa—is very thick. Even if one ignores the overlapping bone plates and intervening matrix, it is clear that the orbital wall is much thicker than in other hominoids, including modern humans and gorillas.

The maxillary sinus extensively pneumatizes the zygomatic bone. It runs laterally far beyond the zygomaticomaxillary suture, whose remains demarcate the medial border of the bone. This lateral extension of the sinus constitutes approximately 20% of the volume of the entire zygomatic bone. The posterior wall of the sinus, which bears the temporal surface, is particularly thick (2.5 mm), a property that is confirmed by the remains of the sinus cavity on the maxilla. The thickness is of interest in light of the paper-thin posterior maxillary sinus wall characterizing modern humans and the African apes.

The Mandible

Preservation

The A.L. 444-2 mandible consists of a nearly complete right corpus, with the root of the ascending ramus and the symphysis. The left corpus is represented by the region below the broken incisors and the mesial wall of the canine alveolus but does not reach the base. Refer to Chapter 2 for details of preservation.

Morphology of the Lateral Aspect

The corpus is vertically deepest anteriorly beneath I_2-P_4, where the basal margin appears mildly swollen. It shallows posteriorly as the basal contour angles superiorly relative to the occlusal and alveolar planes. The lateral prominence is massive and centered 23 mm lateral to the midpoint of M_2. It abruptly flattens into the external corpus below distal M_1, and beneath P_4 it is barely palpable. The strong lateral projection of the prominence gives A.L. 444-2 a greater corpus breadth at the M_2 level than any other mandible in the Hadar sample (see below for discussion of metric variation). The oblique line runs anteriorly (across a missing flake of bone) from the superior aspect of the lateral prominence, and disappears approximately 15 mm below P_4. It does not lead anteriorly to a lateral superior torus. The broad but shallow extramolar sulcus resides high on the corpus and fades out anteriorly at about the M_1 level. In lateral view, when the M_3 is shifted mesially to its correct position in the tooth row, the ascending ramus obscures from one-half to two-thirds of the crown.

The transition from the lateral to the anterior corpus is defined by a bulging C/P_3 jugum. Additional, much less prominent, tooth root juga occur in the alveolar bone at the mesiobuccal corner of the M_1 and P_4. A shallow, but palpable lateral corpus hollow is located anterior to the lateral prominence and posterosuperior to the mental foramen; it is deepest below P_3/P_4, and its lower limit is defined by a poorly developed marginal torus. Platys-

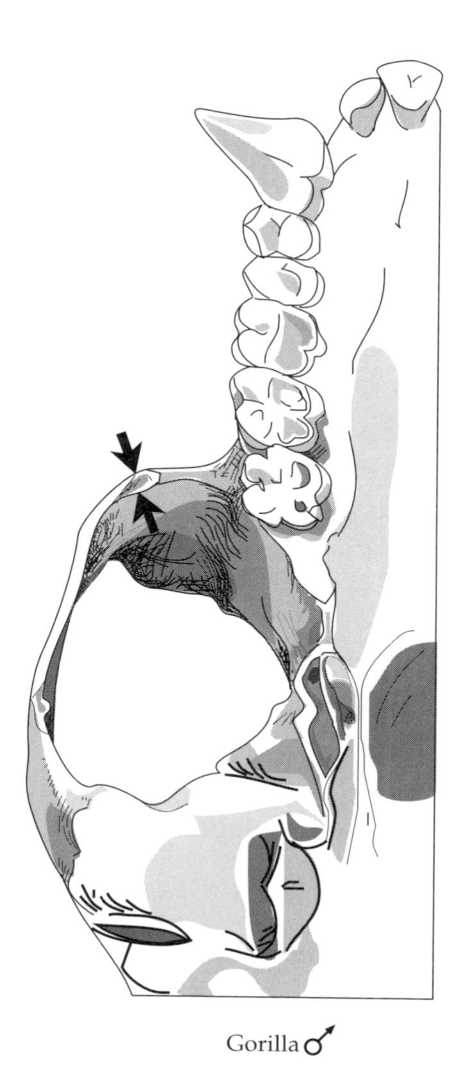

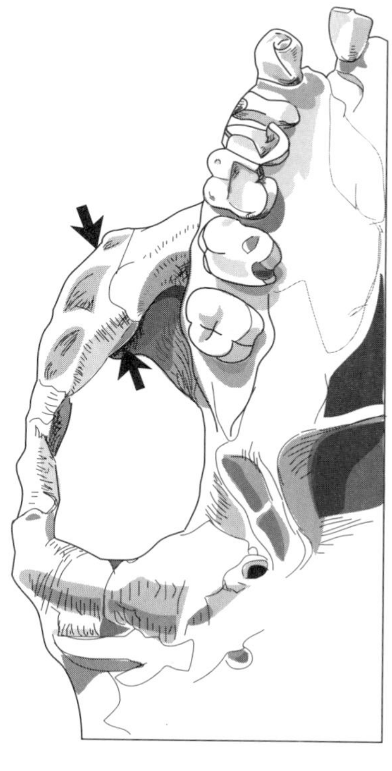

Figure 5.28 Thickness of the masseter muscle origin site in a male gorilla and in A.L. 444-2. The arrows indicate the anterior masseter origin. In spite of the great difference in size between the male gorilla skull and that of A.L. 444-2, the origin of the masseter in the gorilla is much more delicate.

matic striae are evident on the anterior portion of the marginal torus, and split lines are absent.

The mental foramen, an elongate oval in shape and opening laterally and very slightly anteriorly, is located at the anteroinferior margin of the lateral corpus hollow below P_3/P_4. It has a maximum diameter (posteroinferior to anterosuperior) of 3.4 mm and measures 2.0 mm perpendicular to this long axis. It is located approximately at midcorpus height: the foramen's lower margin is 23.7 mm from the base and 21.6 mm from the alveolar margin.

Morphology of the Anterior Aspect

In lateral view the external symphyseal profile is fairly vertical, angling only slightly posteriorly as it approaches the (lost) basal margin. Its surface describes a nearly flat, vertically uneven *S* shape, with a very weak superior concavity and a mildly rounded inferior convexity separated by a low, circular symphyseal tubercle located in the midline just below the midpoint of anterior corpus height. Moderate incisor juga are present in the alveolar bone, reflecting the roots of the slightly procumbent incisors. Shallow anterior mandibular incurvatios appear below the two central incisors. Only a trace of what was a moderately developed basal incisura remains to the right of the midline.

Morphology of the Posterior Aspect

A strongly hollowed postincisive planum slopes steeply posteroinferiorly from the slightly abraded alveolar margin. The superior transverse torus, situated just above midcorpus height, slightly overhangs the anterior wall of the damaged genioglossal fossa, extending posteriorly to the C/P_3 level, where it quickly grades into the medial aspect of the corpus. The inferior transverse torus is mostly lost; it was situated low on the corpus, but well above the basal contour, and extended posteriorly to the distal P_3 level. The genioglossal fossa appears to have been broad and deep; the nature of the supraspinous foramen cannot be ascertained.

Morphology of the Medial Aspect

The alveolar margin is intact at P_3/P_4, but eroded or chipped away in the M_1-M_3 region. The internal alveolar surface is smooth, showing no signs of root juga or a torus mandibularis. A large, rounded alveolar prominence extends anteriorly from below M_3, where cortical bone is flaked away, and vertically deepens anteriorly where it grades into the lower aspect of the superior transverse torus. There is little evidence of the mylohyoid line, probably due to poor preservation. A trace of it can be detected in the intertransverse toral area beneath P_3. Below this, the alveolar prominence grades into a shallow, anteroposteriorly elongate sulcus, constituting the merged anterior and posterior subalveolar fossae, a portion of whose depth is accentuated by slight inward crushing and flaking of the external surface.

Morphology of the Basal Aspect

The corpus base is inflated in dramatic contrast to the hollowed contours of the lateral and medial surfaces above it. The basal margin itself is blunt and weakly everted, with eversion most pronounced posteriorly. Although most of the anterior base is lost, an ill-defined right digastric fossa forms a flattened surface posteriorly. At the posterior edge of this surface sits a spinelike anterior marginal tubercle, which forms the lowest point on the basal contour.

Morphology of the Occlusal Aspect

Although the crack between M_1 and M_2 has displaced the posterior part of the right corpus laterally, reconstruction of the mandible suggests straight, posteriorly divergent premolar-molar tooth rows, with the canine and incisors set in a smooth, anteriorly convex arc. The outline of the internal corpus is fairly narrow but not tightly constricted anteriorly. It widens out posteriorly, giving the inner contour an anteriorly blunt V-shaped appearance in occlusal view. The bone cortex, where exposed at the break distal to M_3, is thickest at the base (5.4 mm), with the lateral and medial walls subequal in thickness (3.1 mm).

Comparative Mandibular Anatomy

The original diagnosis and description of *A. afarensis* (Johanson et al., 1978) emphasized the distinctive anatomy in the mandible of the Pliocene hominins from Hadar and Laetoli, and a mandible, LH 4, was designated the type specimen for the taxon. A total of 32 mandibles of *A. afarensis* were known at that time, with 28 coming from Hadar and the remainder from Laetoli. During the 1990s an additional 27 mandibles and mandible fragments were recovered from Hadar. The 3.4 Myr-old site of Maka, Ethiopia, has yielded an additional five mandibular specimens (White et al., 1993, 2000), and one specimen possibly referrable to this species was found at Koro Toro, Chad, dating to 3.0–3.5 Ma (Brunet et al., 1996).

The distinctive, archaic features of the *A. afarensis* mandible have been profiled in a series of papers (Johanson and White, 1979; White et al., 1981; Johanson et al., 1982a). These features include:

- A posteriorly sloping, rounded, bulbous anterior corpus
- A weak to moderate superior transverse torus
- A low, rounded, basally set inferior transverse torus
- An inferiorly placed, anterosuperiorly opening mental foramen
- A hollowed lateral surface (superior and posterior to the mental foramen), defined anteriorly by a C/P_3 root bulge and posteriorly by a weak lateral prominence
- Generally straight molar/premolar dental rows, but with variations from a slight lateral convexity to a slight lateral concavity
- Ascending ramus arising high on the corpus to form a narrow extramolar sulcus
- High level of sexual dimorphism in size

It is against this background that the morphology of the A.L. 444-2 mandible is now assessed.

As is the case with many other *A. afarensis* mandibles, the anterior corpus of the A.L. 444-2 specimen is, in its inferior portion, rounded and bulbous, but the symphyseal cross section is strikingly vertically oriented. A number of other fossils added to the Hadar mandible series during the 1990s feature a more upright symphyseal axis than is typical of the 1970s sample: A.L. 417-1a, A.L. 437-1, A.L. 437-2, A.L. 438-1g, and A.L. 620-1 (see Figure 5.29). While it remains true that few, if any, mandibles of *A. africanus* have symphyseal axes as inclined as those of A.L. 198-1, A.L. 277-1, A.L. 333w-60, or A.L. 400-1a, the degree of overlap between the mandibles of this species and those of *A. afarensis* is significantly greater than was previously documented. (Undescribed specimens from Sterkfontein may result in revisions of the mandibular morphology of *A. africanus*, which, for reasons of preservation, has been heavily weighted toward the Makapansgat sample.) Thus, the inclination of the symphysis is quite variable in *A. afarensis* (Figure 5.29). In A.L. 444-2 and in one of the smallest Hadar mandibles, A.L. 288-1i, the mandibular symphysis is fairly vertical. More posteriorly inclined symphyses are encountered across the Hadar mandible

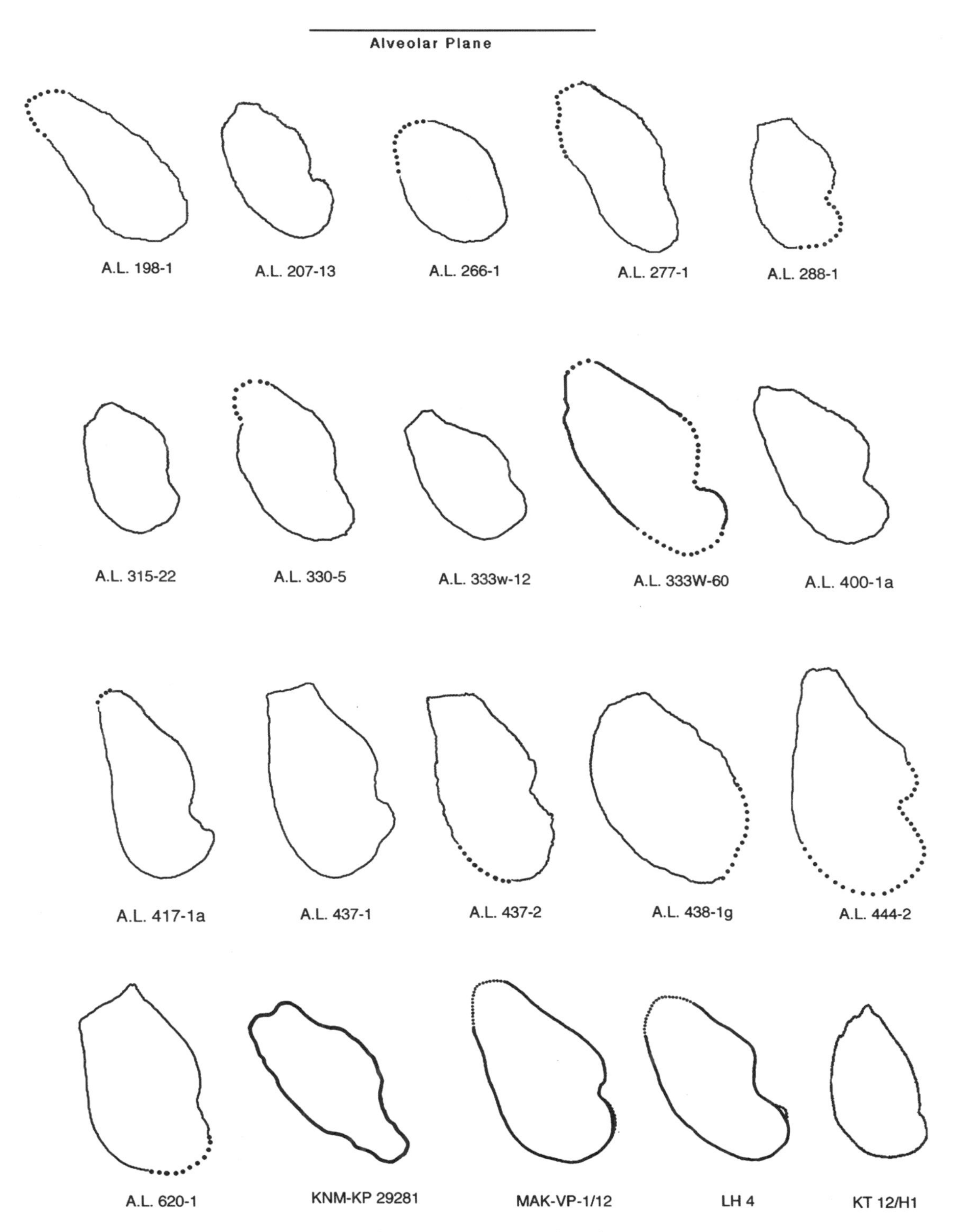

Figure 5.29 Midsagittal cross sections of the anterior corpus of *A. afarensis* mandibles, taken perpendicular to the alveolar plane, revealing the contours and inclination of the symphysis. The dotted outline denotes reconstructed contour. LH 4 and MAK-VP 1/12 sections from White et al. (2000); KT 12/H1 (*A. bahrelghazali*) section from Brunet et al. (1996); KNM-KP 29281 (*A. anamensis*) section from Ward et al. (2001). Bone preservation on both sides of A.L. 444-2 midline gives confidence in depicted outline.

size range (compare A.L. 207-13 with A.L. 333w-60; see Figure 5.29). The inclination of the symphyseal cross section of the MAK-VP 1/12 mandible is best matched by that of Hadar specimens A.L. 330-5 and A.L. 400-1a.

In terms of its symphyseal inclination the *A. afarensis* type mandible, LH 4 falls within the range for the Hadar sample but differs from Hadar homologs in the external midsagittal contour of its anterior corpus. In the Laetoli specimen the inferior part of the contour is not bulbous but is "cut away" to create an arched, continuously retreating profile (Figure 5.29), a configuration shared with mandibles of *A. anamensis* (Ward et al., 2001) but, interestingly, not with mandibles of extant African apes. The difference between the Laetoli specimen and those from Hadar does not appear to relate to the inclination of the symphyseal cross section per se, as Hadar mandibles with the symphysis inclined to a similar degree as LH 4's maintain a more "filled out" external anterior profile (e.g., A.L. 400-1a; see Figure 5.29). We return to the Laetoli/Hadar mandible comparison in Chapter 6.

The basal set of the inferior transverse torus in A.L. 444-2 conforms to the *A. afarensis* pattern. However, two other large mandibles in the 1990s collection (A.L. 437-1 and A.L. 438-1g) have a more elevated inferior torus than does A.L. 444-2 or do many of the mandibles in the 1970s sample (but see A.L. 207-13, a mandible from the 1970s with a relatively high inferior torus). The superior transverse torus in A.L. 444-2 is among the most strongly developed in the sample of *A. afarensis* mandibles, approximating the condition in A.L. 333w-60, A.L. 400-1a, and A.L. 438-1g (Figure 5.29). The size of the superior torus in A.L. 444-2 is all the more striking in light of its relatively vertical postincisive planum.

Typical of *A. afarensis*, lateral corpus hollowing is present, and easily palpable, on the A.L. 444-2 mandible. The position of this hollow varies relative to corpus height in *A. afarensis*. In A.L. 288-1i, A.L. 333w-60, and A.L. 400-1a, the hollow is situated low on the corpus while in A.L. 437-2, A.L. 438-1g, and A.L. 444-2, it is confined to a higher position. Since the mental foramen is always positioned at the anteroinferior corner of the hollow, the position of the hollow is correlated with the height of the foramen above the mandible base (see the following discussion of the relative height of the mental foramen).

The dental arcade of the reconstructed A.L. 444-2 mandible strongly resembles that of other *A. afarensis* mandibles such as A.L. 400-1a, LH 4, and MAK-VP 1/12. The ratio between the internal bialveolar margin distances measured at the canine and M_2 positions, which expresses the degree to which the alveolar rows parallel one another, is 51% in A.L. 444-2, which falls in between the more divergent alveolar arch of A.L. 288-1i (47%) and the more parallel-sided MAK-VP 1/12 (58%), LH 4 (58%), and A.L. 417-1a (62%) mandibles.

In all *A. afarensis* mandibles (and, indeed, in all subsequent hominin species) the canine crown is medially shifted relative to the long axis of the postcanine tooth row, such that the canine tends to be spatially integrated with the anterior dental arch. This configuration contrasts with the more plesiomorphic condition in *A. anamensis*, in which the canine is aligned with the postcanine axis, and the medial curve of the alveolar process commences anterior to it (Ward et al., 2001).

The ascending ramus in A.L. 444-2 arises high on the corpus, as in almost all other *A. afarensis* mandibles. Consistent with the larger size of its lateral prominence, the A.L. 444-2 mandible has a broad extramolar sulcus (buccinator groove), which measures 22 ± 2 mm wide from the midpoint of the occlusal surface of M2 at the level of the postcanine occlusal plane (compared to 16.8 ± 1 mm for A.L. 128-23 and 17.6 ± 1 mm for A.L. 266-1; White and Johanson, 1982). Although not directly measurable, the A.L. 438-1g mandible exhibits an even wider sulcus than does A.L. 444-2.

Neither the height nor the anterior extension of the ramus root on the corpus appears to covary with corpus size in the Hadar sample. In one new large Hadar mandible, A.L. 437-2, the ramus root reaches inferiorly to approximate midcorpus level, as frequently seen in later *Australopithecus* species (White et al., 1981), while in A.L. 444-2, A.L. 438-1g, and A.L. 437-1, also large specimens, the ramus origin sits high on the corpus, as in many other *A. afarensis* mandibles. The root of the ramus extends forward to the mesial M_1 level in A.L. 444-2 and A.L. 438-1g, as it does in the small female A.L. 288-1i. In most *A. afarensis* mandibles the position of the ramal root ranges from middle (e.g., A.L. 128-23, A.L. 330-5, and A.L. 333w-1) to distal M_1 (e.g., A.L. 277-1, A.L. 417-1a, A.L. 437-1, and A.L. 620-1). The enlarged mandible sample from Hadar confirms White et al.'s (2000: 62) recent observation based on the Maka mandibles: "the *A. afarensis* mandibular ramus was vertically disposed, anteriorly placed and robust compared to that in chimpanzees."

Comparative Metrics of *A. afarensis* Mandibles

With the addition of the A.L. 444-2 skull and other fossils from the 1990s Hadar hominin assemblage, the sample of measurable *A. afarensis* mandibles has nearly doubled from 12 to 23 specimens. This increase provides a solid basis for the analysis of metrical variation (and sexual dimorphism) in the *A. afarensis* mandible. Metrical data for Hadar mandibles are presented in Tables 5.13–5.17.

Mandible Corpus Breadth

Data in Table 5.14 show that with the increase in Hadar hominin mandible sample size has come an elevation in

Table 5.13 Miscellaneous Measures of the *A. afarensis* Mandible[a]

Measurement	A.L. 444-2	A.L. 417-1a	A.L. 288-1i	A.L. 400-1a	A.L. 266-1	MAK-VP 1/12	LH 4
Maximum bicondylar breadth	(149)	n/a	108.3	n/a	n/a	(118)	n/a
Minimum bicondylar breadth	(89)	n/a	64.0	n/a	n/a	68.2	n/a
Maximum mandibular length	(152)	n/a	99.8	n/a	n/a	(120)	n/a
Bimental foramen breadth (anterior edge)	(56)	(44.5)	36.1	(44)	(44)	40.3	43.9
Maximum condylar breadth	(32)	n/a	23.1	n/a	n/a	31.7	n/a
Maximum condylar length	n/a	n/a	9.1	n/a	n/a	15.0	n/a
Gonion to gnathion	(118)	n/a	82.3	n/a	n/a	n/a	n/a
Gonion to infradentale	(136)	n/a	90.8	n/a	n/a	n/a	n/a
Gonion to highest condyle	(90)	n/a	53.2	n/a	n/a	n/a	n/a
Bi-internal alveolar margin breadth at:							
I1	n/a	n/a	n/a	n/a	n/a	4.6	n/a
I2	(14)	n/a	n/a	14.0	n/a	11.6	n/a
C	(21)	(18)	16.6	20.8	n/a	20.4	(21)
P3	(31)	(23)	19.4	27.6	26.0	26.7	30.0
P4	(33)	(23)	31.1	22.9	29.4	29.5	35.0
M1	(36)	(25)	32.9	28.5	33.2	31.5	30.0
M2	(41)	(29)	35.0	n/a	n/a	(35)	36.0
M3	(44)	(34)	40.2	41.3	n/a	(43)	(41)

[a]Measurements of the reconstructed condyles of the A.L. 444-2 mandible are based on the preserved dimensions of the cranial base and are considered reliable (±2.0 mm). Other measurements involving the reconstructed ramus of the A.L. 444-2 mandible (based on A.L. 333-108) are for illustration purposes only. Bi-internal alveolar width measurements for A.L. 444-2 and A.L. 417-1a are based on the symmetrical reconstruction about the midline and are considered reliable. Values in parentheses are estimates (±1.0–2.0 mm). n/a, not available. Data for the Maka specimen are from White et al., 2000; for the 1970s Hadar specimens from White and Johanson, 1982; and for LH 4 from White, 1977.

the mean and standard deviation in minimum transverse corpus breadth. This increase is seen at almost every tooth and interdental position. The average increase in mean minimum corpus breadth is 0.7 mm and seems to weigh slightly in favor of tooth positions mesial to P_4 (where it is also the case that 1970s sample sizes were smallest).

Although the A.L. 444-2 mandible's corpus is broad by Hadar *A. afarensis* standards (its breadth dimensions lie on average 3.1 mm above the Hadar *A. afarensis* sample means), this specimen does not feature the broadest corpus in the mandible collection, a status attained by another specimen in the 1990s sample, A.L. 438-1g. Both of these mandibles have particularly high corpus breadths at the M_2 position (Table 5.14), where their massive lateral prominences dramatically inflate the lateral aspect of the corpus.

Mandible Corpus Height

Mean mandible corpus height increases sharply with the addition of specimens recovered in the 1990s (Table 5.14), although the picture is distorted by the very large jump in average corpus height at anterior dental positions (the result of the tiny number of mandibles measurable at these positions in the 1970s collection). Focusing on data for positions distal to the canine, the average increase in corpus height is 1.9 mm and is of greatest magnitude in the premolar region. At almost every tooth position the 1990s collection provides the maximum sample value; in only one case (M_2/M_3) does it also supply a new minimum value.

The A.L. 444-2 mandible corpus is very tall. Its corpus height exceeds the Hadar sample mean at every position where preservation permits comparisons, by between 1.1 and 1.7 SD. The deviation is greatest at the P_3/P_4 and M_1 positions. At 7 of 11 measurable positions, the value for A.L. 444-2 defines the high end of the range of variation for corpus height in the Hadar *A. afarensis* sample.

The distribution of Hadar corpus heights at M_1 is remarkably similar in pattern to that for a sample of 20 gorilla (10 male and 10 female) mandibles, in which two discrete single-sex clusters are separated by a small, intermediate, mixed-sex cluster (Figure 5.30). Only one (female) gorilla mandible falls in the 4-mm range that separates the female cluster from the mixed cluster; in the Hadar distribution not even a single specimen lies in the 3.2-mm interval that separates the low and intermediate corpus height clusters. In the Hadar distribution the "intermediate" cluster ($n = 6$) includes one specimen that we can confidently label as female (A.L. 417-1a, which is associated with a complete maxilla and both upper and lower canines) and two others as male (A.L. 277-1, which contains a canine, and A.L. 620-1, which contains one of

Table 5.14 Mandible Corpus Breadth[a] (mm)

	I1/I1	I1/I2	I2	I2/C	C	C/P3	P3	P3/P4	P4	P4/M1	M1	M1/M2	M2	M2/M3
1990s Hadar sample														
A.L. 198-22	—	—	—	—	—	—	—	—	—	—	21.7	20.1	20.9	23.9
225-8	—	—	—	—	—	—	—	—	—	—	—	*20.0*	21.4	*23.0*
228-2	—	—	—	—	—	—	—	15.9	16.0	15.7	16.3	18.0	—	—
315-22	17.8	16.9	16.9	17.7	17.6	17.5	17.4	16.6	17.3	17.8	19.2	19.7	*20.0*	*21.0*
330-5	18.2	17.6	17.8	17.9	18.7	19.0	19.4	18.0	18.5	18.6	20.9	18.2	19.5	21.9
417-1a	17.7	18.1	18.1	17.3	17.3	17.7	18.9	18.6	18.4	17.7	18.0	17.5	18.4	20.1
432-1	—	—	—	—	—	—	—	—	—	—	—	—	20.3	22.3
433-1a, b	—	—	—	—	—	—	22.1	21.0	*20.3*	19.7	20.2	20.5	20.8	*22.0*
436-1	—	—	—	—	—	—	—	—	—	—	—	—	19.6	22.0
437-1	22.2	22.9	22.7	22.6	23.9	24.0	23.8	22.6	21.2	19.4	20.0	19.8	19.6	22.5
437-2	21.8	21.0	20.9	21.3	22.1	22.2	22.1	22.8	22.2	21.1	22.2	23.0	24.2	*27.0*
438-1	25.5	25.0	25.0	25.0	25.5	25.7	25.3	25.1	*25.0*	23.0	24.7	25.3	28.1	31.9
444-2	*24.0*	23.0	23.2	23.5	23.9	23.6	23.2	21.7	21.1	21.5	23.0	—	30.5	—
582-1	—	—	—	—	—	25.9	25.5	24.2	22.6	21.6	21.4	—	—	—
620-1	22.9	22.6	22.3	22.3	23.3	22.5	21.6	20.2	19.5	19.0	20.5	21.5	22.6	—
766-1	22.7	22.5	22.6	23.0	23.6	—	—	—	—	—	—	—	—	—
1970s Hadar sample														
A.L. 128-23	—	—	—	—	18.2	17.1	*16.8*	16.5	16.6	16.8	18.0	20.4	22.9	—
145-35	—	—	—	—	—	—	21.1	20.0	18.9	*16.8*	21.1	22.1	24.8	—
188-1	—	—	—	—	—	—	—	—	—	—	—	*18.8*	22.3	22.7
198-1	—	—	—	18.2	17.9	17.5	17.2	16.2	15.8	15.6	15.8	16.9	18.1	—
207-13	—	—	—	—	—	—	—	—	*17.4*	*17.3*	*18.1*	*18.8*	*18.4*	—
266-1	*21.1*	22.2	*21.4*	*21.1*	*21.9*	*20.9*	*20.8*	*20.1*	*19.9*	*20.4*	*21.7*	23.7	24.2	—
277-1	—	—	—	—	19.9	19.6	19.3	18.6	17.8	17.5	17.9	18.8	—	—
288-1i	17.3	17.2	17.1	17.0	17.1	16.6	16.8	16.2	16.6	16.8	17.1	—	—	—

311-1		—	—	—	—	*25.7*	*23.6*	*23.4*	*23.3*	*22.0*	*21.7*	—	—	—	—
333w-1a, b		—	—	—	—	—	—	—	19.1	18.9	18.9	19.4	21.5	23.0	—
333w-12		—	18.0	18.6	18.9	19.0	18.5	18.1	17.7	16.8	16.7	17.4	18.3	—	—
333w-32+60		*21.9*	*22.4*	*22.9*	—	*23.4*	*23.7*	*23.1*	*22.3*	*22.0*	22.4	23.6	23.5	23.6	—
333w-46		—	—	—	—	—	—	*19.7*	*18.9*	—	—	—	—	—	—
333-97		—	—	—	—	—	22.4	19.8	—	—	—	—	—	—	—
400-1a		19.2	19.3	19.2	19.5	20.0	20.4	20.1	18.9	18.5	18.5	18.7	—	—	—
MAK 1/12[b]		—	21.3	20.5	20.7	20.8	19.9	19.1	18.2	17.7	17.6	18.7	20.3	20.6	24.7
MAK 1/2[b]		—	—	—	—	—	—	—	—	—	—	19.6	20.2	21.4	23.6
LH 4[c]		19.1				19.0		18.5		17.5	19.1	19.4		22.4	
Statistics for 1970s Hadar sample	Mean	19.9	19.8	19.8	18.9	20.3	20.0	19.7	19.0	18.4	18.3	19.0	20.3	22.2	22.7
	SD	2.1	2.4	2.3	1.5	2.8	2.6	2.2	2.2	2.0	2.2	2.3	2.3	2.5	—
	Min	17.3	17.2	17.1	17.0	17.1	16.6	16.8	16.2	15.8	15.6	15.8	16.9	18.1	—
	Max	21.9	22.4	22.9	21.1	25.7	23.7	23.4	23.3	22.0	22.4	23.6	23.7	24.8	—
	n	4	5	5	5	9	10	12	12	12	12	11	10	8	1
Statistics for combined Hadar sample	Mean	20.9	20.6	20.6	20.4	21.1	21.0	20.7	19.8	19.3	18.9	19.9	20.3	22.1	23.4
	SD	2.7	2.7	2.6	2.6	3.0	3.0	2.6	2.7	2.4	2.2	2.4	2.2	3.2	3.2
	Min	17.3	16.9	16.9	17.0	17.1	16.6	16.8	15.9	15.8	15.6	15.8	16.9	18.1	20.1
	Max	25.5	25.0	25.0	25.0	25.7	25.9	25.5	25.1	25.0	23.0	24.7	25.3	30.5	31.9
	n	13	14	14	14	18	19	22	23	23	23	23	21	21	12
Combined mean –70s mean		1.0	0.8	0.8	1.5	0.8	1.0	1.0	0.8	0.9	0.6	0.9	0.0	–0.1	0.7
Combined min –70s min		0	–0.3	–0.2	0	0	0	0	–0.3	0	0	0	0	0	—
Combined max –70s max		3.6	2.6	2.1	3.9	0	2.2	2.1	1.8	3.0	0.6	1.1	1.6	5.7	—
A.L. 444-2 above combined mean		3.1	2.4	2.6	3.1	2.8	2.6	2.5	1.9	1.8	2.6	3.1	—	8.4	—

[a]Minimum parallel corpus breadth at positions along the toothrow, as defined in White and Johanson, 1982 (p. 526). Italicized values are estimates (±1.0–2.0 mm). Where both sides are present and measureable, the value provided is the average. Dash indicates measurement not possible at the designated tooth or interdental position.

[b]From White et al., 2000.

[c]From White, 1977, and pers. comm.

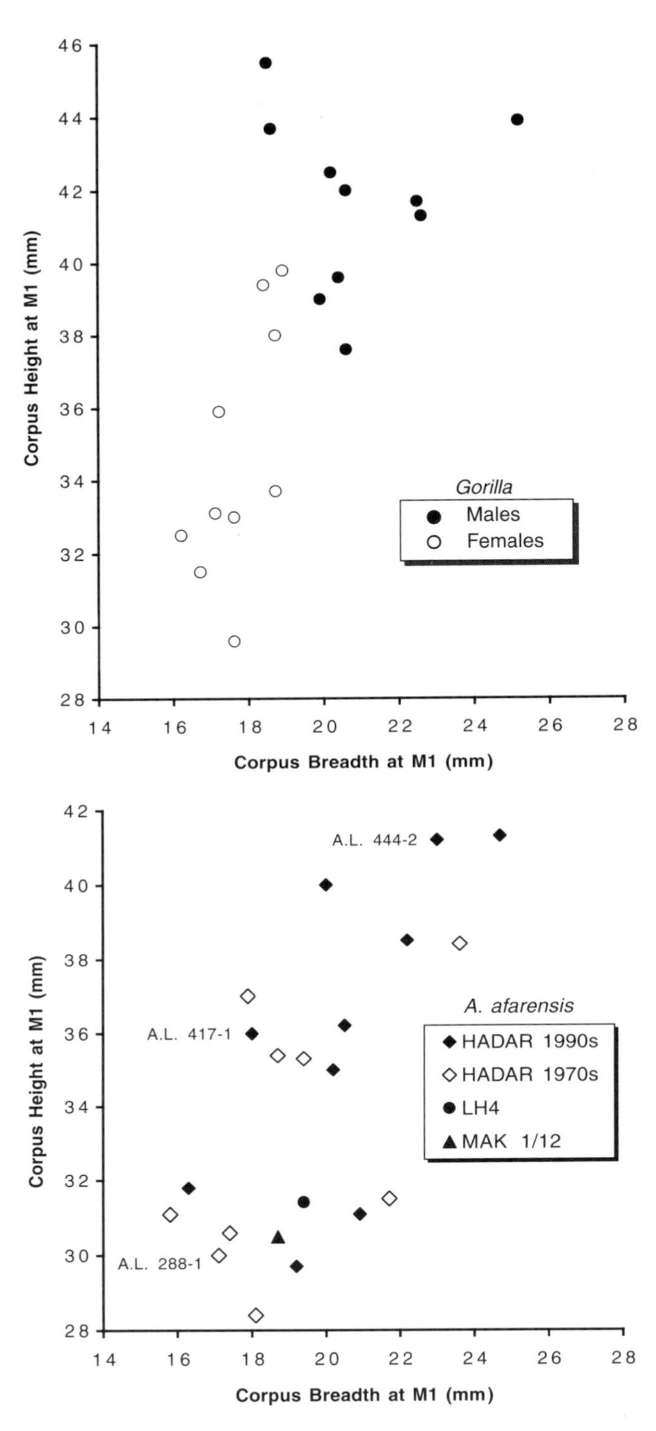

Figure 5.30 Bivariate plots of mandible corpus height and breadth measured at M1 level, in gorillas (*top*) and *A. afarensis* (*bottom*). Data for MAK-VP 1/12 from White et al. (2000).

the largest known Hadar hominin M_3s); the others cannot be assigned to sex with confidence.

Although the pattern of variation represented in Figure 5.30 might suggest a gorilla-like level of sexual dimorphism in *A. afarensis* mandibular corpus height, this impression stems from the fact that four of the five deepest mandibles in the hominin plot originate from the youngest *A. afarensis*-bearing levels in the Hadar stratigraphic sequence, in which Lockwood et al. (2000) documented a significant increase in mandible size (see below for further discussion). Thus, even though these late mandibles appear to belong to male individuals (Lockwood et al., 2000) if the size increase over time is ignored, the actual degree of sexual dimorphism in mandible corpus height within the Hadar sample as a whole would be overestimated.

The A.L. 444-2 mandible corpus progressively deepens anteriorly, with maximum depth values recorded in the canine and incisor regions (Table 5.15). This anterior increase in corpus depth is characteristic of Hadar *A. afarensis* generally, though there are exceptions (for example, A.L. 438-1g and A.L. 437-1 exhibit maximum corpus depth in the premolar region). Anterior deepening of the corpus also characterizes *A. afarensis* type specimen LH 4 and, to a lesser extent, the Maka mandible MAK-VP 1/12 (White et al., 2000). Smaller (female) mandibles with a well-preserved anterior corpus remain relatively rare in the Hadar sample (e.g., A.L. 288-1i and A.L. 417-1a), but on present evidence it does not appear that sexual dimorphism in canine crown (or root) size influences the pattern of corpus depth across the Hadar sample.

Mandible Corpus Shape at M_1

The mandible corpus shape index (breadth/height × 100) calculated at the M_1 position is often used to compare variation in hominoid corpus "robusticity" (e.g., Chamberlain and Wood, 1985; Kimbel and White, 1988b). A higher index value indicates a relatively broad corpus. In interspecific comparisons of early hominins a transversely broad mandible corpus has been interpreted as a functional response to high magnitude loading and axial twisting of the corpus during mastication (e.g., Hylander, 1979, 1988; but for a cautionary note, see Daegling and Hylander, 1998).

For the entire Hadar *A. afarensis* sample the mean shape index value is 57.5% (n = 19 specimens for which both variables can be measured), which is essentially identical to the value for the 1970s sample alone (Table 5.16). The A.L. 444-2 mandible corpus is slightly less robust than the Hadar average, with an index value of 55.8%. Often in *A. afarensis* larger (male) mandibles (e.g., A.L. 277-1 and A.L. 437-1) have a more slender (i.e., relatively deep) corpus than do smaller (female) mandibles,

Table 5.15 Mandible Corpus Height[a] (mm)

	I1/I1	I1/I2	I2	I2/C	C	C/P3	P3	P3/P4	P4	P4/M1	M1	M1/M2	M2	M2/M3
1990s Hadar sample														
A.L. 198-22	—	—	—	—	—	—	—	—	—	—	—	—	34.0	32.8
225-8	—	—	—	—	—	—	—	—	—	—	31.1	28.4	28.1	—
228-2	—	—	—	—	—	—	—	36.0	36.0	34.9	31.8	29.8	—	—
315-22	—	—	—	—	—	31.5	32.0	32.4	33.0	32.0	29.7	31.4	28.0	—
330-5	—	—	—	33.0	—	—	32.0	32.7	31.4	31.7	31.1	31.0	28.3	27.2
417-1a	38.2	—	38.0	36.7	38.0	37.4	39.0	38.6	37.2	36.8	36.0	35.2	32.8	31.8
418-1	—	—	—	—	—	—	—	—	—	—	—	—	36.0	—
433-1a, b	—	—	—	—	—	—	—	—	—	—	35.0	—	—	—
436-1	—	—	—	—	—	—	—	—	—	—	—	—	26.0	23.7
437-1	45.2	43.0	43.0	43.0	44.0	44.5	46.0	45.0	44.0	43.8	40.0	—	—	—
437-2	45.0	45.0	45.0	43.6	43.5	43.5	44.0	44.5	43.4	41.5	38.5	37.2	37.0	—
438-1g	39.6	39.0	39.0	39.0	39.0	39.4	41.0	41.0	42.0	41.4	41.3	37.7	37.1	37.3
440-1	—	—	43.0	43.3	43.0	44.0	—	—	—	—	—	—	—	—
582-1	—	—	—	—	—	—	40.0	41.1	40.5	—	—	—	—	—
444-2	—	—	45.0	44.5	45.0	44.3	44.5	44.3	43.9	43.0	41.2	39.0	37.6	—
620-1	—	43.2	—	—	—	—	40.0	41.0	38.0	37.8	36.2	35.0	34.5	—
1970s Hadar sample														
A.L. 128-23	—	—	31.4	32.4	30.0	31.8	30.0	32.1	—	—	—	—	—	—
145-35	—	—	—	—	—	—	—	32.3	28.0	30.4	27.8	27.0	—	—
188-1	—	—	—	—	—	—	—	—	—	—	—	34.8	34.3	34.2
198-1	—	—	—	—	—	—	—	34.3	32.2	32.4	31.1	31.1	30.8	31.1
207-13	—	—	—	—	—	—	—	—	—	28.0	28.4	27.0	25.3	25.0
266-1	—	—	—	—	—	32.2	31.0	32.3	—	32.2	31.5	30.4	27.6	27.0
277-1	—	—	—	—	—	40.0	41.2	40.8	39.2	38.5	37.0	—	—	—
288-1i	32.5	—	—	—	—	31.7	31.0	31.1	29.3	30.0	30.0	29.1	27.6	26.7
333w-1a, b	—	—	—	—	—	—	—	—	37.5	37.5	35.3	34.0	32.4	—
333w-12	—	—	—	31.4	—	31.0	—	31.8	31.1	31.5	30.6	—	—	—
333w-32+60	45.2	—	—	—	—	41.8	42.6	43.0	40.1	40.5	38.4	38.0	35.4	34.4
333w-46	—	—	—	—	—	36.2	34.5	34.6	—	—	—	—	—	—
400-1a	39.4	36.2	—	37.2	33.6	36.7	34.4	36.0	35.6	35.4	35.4	—	—	—

(continued)

Table 5.15 *(continued)*

		I1/I1	I1/I2	I2	I2/C	C	C/P3	P3	P3/P4	P4	P4/M1	M1	M1/M2	M2	M2/M3
MAK 1/12[b]		—	—	—	—	—	35.8	33.4	34.3	32.5	32.4	30.5	30.1	29.6	30.1
MAK 1/2[b]		—	—	—	—	—	—	—	—	—	—	—	34.6	32.6	31.8
LH 4[c]									38.3	34.7	34.4	31.4	30.0	29.5	
Statistics for 1970s	Mean	39.0	36.2	31.4	33.7	31.8	35.2	35.0	34.8	34.1	33.6	32.6	31.4	30.5	29.7
Hadar sample	SD	6.4	0.0	0.0	3.1	2.5	4.1	5.1	4.0	4.6	4.1	3.7	3.9	3.8	4.1
	Min	32.5	36.2	31.4	31.4	30.0	31.0	30.0	31.1	28.0	28.0	27.8	27.0	25.3	25.0
	Max	45.2	36.2	31.4	37.2	33.6	41.8	42.6	43.0	40.1	40.5	38.4	38.0	35.4	34.4
	n	3	1	1	3	2	8	7	10	8	10	10	8	7	6
Statistics for combined	Mean	40.7	41.3	40.6	38.4	39.5	37.7	37.7	37.2	36.8	35.8	34.2	32.7	31.8	30.1
Hadar sample	SD	4.7	3.6	4.9	5.0	5.4	5.2	5.5	4.9	5.1	4.8	4.2	3.9	4.1	4.4
	Min	32.5	36.2	31.4	31.4	30.0	31.0	30.0	31.1	28.0	28.0	27.8	27.0	25.3	23.7
	Max	45.2	45.0	45.0	44.5	45.0	44.5	46.0	45.0	44.0	43.8	41.3	39.0	37.6	37.3
	n	7	5	7	10	8	15	16	20	18	19	21	17	18	11
Combined mean –70s mean		1.7	5.1	9.2	4.7	7.7	2.6	2.7	2.4	2.7	2.1	1.6	1.3	1.3	0.4
Combined min –70s minimum		0	0	0	0	0	0	0	0	0	0	0	0	0	−1.3
Combined max –70s maximum		0	8.8	13.6	7.3	11.4	2.7	3.4	2.0	3.9	3.3	2.9	1.0	2.2	2.9
A.L. 444-2 above sample mean				4.4	6.1	5.5	6.6	6.8	7.1	7.1	7.2	7.0	6.3	5.8	

[a]Perpendicular (minimum) corpus height at positions along the toothrow, as defined in White and Johanson, 1982 (p. 526). Italicized values are estimates (±1.0-2.0 mm). Where both sides are present and measureable, the value provided is the average. Dash indicates measurement not possible at the designated tooth or interdental position

[b]From White et al., 2000.

[c]From White, 1977.

Table 5.16 Mandible Corpus Shape

		Corpus Breadth (mm)	Corpus Height (mm)	Shape Index[a]
1990s Hadar sample				
A.L. 228-2		16.3	31.8	51.3
315-22		19.2	29.7	64.6
330-5		20.9	31.1	67.2
417-1a		18	36.0	50.0
433-1a, b		20.2	35.0	57.7
437-1		20	40.0	50.0
437-2		22.2	38.5	57.7
438-1		24.7	41.3	59.8
444-2		23	41.2	55.8
620-1		20.5	36.2	56.6
	Mean	20.5	36.1	57.1
	SD	2.4	4.2	5.8
	Min	16.3	29.7	50.0
	Max	24.7	41.3	67.2
	n	10	10	10
1970s Hadar sample				
A.L. 198-1		15.8	31.1	50.8
207-13		18.1	28.4	63.7
266-1		21.7	31.5	68.9
277-1		17.9	37.0	48.4
288-1i		17.1	30.0	57.0
333w-1a, b		19.4	35.3	55.0
333w-12		17.4	30.6	56.9
333w-32+60		23.6	38.4	61.5
400-1a		18.7	35.4	52.8
	Mean	18.9	33.1	57.2
	SD	2.4	3.5	6.5
	Min	15.8	28.4	48.4
	Max	23.6	38.4	68.9
	n	9	9	9
Combined Hadar sample	Mean	19.7	34.7	57.5
	SD	2.5	4.1	5.9
	Min	15.8	28.4	48.4
	Max	24.7	41.3	68.9
	n	19	19	19
MAK-VP 1/12		18.7	30.5	61.3
LH 4		19.4	31.4	61.8

[a]Mandible corpus shape index is calculated as the minimum corpus breadth at M_1/perpendicular corpus height at $M_1 \times 100$.

as has been pointed out previously (Kimbel and White, 1988b; Lockwood et al., 2000) (see Figure 5.31).

If we divide the Hadar sample into "male" and "female" subsets using the gap between the low and intermediate corpus height clusters, but assign A.L. 417-1a to the "female" category, the different corpus shape index supports the impression that "male" Hadar mandibles tend to be less robust than those of females: the index for the "female" sample ($n = 10$) is 58.9% and for the "male" sample ($n = 9$) is 55.5%. Adding MAK-VP 1/12 and LH 4 to the female group (their corpus height dimensions lie squarely within the low corpus height cluster) raises the "female" shape index to 59.4%, slightly enhancing the contrast between the *A. afarensis* subsets. Although the means for the *A. afarensis* subsets fall within 1 SD of each other, only 1 of 10 mandibles in the "male" sample has a shape index value above 60% (A.L. 333w-60), while 6 of 11 mandibles in the "female" sample fall above this mark.

Vertical Position of the Mental Foramen

A relatively low position of the mental foramen—at or below midcorpus—has been cited as a diagnostically primitive feature of the *A. afarensis* mandibular corpus

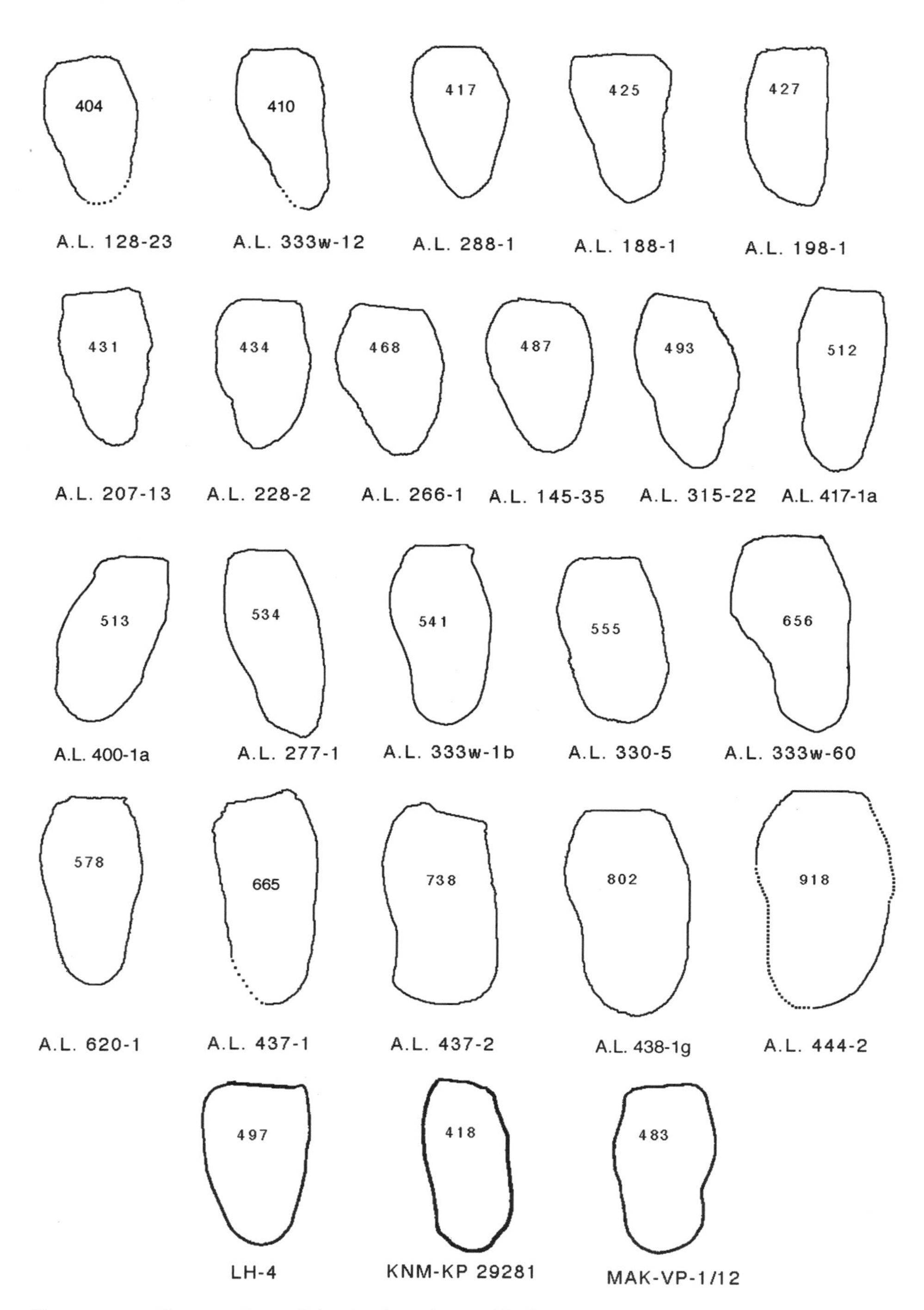

Figure 5.31 Cross sections of the *A. afarensis* mandibular corpus at mid-M1 level, perpendicular to alveolar plane. Dotted outline denotes reconstructed contour. Figures inside the outlines are the cross-sectional areas (mm^2) of the corpus at M1, measured on scanned outlines of sectioned casts on a Macintosh computer using version 1.62 of the public domain NIH Image program (developed at U.S. NIH and available on the Internet at http://rsb.info.nih.gov/nih-image). LH 4 and MAK-VP 1/12 sections from White et al. (2000); KNM-KP 29281 (*A. anamensis*) section from Ward et al. (2001).

(Johanson et al., 1978; Johanson and White, 1979; White et al., 1981). An index expressing the height of the foramen's lower margin above the basal contour as a percentage of the total corpus height in the same coronal plane can be assessed on 10 adult mandibles in the 1990s Hadar sample, bringing the total sample of specimens available for comparison of this index to 20. The mean index value for the combined Hadar sample is 44.2%, which is 1.6% higher than the value for the 1970s sample alone. Thus, the mental foramen occupies a slightly elevated position on the corpus in the 1990s sample. As data in Table 5.17 show, the higher index value is due mostly to large (male) mandibles A.L. 437-1, A.L. 437-2, A.L. 438-1g, and A.L. 444-2, which feature an especially deep basal segment of the total corpus height and thus a superiorly shifted foramen.

Variation and Time in the A. afarensis *Mandible Sample*

Using the specimens available in the 1970s, Kimbel and White (1988b) found size variation in the *A. afarensis* mandible corpus to fall within the range for African great apes species. With the recently augmented Hadar mandible samples at hand, Lockwood et al. (2000) reexamined corpus dimensions (at M_1) using randomization tests to evaluate the significance of differences between coefficients of variation for *A. afarensis* and hominoid comparative samples. In this study, the null hypothesis that the *A. afarensis* mandible corpus is no more variable than that of (single subspecies) samples of African great ape mandibles was rejected in five of six tests (*Pan troglodytes* [$n = 56$]: corpus height, corpus breadth and geometric mean; *Gorilla gorilla* [$n = 60$]: corpus breadth, geometric mean). In contrast, variation in the corpus shape index was found to be within the range for the comparative taxa.

Lockwood et al. (2000) used nonparametric rank correlation and randomization tests to investigate whether anagenetic change within the *A. afarensis* lineage contributed significantly to the elevated levels of size variation in the mandible corpus. They discovered a pronounced temporal trend toward larger corpus size (especially corpus height) within the Hadar stratigraphic sequence, localized in the upper part of the Kada Hadar member, from which A.L. 444-2 and three other large mandibles (A.L. 437-1, A.L. 437-2, and A.L. 438-1g) were recovered. When these four specimens were removed from the analysis, no significant size trend was detected. Lockwood et al. (2000) concluded that the relatively high levels of variation in the *A. afarensis* mandible corpus were due partly to change in size over time. Confirming earlier suggestions of the lack of covariation between corpus size and shape in *A. afarensis* (Kimbel and White, 1988b), the Lockwood et al. study did not find temporally directed change in the corpus shape index. Apparently, ancestral corpus proportions were preserved, as the *A. afarensis* mandible grew larger over time. We return to this topic in Chapter 6.

Dentition

Preservation

The maxillary dentition of A.L. 444-2 is much better preserved than the mandibular dentition. The latter preserves crown and root portions of the left I_1 and I_2 and right I_1 to M_3. The LI_1 is missing the distolabial (DLa) corner, while the RI_1 preserves only the lingual face. The LI_2 preserves only a small portion of enamel at the base of the gingival eminence and a fragment of the labial face that does not join the root. Except for a small portion of enamel at the distolingual (DL) corner, the entire crown of the RI_2 is missing, but the unattached LI_2 preserves crown morphology. The mesiolabial corner of the R_C is intact, while an unjoined flake preserves the lingual face. The crown is broken away on the RP_3, exposing the root canal and leaving only a portion of the distal enamel rim. The heavily worn P_4 is missing most of the buccal face, except for the distobuccal (DB) corner. The RM_1 preserves the buccal face and DB corner of enamel, while the lingual portions of the crown are missing. No enamel portions of the RM_2 and RM_3 crowns are preserved.

The LI^1 is missing a chip of enamel from the DLa corner, as well as a small chip from the cervical edge of the lingual tubercle. On RI^1 a bit of lingual enamel at the lingual tubercle is preserved, as is a portion of the DLa face. While only a small portion of the labial face (near the cervix) is present on the broken L^C, its antimere is present except for the lingual half of the distal face. Of the P^3s, the left one preserves only the cervical portion of the lingual face; otherwise, both crowns are broken away. The LP^4-M^3 tooth crowns are well preserved. On the right side, the buccal half of the P^4 is broken away, and both the lingual face of the M^1 and a small chip of the DL corner of M^2 are lost; the RM^3 crown is intact. Hairline cracks traverse several of the crowns but do not significantly alter crown dimensions. A total of 21 enamel and dentine chips could not be attached to the upper or lower dentition.

Mandibular Dentition

Crown Morphology

First lower incisor. The incisal edge of the LI_1 is labiolingually (LL) compressed, and the basal outline is a mesiodistally (MD) compressed oval. The crown broadens MD occlusally, and in labial view the mesial edge is

Table 5.17 Measurements of the Mental Foramen and its Vertical Position on the Mandible Corpus[a]

		Foramen Height (mm)	Foramen Length (mm)	Foramen to Base (mm)	Foramen to Alveolar Margin (mm)	Height Index (%)
1990s Hadar sample						
A.L. 225-8		1.9	3.0	—	—	—
228-2		2.6	4.8	15.4	20.1	43.4
315-22		1.8	3.0	13.0	21.1	38.1
330-5		2.9	5.5	13.8	19.7	41.2
417-1a		2.2	2.4	15.6	21.5	42.0
433-1a		2.3	3.3	—	17.0	—
437-1		2.6	3.0	19.9	25.0	44.3
437-2		2.8	4.1	22.1	23.2	48.8
438-1g		2.9	4.3	24.3	*20.5*	54.2
582-1		2.4	3.5	*23.0*	21.1	52.2
444-2		2.0	3.4	23.7	21.6	52.3
620-1		2.7	4.9	17.3	23.5	42.4
1970s Hadar sample						
A.L. 128-23		3.1	3.9	14.6	17.4	45.6
145-35		2.9	2.6	—	18.0	—
198-1		3.3	4.0	13.9	18.4	43.0
207-13		—	4.4	13.0	21.0	38.2
266-1		3.4	2.6	14.2	18.4	43.6
277-1		3.1	3.5	16.6	23.0	41.9
288-1i		3.9	3.5	11.5	20.0	36.5
311-1		3.3	3.4	—	26.3	—
333w-1a,b		3.6	4.3	19.0	18.7	50.4
333w-12		3.2	4.1	12.8	19.0	40.3
333w-32+60		3.4	4.2	16.5	24.1	40.6
400-1a		3.9	3.7	16.7	20.1	45.4
MAK-VP 1/12[b]		—	—	15.9	18.8	45.8
Statistics for 1970s Hadar sample	Mean	3.4	3.7	14.9	20.4	42.6
	SD	0.3	0.6	2.3	2.8	4.0
	Min	2.9	2.6	11.5	17.4	36.5
	Max	3.9	4.4	19	26.3	50.4
	n	11	12	10	12	10
Statistics for combined Hadar sample	Mean	2.9	3.7	16.8	20.8	44.2
	SD	0.6	0.8	3.9	2.5	5.0
	Min	1.8	2.4	11.5	17	36.5
	Max	3.9	5.5	24.3	26.3	54.2
	n	23	24	20	23	20
Combined mean −70s mean		−0.5	0.0	2.0	0.4	1.7
Combined min −70s minimum		−1.1	−0.2	0	−0.4	0
Combined max −70s maximum		0	1.1	5.3	0	3.8

[a]Height of the mental foramen is superoinferior; length is anteroposterior. The height index is the vertical height of the foramen's inferior margin above the corpus base expressed as a percentage of the total height of the corpus in the coronal plane of the foramen. Hadar 1970s data from White and Johanson, 1982. Italicized values are estimates (±1.0-2.0 mm).
[b]From White et al., 2000.

fairly vertical, with the distal edge exhibiting a slight distal-occlusal flare. The lingual and labial enamel lines are slightly convex rootward. The mesial and distal enamel lines peak strongly toward the incisal edge. Occlusal wear on the LI_1 exposes a strip of dentine that measures 1.2 mm wide and 4.2 mm long. In labial view of LI_1, occlusal wear appears concave incisally, with topographic highpoints at the mesial and distal margins of the incisal edge. Enamel thickness is 1.1 mm labially and 1.0 mm lingually.

The lingual face has a weak gingival eminence that gives rise to a weak vertical median ridge, which quickly disappears into a slightly concave lingual surface. Both the mesial and distal marginal ridges are only weakly expressed. The chemically pitted labial face is convex vertically and horizontally. A shallow, vertical furrow occurs near the labial enamel line, just mesial to the DLa corner.

Second lower incisor. The gingival eminence on the LI_2 is weak, featureless, and continuous with a slightly concave lingual face. I_2 roots are strongly MD compressed. Incisal wear angles slightly lingually and inferiorly and bears a cupped dentine exposure that is labially convex. A flat mesial interproximal facet (IPF) occurs on the LI_2.

The labial face is slightly convex mesiodistally and more strongly, vertically. Weak hypoplastic lines occur near the root. Both DLa and mesiolabial (MLa) corners are squared off. The labial enamel line is convex downward, and the labial enamel rim is ca. 1.3 mm thick.

Lower canine. The lingual face bears a low, flatly polished, vertical, medial ridge and has a wide gingival eminence. A V-shaped, occlusally widening groove borders the medial ridge distally. A strong posterior cingulum is present.

The tall, mesial half of the labial face is gently convex vertically and strongly convex MD and possesses a shallow, vertical MLa groove. Hypoplastic lines occur just above the upwardly convex labial enamel line. Judged by the shape of the root, crown shape is a MD compressed oval; no distal or mesial IPFs could be discerned. The labial enamel rim is 1.8 mm thick.

Lower third premolar. The long axis of the exposed, MD compressed root is set obliquely to the long axis of the tooth row, its longest dimension running from the mesiobuccal corner to the DL corner. A MD compressed cross section of the root canal is exposed and measures 0.8 mm MD and 1.5 mm buccolingually (BL). The distal IPF is situated lingual to the midline, and the distal enamel rim is 1.2 mm thick. Without radiographs it is not possible to determine if the roots are multiple.

Lower fourth premolar. The crown outline is a rounded square with an abbreviated DB corner. Most morphology is absent due to extensive occlusal wear, which produces a wide, deeply cupped dentine exposure on most of the crown with the exception of a flat, enamel island on the DL quarter of the tooth. The mesial IPF is flat, centrally located at the enamel rim level and measures 1.0 mm wide. The flat, 1.3-mm wide distal IPF, also at the enamel rim level, is set lingual to the crown midline and measures 4.9 mm. Although the enamel is missing from the lingual face, the dentine preserves a MD convex contour. The enamel line appears horizontal mesially and lingually.

Lower first molar. Occlusal wear has exposed a MD elongated trough of cupped dentine, which extends from the thin (1.9 mm) mesial marginal ridge to a thicker (2.3 mm) distal marginal ridge. Wear has left intact the buccal occlusal margin, which is thinner (1.3 mm) at the protoconid and thicker (2.3 mm) at the hypoconid. The occlusal outline is a rounded rectangle. A centrally located mesial IPF is flat and measures 4.9 mm where it is in contact with the distal IPF of the P_4. Only a small crescent of the extreme buccal portion of the distal IPF remains, the rest having been flaked away. The mesial buccal groove is shallow and extends 1.5 mm below the occlusal rim. The distal buccal groove is also an indistinct shallow trough. In occlusal view the buccal face is bilobate, essentially straight MD and convex vertically. The buccal enamel line is chipped, but dips between the two buccal roots. Distally the enamel line is horizontal.

Comparative Mandibular Dentition

Morphological and metrical comparisons of the A.L. 444-2 mandibular dentition with other *A. afarensis* material is severely limited due to heavy dental wear and postmortem loss of crown enamel. Overall, however, the preserved morphology of the I_1, I_2, P_4, and M_1 exhibit no significant differences from other dental remains from Laetoli (White, 1977, 1980), Hadar (Johanson et al., 1982c), and Maka (White et al., 2000). Although crown measurements are limited (Table 5.18), the proportions of anterior to posterior teeth are similar to those seen in other specimens, such as A.L. 400-1a and MAK-VP 1/12 (White et al., 2000). There is no evidence of marked expansion of cheek teeth (P_3-M_3) or significant reduction in the size of the anterior dental battery (I_1-C).

Metrically, the preserved mandibular dentition of A.L. 444-2 tends toward the upper end of the range of variation for the known *A. afarensis* sample. The 7.1 mm MD dimension of the I_1, is above the mean of 6.5 mm, but well below the 8.0 mm long LH 2 I_1. The 7.2 mm BL dimension of the I_1 is just below the mean of 7.4 mm for this tooth in *A. afarensis*. The 8.8 mm BL measurement for the I_2 in A.L. 444-2 exceeds the mean of 7.8 mm for this tooth and defines the upper limit for this dimension in *A. afarensis*. The MD lengths of both the P_4 and M_1 are at the upper end of the known ranges for these teeth. Of the 23 *A. afarensis* P_4s for which MD measurements are available, the 11.4 mm length of the A.L. 444-2 P_4 is matched only in A.L. 582-1, while the 14.6 mm length of the M_1 is matched in A.L. 241-1 and is exceeded by the 14.8-mm long M_1 in A.L. 440-1.

Metric Features of the A. afarensis *Adult Mandibular Dental Sample*

With the addition of the fossils recovered during the 1990s, the Hadar dental sample contains 135 nonantimeric

Table 5.18 Catalog of Hadar Mandibular Dental Measurements[a] (mm)

Specimen	MD Length	BL Breadth	Comments[b]
I/1			
A.L. 333w-58	6.2	7.6	L avr./B est./Lw
A.L. 333w-9a,b		7.3	
A.L. 400-1a	5.6	7.5	Lw
A.L. 440-1	6.5		
A.L. 444-2	7.1	7.2	L+B avr./L corr./B est.
A.L. 582-1	5.6	6.9	
I/2			
A.L. 198-18	6.2	6.7	L+Bw
A.L. 333w-58	7.2	8.2	Lw
A.L. 333w-9a,b	6.1		L corr.
A.L. 400-1a	6.9	7.7	L+B avr./B est.
A.L. 437-2	5.0	8.7	L vw
A.L. 444-2		8.8	B est.
/C			
A.L. 128-23	7.5	8.9	
A.L. 198-1	8.9	8.8	L corr.
A.L. 277-1		11.8	B est.
A.L. 333-103		10.3	
A.L. 333-90		9.7	
A.L. 333w-10		12.0	
A.L. 333w-58	9.5	12.4	
A.L. 366-6	8.2		
A.L. 400-1a	7.8	9.3	L avr./L corr./Lsw
A.L. 417-1a	7.6		
A.L. 437-2	9.3	11.5	L corr.
A.L. 440-1	9.0	10.4	
A.L. 487-1f	9.2	11.4	

Specimen	MD Length	BL Breadth	Comments
P/4			
A.L. 128-23	7.7	10.0	
A.L. 145-35	9.5		L corr.
A.L. 176-35	10.7	10.6	
A.L. 198-1	8.9	9.8	L corr.
A.L. 207-13	9.2	10.0	L corr.
A.L. 228-2	9.6	10.9	L corr.
A.L. 266-1	9.7	10.7	L avr. corr./B avr.
A.L. 277-1	10.4	11.8	L. corr.
A.L. 288-1i	8.2	10.5	L corr.
A.L. 330-5	8.7	10.4	
A.L. 330-7	10.5	12.1	
A.L. 333-44	11.1	11.2	L corr.
A.L. 333w-1	9.5	10.5	L avr./B avr.
A.L. 333w-32, 60	9.5	12.8	
A.L. 400-1a	9.8	11.2	L avr. corr./B avr.
A.L. 417-1a	8.6	11.2	L corr.
A.L. 433-1a	9.5		L corr.
A.L. 443-1	10.8	11.7	
A.L. 444-2	11.4		L corr.
A.L. 582-1l	11.4	11.8	L avr. corr./B est.
M/1			
A.L. 128-23	11.2	11.1	
A.L. 145-35	13.0	13.4	
A.L. 198-22	12.8		L corr.
A.L. 198-1	10.1		L corr.
A.L. 200-1b	13.0	12.5	L corr.
A.L. 228-2	12.8	12.0	L corr.
A.L. 241-14	14.6	13.5	B est.
A.L. 266-1	12.7	11.9	L avr. corr./B avr.

Specimen	MD Length	BL Breadth	Comments
M/2			
A.L. 128-23	12.1	12.5	
A.L. 145-35	15.4	14.2	
A.L. 188-1	15.4	15.2	L corr.
A.L. 198-1	12.4	12.4	L corr.
A.L. 207-13	13.2	12.5	L corr.
A.L. 225-8	12.4	11.1	L corr.
A.L. 266-1	13.0	14.0	
A.L. 277-1	15.1	14.5	L corr.
A.L. 288-1i	13.2	12.2	L corr.
A.L. 330-5	12.7	12.8	
A.L. 333-74	14.0		L corr.
A.L. 333w-1	13.7	12.8	L+B avr/L corr.
A.L. 333w-27	15.4	14.1	L corr./B est.
A.L. 333w-32, 60	14.5	14.6	
A.L. 333w-48	12.6	12.1	B est.
A.L. 333w-57	14.3	12.1	L corr.
A.L. 333w-59	14.0	14.4	L corr.
A.L. 400-1a	15.0	14.6	L+B /L corr.
A.L. *417-1a, b*	13.2	13.1	L+B avr./L corr.
A.L. 418-1	16.5		L corr.
A.L. 437-1	16.1	13.9	L corr./B est.
A.L. 437-2	15.6		L corr.
A.L. 438-1	16.0		L corr.
A.L. 440-1	15.8	13.8	L corr.
A.L. 443-1	15.2	14.2	
M/3			
A.L. 188-1		14.9	
A.L. 198-1	14.6	12.1	L corr.
A.L. 207-17	13.4	11.3	

P/3				A.L. 277-1	12.5		L corr.	*A.L. 225-8*	15.0	13.8	
A.L. 128-23	8.2	9.6	L corr.	A.L. 288-1i	12.4	11.0	L corr.	A.L. 266-1	15.8	13.7	L corr; B est.
A.L. 198-1	9.4	9.5	L corr.	*A.L. 315-22*	13.4		L corr.	A.L. 288-1i	14.2	12.2	L+B avr./L corr.
A.L. 207-13	8.8	9.5	L corr./B est.	*A.L. 330-5*	12.4	12.1	L corr.	*A.L. 330-5*	13.7	12.7	
A.L. 266-1	9.4	10.7	L avr. corr./B avr. est.	*A.L. 330-7*	13.7	13.4	L corr.	A.L. 333-74	14.2	13.8	L corr.
A.L. 277-1	9.6	12.3	L corr.	A.L. 333-74	13.5	13.5	L corr.	A.L. 333W-32, 60	14.2	14.3	L+B avr./B est.
A.L. 288-1i	8.4	10.0	L avr. corr.	A.L. 333W-1	13.1	12.2	L avr. corr./B avr.	A.L. 333W-57	14.4	12.5	
A.L. 311-1	8.5	10.8	L corr.	A.L. 333W-12	13.1	12.7	L corr.	A.L. 333W-59	14.0	13.1	B est.
A.L. 315-22	7.9	8.9		A.L. 333W-32, 60	13.2	13.2	L corr.	A.L. 400-1a	15.2	13.7	L+B avr./L corr
A.L. 333-10	9.8	11.1	L corr.	A.L. 400-1a	13.2	12.6	L avr. corr./B avr.	A.L. 411-1	15.0		L corr.
A.L. 333W-1	9.7	10.3	L avr./B avr.	A.L. 411-1	12.5		L corr.	*A.L. 417-1a, b*	15.4	13.3	L+B avr./L corr.
A.L. 333W-32, 60	9.0	12.6		*A.L. 417-1a*	12.4	11.9	L corr.	*A.L. 437-1*	16.5	13.6	L corr.
A.L. 333W-46	9.6	9.9	L corr.	*A.L. 440-1*	14.8	13.1	L corr.	*A.L. 462-7*	16.6	14.2	B est.
A.L. 333W-58	9.4	10.4	L corr.	*A.L. 444-2*	14.6		L corr.	*A.L. 438-1*	16.5		L corr.
A.L. 400-1a	9.8	11.2	L avr. corr.					*A.L. 465-5*	14.0		L corr./est.
A.L. 417-1a	8.9	10.8						*A.L. 487-1a*	17.2		L corr.
A.L. 438-2	8.1	9.5						*A.L. 620-1*	17.4	15.3	L corr.
A.L. 440-1	8.9	10.7	L est.								
A.L. 582-1	11.4	10.0	L corr./B est.								
A.L. 655-1	9.7	10.6									

MD, mesiodistal; BL, buccolingual

[a]Includes all specimens recoverd 1974–1999. Italicized specimen numbers denote Hadar teeth added during 1990–1999. For specimens in which both left and right teeth are present and measureable, the average value is given.

[b]L, length; B, breadth; avr, average; corr, corrected; est, estimated; sw, slightly worn; w, worn; vw, very worn.

adult lower teeth. Even though the mandibular dentition in the A.L. 444-2 skull is not well preserved, it is worthwhile to summarize some of the main metrical features of the newly enlarged *A. afarensis* mandibular dental collection (see Tables 5.18 and 5.19). All of the following comments refer to the number of *individuals* rather than to the number of *teeth*, and sample sizes refer to the number of teeth on which one or both standard crown dimensions can be measured or estimated.

Lower central incisor. Four measurable teeth, representing three individuals, have increased the *A. afarensis* sample to six individuals. The A.L. 444-2 skull supplies the largest Hadar I_1, but the unerupted crown from LH 2 remains the largest I_1 (labiolingually and especially mesiodistally) in the *A. afarensis* hypodigm.

Lower lateral incisor. The Hadar sample numbers six individuals, with the addition of two specimens during the 1990s. Labiolingually, the A.L. 444-2 tooth is the largest in the species' hypodigm, barely exceeding the breadth of the tooth associated with Hadar mandible A.L. 437-2.

Lower canine. The Hadar lower canine sample has been augmented by four teeth, bringing the sample from the site to 13 individuals. No Hadar tooth approaches the LH 3 canine in its extremely long mesiodistal diameter (White, 1985; Lockwood et al., 2000), which, uniquely in the hypodigm, substantially surpasses its labiolingual breadth. Lower canine crown size in *A. anamensis* is approximately equivalent to that in *A. afarensis* (see Ward et al., 2001).

Lower third premolar. Six teeth have been added to the Hadar sample, which now totals 19 nonantimeric specimens. The statistical profile of the P_3 sample remains essentially unchanged from the 1970s (White et al., 1981). However, as discussed by Lockwood et al. (2000), there is evidence of a significant temporal trend in the *A. afarensis* P_3 sample, with the older Laetoli P_3s ($n = 5$) standing apart from their Hadar (and Maka) homologs in their mesiodistally expanded crown diameters. As a result, variation in P_3 length (as judged by the CV) is much higher for the combined *A. afarensis* sample (10.9) than for the Hadar series alone (8.9; Table 5.19) (see Lockwood et al., 2000). The crown length of the Laetoli P_3s averages 1.5 mm greater than those from Hadar, while the crown breadth is essentially the same in the two site samples. Only one Hadar tooth, A.L. 582-1, has a mesiodistal dimension greater than 10 mm, while this is true of all of the Laetoli P_3s. Although overlapping with the upper end of the Hadar range, the *A. anamensis* P_3s ($n = 6$ individuals) also appear, on average, to have a relatively enlarged mesiodistal dimension (Ward et al., 2001).

Lower fourth premolar. There are 20 nonantimeric teeth in the Hadar sample, which includes nine specimens recovered in the 1990s. The Hadar sample includes both the smallest (A.L. 288-1i) and the largest (A.L. 582-1) P_4 in the *A. afarensis* hypodigm, and the statistical profile is little changed from that for the 1970s sample. Based on the sample CV (Table 5.19), P_4 mesiodistal length is the most variable dimension in the Hadar adult lower postcanine dentition. However, in contrast to P_3, there is no indication of a temporal trend in P_4 crown length or overall size within the Hadar sample or when the Laetoli and Maka teeth are incorporated into the analysis (Lockwood et al., 2000).

Lower first molar. Eight new teeth bring the Hadar sample total to 22 individuals. Again, this sample includes both the smallest (A.L. 128-23) and largest (A.L. 241-1) *A. afarensis* M_1s. All of the new Hadar teeth fall in size and shape between these extremes. As pointed out by White (1985), the Laetoli teeth ($n = 4$) cluster near the high end of the *A. afarensis* range for length and breadth, probably due to sampling bias.

Lower second molar. The Hadar sample size has been increased to 25 individuals, with nine new M_2s recovered during the 1990s fieldwork. As for P_4 and M_1, Hadar supplies both the largest (A.L. 188-1 and probably A.L. 418-1, for which only length can be measured) and smallest (A.L. 128-23) M_2s in the *A. afarensis* hypodigm. New Hadar specimens have done little to alter the statistical profile for this tooth, which remains the most variable of the three lower molars in *A. afarensis.* There is no hint of a temporal trend in M_2 size with the addition of the small Laetoli ($n = 3$) and Maka ($n = 3$) samples to the hypodigm.

Lower third molar. Ten new additions to the Hadar collection double the number of M_3s to 20. These include several impressively large teeth that surpass the largest M_3 in the 1970s *A. afarensis* hypodigm (LH 4). The new Hadar collection also includes the smallest *A. afarensis* M_3 (A.L. 207-17). With these additions to the margins of the range of variation have come upward jumps in CV for crown length and breadth relative to 1970s levels, but these values remain within the bounds expected for large hominoids (Lockwood et al., 2000).

Maxillary Dentition

Crown Morphology

Upper first incisor. Incisal wear is essentially horizontal and exposes a slightly cupped rectangular dentine exposure measuring 2.3 mm LL and 8.4 mm MD. The slight gingival eminence gives rise to a slightly concave, polished, featureless lingual surface. The labial surface is

Table 5.19 Summary Statistics for the *A. afarensis* Mandibular Dentition[a] (mm)

			Hadar only	
	MD Length	La(B)L Breadth	MD Length	La(B)L Breadth
I/1				
Mean	6.5	7.4	6.2	7.3
SD	0.9	0.3	0.6	0.3
CV	14.3	4.0	10.3	3.8
Max	8.0	7.7	7.1	7.6
Min	5.6	6.9	5.6	6.9
n	6	6	5	5
I/2				
Mean	6.3	8.0	6.3	8.0
SD	0.8	0.7	0.9	0.9
CV	13.0	9.0	13.6	10.7
Max	7.2	8.8	7.2	8.8
Min	5.0	6.7	5.0	6.7
n	7	7	5	5
/C				
Mean	9.0	10.5	8.6	10.6
SD	1.1	1.1	0.8	1.3
CV	12.7	10.6	9.1	12.3
Max	11.7	12.4	9.5	12.4
Min	7.5	8.8	7.5	8.8
n	12	15	9	11
P/3				
Mean	9.5	10.5	9.2	10.4
SD	1.0	0.7	0.8	0.8
CV	10.9	6.9	8.9	7.9
Max	12.6	12.6	11.4	12.6
Min	7.9	8.9	7.9	8.9
n	25	25	19	19

			Hadar only	
	MD Length	BL Breadth	MD Length	BL Breadth
P/4				
Mean	9.8	11.0	9.7	11.0
SD	1.0	0.8	1.0	0.8
CV	10.3	7.1	10.7	7.5
Max	11.4	12.8	11.4	12.8
Min	7.7	9.8	7.7	9.8
n	24	21	20	17
M/1				
Mean	13.1	12.6	13.0	12.5
SD	1.0	0.8	1.0	0.8
CV	7.3	6.1	8.0	6.5
Max	14.8	13.9	14.8	13.5
Min	10.1	11.0	10.1	11.0
n	28	22	22	16
M/2				
Mean	14.4	13.4	14.2	13.4
SD	1.3	1.0	1.3	1.1
CV	8.7	7.3	9.3	8.3
Max	16.5	15.2	16.5	15.2
Min	12.1	11.1	12.1	11.1
n	31	27	25	21
M/3				
Mean	15.1	13.4	15.1	13.4
SD	1.2	1.0	1.2	1.1
CV	7.8	7.3	8.1	7.9
Max	17.4	15.3	17.4	15.3
Min	13.4	11.3	13.4	11.3
n	23	20	19	16

[a]Sample includes Laetoli, Maka, and Hadar specimens recovered through the 1999 field season. Maka data from White et al., 2000. Mean and SD are shown as rounded values; CV was calculated using the raw values.

convex vertically, but more so MD, with a square mesial corner. The chemically pitted labial face is polished, with a centrally placed chip near the cervical margin. Both the lingual and labial enamel lines are slightly convex upward, whereas mesial and distal enamel lines are concave rootward. Mesial and distal IPFs, located at the incisal level, are only slightly concave.

Upper canine. Occlusal outline is a slightly MD compressed oval. Apical wear sloping upward and inward from the labial edge exposes a large, deeply cupped, diamond-shaped dentine island, which tapers mesially and more gradually distally, and at its midpoint measures 3.8 mm wide. Enamel thickness measures 1.4 mm lingually and 1.6 mm labially. The labial face, moderately convex vertically, is strongly convex MD. This face is polished and somewhat uneven, with a number of striae. The lingual face is polished and only slightly concave near the apex. Although damaged, the gingival eminence appears to have been only moderately developed. Lack of a mesial IPF indicates that an I^2/C diastema was present. Only a small portion of the distal IPF is discernable on the preserved buccal enamel rim. There is a hint of a mesial lingual groove, which is not well demarcated from the weak mesial marginal ridge. The mesial enamel line is strongly convex apically and is flanked by prominent enamel borders that converge apically into a strong mesial cingulum.

Upper third premolar. Slightly concave occlusal wear obliterates any surface morphology. The lingual face is MD pinched and strongly convex. The lingual enamel line is slightly convex rootward. The tooth is two rooted.

Upper fourth premolar. The occlusal outline is a MD compressed oval with its longest MD dimension in the lingual half. The occlusal surface is heavily worn (more on the right side) with a large, oval-shaped, deeply concave dentine exposure on the protocone measuring 3.5 mm BL and 4.2 mm MD. Dental enamel thickness, measured on the lingual occlusal rim is 2.4 mm. An irregular pit (chemically eroded?) occurs on the buccal occlusal rim. The metacone, smaller than the protocone, is slightly peaked, with flattening occlusal wear and bearing BL oriented striae, on the distal and mesial slopes. The buccal and lingual faces are strongly convex MD and slightly vertically. Weak buccal and lingual basal bulges are obvious. There are several pits on the buccal face and very faint mesial and distal buccal grooves. A weak, vertical mesiolingual groove creases the lingual face. The mesial and distal enamel lines are horizontal, while the lingual and buccal ones are convex rootward. In buccal view the crown tapers rootward. Located slightly buccal to the crown midline is a flat, distal IPF about 5.3 mm BL. It is angled, MB to DL, to the long axis of the tooth row. A slightly concave mesial IPF is situated at the midpoint of the tooth and measures 5.5 mm. Pinpoint-sized chemical pits occur on the lingual face and the lingual half of the occlusal surface.

Upper first molar. The occlusal outline is a rounded square with a slightly bilobate buccal face and a straighter lingual face. BL oriented fine wear striae are discernable on the deeply worn occlusal surface that exposes a large dentine area, measuring 9.9 mm MD and 7.9 mm BL, extending from the lingual occlusal edge buccally over about two-thirds of the occlusal surface. The enamel thickness measured on the lingual occlusal rim is 1.0 mm. A small, 1.1-mm round dentine exposure occurs on the flatly worn paracone, and an even smaller one occurs on the flatly worn metacone. The flat mesial IPF measures 5.4 mm BL. The flat distal IPF, situated slightly lingual of the crown midline, measures 5.6 mm. The paracone is slightly larger than the metacone. The two buccal cusps appear to be more mesially positioned relative to the two lingual ones. A weak groove occurs at the cervical margin on the lingual surface. The buccal occlusal groove is shallow, but it indents the occlusal margin before continuing three-quarters of the way up the chemically pitted buccal face. A weak DB groove creases the buccal face of the metacone. The lingual enamel line is horizontal, and the buccal enamel line peaks between the root lobes.

Upper second molar. The occlusal outline is a rounded square, but with a reduced BL dimension distally. Occlusal wear has polished and obliterated nearly all of the occlusal groove and fovea pattern. On the RM^2 a large, deep 4-mm dentine exposure occurs on the protocone. Lingual dentine exposures are larger on the RM^2. Cupped dentine exposures also occur on the hypocone, 2.7 mm, and on the metacone, 1.2 mm. Fine BL oriented wear striae occur on the occlusal surface. Occlusal, buccal, and lingual surfaces bear chemical pits. The two buccal cusps are situated mesial to the lingual ones; the paracone is the most pointed. The protocone is the largest, followed by the paracone and then the subequal hypocone and metacone. The buccal occlusal groove is shallow, but it indents the occlusal margin before continuing on the buccal face nearly up to the enamel line. The lingual groove is absent on the occlusal surface and broader and less distinct on the lingual face. The slightly concave 5.7-mm mesial IPF is located just buccal to the crown midline. The flat, distal IPF, situated on the crown midline, measures 5.9 mm but does not reach the occlusal margin like the mesial one. Both lingual and buccal enamel lines appear horizontal.

Upper third molar. The M^3 is the largest of the molars and the M^1 the smallest. The occlusal outline is a rounded trapezoid that tapers distally. Antemortem chips occur on the buccal face of the paracone and metacone of LM^3. All exposed enamel surfaces, especially on the LM^3, are chemically pitted.

Although occlusal wear polishes the entire surface, and fine BL oriented striae are evident, some morphological detail remains. The occlusal surface is faceted from wear on the ML slope of the protocone, the lingual slope of the paracone and metacone, and the MB slope of the hypocone. The mesial marginal ridge is flattened and polished, but the distal marginal ridge forms a definitive boundary for the posterior fovea (especially on the LM^3). The buccal cusps are situated mesial to the lingual ones. The cusp tips appear to be situated close to the occlusal margins. The largest cusp is the protocone, followed by the paracone, then the hypocone, and finally the metacone. The fovea anterior is broad and deep, and it encroaches more onto the lingual slope of the paracone than onto the buccal slope of the protocone. A MD running groove is obvious on the LM^3 and creases the enamel crest that connects the protocone and the paracone. The mesial longitudinal groove indents a low, weak crista obliqua and continues distally through the fovea posterior to notch the distal marginal ridge. The broad, deep posterior fovea is situated buccal to the crown midline. The lingual occlusal groove is deep, notches the occlusal rim, and continues onto the lingual face where it is weakly impressed. Immediately mesial to the lingual groove, on the lingual face, is an indistinct wrinkle, probably a manifestation of a Carabelli's trait. Both the buccal and lingual faces are MD and vertically convex. The shallow buccal occlusal groove notches the occlusal rim and continues up the buccal face as a shallow, broad groove. The distal, buccal, and lingual enamel lines are horizontal. A centrally placed mesial IPF, measuring 5.6 mm, does not reach the occlusal level.

Comparative Maxillary Dentition

The morphology of the *A. afarensis* maxillary dentition is not well represented by the heavily worn teeth of A.L. 444-2, but it presents no obvious departures from previously described morphology (White et al., 1981; Johanson et al., 1982c). Indeed, this is true of the 1990s maxillary dental sample as a whole. Nevertheless, the A.L. 444-2 maxillary tooth size is larger than the mean values for comparable teeth assigned to *A. afarensis* (with the exception of the MD dimension of I^1). In fact, the MD dimensions of the P^4, M^2, M^3, and La (B)L dimensions of the I^1 and P^4-M^3 define the high end of the range for these teeth in the hypodigm. As a result, and because the A.L. 444-2 canine size is not unusually large (it is matched or exceeded in LaL breadth by A.L. 333-1, A.L. 333w-2, A.L. 333x-3, and LH 3), its canine:postcanine tooth size ratio is lower than in the few other *A. afarensis* specimens in which the ratio can be calculated. It is, of course, nothing like the extreme disproportion seen in later *Australopithecus* dentitions (Robinson, 1956; Tobias, 1967; White et al., 1981; Suwa, 1989).

Although little useful morphology remains on the A.L. 444-2 dentition, the new Hadar collection includes maxillary dental elements that amplify previous descriptions of *A. afarensis* dental morphology, especially that of the more diagnostic anterior teeth (White et al., 1981). Four adult maxillary canines have been added to the Hadar hominin sample as the result of 1990s fieldwork (see Table 5.20). Two of these are fairly heavily worn (the teeth associated with the A.L. 444-2 and A.L. 417-1d maxillae), but two isolated specimens (A.L. 487-1c and A.L. 763-1) are relatively unworn, permitting comparative evaluation. Crown dimensions (Table 5.20) suggest that A.L. 487-1c is from a male individual (which is also indicated by the size of the associated lower canines, postcanine teeth and jaw fragments); morphologically, it compares favorably with the canines from A.L. 333-2 (worn) and LH 6 (relatively unworn). The A.L. 763-1 canine is smaller (Table 5.20) and is most likely from a female individual. This tooth is most like A.L. 200-1a morphologically; A.L. 417-1d and LH 5 represent more advanced wear stages in morphologically equivalent teeth.

Neither A.L. 487-1c nor A.L. 763-1 evinces mesial interproximal facets, implying the presence of I^2/C diastemata, which are also found in A.L. 444-2 but not in A.L. 417-1d. While it may be that their relatively unworn state implies that these teeth had not yet at the time of death come into interproximal contact with the I^2, distal interproximal facets are visible on both specimens. Thus, we interpret the lack of a mesial facet as a sign that a diastema would have been maintained. The diastemata in these new maxillary dentitions increase the frequency given by White et al. (1981), who reported that two of six *A. afarensis* individuals in the 1970s sample showed this feature in the maxillary tooth row. With the new Hadar specimens included, the frequency rises to 50% (5/10).

In their comparison of *Australopithecus* dentitions White et al. (1981) noted that the labial (and lingual) crown profile of *A. afarensis* maxillary canines is more often asymmetric than in other species, including *A. africanus*. This asymmetry is due to the differential heights of the mesial and distal crown shoulders along the cervicoapical axis, with the mesial shoulder lower (more apical) than the distal, resulting in a long distal and a short mesial occlusal edge. While this is best seen in relatively unworn teeth (such as A.L. 333x-3, A.L. 400-1b, A.L. 487-1c, and A.L. 763-1), it is often detectable even in worn specimens (such as A.L. 200-1a). Laetoli upper canines are more symmetrical than their Hadar counterparts (see, for example, LH 3 and LH 6), with more cervically positioned mesial shoulders.

Unworn or little worn maxillary canines of *A. africanus* usually are more symmetrical (e.g., Sts. 52a, Stw. 151, Stw. 183, Stw. 287, Stw. 369, and Stw. 410), as in robust species of *Australopithecus* and later species of *Homo*, though as noted by White et al. (1981) asymmetrical canines do occur in *A. africanus*. It is worth pointing out that those Sterkfontein specimens whose taxonomic position is uncertain—Stw. 252 and Stw. 498—feature symmetric canines like that seen in *A. africanus*. That these teeth are quite large but symmetric disputes Strait et al.'s (1997) claim of character redundancy due to a correlation between large canine size and crown asymmetry.

Ward et al. (2001) have noted that the single unworn maxillary canine of *A. anamensis* (KNM-KP 35839) has a symmetrical crown profile when compared to Hadar homologs. This symmetry is unlike that seen in *A. africanus*, however. In the *A. anamensis* tooth, the mesial and distal shoulders are both positioned very close to the cervical margin of the crown, where they merge with a prominent lingual basal tubercle (see Ward et al., 2001: 344–347). This gives the *A. anamensis* tooth a more apelike lingual profile, with subequally long mesial and distal occlusal edges that converge toward the crown apex (although no honing function has been suggested for this tooth). In contrast, the crown shoulders are more apically positioned, lying an equal distance below the cervical margin, in *A. africanus* maxillary canines, with approximately half the crown's height projecting below them in lingual view.

Although large and bearing thin, apelike, enamel caps, the two described maxillary canines of *Ard. ramidus* are distinctly hominin-like in their crown symmetry, with apically shifted crown shoulders yielding "a low, blunt canine tooth relative to more projecting ape canines" (White et al., 1994: 308). In this respect, the *Ardipithecus* canines appear less apelike than those of *A. anamensis*, although it is important to reiterate that in none of these early hominins does canine morphology suggest functional equivalence to the shearing canine in the sectorial C/P_3 complex of the great apes.

A variety of maxillary canine morphologies appear among very early hominin species. This diversity extends within species, if the symmetrical canine crowns in the small Laetoli sample (n = 3 individuals) accurately reflects

Table 5.20 Catalog of Hadar Maxillary Dental Measurements[a] (mm)

Specimen	MD Length	BL Breadth	Comments[b]	Specimen	MD Length	BL Breadth	Comments	Specimen	MD Length	BL Breadth	Comments
I1/				P3/				M2/			
A.L. 198-17a+b	9.0	7.1	L vw	A.L. 199-1	7.5	11.3	L corr.	A.L. 199-1	12.1	13.4	L corr.
A.L. 200-1a	10.9	8.4	L+B avr./L sw	A.L. 200-1a	8.8	12.4	L avr./L corr.	A.L. 200-1a	13.5	14.9	L+B avr.
A.L. 293-3		8.1		A.L. 333-1	8.8	12.3	L+B avr./L corr.	*A.L. 417-1d*	13.2	14.7	L avr. corr./B avr.
A.L. 333x-20	10.8	8.6		A.L. 333-2	9.0	12.2	L+B avr./L corr.	*A.L. 442-1*	12.3	13.9	L corr./B est.
A.L. 333x-4		8.6		*A.L. 417-1d*	8.4	11.7	L avr. corr.	*A.L. 444-2*	14.1	15.8	L+B avr./L corr./B est.
A.L. 444-2	10.5	9.7	L corr./B est.	*A.L. 486-1*	9.1	12.6		*A.L. 486-1*	13.6	14.9	
A.L. 486-1	10.6	8.4						*A.L. 651-1*	12.6	14.3	L corr./B est.
				P4/				*A.L. 770-1a*	13.5	15.2	L corr. avr.
I2/				*A.L. 125-11*	9.2		L corr.				
A.L. 198-17a+b	6.7	6.2	L vw	A.L. 199-1	7.6	11.1	L corr.	M3/			
A.L. 200-1a	7.3	7.2	L+B avr./L sw	A.L. 200-1a	8.4	12.2	L+B avr.	A.L. 161-40	11.8	13.4	L corr.
A.L. 249-26	8.1			A.L. 333-1	9.1	12.4	L corr.	A.L. 199-1	11.4	13.1	L corr./B est.
A.L. 333-2		7.1		A.L. 333-2	9.5	11.7	L+B avr./L corr.	A.L. 200-la	14.3	15.1	L+B avr.
A.L. 333w-28		6.5		A.L. 333w-42	9.5	12.6	L. corr.	A.L. 333x-1	13.6	15.5	
A.L. 333x-2, 17	8.2	7.9	L w	*A.L. 417-1d*	8.5	12.5	L avr. corr./B avr.	*A.L. 417-1d*	13.0	14.8	L avr. corr./B avr.
A.L. 486-1	7.6	7.3		*A.L. 423-1*	8.7	12.1	L corr.	*A.L. 444-2*	14.8	16.3	L+B avr./L corr.
A.L. 417-1d	6.6	7.3	L corr.	*A.L. 444-2*	10.8	14.5	L avr. corr.	*A.L. 486-1*	13.3	15.1	
				A.L. 486-1	9.4	12.8		*A.L. 651-1*	12.0	13.3	L corr.
C/				*A.L. 651-1*	8.1	12.2	L corr./B est.	*A.L. 770-1a*	12.7	15.4	L+B avr./L corr.
A.L. 199-1	8.9	9.3	L corr.	*A.L. 699-1*	9.5						
A.L. 200-1a	9.5	10.9	L+B avr./L sw	*A.L. 770-1a*	9.1	12.1	L corr./B est.				
A.L. 333-1	10.0	12.4	L corr.								
A.L. 333-2	9.8	10.9	L+B avr./B est./L w	M1/							
A.L. 333w-2	10.1	11.9		*A.L. 125-11*	11.5		L corr.				
A.L. 333x-3	10.4	11.5		A.L. 199-1	10.8	12.0	L corr.				
A.L. 400-1b	9.2	10.3		A.L. 200-1a	12.2	13.3	L+B avr./L corr.				
A.L. 417-1d	9.4	9.9		*A.L. 309-8*		12.9					
A.L. 444-2	10.4	11.5		A.L. 333-86	11.9	12.2	B est.				
A.L. 487-1c	10.3	10.8		*A.L. 417-1d*	12.1	13.3	L avr. corr./ B avr.				
A.L. 763-1	8.8	10.4		*A.L. 423-1*	11.9		L corr.				
				A.L. 444-2	13.5	15.0	L avr. corr.				
				A.L. 486-1	12.8	14.4					
				A.L. 651-1	10.5		L corr.				

[a]Includes all specimens recoverd 1974–1999. Italicized specimen numbers denote Hadar teeth added during 1990–1999. For specimens in which both left and right teeth are present and measureable, the average value is given.

[b]L, length; B, breadth; avr, average; corr, corrected; est, estimated; sw, slightly worn; w, worn; vw, very worn.

Table 5.21 Summary Statistics for the *A. afarensis* Maxillary Dentition[a]

			Hadar only					Hadar only	
	MD Length	La(B)L Breadth	MD Length	La(B)L Breadth		MD Length	BL Breadth	MD Length	BL Breadth
I1/					P4/				
Mean	10.6	8.4	10.4	8.4	Mean	9.1	12.4	9.0	12.4
SD	0.8	0.7	0.7	0.7	SD	0.7	0.8	0.8	0.8
CV	7.8	8.5	6.7	8.5	CV	7.4	6.2	8.5	6.5
Max	11.8	9.7	10.9	9.7	Max	10.8	14.5	10.8	14.5
Min	9.0	7.1	9.0	7.1	Min	7.6	11.1	7.6	11.1
n	6	8	5	7	*n*	18	12	13	11
I2/					M1/				
Mean	7.5	7.2	7.4	7.1	Mean	12.1	13.4	11.9	13.3
SD	0.6	0.6	0.6	0.5	SD	1.0	0.9	0.9	1.0
CV	7.5	7.8	8.4	7.4	CV	8.2	6.6	7.3	7.6
Max	8.2	8.1	8.2	7.9	Max	13.8	15.0	13.5	15.0
Min	6.6	6.2	6.6	6.2	Min	10.5	12.0	10.5	12.0
n	8	9	6	7	*n*	15	12	9	7
C/					M2/				
Mean	9.9	10.9	9.7	10.9	Mean	13.0	14.7	13.1	14.6
SD	0.7	1.0	0.6	0.9	SD	0.6	0.6	0.7	0.7
CV	7.1	8.8	5.8	7.9	CV	4.5	4.1	5.0	4.8
Max	11.6	12.5	10.4	12.4	Max	14.1	15.8	14.1	15.8
Min	8.8	9.3	8.8	9.3	Min	12.1	13.4	12.1	13.4
n	14	14	11	11	*n*	11	12	8	8
P3/					M3/				
Mean	8.7	12.4	8.6	12.1	Mean	12.5	14.4	13.0	14.7
SD	0.5	0.6	0.5	0.4	SD	1.2	1.1	1.1	1.1
CV	5.7	4.8	6.3	3.7	CV	9.9	7.5	8.3	7.3
Max	9.3	13.4	9.1	12.6	Max	14.8	16.3	14.8	16.3
Min	7.5	11.3	7.5	11.3	Min	10.9	13.0	11.4	13.1
n	9	9	6	6	*n*	12	12	9	9

[a]Sample includes Laetoli, Maka, and Hadar specimens recovered through the 1999 field season. Maka data from White et al., 2000. Mean and SD are shown as rounded values; CV was calculated using the raw values.

the mean condition of the population from which it was drawn. Reduction of this diversity appears to have occurred subsequent to *A. afarensis*; in many respects the maxillary canine morphology of *A. africanus* is much more like that of robust *Australopithecus* and *Homo* than it is like that of *A. afarensis*. This is in spite of the fact that *A. africanus* retains the relatively large canine crown size present in its putative ancestor, *A. afarensis*.

Metric Features of the A. afarensis *Adult Maxillary Dental Sample*

Compared to the lower dentition, the adult maxillary dental sample remains underrepresented in *A. afarensis* (no doubt owing to the differential taphonomic survivorship of the mandible vs. the cranium). In no case have additions from the 1990s Hadar fieldwork substantively altered the statistical profile based on the 1970s sample (Tables 5.20 and 5.21). With the exception of A.L. 444-2, which has been discussed here, the only noteworthy specimen is A.L 486-1 (from the Denen Dora Member), whose postcanine tooth size tends to slightly exceed the range for 1970s Hadar specimens (Table 5.20). These two specimens supply M^1s that substantially reduce the crown size discrepancy between the 1970s Hadar and Laetoli M^1 samples noted by White (1985).

Lockwood et al. (2000) observed in *A. afarensis* a statistically significant trend toward M^3 crown enlargement over time. Mesiodistal elongation is the predominant change, though whether this can be attributed to intersite differences between Laetoli (which has relatively broad M^3s) and Hadar, as suggested by White (1985), or can be extrapolated across the Hadar Formation is unclear, given the still small sample sizes (Lockwood et al., 2000: 32).

6

Implications of A.L. 444-2 for the Taxonomic and Phylogenetic Status of *Australopithecus afarensis*

Morphology of the A.L. 444-2 Skull: Summary of the Major Features

The Skull as a Whole

A.L. 444-2 is the first specimen to preserve the cranium and mandible of a single adult individual of *A. afarensis*. Pairing this specimen with A.L. 417-1, which includes a mandible and maxilla, enables us to compare comprehensively the craniofacial morphology of male and female individuals of the species for the first time.

The occluded mandibles and maxillae of A.L. 444-2 and A.L. 417-1 reveal a distinctive hominoid snout contour, combining a strongly inclined, convexly sloping nasoalveolar clivus with a relatively upright mandibular symphysis, a straight to slightly rounded anterior symphyseal outline, and an anteriorly placed gnathion.

Both *A. afarensis* specimens feature a very deep mandibular corpus, whose height occupies close to 70% of the orbitoalveolar height of the face. In the African great apes, this value ranges from 36% to 54%, and in modern humans, it is 66%. The high value in humans is due to a short orbitoalveolar region rather than to a deep mandible. *A. afarensis* appears to share a relatively deep corpus with *A. robustus* (the only robust species in which the feature can be determined for a single individual) but not with *A. africanus*.

Relative to the calvarial length, the A.L. 444-2 braincase height is apelike, falling between the tall modern human braincase and the low braincase of *A. boisei* and *A. aethiopicus*. In *A. africanus* (Sts. 5) and *H. habilis* (KNM-ER 1813) the relative braincase height is like that of A.L. 444-2 and the great apes. According to Le Gros Clark's (1950) index expressing the height of the calvaria above the roof of the orbit as a percentage of total calvarial height, Sts. 5 and KNM-ER 1813 have tall, "humanlike" braincases, whereas A.L. 444-2, *A. boisei*, *A. aethiopicus*, and the African great apes group together with low braincases.

In contrast to the rounded, nearly circular midsagittal outline of the chimpanzee calvaria, the posterior parietal/occipital arc in A.L. 444-2 is steep and deviates anteriorly from the circle. This is also true of the *A. boisei* calvaria.

As expected from the calvarial height comparison, the slope of the A.L. 444-2 frontal squama is smaller than that of *A. africanus* and *H. habilis*. It is slightly greater than that of chimpanzees and gorillas and is similar to that of *A. boisei* and *A. aethiopicus*.

Due to the low position of lambda, the angle of the parietal chord (br–la) to the FH is more pronounced in A.L. 444-2 than in the great apes, in which lambda occupies an elevated position on the calvaria. The Hadar specimen is similar to *A. boisei* and *A. aethiopicus* in this regard; with a high lambda, *A. africanus* and *H. habilis* tend to resemble the human condition.

As in all hominins, inion approximates the FH in A.L. 444-2. The combination of a low lambda and low inion is also typical of *A. robustus*, *A. boisei*, and *A. aethiopicus* but not *A. africanus* or *Homo*, in which lambda is higher. The great ape pattern, in which both inion and lambda are above FH, is not found in any hominin.

The nuchal plane (i–o) forms a very low angle to the FH in A.L. 444-2, whereas this angle is much steeper in some other *A. afarensis* crania (e.g., A.L. 162-28 and A.L. 439-1). A matching degree of variation is seen in *A. boisei* (e.g., OH 5 vs. KNM-ER 23000), but no hominin skull approaches the nuchal plane steepness seen in the great apes.

In both A.L. 417-1 and A.L. 444-2 a smaller percentage of palate length projects anterior to the coronal plane of sellion than in the great apes. The *A. afarensis* values overlap with those for *A. africanus*, which is highly variable in this regard, but those for *A. robustus* and *A. boisei* are much smaller still, reflecting the retruded position of the snout in these species. However, *A. aethiopicus* is notable for the most forwardly protruding snout of any hominin, exceeding even the great ape means. The "palate projection" index mirrors the array of angular values of prognathism among the great apes and fossil hominins.

In A.L. 444-2, the upper segment of the face (nasion–nasospinale) is less prognathic (i.e., it is more upright) than in the great apes. This difference lies behind the generally "hominin-like" appearance of the *A. afarensis* face, which is subsequently elaborated in structurally dissimilar ways in *Homo* and late robust *Australopithecus* species. The *A. aethiopicus* specimen KNM-WT 17000 is the most apelike of any hominin in this relationship.

Indices of the masticatory system—projection of the palate relative to the coronal plane of sellion, projection of the palate relative to the masseter's origin, and projection of the masseter's origin relative to the coronal plane of sellion (Rak, 1983)—for A.L. 444-2 indicate a reduction in the anteroposterior distances that separate these landmarks relative to the great ape configuration. *A. afarensis* resembles *A. africanus* in these indices, and both stand in strong contrast to the later robust *Australopithecus* species, in which the anteroposterior compression is extreme (again, *A. aethiopicus* appears anomalously apelike). However, the summary "overlapping" index, which expresses the amount of the palate length's overlap of the distance between the articular eminence and the masseter's origin as a percentage of the total "length of the masticatory system"—the distance between the articular eminence and prosthion—is low in A.L. 444-2 and is thus the most apelike of any hominin skull, including that of *A. aethiopicus*.

The squamosal suture extends to a very high position on the braincase of A.L. 444-2, but the amount of overlap of the temporal squama on the parietal bone is modest, as reflected in the absence of extensive striae parietalis that accompanies the extreme (and unique) overlap of these bones in *A. boisei*.

Although A.L. 444-2 shares with robust *Australopithecus* crania a large bizygomatic width in relation to cranial size (as judged by biorbital breadth), the *A. afarensis* cranium features a less expanded temporal foramen due to its sagittally oriented (nonflaring) zygomatic arches and broad postorbital region.

In superior view, the calvarial shape in A.L. 444-2 is an elongate oval, much as in chimpanzees and *A. africanus*. This contrasts with the teardrop shape of the *A. boisei* calvaria produced by the combination of a narrow postorbital region, a great biporial saddle breadth, and, especially, a short anteroposterior distance between these widths.

A.L. 444-2 is similar to all other *Australopithecus* crania in the lateral deviation from the sagittal plane of the lateral margin of the zygomatic bone's frontal process. Thus, in contrast to the great apes and *Homo*, in which the lateral margin of the process is aligned with the sagittal plane, the facial mask tapers superomedially in the *Australopithecus* crania.

The low, curved inferior margin of the maxilla's zygomatic process positions the anterior origin of the masseter muscle (the zygomatic tubercle) at a lower position on the face in A.L. 444-2 (and other *A. afarensis* crania, such as A.L. 333-1 and A.L. 417-1d) than in *A. africanus*, *A. robustus*, and *A. boisei*, where the inferior margin of the process is straight and strongly angled superolaterally; *A. afarensis* shares the low anterior masseter origin with *Homo* and the great apes.

Despite the massiveness of the A.L. 444-2 peripheral facial skeleton, the biconvex subnasal region, sharp lateral margins of the nasal aperture, deep canine fossae, and flat, relatively vertical infraorbital plate conform to the distinctive, generalized facial pattern observed in other, more fragmentary, *A. afarensis* remains. This species' primitive facial topography appears also to be expressed, at least in part, in the 2.5-Myr-old species *A. garhi*, but other *Australopithecus* species show different degrees of specialization in circumnasal, infraorbital, and zygomaticomaxillary topography.

As deduced from the A.L. 417-1 face and the A.L. 438-1b frontal fragment, the interorbital region is very narrow in *A. afarensis* (constituting less than 50% of the orbit's breadth), which is not observed otherwise in *Australopithecus* or in the African great apes. The narrowness of the interorbital region extends inferiorly to the nasal aperture, which, in relation to orbital width, is narrower in *A. afarensis* than in any other *Australopithecus* species.

The occipital profile of the A.L. 444-2 calvaria is bell shaped, as it is also in A.L. 333-45, but does not exhibit the strongly inclined lateral walls common in *A. boisei* (and also apparent in *A. aethiopicus*). The *A. afarensis* calvaria on the interporial plane is higher than in robust *Australopithecus*, though relative to interporial width, the crania of *A. africanus* and early *Homo* are taller still, with more vertical lateral walls.

In A.L. 444-2 the maximum posterior calvarial width is measured at the level of the mastoid/supramastoid crests (as in A.L. 333-45) from which the straight lateral contour of the pars mastoidea angles sharply inferomedially beneath the cranial base. This is the primitive hominoid pattern shared, albeit on a reduced scale, by *A. africanus*. In robust *Australopithecus* species the maximum breadth of the posterior calvaria occurs lower on the pars mastoidea, where the bulbous lateral contour corresponding to the mastoid crest protrudes laterally beyond the margin of the supramastoid crest above, and the mastoid's inflection beneath the cranial base is comparatively weak.

The nuchal area of the cranial base is anteroposteriorly abbreviated in relation to the posterior (bisupramastoid) width of the base in A.L. 444-2 (and A.L. 333-45), a morphology shared with robust *Australopithecus* crania. All other hominoids, including *A. africanus* and *H. habilis*, show greater relative posterior extension of the nuchal area. This difference may help explain the greater relief of muscle origin scars on the nuchal plane of many of the larger *A. afarensis* and robust *Australopithecus* crania.

The center of the cranial base, including the occipital condyles, is elevated in relation to the more peripheral basal structures, giving the base a coronally concave profile in *A. afarensis* and all other hominins. In the apes, the condyles occupy the most inferior position in a coronally convex basal profile.

The foramen magnum is far forward on the A.L. 444-2 cranial base such that basion lies anterior to the bitympanic line, on the coronal plane of the carotid canals. This is associated with a reduced anterior cranial base length compared to the great apes, in which basion is situated posterior to the bitympanic line. Among other hominins, only in robust *Australopithecus* species does basion typically surpass the bitympanic line *and* approximate the bicarotid canal line (although this may also be true of the otherwise highly plesiomorphic cranial base of early Pliocene *Ardipithecus ramidus*).

Associated with the reduced anterior cranial base length in A.L. 444-2 (as well as other hominins), the carotid canals are more anteriorly placed, which yields a tympanic element that is strongly angled to the coronal plane. On the great ape cranial base, the lateral margin of the tympanic and the carotid canals are coronally aligned, as is, therefore, the tympanic element itself.

As a consequence of the forward migration of the carotid canals and the angulation of the tympanic element, the external aspect of the petrous pyramid is less sagittally oriented in A.L. 444-2 than in the great apes. In this feature, *A. afarensis* occupies the middle ground between the more sagittally oriented petrous of the great apes and *A. africanus*, and the more coronally oriented petrous of modern humans and robust *Australopithecus* species.

The new Hadar cranium is like that of many other hominoids (including other *A. afarensis* specimens) in the moderate posteromedial-anterolateral angulation of the long axis of its articular eminence (seen in basal view). Crania of *A. boisei* uniquely exhibit a strongly angled eminence, which, given its great mediolateral extension, positions the articular tubercle at its lateral end and the entoglenoid process at its medial end on two quite distant coronal planes.

Posteriorly approximated temporal lines; a relatively small "bare area" of the occipital; extensive, laterally prominent compound temporonuchal crests that traverse asterion; and the asterionic notch sutural articulation all support the inference of strongly developed posterior fibers of the temporalis muscle in A.L. 444-2. This pattern of morphology reiterates that seen in other, more fragmentary remains of *A. afarensis* (including four other new specimens), which contrasts with the emphasis on the anterior temporalis fibers that is more common in other *Australopithecus* species—notwithstanding documented variation in the extent of the sagittal crest in *A. boisei*.

A. afarensis has a higher frequency of sagittal and compound temporonuchal crests in small, presumably female, crania than in any other well-documented hominin taxon. The cresting pattern is among the most apelike aspects of *A. afarensis* cranial morphology.

The pattern of bony sulci on the endocranial aspect of the occipital bone, as well as the endocranial cast itself, indicate a dominant, but asymmetrically developed, occipital-marginal venous outflow track in A.L. 444-2. Combined with other new fossil evidence from Hadar, the incidence of this presumably specialized venous drainage pattern in *A. afarensis* is currently 89% (eight of nine individuals), a high frequency shared with *A. robustus* and *A. boisei*. In the smaller *A. africanus* sample the incidence of this pattern is 29% (two of seven individuals), which matches the highest frequency so far documented in any sample of a modern human population.

The Endocast

Several methods of reconstructing the deformed and incomplete A.L. 444-2 endocast yield an endocranial capacity estimate of 550 ml. The average endocranial capacity for three *A. afarensis* specimens is 478 ml, which falls between the means for *A. africanus* (452 ml at minimum, $n = 7$) and *A. boisei* (513 ml, $n = 4$).

Surface detail is very difficult to discern on the endocast, but some shape relationships are informative in a comparative context. The A.L. 444-2 endocast resembles those of *A. robustus* and most *A. boisei* (but not *A. aethiopicus*) in the strong degree of forward tucking of the cerebellum beneath the occipital lobes of the cerebrum and in the roughly triangular shape and flattened posterolateral surface of the cerebellar lobes. However, the cerebellar lobes of the A.L. 444-2 endocast are very small compared to those of all other hominins, including *A. africanus*.

The anterior poles of the frontal lobe are full and rounded in A.L. 444-2, as in apes and *A. africanus*, and are not strongly tapered anteriorly as in robust *Australopithecus*. The unusual narrowing of the frontal lobes in the robust endocast most likely relates to the distinctive shape of their calvaria in superior aspect.

The Disarticulated Skull and the Dentition

The frontal bone, previously poorly represented in *A. afarensis*, shows distinctive morphology in several areas. The breadth across the postorbital constriction is absolutely and relatively large compared to that of any other *Australopithecus* frontal, except the 3.9-Myr-old Belohdelie specimen. As demonstrated by A.L. 444-2 and another new Hadar frontal fragment, A.L. 438-1b, the squama is low but flat to mildly convex in sagittal cross section, lacking any sign of the concave frontal trigone that characterizes robust *Australopithecus* crania. No known *Australopithecus* frontal features the "torus and sulcus" morphology of the African great apes.

In common with other *Australopithecus* frontals, the supraorbital element is barlike and inclined posterosuperiorly in a lateral view of A.L. 444-2. The temporal line occupies the topographically highest point on the supraorbital surface rather than truncating the bar posteriorly as is usually the case in robust *Australopithecus* frontals. In the African great apes, the temporal line is lower than the supraorbital surface.

The anterior outline of the A.L. 444-2 supraorbital element in superior view is straight and coronally aligned, with a slight anterior deviation from this line at the lateral extremity of the supraorbital bars (also inferred for A.L. 438-1b) that creates a bulbous superolateral corner. The common morphology in other *Australopithecus* species is for the supraorbital profile to retreat posterolaterally, leaving the medial part of the supraorbital structures more anterior than the lateral part.

The medial to lateral gradient in vertical supraorbital thickness and the reconstruction of the missing sections of the frontal's zygomatic process permit the inference of a "squared off" supraorbital corner in A.L. 444-2 in frontal view. Other *Australopithecus* frontals feature an inferolaterally sloping corner and usually display the opposite supraorbital thickness gradient. Gorilla frontals have a square supraorbital corner, whereas chimpanzees are highly variable in this regard, and both ape species commonly exhibit a medial to lateral increase in supraorbital thickness.

The A.L. 444-2 frontal squama is thicker than in other *Australopithecus*, early *Homo*, and great ape crania. A thick frontal squama also characterizes the Belohdelie specimen.

The parietal bone's sagittal chord (br–la) in A.L. 444-2 is very long, greater than in any measurable cranium of *A. africanus* or *A. boisei* and within the range for early *Homo*. The long sagittal dimension is coupled with a much shorter coronal dimension to create a markedly rectangular parietal bone, which contrasts with the squarer parietal in the great apes.

As for the frontal bone, the parietal bone of A.L. 444-2 (and of *A. afarensis* generally) is thick, especially in the vicinity of bregma, in comparison to other early hominins and extant hominoids.

In A.L. 444-2 the base of the temporal squama flares laterally to create a nearly horizontal shelf in the mastoid region. This morphology is shared with other large *A. afarensis* crania and is more extreme than in other *Australopithecus* species.

Mastoid pneumatization proliferates far forward in the temporal squama of A.L. 444-2, and of other *A. afarensis* crania, thickening the base of the squama to a greater degree than in any other *Australopithecus* species. Only in the great apes is an equivalent degree of pneumatization of the squama regularly encountered.

A. afarensis is like many other hominoids in the close proximity of the anteriorly inflected mastoid process tip to the coronal plane of porion (in lateral view). Only *A. boisei* departs from this common configuration, with a weakly inflected mastoid tip usually situated approximately halfway between the coronal planes of porion and asterion.

A.L. 444-2 reiterates the primitive basal mastoid morphology of *A. afarensis*: a relatively modestly projecting tip; a shallow, triangular digastric fossa that weakly impinges on the posterolateral mastoid face; and a superimposed occipitomastoid crest and suture. In these features *A. africanus* resembles *A. afarensis*, but *A. robustus* and *A. boisei* differ in the anteroposteriorly elongate digastric fossa, which extends almost to the rear limit of the calvaria in basal view. This pattern is common in later *Homo* and is also found in some early crania of this genus.

Although the mediolateral breadth of the A.L. 444-2 mandibular fossa is as large as those of some *A. robustus* and *A. boisei* crania, its fossa is substantially shorter anteroposteriorly. In fact, *A. afarensis* has the shortest relative mandibular fossa length of any *Australopithecus* species.

An extensive, horizontal preglenoid plane; weak articular eminence; inferiorly pointing entoglenoid process; and vertically shallow, "tubular" tympanic element residing completely posterior to a large postglenoid process are all primitive, chimpanzee-like, features of the *A. afarensis* glenoid region. The anatomy of the A.L. 444-2 skull—with its articular eminence that is more convex on the sagittal plane, its steeper preglenoid plane, and its more upright (though still "tubular," i.e., not platelike) tympanic element—is slightly derived toward the humanlike condition found in some *A. africanus* crania. *A. boisei* stands apart from all other hominoids in its highly modified, autapomorphic, glenoid region.

The tubular form of the tympanic element in A.L. 444-2 and other *A. afarensis* crania is common in the great apes but not normally encountered in other *Australopithecus* species except *A. anamensis*. In all subsequent species of this genus, as well as in *Homo*, the tympanic is platelike—with variable development of the petrous crest and vaginal process of the styloid, more vertically oriented, and more closely approximating the coronal plane of the postglenoid process. In *A. boisei* uniquely, the platelike tympanic tends to incline posteriorly on the mastoid process, sealing the primitively "open" tympanomastoid groove of *A. afarensis* and most *A. africanus* temporals. *A. africanus* is unique in the "paddle" shape of its tympanic plate and in the morphology of the eustachian process at its medial termination.

The A.L. 444-2 occipital bone exemplifies the common *A. afarensis* morphology: a short, broad squama with a low occipital plane and long nuchal plane (i.e., a high occipital scale index). Among the species of *Australopithecus*, only in *A. afarensis* do crania inferred to be those of females have a relatively long nuchal plane, a characteristic of female great ape crania.

Due to its relatively horizontally oriented nuchal plane, the occipital squama of A.L. 444-2 is more strongly flexed across the occipital plane/nuchal plane transition than in other *A. afarensis* crania, in which the nuchal plane is steeply inclined and thus more closely aligned with the occipital plane. The A.L. 444-2 condition resembles that encountered most often among later hominins. However, another new, large *A. afarensis* occipital (A.L. 439-1), positioned close in time to A.L. 444-2, shows the steep nuchal plane and weakly flexed squama that are more typical of the species. A similar degree of variation in these features is seen in the *A. boisei* sample

Larger *A. afarensis* occipital bones, including that of A.L. 439-1 and A.L. 444-2, display a coronally flat nuchal plane, whereas smaller occipitals—presumed females—have a coronally convex nuchal plane. *A. afarensis* is the only hominin species to show the latter condition, which is the usual one in chimpanzees and female gorillas.

Sharp marginal crests defining a nasal aperture that forms a single plane on the face are shared by A.L. 444-2 and all other *A. afarensis* maxillae except the Laetoli specimen Garusi I, which, in its rounded lateral margins and indistinct transition across the entrance to the nasal cavity, resembles chimpanzees.

The nasal cavity morphology of A.L. 444-2 matches that observed in every other *A. afarensis* maxilla: a "stepped" nasal cavity floor, an elevated intranasal platform intervening between the inferior nasal margin and the incisive fossa, and an (inferred) high position of the anterior tip of the vomer that inserts at the posterior end of the incisive crest, well posterior to the anterior nasal spine. Some components of this morphological package appear in some *A. africanus* maxillae, but no specimen has all of them. In all robust *Australopithecus* species, including *A. aethiopicus*, the nasal cavity floor is unstepped and the nasal spine, anterior vomeral insertion, and incisive fossa are merged within the nasal cavity.

As exemplified by A.L. 444-2, *A. afarensis* primitively retains the long, narrow palate shape (as measured internally) common to all *Australopithecus* species. The palate is more parallel-sided than in later *Australopithecus*, but not as much as in the earlier *A. anamensis*.

The A.L. 444-2 palate is shallow and flat between orale and staphylion. This is the usual condition for *A. afarensis* palates, which differ from the deeper palates of later *Australopithecus* species and *Homo*, in which the premaxillary and maxillary palatal planes are set at a strong angle to one another. The palatal morphology of one new Hadar maxilla, A.L. 417-1d, resembles this more derived configuration.

The main division of the maxillary sinus floor is posterior in A.L. 444-2 and other *A. afarensis* maxillae, as it is in *A. africanus* and the great apes. In contrast, the main sinus floor divide is usually anterior in robust *Australopithecus* and *Homo*.

The reconstructed shape of the joined nasal bones in A.L. 444-2 and A.L. 417-1d is that of an hourglass, which is also the shape of the nasal bones in the A.L. 333-105 *A. afarensis* juvenile. No known *A. afarensis* cranium has the inferiorly tapering nasal bone shape often encountered in robust *Australopithecus* species.

Despite the apparent massiveness of the zygomatic bone in A.L. 444-2, and the thick, liplike, heavily scarred origin site for the masseter muscle, the body of the bone is actually quite thin anteroposteriorly. Its vertical temporal surface bears a distinctive central hollow, which is also present in other *A. afarensis* zygomatics, marking the thinnest area of the bone.

The corpus of the A.L. 444-2 mandible is among the largest (in depth and breadth) in the *A. afarensis* hypodigm. However, the proportions of the corpus do not waver from the relatively slender average condition for *A. afarensis* mandibles. The substantially augmented Hadar mandible sample supports previous suggestions that corpus size and shape do not covary in this species.

In its straight, inferiorly bulbous external contour, the anterior mandibular corpus of A.L. 444-2 aligns with that of many other *A. afarensis* mandibles. However, this specimen, as well as several other newly recovered mandibles from Hadar, displays a more upright symphyseal axis than is typical of the *A. afarensis* hypodigm from the 1970s. The Laetoli type specimen LH 4 resembles some Hadar mandibles in its oblique symphyseal axis but is more similar to the mandibles of *A. anamensis* than to those of Hadar hominins in its curved, retreating external anterior corpus profile.

Where preservation permits assessment of their dimensions, the upper and lower teeth of A.L. 444-2 fall at or near the high end of the size range for the *A. afarensis* hypodigm. However, additions to the Hadar hominin assemblage since the 1970s do not significantly affect previously documented statistical profiles (mean and standard deviation) for tooth size and shape.

A. afarensis maxillary canines tend to be strongly asymmetric in lingual and labial views, with mesial and distal crown shoulders positioned at different heights relative to the crown apex. Laetoli upper canines are slightly more symmetric, but not as symmetric as in *A. africanus* teeth, which resemble all later hominins in this respect.

Laetoli and Hadar site samples of *A. afarensis* differ in the shape of the lower third premolar. The mesiodistally expanded Laetoli teeth are easily distinguished in the context of the now substantially enlarged Hadar P_3 sample size. The sole Maka P_3 stands with its Hadar homologs in this respect.

Taxonomic and Phylogenetic Status of *A. afarensis*

More than 22 years ago *A. afarensis* was characterized as a "basal, undifferentiated" species that "retain[ed] hints of a still poorly known Miocene ancestor" (Johanson and White, 1979: 328, 325), perceptions built chiefly on the distinctly apelike morphology pervading the hominin teeth and jaws from Hadar and Laetoli. Indeed, little in the then known remains of *A. afarensis* countered the impression that this species was perfectly primitive relative to geologically younger hominin taxa. Although few commentators have contested this characterization of *A. afarensis* crania, teeth, and jaws as prevailingly plesiomorphic vis-à-vis those of later hominin species, arguments concerning the taxonomic homogeneity of the species' hypodigm and, perforce, its phylogenetic affinities have been aired. The new fossil material from Hadar is directly relevant to several of these taxonomic and phylogenetic issues.

In this chapter we revisit the taxonomic and phylogenetic status of *A. afarensis* in light of the A.L. 444-2 skull and other new cranial, mandibular and dental specimens recovered at Hadar since the 1970s. We direct our discussion toward answering three questions:

1. How do the new Hadar fossils affect the debate over the taxonomic unity of the *A. afarensis* hypodigm?
2. Does the new material affect the characterization of the *A. afarensis* skull and dentition as wholly primitive relative to later hominin taxa?
3. How does the new material contribute to the debate over the phylogenetic relationships among early hominin species?

Taxonomic Unity of *A. afarensis*

The Hadar hominin sample spans an impressive range of size and morphology, a fact that has led a number of investigators to the conclusion that *A. afarensis* should be subdivided into more than one species-level taxon (e.g., Olson, 1981, 1985; Senut, 1983; Zihlman, 1985; Falk, 1988; Groves, 1989). The multiple species taxonomic solution for the Hadar sample has always been tenuous, however, because in anatomical regions where morphological (usually qualitative) distinctions have been drawn, sample sizes have been very small (e.g., femur and cranium), but where sample sizes are adequate and excessive amounts of size variation have been claimed (e.g., elements of the dentition and mandible corpus), it has been possible to show that whereas size variation in the hypodigm approximates the upper limit of size variation within the largest, most dimorphic of the extant hominoid species (gorilla and orangutan), it does not exceed those limits with statistical significance (e.g., Cole and Smith, 1987; Kimbel and White, 1988b; Lockwood et al., 1996). Hinging on the outcome of such nuanced taxonomic questions are decisions about the degree of sexual size dimorphism attained by the Hadar hominins, and, in turn, inferences about life-history characteristics permitted by those decisions (e.g., Lovejoy, 1981; Lovejoy et al., 1989).

Further complicating the issue of the taxonomic unity of *A. afarensis* is the fact that the aggregate time depth of the Hadar and Laetoli samples is on the order of 700,000 years. It is axiomatic in paleontology that time provides a "vector" of variation that has no homolog in the neontological realm, where geography and sexual dimorphism are the major factors in skeletal variation.

Time is implicated in the production of morphological variation in at least two important ways. First, because species taxa constitute lineages, morphological change can accumulate over the duration of a species' temporal span (anagenesis). Whether there is evidence of anagenetic morphological change and how significant it is in relation to the amount of morphological change associated with the origin of a lineage (cladogenesis) are empirical questions that can be posed meaningfully only in cases in which the fossil record is densely sampled over a well-calibrated stratigraphic record devoid of major gaps. An example of this type of variation is the increase in size of the *A. afarensis* mandibular corpus within the Hadar Formation, which we have discussed (see also Lockwood et al., 2000).

The second temporal source of morphological variation we may call "phylogenetic polymorphism," which is the presence in a species of more than one state of a character due to the phylogenetically "transitional" status of the character. In essence, the variation observed is a snapshot of a character in the process of a state change, and the presence of the polymorphism in a species is a function of the species' locus on the phylogenetic tree in relation to that process. An example of this type of variation is the P_3 morphology of the Hadar hominins. Within the Hadar sample we find "bicuspid" P_3s with a well-developed lingual cusp (metaconid) and relatively closed anterior fovea, which are derived states, and "unicuspid" P_3s with only a "swollen lingual ridge" (White et al., 1981: 459) emanating from the protoconid apex and a more open anterior fovea, which are primitive states. These morphologies do not map onto different size groups within the sample (or any other morphological groupings, for that matter), so the alternate expressions of P_3 morphology do not appear to be a function of sexual dimorphism. Not implausibly, the two P_3 states have been used to divide the Hadar sample taxonomically (e.g., Olson, 1981; Coppens, 1983). However, an equally valid suggestion is that *A. afarensis* may be expected to express both states if the C/P_3 complex was changing (under the influence of natural selection) from the ancestral sectorial form to the derived nonsectorial form during middle Pliocene hominin evolution (see Greenfield, 1990). This assumption is borne out by three observations: (1) the P_3 is entirely unicuspid in the earlier, potentially ancestral *A. anamensis* (Ward et al., 2001); (2) the P_3 is bicuspid in all hominins appearing in the fossil record subsequent to *A. afarensis* (White et al., 1981); and (3) although demonstrably apelike in form, the unicuspid *A. afarensis* P_3 was not functionally sectorial (i.e., there is little or no honing and occlusal wear is largely apical; Greenfield, 1990). If the C/P_3 complex was undergoing a selection-driven transformation, we may expect to see a paleontological pattern of vectored change in the relative frequencies of the ancestral and derived character states over time, with ultimate fixation of the derived state. It also implies that we may expect to find evidence of frequency change within the *A. afarensis* lineage itself. Although there is no evidence for such change within the Hadar sequence, in the time interval between Laetoli and Hadar the P_3 does become significantly less mesiodistally elongated (Lockwood et al., 2000), which is also a derived hominin condition. Thus, "phylogenetic polymorphism" can explain attenuated intraspecific variation in a character whose primitive state subsequently becomes lost in sister taxa, with attendant reduction or loss of variation.[1]

Statistical Assessment of Quantitative Variation

Armed with the large samples of Hadar hominin mandibles and teeth acquired since 1990, the one-species hypothesis was retested by Lockwood et al. (2000), who were able to assess the contribution temporal change makes to the overall profile of metrical variation within *A. afarensis*. Using nonparametric rank correlation and randomization tests to assess statistical significance, they found that a trend toward larger mandible corpus size (whether measured as corpus breadth or corpus height, or the geometric mean of the two)—focused on the upper Kada Hadar Member sample (ca. 3.0 Ma) that includes A.L. 444-2—contributed significantly to the high level of variation in this parameter relative to that observed in samples of extant great ape mandibles. This result is underscored by the fact that when the mandible sample was treated as a single analytical unit, corpus breadth and height were found to be significantly more variable than in large reference samples of chimpanzees and gorillas, unless the geologically younger sample was removed from the analysis (Lockwood et al., 2000).

Clearly, a failure to take into account the effect of anagenetic change on size variation could lead to a rejection of the null hypothesis that *A. afarensis* mandibular corpus size varies as expected for a single extant hominoid species, and thus, inappropriately in our view, to the conclusion that two species are represented in the mandible sample. The fact that neither mandible corpus shape nor other aspects of corpus morphology consistently distinguish the late sample of large mandibles from those of earlier (ca. 3.4–3.2 Ma) Hadar levels is consistent with the interpretation of this variation as occurring within a single-species lineage.

Significant temporal trends were also observed by Lockwood et al. (2000) in the P_3 (mesiodistally longer), the M^3 (larger), and the lower canine (labiolingually broader), although, in contrast to the mandible, these dental trends did not significantly affect the within-sample level of size variation for these parameters, which was not excessive relative to that found across the spectrum of hominoid comparative samples. In an examination of dental size coefficients of variation using bootstrapping methodology, Lockwood et al. (2000) found the number of rejections of the one-species hypothesis ($p < .05$) out of 24 comparisons (length and breadth of upper and lower C–M3) to depend on the comparative sample employed as a reference, with a range of rejections varying between 10 and 0. The fewest rejections involved comparisons with gorillas (2 of 24), orangutans (1 of 24) and a pooled two-species sample of bonobos and chimpanzees (0 of 24). Lockwood et al. (2000: 46) concluded that "levels of variation within *A. afarensis* best compare to those of the most sexually dimorphic of living hominoids, gorillas and orang-utans" and that the observed temporal trends were reasonably interpreted as anagenetic (i.e., intraspecific).

While the Lockwood et al. (2000) study supports the taxonomic unity of the *A. afarensis* hypodigm, it qualifies previous claims of morphological stasis within the species (Johanson and White, 1979; White et al., 1993; Kimbel et al., 1994; see below). These claims focused explicitly on the temporal stability of diagnostic morphological skull characters rather than on metric trends within a well-diagnosed species lineage. We take up the question of variation in qualitative morphological characters next.

Qualitative Assessment of Variation

The new skull material from Hadar adds information on the range of variation in some of the diagnostic characteristics of *A. afarensis*. Enlarged ranges of variation are observed in some characters of the temporal and occipital bones, as well as the maxilla and mandible.

Temporal bone. Previous characterizations of *A. afarensis* cranial base morphology, influenced chiefly by the A.L. 333-45 calvaria, emphasized the primitive nature of the mandibular fossa: "shallow, bounded by only a hint of articular eminence" (White et al., 1981: 456; see also Johanson and White, 1979: 323; Johanson et al., 1982a: 383; Kimbel et al., 1984: 374). Craniofacial fragment A.L. 58-22 was described as having a "shallow mandibular fossa" with a "low articular eminence" and "irregular, perhaps partly degenerative topography" (Kimbel et al., 1982: 455). The mandibular fossa of the new skull A.L. 444-2 expands the range of variation in this suite of features. Its articular eminence, while certainly quite low by comparison to *A. boisei* or *Homo*, is strongly delimited and set off from the roof of the mandibular fossa, which adopts the form of an anteroposteriorly constricted groove. This contributes to a greater topographic disparity between the summit of the eminence and the floor of the fossa than in other *A. afarensis* temporal bones (i.e., A.L. 166-9 and A.L. 333-45).

The chimpanzee-like horizontal orientation, tubular sagittal cross section, and undifferentiated crista petrosa of the tympanic element has also played a prominent role in conceptions of a primitive *A. afarensis* cranium (Johanson and White, 1979: 323; White et al., 1981: 456; Kimbel et al., 1984: 374). In A.L. 444-2 the crista petrosa remains poorly defined and the tympanic cross section is tubular (as opposed to the sharp crista petrosa and cone-shaped cross section of *A. africanus* tympanics), but the tympanic element as a whole is more upright than in A.L. 166-9 or A.L. 333-45. This tympanic element does approximate the orientation of that in temporal fragment A.L. 333-84, however.

The overall effect of the A.L. 444-2 skull is to shift the mean expression of these cranial base attributes toward the derived spectrum of the *Australopithecus* morphocline. While it remains the case that the *A. afarensis* temporal bone is more primitive than that of *A. africanus*, the question naturally arises whether the augmented variation signals taxonomic diversity. The answer to this question is not straightforward, however, as it is complicated by the evidence for the temporal trend toward larger skull size, which was adduced by Lockwood et al. (2000). A comparison of chimpanzees and gorillas may be illustrative. As we noted in the discussion of the temporal bone, in the gorilla the topography of the enlarged glenoid region is accentuated by a higher articular eminence and a more upright tympanic element than in the chimpanzee. Although the cause of this difference needs to be investigated thoroughly, one plausible suggestion is that this increase in joint complexity is related to the larger body size and joint cavity size in the gorilla. Although the differences between these hominoids are interspecific, differences due to an anagenetic trend toward larger size might be expected to mimic those correlated with an interspecific "static" difference in body size. Further fossil evidence will need to be accumulated to investigate this issue satisfactorily.

Occipital bone. The collection of new hominin occipital bones from Hadar introduces new morphological variation into the sample while also reiterating the diagnostically primitive morphological pattern previously described for this anatomical region of the *A. afarensis* cranium. The range of variation in the occipital bone sample is fully encompassed by new undoubtedly male specimens A.L. 439-1 and A.L. 444-2, both of which derive from the youngest *A. afarensis*–bearing stratigraphic horizons in the Hadar Formation, rendering the temporal trend toward larger size irrelevant to the interpretation of their differences. In A.L. 439-1 the nuchal plane is rounded sagittally, steeply angled to the horizontal, and separated from the occipital plane by a posterosuperiorly protruding compound T/N crest. This is the apelike morphological pattern described for all of the occipital specimens from the 1970s Hadar sample (A.L. 162-28, A.L. 333-45, and the uncrested A.L. 288-1a), as well as for KNM-ER 2602 from Koobi Fora, Kenya (Kimbel et al., 1984; Kimbel, 1988). The morphological details discernible on the small occipital fragments A.L. 224-9, A.L. 427-1b, and the subadult A.L. 444-1 agree with this pattern. In contrast, A.L. 444-2 possesses a much more horizontal, relatively flat nuchal plane from which the compound T/N crest is vertically suspended inferiorly.

The degree of difference between these two Hadar specimens is matched by differences long known to characterize the *A. boisei* sample, as seen in a comparison of the occipital bones of the adult male crania KNM-ER 406 and OH 5 (e.g., White et al., 1981; Kimbel et al., 1984; Wood, 1991b; Brown et al., 1993). The latter specimen shows a horizontal nuchal plane with an inferiorly suspended nuchal crest, contrasting with the steeper nuchal plane and upwardly directed nuchal crest in the Koobi Fora cranium (and also, apparently, the Konso specimen KGA 10-525, judging from the photographs in Suwa et al.,

1997). Very few observers think that these crania belong to distinct species, and despite the enhanced range of variation resulting from the addition of the A.L. 444-2 cranium to the Hadar sample, taxonomic splitting does not appear to be warranted on these grounds.

Maxilla. Notwithstanding overall size differences, maxillary morphology in A.L. 444-2 and other new additions to the Hadar sample (A.L. 413-1, A.L. 417-1d, A.L. 427-1a, A.L. 442-1, and A.L. 486-1) is similar to that encountered in the 1970s sample of *A. afarensis* (A.L. 199-1, A.L. 200-1a, A.L. 333-1, and A.L. 333-2), especially in terms of the posteriorly positioned zygomatic process, the well-developed canine fossa, the prognathic, biconvex subnasal (nasoalveolar) region, and the distinctive complex of primitive nasal cavity relationships described above. However, when compared to the rest of the sample, the palate of A.L. 417-1d is very deep and its premaxillary portion is more strongly flexed inferiorly at the palatal incisive foramen (the palate depth range is 5.5 mm). As discussed above, this creates a vaulted coronal cross section and a distinctly flexed midsagittal cross section that is unusual for *A. afarensis* palates (Johanson and White, 1979: 323; White et al., 1981: 452; Kimbel et al., 1984: 374).

Ranges of variation in palatal depth within modern human populations can reach at least 10 mm (de Villiers, 1968), and a similar relative degree of variation characterizes the *A. africanus* palate sample as well. Unfortunately, due to breakage or distortion, the *A. africanus* sample includes very few measurable palates, but a comparison of the visibly shallow Sts. 53 and MLD 9, on the one hand, and the much deeper Sts. 5 and Sts. 52a, on the other, reveals a considerable variety of palate depths within this species—certainly at least as much as within the Hadar hominin sample, although the *A. afarensis* palate is still, on average, shallower than that of the South African taxon. It is also noteworthy that the shallower palates of *A. africanus* are the same ones in which the flexion of the premaxillary portion is weak. Thus, while Sts. 53 and MLD 9 exhibit weak anterior flexion, in Sts. 5 and Sts. 52a the flexion is marked. The A.L. 417-1d palate likewise exhibits the combination of relatively great depth and anterior flexion. Thus, while this Hadar specimen hints at a taxonomically distinctive pattern of morphology, examples from the *A. africanus* hypodigm prevent an unequivocal finding of taxonomic diversity within the Hadar cranial sample. The large number of craniofacial similarities held in common by A.L. 417-1d and other Hadar maxillae of *A. afarensis* reinforce this conclusion.

Mandible. Apart from the larger corpus size of the temporally late mandibles in the Hadar Formation, the newly acquired Hadar hominin mandibles replicate the morphological pattern that was well documented in the 1970s sample from the site (White et al., 1981; White and Johanson, 1982), and thus there is little in them to suggest taxonomic diversity within the Hadar sample. Several attributes of the geologically late mandibles can be linked to large corpus dimensions, such as the slightly higher average position of the mental foramen (due to a deeper basal segment of the corpus) and a wider extramolar sulcus in some of these jaws (A.L. 438-1g and A.L. 444-2). Although these large late mandibles also tend to have more vertically inclined symphyseal cross sections than was usually seen in the 1970s Hadar sample, this apparently does not relate to their temporal position, as other recently recovered mandibles from earlier Hadar time horizons also have relatively vertical symphyseal axes (e.g., A.L. 417-1a from the Sidi Hakoma Member and A.L. 620-1 from the Denen Dora Member). The new finds create a more derived average symphyseal orientation for the Hadar *A. afarensis* sample.

Hadar and Laetoli Site Samples: Do They Represent the Same Hominin Species?

In response to questions concerning the validity of pooling the Hadar and Laetoli samples into a single hominin species (e.g., Tobias, 1980), White (1985) addressed the dental metrical and morphological differences between these site samples. Although he identified differences in postcanine tooth size between the collections, most such cases were attributed to size bias in the much smaller Laetoli sample. White (1985: 149) concluded that "the dental element-by-element comparison . . . reinforces this pooling, as surprisingly few metric or morphological characters are found to differentiate the site samples. The total range of variation that results from a pooling of Hadar and Laetoli teeth is fully consistent with ranges universally documented for other early and modern hominoid taxa."

Suwa (1990) subsequently investigated premolar morphology in detail and described the more primitive form (asymmetric, buccal dominance of the crown) of the Laetoli maxillary P^3 sample in relation to that from Hadar. He also noted that the derived, "molarized" morphology of the Hadar A.L. 333 maxillary and mandibular premolars is primarily responsible for this intersite difference, and he suggested that the divergent A.L. 333 premolar morphology might indicate either an evolutionary/taxonomic distinction or a sampling artifact. There has been little follow-up to these studies, or comparative investigations of nondental elements, due mostly to the frustratingly small sample from Laetoli.

Interest in the relative evolutionary positions of the Laetoli and Hadar samples has been rekindled with the discovery of *A. anamensis*, a ca. 3.8 to 4.2 Myr-old species whose mandibular, maxillary, and dental anatomy is significantly more plesiomorphic than that of *A. afarensis*

(M. Leakey et al., 1995, 1998; Ward et al., 1999, 2001). Leakey et al. (1995: 568) state that "*A. anamensis* can be readily distinguished from the younger Hadar sample, but has closer affinities with the older Laetoli specimens," and they provide the following list of dental similarities between *A. anamensis* and the Laetoli sample: "very large size of the canines, the accessory distal cuspules of the lower canines, the exaggerated flare on the molars, the tapering of the upper molars, and the vertically implanted upper canines." Elsewhere in the same article, Leakey et al. draw attention to affinities between the *A. anamensis* maxilla KNM-KP 29283 and the Garusi I maxilla from Laetoli (the vertical implantation of the canine root and the more upright plane of the nasal aperture). Next, we reevaluate Garusi I in light of the new additions to the Hadar maxillary sample.

Garusi I and the Hadar Maxillary Sample

The Garusi I maxillary fragment preserves the only adult facial morphology in the Laetoli sample of *A. afarensis*. Although broken anteriorly, when viewed in median cross section, the premaxillary region is horizontally inclined, and the very shallow anterior and postcanine palatal surfaces are coplanar. On the facial surface, the nasoalveolar and nasocanine contours are distinctly angled relative to one another, the former jutting out beyond the substantial canine alveolus, whose labial plate comprises the latter. Thus far, Garusi I can be described as "typical" of *A. afarensis* maxillae. However, Leakey et al. (1995: 566) drew attention to ways in which they believe Garusi I resembles *A. anamensis* maxillae more than it does *A. afarensis* maxillae from Hadar: (1) its palate is narrower and shallower than the Hadar palates; (2) its (missing) canine was more vertically implanted, resulting in a more vertical nasocanine contour (sensu Kimbel et al., 1984), than in Hadar maxillae.

New Hadar specimens permit testing of these claims. Our data indicate that internal palate breadth measured at P^3 for Garusi I (cast measurement) and Hadar specimens A.L. 417-1d and A.L. 442-1 are essentially the same, at 24 mm. Accordingly, the Garusi I palate does not appear unusually narrow when compared to the Hadar *A. afarensis* sample. However, judging from the size of its premolar crowns and the diameter and height of its canine alveolus, Garusi I is probably male, whereas the two small Hadar specimens are almost certainly females. Thus, the Garusi palate may be narrow compared to Hadar homologs of similar dental size (e.g., A.L. 333-1, in which the internal palate breadth at P^3 is ca. 30 mm).

The absence of a distinct transition between the lingual wall of the alveolar process and the palatal surface at midline—the basis for the extreme shallowness of the Garusi I palate—is matched in Hadar specimens A.L. 200-1a, A.L. 333-1, A.L. 333-2, and A.L. 486-1. In other Hadar maxillae, the wall of the alveolar process is taller opposite the premolars, and, consequently, the transition to the palatal surface is more steeply graded (e.g., A.L. 417-1d). With respect to palate depth, the Garusi maxilla does not stand apart from the Hadar sample.

We also find little to distinguish Garusi I in the comparison of midfacial verticality. Although some Hadar maxillae show a more posteriorly inclined nasocanine contour (A.L. 199-1, A.L. 200-1a, A.L. 427-1a, and A.L. 486-1), many do not, showing instead a more upright rise of the lateral nasal region, similar to that in Garusi I and *A. anamensis* specimen KNM-KP 29283 (A.L. 333-1, A.L. 333-2, A.L. 417-1d, and A.L. 442-1).

The morphology around the lower part of the nasal aperture differs between Garusi I and the Hadar maxilla A.L. 200-1a, as previously noted by Puech et al. (1986). In fact, in all Hadar maxillae with the relevant area preserved, the bony surface immediately lateral to the nasal aperture is a thin, flat to slightly convex plate whose horizontal cross section tapers medially to a sharp crest, which is the lateral margin of the nasal aperture. The morphology of this region is not directly influenced by the canine root (even when this root is large, as, for example, in A.L. 333-1 and A.L. 444-2), which is a topographically distinct entity in the midface (e.g., A.L. 200-1a, A.L. 333-1, A.L. 417-1d, and A.L. 427-1a). Inferiorly, the lower margin of the nasal aperture in Hadar maxillae usually is a well-defined, sometimes raised ridge that demarcates the subnasal surface from the anterior floor of the nasal cavity. As described in Chapter 5, the nasal aperture in *A. afarensis* is bounded by a continuous rim constituting the thin lateral and inferior margins that meet in sharp inferolateral corners.

The Garusi I specimen does not share this morphological pattern. Here, the bony surface lateral to the nasal aperture is a stout column with a tightly curved horizontal cross section that passes medially into the nasal cavity without interruption by a distinct margin. In contrast to the solid anterior pillar in the face of *A. africanus* (Rak, 1983), however, the bony column alongside the nasal aperture in Garusi I is hollow, faithfully reflecting the (considerable) size of the missing canine root. Thus, in contrast to Hadar maxillae, the canine root in Garusi I directly shapes the morphology of the face lateral to the nasal aperture, as it does in many chimpanzees. At the lower margin of the nasal aperture in Garusi I the convex subnasal surface grades insensibly into the anterior floor of the nasal cavity; there is no hint of a marginal crest separating these surfaces from one another. Although some Hadar maxillae lack the raised marginal crest inferiorly, there is almost never a problem deciding where the subnasal surface ends and the nasal cavity floor begins. This decision is impossible in Garusi I. The nasal aperture/cavity morphology of Garusi I is very similar to that of *A. anamensis*.

In sum, Garusi I differs from the Hadar maxillary sample in its (relatively) narrow palate (at least in the premolar region) and the morphology of the nasal aperture. It does not appear distinct in palate depth (again, in the premolar region) or in the verticality of the midface. In the context of the small sample of *A. afarensis* maxillae known in the 1970s the significance of Garusi I's morphological distinctions was difficult to evaluate. However, with the enlarged Hadar sample and the discovery of the more plesiomorphic *A. anamensis* maxillae, these distinctions may be evolutionarily significant. We return to this subject below.

LH 4 and the Hadar Mandible Sample

In terms of corpus size, shape, cortical contouring, and dental arch form, LH 4, the type specimen of *A. afarensis*, closely resembles the Hadar mandibles attributed to this taxon (e.g., A.L. 400-1a). The morphology of the Laetoli mandible does differ from that of the Hadar sample in at least one important respect, however. In the Hadar mandibles that preserve the symphyseal cross section intact, as well as in the Maka mandible MAK-VP 1/12, the external contour is usually straight, though receding, with a filled out (slightly bulbous) basal segment. As shown in Chapter 5, this cross-sectional shape is independent of the inclination of the symphyseal axis. In contrast, the anterior outline of the symphyseal cross section in the Laetoli mandible is strongly convex and "cut away" inferiorly so that the most inferior point on the basal contour is more posterior, closer to the posterior boundary of the cross section, than in Hadar mandibles. While some Hadar mandibles hint at the Laetoli morphology (A.L. 207-13, A.L. 330-5, and possibly the incomplete A.L. 333w-60), none show it as fully developed as LH 4. Moreover, the LH 4 morphology closely resembles that observed in the several mandibles of *A. anamensis*, whose symphyseal cross section has been characterized by Ward et al. (2001: 334) as "unique among hominoids and fossil hominids in its smooth convexity across this contour." With its strongly inclined symphyseal axis and convex, retreating anterior corpus, LH 4 presents a morphological package so far unmatched in any Hadar mandible. These differences between Laetoli and Hadar adult mandible morphology are already detectable in early ontogenetic stages of jaw growth. In the LH 2 mandible (with M_1 erupted) the external symphyseal contour flattens and retreats basally (much as in LH 4), whereas in A.L. 333n-1, a recently discovered Hadar mandible (with M_1 unerupted), the external symphyseal surface is clearly vertical all the way down to the basal margin in the midline of the anterior corpus.

Laetoli Dental Sample

The Laetoli hominin dental sample, as small as it is, overlaps the Hadar dental sample in most comparable aspects of morphology and size. They are not identical, however (White, 1985). We have already discussed the significantly longer mesiodistal dimension of the Laetoli P_3s (see Lockwood et al., 2000), which in this respect tend to resemble the *A. anamensis* P_3s. In contrast, the occlusal morphology of the Laetoli P_3s is more like that of the Hadar sample: the metaconid is well developed in four of five known Laetoli teeth (White, 1985), whereas *A. anamensis* P_3s are uniformly unicuspid. Ward et al. (2001) have observed that, compared to Hadar homologs (A.L. 333x-3 and A.L. 400-1b), relatively unworn maxillary canines from Laetoli (LH 3 and LH 6) are more symmetrical in lingual view due to the similar cervicoincisal positions of the mesial and distal crown shoulders. In the Hadar teeth the mesial crown shoulder is much closer to the crown tip than is the distal one. The Laetoli teeth resemble the single known unworn upper canine of *A. anamensis*, KNM-KP 35839, in this feature (Ward et al., 2001; see Chapter 5 for further discussion).

In sum, the Laetoli hominin sample deviates from the Hadar sample in three important areas: the maxillary canine/lower third premolar, the circum-nasal region and nasal cavity, and the anterior mandibular corpus. In these respects the morphology of the Laetoli sample is prefigured in the temporally antecedent sample of jaws and teeth of *A. anamensis*. Because the snout has been the site of considerable morphological transformation during hominin phylogeny, we view the anatomy of the Laetoli sample as evolutionarily significant.

Could the Laetoli hominins belong to the same species as the Kanapoi and Allia Bay specimens that constitute the hypodigm of *A. anamensis*? Although the Laetoli sample size is small, we believe it is significant that the available specimens exhibit a morphology that is markedly similar to that found in the earlier *A. anamensis* material. Given their overlap with the Hadar fossils in other respects, however, the Laetoli hominins are morphologically and temporally intermediate between *A. anamensis*, on the one hand, and the Hadar *A. afarensis* sample, on the other. This finding can be interpreted in alternative ways:

1. The Laetoli sample may represent a hominin population that was in a phyletically intermediate position on a lineage connecting *A. anamensis* to *A. afarensis*.
2. The Laetoli hominins sample a population that exhibits intermediate states in some characters but represents a distinct clade in a sister-species relationship with *A. afarensis* and/or subsequent taxa.

The phylogenetic analyses of the relevant fossils that might permit choosing between these alternatives have yet to be carried out, although, given the small samples currently available from Kanapoi, and especially Allia Bay and Laetoli, such resolution may not presently be feasible. While the recognition of species taxa is independent of, and logically prior to, phylogenetic analysis, we would be reluctant to divide the Laetoli and Hadar samples taxonomically until the results of detailed comparative studies are in hand. Nevertheless, we predict that the evidence of a phylogenetically more differentiated Hadar sample in relation to that from Laetoli will be of paramount importance in interpretations of mid-Pliocene hominin evolution.

Phylogenetic Position of *A. afarensis*

To investigate the phylogenetic position of *A. afarensis* in the hominin clade, we performed a cladistic analysis using (1) MacClade 4 (Maddison and Maddison, 2000) to enter data and visually examine patterns of character change and (2) PAUP 3.0s+1 (Swofford, 1991) to search for parsimonious arrangements of taxa. We employed 82 characters drawn from our comparative observations on the *A. afarensis* skull and dentition presented in this study, and assigned states to eight hominin species that comprised the taxonomic units in the cladistic analysis (*A. afarensis*, *A. africanus*, *A. aethiopicus*, *A. robustus*, *A. boisei*, *H. habilis*, *H. rudolfensis*, *H. erectus*). Outgroup taxa consisted of the extant chimpanzee and gorilla, which were treated as a single taxon for purposes of the cladistic analysis.[2]

We did not include *Ard. ramidus*, *A. anamensis*, *Kenyanthropus platyops* or *A. garhi* in this analysis. While this omission will undoubtedly result in an oversimplified set of results, too little information is available on the morphology of these taxa (due either to preservation bias or to lack of published details) to warrant their inclusion in an analysis that, to a significant extent, emphasizes the morphology of the skull as a whole. Published information on *Ard. ramidus* (White et al., 1994) includes glimpses of basal hominin dental and cranial base morphology, but does not (so far) appear to influence the hypothesized relationship(s) of *A. afarensis* to subsequent taxa. The fully described sample of *A. anamensis* remains (Ward et al., 2001) is dominated by mandibles and teeth whose morphology is almost entirely plesiomorphic in characters that distinguish it from that of *A. afarensis*. The morphology of the Kanapoi *A. anamensis* maxilla has intriguing implications for the status of the Laetoli hominins (see above), but a more complete representation of *A. anamensis* cranial morphology is necessary before this taxon can perform a useful role in comprehensive phylogenetic analyses. In any event, as with *Ard. ramidus*, there is nothing in the known morphology of *A. anamensis* that would alter the position of *A. afarensis* relative to later hominins.

The omission of *A. garhi* and *K. platyops*, whose type specimens are crania with associated teeth bearing significant amounts of morphological detail, is more egregious. Both taxa have been hypothesized as potential ancestors to one or more species of early *Homo* (Asfaw et al., 1999; M. Leakey et al., 2001). While the test of these hypotheses must await full publication, potential implications for the role of *A. afarensis* in hominin phylogeny are significant. Thus, we view our results as tentative and liable to revision and consider our analysis to be a starting point for future, more inclusive, examinations of hominin phylogenetic relationships.

Of the 82 characters extracted from our comparative observations, 2 represent the skull as a whole, 6 are from the mandible, 62 are from the cranium, 3 are from the endocast, and 9 are from the dentition. Of the 62 cranial characters, 37 represent the calvaria (cranial vault and base), and 25 represent the face (including the supraorbital region). Table 6.1 provides a complete list of the characters and character states used in the analysis.

The arguments for and against the use of computer-based cladistic analysis have been well rehearsed, and we are consequently under no illusions about the strength and weaknesses of this method of reconstructing phylogeny, especially at the species level, where homoplasy is known to exert a significant confounding influence on tree resolution. The realization that high levels of hard-tissue homoplasy make it difficult to reconcile morphology-based with highly corroborated molecule-based phylogenies of the catarrhine primates have led some investigators to suggest that cladistic analysis is unlikely to provide reliable estimates of hominin phylogenetic relationships, which depend entirely on skeletal attributes (e.g., Collard and Wood, 2000; but see Deane and Begun, 2002). Others have suggested that either functional or developmental linkage (or both) among characters renders parsimony-based cladistic analysis based on morphological details untenable in principle (e.g., Asfaw et al., 1999; Lovejoy et al., 1999; McCollum and Sharpe, 2001). As has been pointed out many times, cladistic analysis assumes that characters contribute independent information about descent to a parsimonious arrangement of taxa. If characters change states together due to functional or developmental linkage, then the assumption that the characters are mutually reinforcing signals of common descent is violated.

The problem of character correlation, of course, is not a problem only for cladistic analysis; it is a problem for many aspects of comparative biology. Lately it is attracting much attention in human evolutionary studies. We wrote about this issue in relation to the use of phylogenetic criteria in species identification (Kimbel and Rak, 1993: 471) and mentioned cases (such as the asterionic region of the hominoid cranium) in which "sound biologi-

Table 6.1 Characters and States Used in the Cladistic Analysis[a]

Character No.	Character/States
1	Rel. mandible corpus depth 0: shallow 1: deep 2: shallow (short face ht)
2	position of gnathion 0: posterior 1: anterior
3	ant. calvarial contour 0: deviates (torus) 1: deviates (no torus) 2: aligned
4	post. calvarial contour 0: aligned with circle 1: deviates from circle
5	rel. height nuchal area 0: high 1: low
6	rel. height vertex 0: moderate 1: low 2: high
7	rel. foramen magnum position 0: posterior 1: intermediate 2: anterior
8	lower facial profile 0: angled 1: aligned
9	rel. palate projection 0: high 1: intermediate 2: low 3: low (pal reduced)
21	infraorbital plate orientation 0: vertical 1: sloped 2: extremely sloped
22	infraorb. topography 0: flat 1: maxillary trigon 2: nasomaxillary basin
23	canine fossa 0: present 1: groove 2: fossula 3: absent
24	subnasal contour 0: biconvex 1: straight 2: straight, guttered
25	interorbital width 0: broad 1: narrow
26	calvarial height/breadth 0: low 1: very low 2: high
27	rel. base width 0: narrow 1: wide
28	post. base shape 0: moderate 1: squat 2: elongate
29	sag+cmpd t/n crest, m=f 0: yes
41	posterior temporal squama 0: flattened 1: vertical
42	temporal pneumatization 0: extensive 1: reduced
43	zygomatic proc. root 0: undivided 1: divided
44	max. lat. projection mastoid 0: high 1: low
45	mastoid tip position 0: anterior 1: posterior
46	mastoid face 0: single posterolateral 1: discrete post and lat 2: single lateral
47	digastric notch 0: confined 1: posteriorly extended
48	mandibular fossa depth 0: shallow 1: intermediate 2: deep
49	articular eminence summit 0: posterior 1: anterior
50	articular eminence form 0: single plane 1: med-lat twisted
63	dental arch shape 0: posteriorly convergent 1: intermediate 2: posteriorly divergent
64	maxillary sinus division 0: posterior 1: anterior
65	zygomatic temp. surface 0: flat 1: deeply excavated
66	zygo. frontal proc. lat. margin 0: vertical 1: laterally divergent
67	superior orb. fissure 0: foramen 1: laterally extended
68	mand. corpus. robusticity 0: low 1: intermediate 2: high
69	mental foramen position 0: very low 1: low 2: intermediate 3: high
70	symphyseal axis 0: strongly inclined 1: intermediate 2: upright
71	ramus root ant. position 0: posterior 1: intermediate 2: anterior

10 rel. masseter projection
0: low
1: intermediate
2: high

11 midfacial prognathism
0: high
1: intermediate
2: low

12 nasoalveolar inclination
0: high
1: intermediate
2: low

13 position zygomatic angle
0: below orbit
1: at orbit
2: above orbit

14 shape of squamosal suture
0: flat
1: arched

15 squamosal suture overlap
0: low
1: high

16 rel. ht. squamosal suture
0: low
1: intermdiate
2: high

17 zygomatic arch alignment
0: sagittal
1: deviant

18 rel. masseter height
0: low
1: high

1: no
2: absent

30 position temp. emphasis
0: posterior
1: intermediate 1
2: intermediate 2
3: anterior

31 nuchal crest emphasis
0: lateral
1: medial
2: absent

32 venous sinus pattern
0: transverse
1: marginal

33 cerebellum position
0: not tucked
1: tucked

34 anterior pole shape
0: rounded
1: beaked

35 calvarial bone thickness
0: thin
1: thick
2: very thick

36 supraorbital corner
0: angled
1: rounded

37 supraorbital thickness gradient
0: med. to lat.
1: lat. to med.

38 supraorbital surface
0: undivided
1: divided

51 postglenoid process size
0: large
1: small

52 postglenoid/tympanic relation
0: separate
1: variable
2: merged

53 tympanic form
0: tubular
1: platelike

54 tympanic orientation
0: horizontal
1: vertical
2: posteriorly inclined

55 crista petrosa
0: absent-weak
1: moderate-strong

56 vaginal process
0: absent
1: present

57 lateral tympanic extension
0: medial to saddle
1: lateral to saddle

58 nuchal plane form
0: transversely convex
1: variable
2: flat

59 nasal cavity form
0: stepped
1: smooth

72 ramus root vert. position
0: high
1: intermediate
2: low

73 lateral corpus hollow
0: very strong
1: moderately strong
2: weak
3: absent

74 relative postcanine size
0: small
1: medium
2: medium-large
3: large
4: very large

75 canine size
0: large
1: reduced
2: small
3: very small

76 P/3 cusp number
0: one
1: one-two
2: two

77 postcanine cusp position
0: marginal
1: intermediate
2: convergent

78 P_3/ mb line ext.
0: always
1: frequent
2: rare
3: absent

(continued)

Table 6.1 *(continued)*

Character No.	Character/States			
			60 palate depth	79 upper canine lingual shape
19	bizyg. width (orbital plane)		0: shallow	0: asymmetric
	0: low	39 frontal squama form	1: intermediate	1: more symmetric
	1: intermediate	0: sulcus	2: deep	2: symmetric
	2: high	1: flat		
		2: concave	61 palatine process	80 dm/1 shape
20	nasal aperture margins		0: ± horizontal	0: bl narrow
	0: sharp or dull-large C	40 rel. postorbital breadth	1: steep posterior angle	1: bl broad
	1: sharp-small C	0: large		2: molarized
	2: dull-small C	1: small	62 palate shape	
	3: sharp, everted		0: narrow	81 dm1/ mesial profile
			1: broad	0: mmr absent, prd ant., fa open
				1: mmr slight, prd ant., fa open
				2: mmr thick, prd=md, fa closed
				82 occlusal wear
				0: strong ant. postcanine gradient
				1: weak ant. postcanine gradient

[a]The transformation of the following characters was treated as ordered in the analysis: 7, 9–13, 16, 19, 21, 30, 48, 60, 63, 69–77, 80.

cal judgment in the definition of characters" can lead to a degree of control over character correlation. Lieberman (1995, 1997, 1999, 2000) has extensively discussed the problem of developmental and functional linkages for character definition in phylogenetic analysis of skeletal morphology, and Lovejoy et al. (1999, 2000) have stressed the importance of the cascading phenotypic effects of altered gene expression in primordial tissue fields (such as the limb bud) for pinpointing the genetic source of adaptive and phylogenetic morphological change. However, for the argument against character segregation to have force, hypotheses of character correlation must be tested. Strait (2001) recently tested the independence of morphological characters in the hominin cranial base and found less correlation than predicted by hypotheses of structural and functional linkage. In addition, some key fossils show a mixture of characters that appears at odds with notions about tight functional integration in the hominin craniofacial complex—a phenomenon that may turn out to be more common than generally appreciated (see also Collard and Wood, 2000). The highly mosaic craniofacial morphology of the KNM-WT 17000 cranium of *A. aethiopicus* is one example, and the early *Homo* cranium KNM-ER 1470, whose facial skeleton combines putative, conflicting synapomorphies of both *Homo* and robust *Australopithecus* species, is another. The greatly expanded postcanine occlusal area combined with plesiomorphic maxillary anatomy in *A. garhi* cranium BOU-VP 12/130 (Asfaw et al., 1999), and of a small molar dentition with derived maxillary anatomy in *K. platyops* cranium, KNM-WT 40000 (M. Leakey et al., 2001), would appear to disconfirm the hypothesis that hominin megadonty is tightly linked functionally to apomorphic shifts in the position of the mandibular adductors (McCollum and Sharpe, 2001). Our own comparative observations of early hominin facial and basicranial regions, which form the body of this study, support the idea of significant independent variation across taxa in the components of classically construed functional cranial units. We thus share the view of Collard and Wood (2001), Strait (2001), and others that, in the absence of "hard" evidence for significant character correlation, there is little reason to be repelled by computer-based cladistic analysis as one method by which to infer the phylogenetic content of large data sets. As in any science, the results are amenable to testing, corroboration or refutation, and modification, if warranted.

Among the decisions facing those who elect to carry out a computer-based cladistic analysis is whether or not to "weight" characters relative to one another a priori, based on their presumed level of phylogenetic informativeness. For example, it is sometimes said that because the masticatory apparatus is involved in a vital function, selection, acting on fundamentally similar genetic backgrounds, is likely to have produced multiple independent examples of similar, if not identical, masticatory adaptations among closely related hominin species—and that therefore characters representing the masticatory apparatus are more likely to display homoplasy than, say, characters of the allegedly more conservative cranial base. According to such arguments, one might be persuaded to assign higher weight to cranial base characters than to masticatory characters, or to discount the validity of proposed cladistic relationships that rely heavily on masticatory characters. We are skeptical of this proposition, not least because it has yet to be demonstrated that the hominin masticatory apparatus is indeed more prone to homoplasy, and thus a less reliable guide to phylogeny, than the cranial base or other regions of the skull. In fact, a recent explicit examination of this question concluded that it is not (Collard and Wood, 2001). Therefore, in our analysis we assign equal weight to all characters irrespective of their anatomical region of origin.

Another, somewhat more subtle way to weight characters a priori is to order them, which is tantamount to specifying a model of change for characters with more than two states. Ordering a character imposes a model of linear change between successive character states; changes that "jump" intermediate states are more costly (i.e., the number of steps is equal to the absolute value of the difference between state numbers; Maddison and Maddison, 2000) than changes between adjacent states. For unordered characters, change between any two states is permitted without additional cost, irrespective of state number (i.e., all changes count as one step). In our analysis, we considered each character and employed the reasonable assumption of ordered change for some quantitative characters (e.g., heights and widths) and for characters for which there are states coded as "intermediate." This means that the transformation between extreme states requires passing through all intermediate states. In the end, we imposed ordered change on 24 of the 41 characters (59%) with three or more states; these characters are designated in Table 6.1. We compared results using our selected ordering with those obtained using "all ordered" and "all unordered" options. Although tree lengths are of course changed when these assumptions are selected, the topology of the most parsimonious trees does not.

Results

Our results substantiate the basal position of *A. afarensis* in hominin phylogeny (Figure 6.1; see Table 6.2 for distribution of character states). In every one of the top 25 parsimonious cladograms found using the branch-and-bound search algorithm in PAUP, *A. afarensis* is the sister taxon to the other ingroup taxa. This is the case whether characters are treated as selectively ordered, en-

Table 6.2 Distribution of Character States among Hominin Taxa

	1	2	3	4	5	6	7	8	9	10	11	12	13	14	15	16	17	18	19	20
Chimp/gorilla	0	0	0	0	0	0	0	0	0	0	0	0	0	0	0	0	0	0	0	0
A. afarensis	1	1	1	1	1	0	2	0	1	1	1	0	1	1	0	2	0	0	2	1
A. africanus	0	1	2	0	1	2	1	1	1	1	1	1	1	1	0	1	0	1	0	2
A. aethiopicus	?	1	1	1	1	?	1	1	0	2	0	0	2	1	1	1	0	1	2	2
A. robustus	1	1	1	?	1	1	2	1	2	2	2	1	1	1	0	?	0	1	1	2
A. boisei	1	1	1	1	1	1	2	1	2	2	2	1	2	1	1	2	1	1	2	2
H. habilis	?	1	0	0	1	2	1	0	3	1	1	1	1	1	0	1	0	0	0	3
H. rudolfensis	?	1	2	0	1	2	?	1	3	2	2	2	1	1	0	1	0	0	0	3
H. erectus	?	1	0	0	1	2	1	0	3	1	2	2	1	1	0	1	0	0	0	3

	21	22	23	24	25	26	27	28	29	30	31	32	33	34	35	36	37	38	39	40
Chimp/gorilla	0	0	0	0	0	0	0	0	0	0	0	0	0	0	0	0	0	0	0	0
A. afarensis	0	0	0	0	1	0	1	1	0	0	0	1	1	0	1	0	0	0	1	0
A. africanus	1	0	1	1	0	2	0	0	1	2	2	0	0	0	0	1	1	1	1	1
A. aethiopicus	2	1	2	1	0	0	1	1	0	1	1	1	0	?	0	1	1	0	2	1
A. robustus	1	1	2	2	0	?	1	?	1	3	1	1	1	1	0	1	1	0	2	1
A. boisei	2	2	3	2	0	0	1	1	1	3	1	1	1	1	0	1	1	0	2	1
H. habilis	0	0	0	0	0	2	1	2	2	2	2	0	0	0	0	1	1	1	0	0
H. rudolfensis	1	0	3	1	0	?	1	?	2	2	2	0	0	0	0	1	1	0	1	0
H. erectus	0	0	0	0	0	2	1	2	2	2	2	0	?	0	2	1	1	1	0	0

	41	42	43	44	45	46	47	48	49	50	51	52	53	54	55	56	57	58	59	60
Chimp/gorilla	0	0	0	0	0	0	0	0	0	0	0	0	0	0	0	0	0	0	0	0
A. afarensis	0	0	0	0	0	0	0	1	0	0	0	0	0	0	0	0	0	0&1	0	1
A. africanus	1	1	0	0	0	0	0	1	0	0	0	0&2	1	1	1	0	0	2	0	1
A. aethiopicus	0	1	0	1	0	0	?	0	0	0	0	0	1	1	1	1	0	2	1	1
A. robustus	1	1	1	1	0	1	1	2	0	0	1	2	1	1	1	1	1	2	1	2
A. boisei	1	1	1	1	1	2	1	2	1	1	1	2	1	1&2	1	1	1	2	1	2
H. habilis	1	1	0	0	0	0&2	0	2	0	0	1	2	1	1	1	1	0	2	0	2
H. rudolfensis	1	1	0	0	0	?	?	2	0	0	1	?	1	1	1	1	0	2	0	2
H. erectus	1	1	0	0	0	2	1	2	0	0	1	2	1	1	1	1	0	2	0	2

	61	62	63	64	65	66	67	68	69	70	71	72	73	74	75	76	77	78	79	80
Chimp/gorilla	0	0	0	0	0	0	0	0	0	0	0	0	0	0	0	0	0	0	0	0
A. afarensis	0	0	1	0	1	1	0	1	1	1	1	1	1	1	1	1	0	1	0&1	1
A. africanus	0	0	2	0	0	1	0	1	2	1	1	1	2	2	1	2	1	2	1	1
A. aethiopicus	0	0	2	?	0	1	1	2	?	2	?	?	3	?	?	?	?	?	?	?
A. robustus	1	0	2	1	0	0	1	2	2	2	1	1	3	3	2	2	2	3	2	2
A. boisei	1	0	2	1	0	1	1	2	3	2	2	2	3	4	3	2	2	3	2	2
H. habilis	0	1	2	1	0	0	?	1	2	2	1	1	2	1	2	2	0	?	1	1
H. rudolfensis	0	1	?	1	0	0	?	1	2	2	1	1	2	1	1	2	0	?	1	1
H. erectus	0	1	2	1	0	0	1	1	2	2	1	1	2	1	2	2	0	?	1	1

	81	82
Chimp/gorilla	0	0
A. afarensis	1	0
A. africanus	1	1
A. aethiopicus	2	?
A. robustus	2	1
A. boisei	2	1
H. habilis	?	1
H. rudolfensis	?	1
H. erectus	1	1

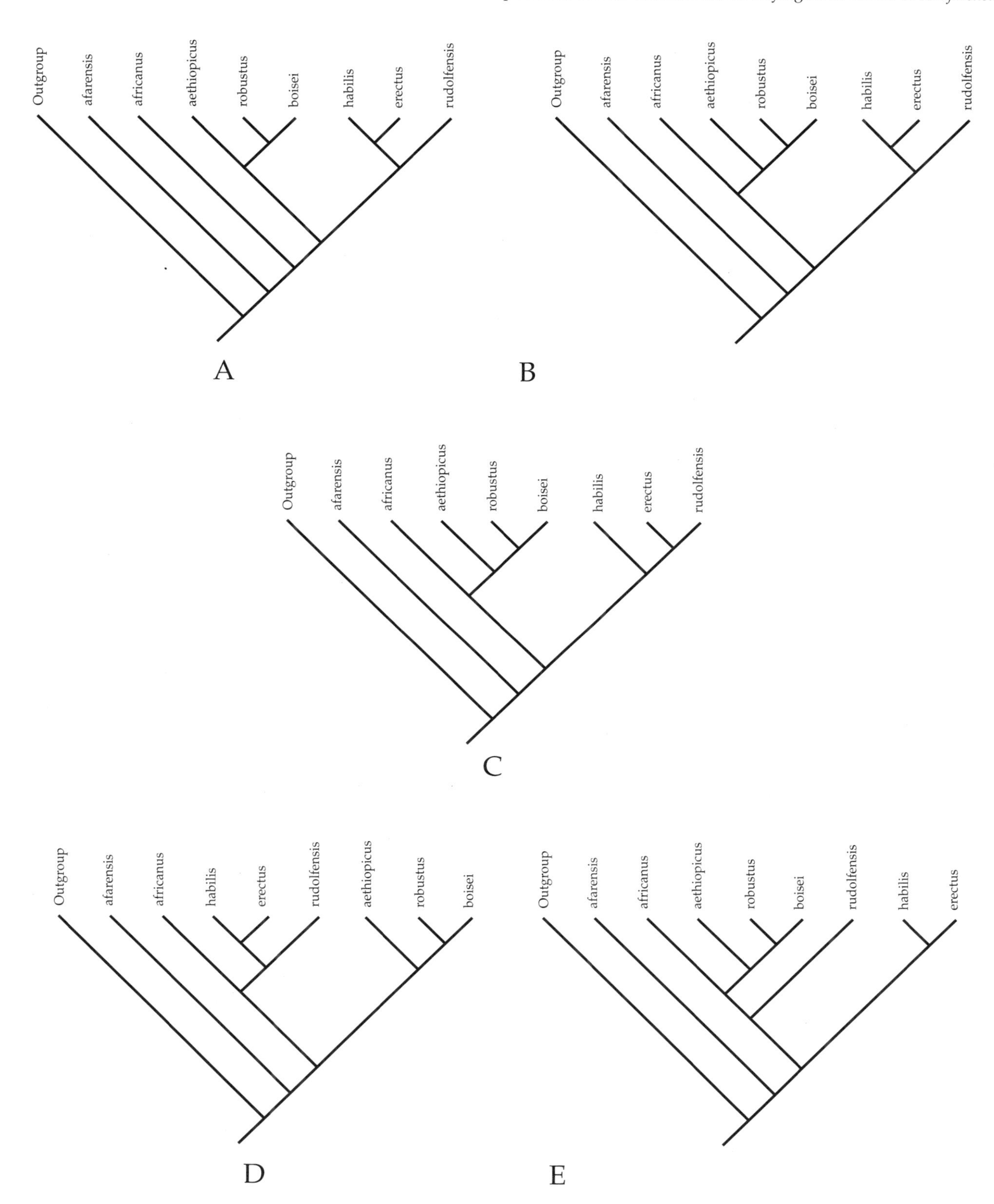

Figure 6.1 The five most parsimonious trees (A-E) found by PAUP's branch-and-bound algorithm. These trees differ by only four steps and a CI of 0.008. However, in each one, *A. afarensis* is basal to the other ingroup taxa included in the analysis.

tirely ordered, or entirely unordered. Two cladograms share the status as most parsimonious (Consistency Index [CI] = 0.739). One (A in Figure 6.1) locates *A. africanus* as the sister taxon to the remaining ingroup taxa, which are divided into monophyletic robust *Australopithecus* and early *Homo* clades. Within the robust clade, *A. aethiopicus* is sister to an *A. robustus* + *A. boisei* clade, and *H. rudolfensis* is sister to an *H. habilis* + *H. erectus* clade. In the other (B in Figure 6.1), *A. africanus* is the basal taxon of a monophyletic robust *Australopithecus* clade, which itself is the sister to the early *Homo* clade. The next three shortest cladograms (C–E) differ from the first two by only one or two steps (CI = 0.735–0.731). Cladogram C (Figure 6.1C) is similar to B except that *H. habilis* is sister to *H. rudolfensis* + *H. erectus*. The last two, cladograms D and E (Figure 6.1 D, E), are equally parsimonious: in D, *A. africanus* is sister to the *Homo* clade, within which *H. rudolfensis* is sister to *H. habilis* + *H. erectus*, whereas in E, *H. rudolfensis* is sister to a clade comprising *A. africanus* and the three robust species.

The strict consensus trees derived from the top three and five cladograms differ in only one respect. Although both of them retain *A. afarensis* as the basal hominin taxon and preserve the monophyly of *A. aethiopicus* + *A. robustus* + *A. boisei* (thus supporting the generic designation of this clade as *Paranthropus*), the strict consensus of the three most parsimonious trees also upholds the monophyly of the three early species of *Homo*, whereas the strict consensus of the five most parsimonious trees does not (Figure 6.2).

Implications for Phylogenetic Hypotheses

Status of A. afarensis

Our finding that *A. afarensis* is the sister taxon to all subsequent hominin taxa corroborates the results of most previously executed cladistic analyses (Skelton et al., 1986; Chamberlain and Wood, 1987; Skelton and McHenry, 1992; Lieberman et al., 1996; Strait et al., 1997; Wood and Collard, 1999). This result is supported by both qualitative (associated with the work of Skelton and colleagues, Strait and colleagues, and the present analysis) and quantitative (associated with Wood and colleagues) data sets. Our results find no support for the hypothesis that *A. afarensis* is sister to the robust *Australopithecus* species, as proposed by Olson (1981, 1985) and Falk (e.g., 1988), which drew on a small number of characters of the basicranium and the endocranial venous sinuses. Based on published evidence (Asfaw et al., 1999), we think that *A. afarensis* will most likely retain its basal position vis-à-vis *A. garhi*, but we are less certain about what to expect when *K. platyops* becomes better known.

Plesiomorphic skull characters of A. afarensis. The morphological attributes of the *A. afarensis* skull that are likely to have been present in the basal hominin species are weighted toward the maxilla, the inferior aspect of the temporal and occipital bones, and the pattern of cranial cresting (and its morphological correlates). The following character states are interpreted as primitive in *A. afarensis* but as derived in subsequent *Australopithecus* species:

1. Subnasal region convex in sagittal and coronal planes (independent nasocanine and nasoalveolar contours)
2. Strong canine fossa
3. Lowest point on root of zygomatic process located posteriorly, above M^1
4. Low, arched zygomaticoalveolar crest
5. Infraorbital plate with occasional weak transverse buttress
6. Palate narrow, usually shallow, with flat to moderately angled premaxillary segment
7. I^2/C diastemata common
8. Tympanic element with tubular cross section and absent or weak petrous crest
9. Temporal squama extensively pneumatized by mastoid cellularization
10. Occipital squama usually with steep, sagittally and transversely convex planum nuchale
11. Posteriorly accentuated sagittal crests and compound temporal/nuchal crests common in males and present in some females; concomitant horizontal flaring of parietal bone and temporal squama posterolaterally, with resulting development of asterionic notch sutural articulation
12. Barlike supraorbital torus with coronally oriented anterior margin and vertical thickness increasing laterally.

The cranial cresting pattern also describes *A. aethiopicus* (based on its only known cranium, KNM-WT 17000), while *A. garhi* is said to retain several of these plesiomorphic states, to the exclusion of subsequent *Australopithecus* species (Asfaw et al., 1999: Table 1; specifically numbers 1, 2, 4, and 6). Assuming these characterizations are borne out by further discoveries, the *A. afarensis* skull will appear less primitive in the context of post-3.0-Myr-old hominins than was previously the case.

Autapomorphies of the A. afarensis *skull.* The *A. afarensis* skull possesses very few apomorphic characters that are unknown in other early hominin species. The following states are probable autapomorphies:

1. Interorbital and nasal aperture breadths narrow (probably linked structurally)
2. Postorbital region of the frontal bone transversely broad
3. Temporal face of zygomatic bone deeply excavated

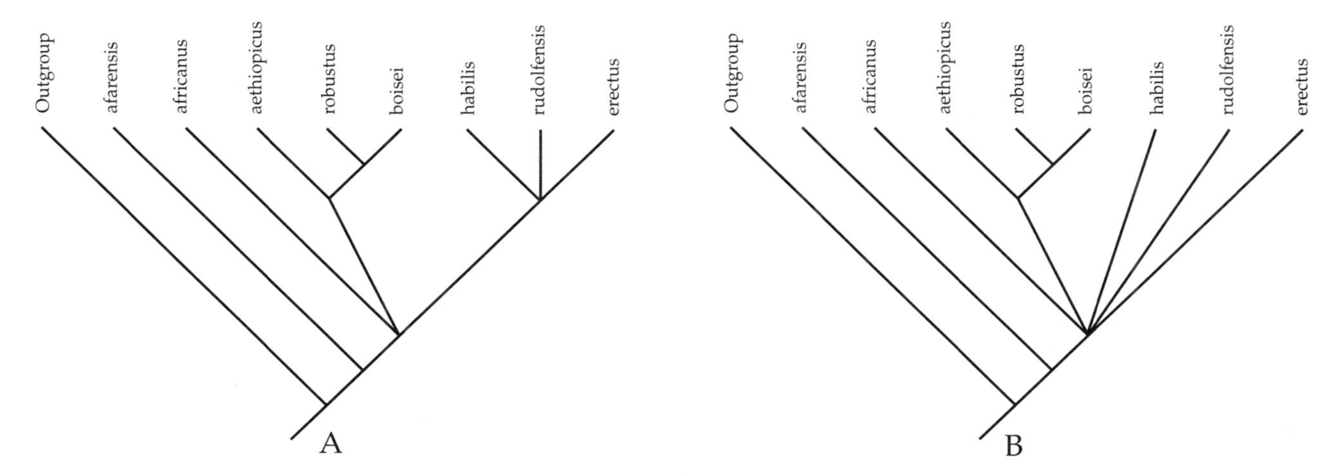

Figure 6.2 The strict consensus of the three (A) and five (B) shortest trees found by PAUP's branch-and-bound algorithm. In the consensus of the three shortest trees, the monophyly of the robust *Australopithecus* and *Homo* clades is preserved, but the latter is lost in the consensus of the five shortest trees.

4. Bones of the calotte (frontal, parietals, and occipital) generally thick

The A.L. 444-2 skull allows us to identify two unusual features of this species' cranium: the broad postorbital region and the very thick cranial vault bones. While a broad postorbital region is also characteristic of the genus *Homo*, we think it unlikely that this similarity is a synapomorphy indicating a unique relationship with *A. afarensis* (as our phylogenetic results confirm). Asfaw et al. (1999: Table 1) contend that *A. africanus* shares the same state of postorbital constriction ("moderate") as do *A. afarensis* and early *Homo*, but our data show that, whether gauged absolutely or relatively, *A. africanus* has a much narrower frontal.[3] Importantly, however, a broad postorbital region is also found in the 3.8-Myr-old Belohdelie (BEL-VP 1/1) frontal, which shares with *A. afarensis* the additional unusual character of high cranial vault thickness. With few exceptions, *Australopithecus* and earliest *Homo* crania have thin—sometimes exceedingly thin—vault bones, which is also the condition in the great apes. The combination of these unusual conditions in BEL-VP 1/1 strengthens its taxonomic assignment to *A. afarensis* (cf. Asfaw, 1987), notwithstanding its remote temporal location relative to the Hadar sample. Of course, if it turns out that the still unknown frontal bone of the more ancient *A. anamensis* also expresses these characters, then the assignment of the Belohdelie specimen may once again be equivocal; but that remains hypothetical at this stage of our knowledge.

Synapomorphic skull characters of A. afarensis *and all subsequent hominin taxa.* The new fossil sample from Hadar augments the list of apomorphic characters that *A. afarensis* shares with other hominin taxa, which implies that the current portrait of the *A. afarensis* skull is less apelike than the one based on the 1970s collections from Hadar and Laetoli (Johanson and White, 1979; White et al., 1981; Rak, 1983; Kimbel et al., 1984). The following apomorphies are shared with subsequent hominin species:

1. Anterior position of gnathion in the facial triangle (due to more upright anterior mandibular corpus)
2. Reduced upper midfacial prognathism (sellion–nasospinale angle)
3. Strongly arched squamosal suture
4. More coronal orientation of petrous element
5. Distinct mastoid tip with circumscribed digastric fossa
6. Rectangular (long sagittal dimension) parietal bone shape
7. Posteriorly located parietal eminence
8. Low position of inion in the midsagittal arc of the calvaria
9. Anteriorly positioned foramen magnum
10. Short anterior cranial base
11. Transversely thick mandibular corpus
12. Lower canine set more medial to P_3 in mandibular arch curvature
13. Vertical, anteriorly positioned mandibular ramus
14. Bicuspid P_3s common

The states for some, but not all, of these characters can be determined for either *Ard. ramidus* or *A. anamensis*. The cranial base traits (nos. 4, 5, 8, and 9) are unknown for *A. anamensis*, but the *Ard. ramidus* cranial base ARA-VP 1/500 shows that they had already evolved in the hominin clade by 4.4 Ma (White et al., 1994: Fig. 1). This raises the possibility that these states are derived at the base of the hominin clade even though *A. africanus* displays apparently more apelike conditions in the position of the foramen magnum and the orientation of the petrous element (Dean and Wood, 1982). The *A. anamensis* mandibles are described as apelike in corpus proportions (slender) and dental arch shape (canines are more lateral in a narrower arch) compared to *A. afarensis* (Ward et al., 2001), but both of these species show a derived reduction of upper midfacial prognathism.

Apomorphies of the sister group of A. afarensis. Apomorphies distinguishing *A. africanus*, robust *Australopithecus* and *Homo* from the last common ancestor they shared with *A. afarensis* span the skull and dentition:

1. Medial position of the nuchal crest's maximum development
2. Uniformly flat nuchal plane
3. Enhanced topography of the glenoid region of the temporal bone, especially of the platelike tympanic element
4. Reduction of temporal squama pneumatization
5. Posterolaterally receding supraorbital torus and lateral to medial vertical thickness gradient
6. Flattened or inflated topography of the mandibular corpus' lateral aspect
7. Higher position of the mental foramen on the mandibular corpus
8. Posteriorly divergent shape of the maxillary dental arch
9. More derived morphology of the upper and lower anterior premolars and the upper canine
10. More uniform postcanine occlusal wear gradient

Many of these characters have long been known to distinguish the plesiomorphic anatomy of the *A. afarensis* skull and teeth from that of other hominins (e.g., White et al., 1981; Rak, 1983; Kimbel et al., 1984), but the distinctive morphology of the supraorbital region, available for the first time in A.L. 444-2, is an important addition to the list.

Some features previously considered apomorphic for the sister group to *A. afarensis*—for example, the more horizontally oriented nuchal plane, the increased depth of the palate, and the uprightness of the anterior mandibular corpus—may not so clearly be unique to that clade in light of the polymorphism introduced by the recent Hadar discoveries described here, which have expanded the range of variation in these features to envelop the more derived end of their morphoclines. Nevertheless, the average conditions of these characters remain distinctive for the hypodigm.

Apomorphies of A. afarensis *and* A. africanus. Although a few workers have questioned the specific distinctiveness of *A. afarensis* vis-à-vis *A. africanus* (e.g., Boaz, 1979; Tobias, 1980), the discovery of A.L. 444-2 at Hadar and Stw. 505 at Sterkfontein make such a conclusion extremely unlikely. The latter specimen is the largest known cranium attributable to *A. africanus*, just as A.L. 444-2 is the largest known *A. afarensis* cranium; each has an endocranial capacity well in excess of 500 cc (on the endocranial capacity of Stw. 505, see Conroy et al., 1998; Lockwood and Kimbel, 1999). In its derived facial, frontal bone, temporal bone, and cranial crest morphology, the Stw. 505 cranium closely matches smaller *A. africanus* crania from Sterkfontein and Makapansgat (Lockwood and Tobias, 1999), thereby sustaining the morphological gap between the crania of this species and those of *A. afarensis*.

Our study reveals no synapomorphies in the skulls of these two species that are not also shared with other taxa. Where *A. afarensis* is apomorphic (e.g., short, wide cranial base; occipital-marginal venous drainage; anteriorly tucked cerebellar lobes), *A. africanus* is relatively plesiomorphic; where *A. africanus* is apomorphic (e.g., circumnasal anterior pillars; division of supraorbital element into distinct supraorbital and superciliary entities; loss of compound temporonuchal crest; platelike tympanic element with strong petrous crest; bicuspid mandibular third premolar), *A. afarensis* is relatively plesiomorphic. What the crania of these species do share is a relatively large number of plesiomorphies relative to those of robust *Australopithecus* and early *Homo* crania. The most obvious of these are a highly prognathic maxilla (also present in *A. aethiopicus*) and several plesiomorphic temporal bone features (e.g., wide tympanomastoid sulcus, absence of vaginal process, coronal offset of the tympanic element and the large postglenoid process). To these we may add the posteriorly divided maxillary sinus floor (Kimbel et al., 1997) and the foramen-like shape of the superior orbital fissure (Rak et al., 1996).

Homoplasic similarity between A. afarensis *and robust* Australopithecus. As we have noted, some accounts of *A. afarensis* cranial morphology have identified character states this species shares uniquely with one or more species of robust *Australopithecus* (Olson, 1981, 1985; Falk, 1988; see also Kimbel, 1984; Kimbel et al., 1984, 1985). Olson's (1981, 1985) study focused on the basal mastoid region of A.L. 333-45, whose strongly pneumatized mastoid process and coextensive occipitomarginal crest and suture, were thought by him to indicate apomorphic conditions shared with *A. robustus* and *A. boisei*. Kimbel et al. (1985) argued that Olson's polarity determinations were incorrect, that these states were in fact primitive, and that only in the marked inferior projection of the mastoid process tip (a hominin synapomorphy) did the mastoid morphology of A.L. 333-45 differ from that usually encountered in the great apes. The A.L. 444-2 skull reinforces these conclusions. Falk and Conroy (1983) and Kimbel (1984) described the pattern of cranial venous outflow as read from the calvarial remains of *A. afarensis*. The very high frequency of enlarged occipital-marginal (OM) sinuses observed in the 1970s Hadar cranial sample matched that of *A. robustus* and *A. boisei*, leading Falk and Conroy to infer a phylogenetic link among these taxa. Although Kimbel (1984) disputed this phylogenetic conclusion based on functional considerations, the recently enlarged Hadar sample confirms that the OM sinus system is fixed at a high frequency in *A. afarensis*. Whether the sharing of this character state by *A. afarensis* and robust *Australopithecus* species is a sign of descent from a common ancestor depends on the outcome of a phylogenetic analysis of all available characters. (It also depends on whether the OM sinus system is a derived or primitive state; inference from ontogenetic development suggests it could be primitive. See Kimbel, 1984.) Our more comprehensive analysis fails to support a direct phyletic link between *A. afarensis* and the robust species of *Australopithecus*. Therefore, derived similarities between the two groups must be interpreted as homoplasies.

Our study of A.L. 444-2 adds significantly to the list of nonhomologous similarities among *A. afarensis*, *A. aethiopicus*, *A. robustus*, and *A. boisei*:

1. Posterior "segmentation" of calvarial contour (deviation from circular contour)
2. Broad, short cranial base
3. Relatively short upper occipital scale (produced by a low lambda)
4. Foramen magnum placed far anterior on the cranial base
5. Occipital-marginal sinus system common
6. Cerebellum tucked under the occipital pole of the cerebrum
7. High percentage of total facial height occupied by mandible corpus height

None of these similarities appears to implicate the highly derived masticatory system of the robust *Australopithecus* species. Although the relatively deep mandible corpus is derived for hominins, in *A. afarensis* it is associated with a generally slender (plesiomorphic) corpus cross section, whereas the cross-sectional shape of the mandible body in robust *Australopithecus* is influenced chiefly by tremendous transverse corpus width, which is often interpreted as part of their specialized masticatory package. All of the other derived similarities in this list relate to the shape of the calvaria, especially of its posterior segment and base, and of the configuration of the posterior part of the brain endocast. Dean (1986) previously commented on the likelihood of a structural link between changes in the conformation of the posterior part of the brain (including the relative position of the cerebellum) and the cranial base. His discussion focused on a set of apparently derived similarities postulated to link *Homo* and robust *Australopithecus* species, but here we provide evidence from a more global phylogenetic analysis that they are liable to convergence. Nevertheless, if our results are correct, then the *Homo* and robust *Australopithecus* clades are more closely related to one another than to *A. afarensis*, and Dean's (1986) scenario (which also invokes synapomorphies of dental development) is a reasonable one for later Pliocene hominins.

Status of A. africanus

Although the many distinctions between the skulls of *A. afarensis* and *A. africanus* are unmistakable, the phylogenetic role of *A. africanus* nevertheless remains enigmatic. In the top five cladograms, this species shifts among three different positions, and it is, consequently, one of the least stable taxa in the analysis (as reflected also in the strict consensus cladogram). In the two shortest, equally parsimonious trees, *A. africanus* is either the sister taxon to all hominins other than *A. afarensis* (as suggested by Strait et al., 1997; see above) or it is the sister to the three robust *Australopithecus* species (as originally envisioned by Johanson and White, 1979; see also White et al., 1981; Rak, 1983; Chamberlain and Wood, 1987). Synapomorphies supporting the latter tree include the general increase in postcanine (especially molar) tooth size, the relatively high masseter muscle origin on the zygomatic, the frequent groovelike transformation of the canine fossa, and more convergent molar cusp apices—all features that emphasize the specialized aspects of the *A. africanus* masticatory system (White et al., 1981; Rak, 1983; Kimbel et al., 1984). Whichever of these hypotheses is more accurate, it is the case that the synapomorphies of the *A. africanus* skull and dentition overshadow the derived similarities between *A. afarensis* and the monophyletic robust *Australopithecus* group.

We do not attempt to draw a phylogenetic conclusion about *A. africanus* here, but we do point out that in recent years there has been a tendency to view the hypodigm as taxonomically diverse (Clarke, 1988, 1994; Kimbel and White, 1988b; Kimbel and Rak, 1993; Moggi-Cecchi et al., 1998; Lockwood and Tobias, 2002). However, a consensus has not developed on how, or even whether, the hypodigm ought to be divided (for a brief review, see Wood and Richmond, 2000), so the phylogenetic position of the hominins from Sterkfontein and Makapansgat (as well as Taung) will continue to be problematic. If there is taxonomic diversity within the hypodigm, morphological distinctions in the skull and dentition will be quite subtle.

Monophyly of the robust Australopithecus *species*

The monophyly of the three robust species—*A. aethiopicus, A. robustus*, and *A. boisei*—is well supported by our analysis (see also Strait et al., 1997). The fact that the strict consensus cladogram does not resolve the relationship between *A. africanus* and *A. aethiopicus* is due to the uncertainty surrounding *A. africanus*. In addition, because the consensus tree preserves a sister relationship between *A. robustus* and *A. boisei*, it is less likely that the hypothesis of a uniquely close relationship between the eastern African species *A. aethiopicus* and *A. boisei* is true; this alternative hypothesis was suggested after the discovery of KNM-WT 17000 (Walker et al., 1986; Walker and Leakey, 1988; Kimbel et al. 1988; Suwa et al., 1996; Asfaw et al. 1999). Likewise, there is no apparent support in this analysis for Skelton and McHenry's (1992) most parsimonious cladogram placing *A. aethiopicus* as the sister to *A. africanus, A. robustus + A. boisei*, and *Homo*, or for the shortest tree in Lieberman et al. (1996), in which *A. aethiopicus* is sister to an *A. robustus + A. boisei* clade, which itself is sister to a clade comprising *A. africanus + Homo*.

Although the apomorphic masticatory system of the robust *Australopithecus* species has traditionally formed the bulk of the evidence in favor of the monophyletic arrangement of these taxa, its presumed high adaptive value has focused attention on the likelihood of its having evolved convergently in separate East and South African robust lineages (e.g., Skelton et al., 1986; Strait et al., 1997; for a demonstration that the hominin masticatory system is unlikely to be more prone to homoplasy than the other cranial regions are, see Collard and Wood, 2001). While our results suggest that the masticatory apparatus is indeed apomorphic for the robust clade, the highly mosaic facial and palatal morphology of *A. aethiopicus* cranium KNM-WT 17000 confines the masticatory synapomorphies of the robust clade to the infraorbital and adjacent zygomaticomaxillary regions, in addition to the postcanine dentition and mandible corpus. This is a significantly trimmed list of characters compared to that historically provided in descriptions of the specialized robust *Australopithecus* masticatory system, which prominently featured reduction of upper midfacial prognathism and retraction of the palate (e.g., Rak, 1983; Kimbel et al., 1984). Perhaps one of the more important messages of KNM-WT 17000 concerns how little we *actually know* about patterns of functional linkage in the evolutionary transformation of the hominin skull. Clearly, however, the functional complexes that describe entire clades cannot be inferred merely from the co-occurrence of character states in the most apomorphic (terminal) taxa in those clades.

In our analysis additional support for the hypothesis of robust monophyly comes from the low position of the calvarial vertex, the concave frontal squama (trigonum frontale), the low position of the maximum lateral mastoid projection, and a smooth nasal cavity floor with merged anterior nasal spine and vomeral insertion within the nasal cavity.

Monophyly of Early Homo

The strict consensus cladogram derived from the three shortest trees upholds the monophyly of the three early species of the genus *Homo*, although the relationships within the clade are left unresolved. Four of the five shortest trees retain these species as a monophyletic group. We qualify this result with the observation that addressing the phylogenetic relationships within *Homo* was not an objective of our study; therefore, we did not include some of the characters that might assist in resolving those relationships with greater clarity (for example, cranial capacity), but which would not be relevant to questions regarding the phylogenetic position of species within *Australopithecus*. Nevertheless, we think our data do bear on the issue in light of suggestions that *Homo* may not be monophyletic if it includes the basal species *H. habilis* and *H. rudolfensis* (Lieberman et al., 1996; Wood and Collard, 1999; Collard and Wood, 2001), with one or both taxa potentially affiliated cladistically with *Australopithecus sensu lato*.

Synapomorphies linking the three *Homo* species include reduced palatal (subnasal) prognathism, everted nasal aperture margins, and a relatively broad palate (for further discussion, see Kimbel et al., 1997). In only one of the five shortest trees is one of the conventionally assigned early *Homo* species, *H. rudolfensis*, affiliated cladistically with *Australopithecus* (Figure 6.1D). In this tree, which is one of two least parsimonious of the shortest five trees, *H. rudolfensis* is the sister to a clade comprising *A. africanus* plus a monophyletic robust *Australopithecus* clade. Not surprisingly, characters supporting this tree are concentrated in the infraorbital and peripheral parts of the face. In both of our shortest trees, *H. rudolfensis* is the sister to the *H. habilis + H. erectus* clade. These results suggest that caution is warranted regarding Wood and Collard's (1999) arguments for transferring *H. habilis* and *H. rudolfensis* to the genus *Australopithecus*.

***Australopithecus afarensis* in Human Evolution**

Our comparative study of the skull of Hadar specimen A.L. 444-2 confirms the basal status of *A. afarensis* in Middle to Late Pliocene hominin evolution. In its cranium, mandible, and dentition it remains highly plesiomorphic relative to geologically younger hominin taxa. However, the focus on the primitive aspects of *A. afarensis* morphology can obscure the fact that structurally this species is far removed from conditions expected in the last common ancestor of all hominins. This realization derives from two compatible sources: (1) the expanded range of morphological variation in *A. afarensis* introduced by A.L. 444-2 and other skull remains recovered during the second phase of fieldwork at Hadar beginning in 1990, which capture some derived states commonly associated with the species' geologically younger sister group; (2) the discovery of geologically older taxa (*Ard. ramidus*, *A. anamensis*) with substantially more plesiomorphic skulls and teeth.

Good Darwinians might expect exactly such a finding of structural and temporal intermediacy, but converting these expectations into a reliably stable phylogenetic topology is, as abundant experience shows, much less straightforward. Especially illustrative is the recent discovery in Kenya of *K. platyops*, whose known occurrence, at 3.5 Ma, is approximately 100,000 years older than the oldest Hadar specimens and contemporary with the Laetoli sample of *A. afarensis*. We are impressed by the strength of the anatomical evidence for the specific distinctiveness of the *K. platyops* material, especially in the context of the A.L. 444-2 skull, but how this evidence will be integrated into existing phylogenetic hypotheses is unclear. Described as more derived than *A. afarensis* in its facial morphology (M. Leakey et al., 2001), *K. platyops* is potentially a sister species to some of the younger taxa included in the sister group of *A. afarensis* in our phylogenetic analysis.

Increasing the sample of fossils in the 3.4–3.7 Ma time period will be important to clarifying the status of these Middle Pliocene hominins. Recently initiated fieldwork immediately to the south of Hadar in the Dikika area, where fossil-rich sediments of the lower Sidi Hakoma and underlying Basal members of the Hadar Formation (> 3.4 Ma) are already yielding new hominin fossils, may contribute in this regard (Zeresenay Alemseged, personal communication).

Central to future thinking about the role of *A. afarensis* in early hominin evolution is the status of the Laetoli sample. We have outlined ways in which the Laetoli hominins differ from the younger Hadar fossils. Although some of these differences have been recognized since *A. afarensis* was first diagnosed, their significance has been enhanced in the wake of the discovery of Early Pliocene (ca. 4.0 Ma) *A. anamensis* (Ward et al., 2001) and by the expanded range of variation within the Hadar sample itself. Both the Laetoli and Kanapoi/Allia Bay (*A. anamensis*) hominin collections are dominated by jaws and teeth, and it is in the snout and canine/lower premolar complex that the Laetoli specimens tend to affiliate with the older material. A significantly expanded sample from the 3.5–3.7-Myr-old upper Laetolil Beds would be enormously helpful in substantiating the extent and anatomical distribution of the morphological differences supporting the Laetoli site sample's intermediate position between the Hadar and Kanapoi/Allia Bay samples (taphonomic factors and limited sediment exposure at the Laetoli site work against the prospect of large numbers of new fossils from the site, however).

Whatever the systematic consequences of the Laetoli sample's temporal and morphological intermediacy may be, one issue of great importance is the glimpse the Kanapoi/Allia Bay, Laetoli, Maka, and Hadar specimens afford of evolutionary change in the front of the jaws and the anterior dentition, particularly the canine/lower premolar complex. A key anatomical region distinguishing apes from humans, its early transformation is documented in unprecedented detail in the gnathic and dental remains sampled from this temporal sequence.

Anagenetic evolution is implicated in the trend toward large size of the mandible (and, presumably, cranium) in the Hadar sample of *A. afarensis* (Lockwood et al., 2000), exemplified by the geologically young but massive A.L. 444-2 skull, and explains the very high level of size variation in the Hadar mandible series, which exceeds that observed even in the most dimorphic extant hominoid species. This example from *A. afarensis* cautions against the common practice of comparing variation in time-transgressive samples with that in modern analog taxa. A failure to account for the effects of temporally vectored change on the profile of variation in fossil species may yield elevated Type I error rates in the assessment of taxonomic diversity and thus open the door to incorrect inferences about past levels of sexual dimorphism and life-history strategies. Unfortunately, there are currently very few fossil hominin species as densely sampled over a well-calibrated stratigraphic record as *A. afarensis* (*A. boisei* is another example; Wood et al., 1994, and Lockwood et al., 2000), so in many cases it will be difficult to sharply delineate the relative influence of different sources of variation on overall levels of variation.

In spite of the increase in the size of the mandible over time, the temporal persistence of diagnostic skull and dental characters in the late *A. afarensis* sample from Hadar argues for maintaining a single-species interpretation of variation documented at the site. Postcanine tooth size variation is consistent with this interpretation, although it is as high as it is in the largest, most dimorphic hominoid species, the gorilla and orangutan (Lockwood et al., 2000). According to the single-taxon hypothesis, the variation in Hadar P_3s is interpretable as "phylogenetic polymorphism," variation in a taxon due to a character's "transitional" state in a particular part of the phylogenetic tree.

The fact that a general increase in postcanine tooth size does not accompany the trend toward large mandibles late in the Hadar sequence suggests that the documented anagenetic change in the *A. afarensis* lineage does not contain information about morphological transformations that forecast relationships between *A. afarensis* and potential descendant taxa. A similar message is contained in the mandible corpus data, which reveal no tendency for large, late Hadar mandibles to be more robust in corpus proportions than earlier mandibles were. These Hadar data therefore provide no support for phylogenetic hypotheses that posit a direct ancestor–descendant relationship between *A. afarensis* and the East African megadont species *A. aethiopicus*, which appears (stratigraphically speaking) with a number of diagnostic robust *Australopithecus* postcanine dental specializations already intact no later than 2.7 Ma (or possibly earlier), some 300,000 years after the latest known *A. afarensis* specimens (Suwa et al., 1996). Instead, the trends observed in late *A. afarensis*, while biologically significant, do not appear to involve aspects of skull and dental morphology marked by subsequent Pliocene phylogenetic and adaptive transformations. This is the most important meaning of "stasis" in the Hadar record of *A. afarensis*. Additional fossils from the 2.7 to 3.0 Ma time period in eastern Africa will determine whether this stasis in *A. afarensis* terminates in extinction or speciation, or yields to phyletic transformation.

At the moment, the putative phylogenetic relationships between *A. afarensis* and either *A. aethiopicus* or *A. garhi* (or both) hang on "evidence" from symplesiomorphic cranial morphology and the Late Pliocene temporal position of the latter two taxa in the East African stratigraphic record. Cladistic evidence for a sister-species relationship between *A. afarensis* and either of these taxa does not at present categorically exclude an intermediate phylogenetic role for the South African species *A. africanus*, and yet, as our results (and that of other authors) show, the phylogenetic position of *A. africanus* (at least as conventionally delineated) is itself highly problematic. Relative to *A. afarensis*, the species *A. africanus*, *A. aethiopicus*, and *A. garhi* each exhibit masticatory modifications, including postcanine megadonty, reminiscent of the Late Pliocene exaggerations of *A. robustus* and *A. boisei* morphology. That such traces surface in the *Homo* clade as well (i.e., in *H. rudolfensis*) reinforces the impression that African Pliocene hominin evolution after *A. afarensis* (<3.0 Ma) was substantially concerned with producing different ways to contend functionally with a changing dietary resource base.

How these changes map on to phylogenetic hypotheses remains an outstanding problem area for paleoanthropologists. Although our results support the idea of a single robust *Australopithecus* clade, the origin of the clade remains obscured in the highly mosaic mix of primitive and derived dental and craniofacial characters observed in the still sparse fossil record between 3.0 and 2.5 Ma. Relative to *A. afarensis* conditions, anomalously primitive characters in the calvaria (especially the cranial base) of *A. africanus* suggest some skepticism is appropriate regarding the necessity of seeing any of the species currently known to fall in this time period as ancestral to known latest Pliocene hominins.

The Hadar sample of *A. afarensis* provides an unusually detailed portrait of hominin evolution during a period characterized by remarkable transformation and diversification in skull and dental morphological patterns. On both ends of the time period encapsulated by the Hadar sample there remain open questions about hominin variation, phylogeny, and adaptation. Progress in answering these questions will certainly come from detailed study and comparison of existing Pliocene collections, but only new fossil evidence can fill the remaining gaps in our knowledge base. Results of ongoing fieldwork across the African continent will no doubt continue to narrow these gaps. The ultimate significance of the extensive and still growing sample of skulls and teeth of *Australopithecus afarensis* accumulated over more than a decade of fieldwork lies in the rich source of data and insight it will provide as fresh hypotheses emerge and seek confirmation against what is already known.

Notes

CHAPTER 1

1. We think that the mandible fragment from the site of Bar el Gahzal, Chad (Brunet et al., 1996), very likely represents this species as well, but as we have not examined the original specimen, we do not include it in the *A. afarensis* hypodigm for the purposes of this research.

2. The identification by M. Leakey et al. (2001) of 3.5 Myr-old *Kenyanthropus platyops* may push the evidence for hominin lineage diversity back into the Middle Pliocene. We refer to this proposal in Chapter 6.

3. The nomenclature of the taxon has also been controversial (see Groves, 1999). Technically *Australopithecus afarensis* Johanson 1978 is the replacement name for *Meganthropus africanus* Weinert 1950, a junior secondary homonym of *Australopithecus africanus* Dart 1925. Recently, Groves (1996) successfully petitioned the International Commission on Zoological Nomenclature to conserve the specific name *afarensis* Johanson 1978 by suppressing *africanus* Weinert 1950. When *afarensis* is removed from the genus *Australopithecus, Praeanthropus afarensis* (Johanson 1978) is the correct binomen. (See Opinion 1941, *Bull. Zool. Nomen.* 56(3): 223–224, 1999.)

CHAPTER 3

1. Rather than using total facial height, we chose the orbitale–alveolar plane dimension because these structures are more available in the fossil record and because we wished to exclude the height of the mandibular corpus itself from the measurement. (The corpus is included in total facial height.) Nevertheless, the percentage of total facial height occupied by the corpus offers a similar picture, as can be seen in the percentage values appearing at the bottom of Figure 3.3.

2. Notably, there is not a dramatic difference between the vertical distance from the FH (the bottom of the orbit) to vertex and the distance from the roof of the orbit to vertex, according to Le Gros Clark's method, as the size of the orbit is our general yardstick for calvarial size.

3. The vertically short orbits in Sts. 5 contribute to the difference between the height of the calvaria above the FH (orbitale) and the height calculated by means of the Le Gros Clark system.

4. This ascent of inion reduces the total arc length, and thus the measurements of the individual segmental arcs acquire greater weight in the calculations.

5. Encrustation or a strange ossification (of the apical ligament) in the region of basion causes the elimination of a saddle that is typically seen between the two occipital condyles and hence brings about an unnatural descent of basion in Sts. 5. If this encrustation were ignored, the midsagittal cross section of the basisphenoid would look quite natural by any standard. It would be straight and thus cancel out the drop of basion. With this correction, the foramen magnum would demonstrate an inclination of +14°.

6. This point does not necessarily coincide with opisthocranion.

7. Here, too, the exocranial structures influence the index value, although the effect is negligible in comparison to that of prognathism. Furthermore, the effect of the posterior exocranial structures (such as the nuchal crest) balances out that of the anterior structures (such as a prominent glabellar mass). Still, note the possible effect of the enormous external occipital protuberance in A.L. 444-2 on its index value, seen in Figure 3.19.

8. Although the increased brain volume in modern humans barely affects the position of glabella, it does show considerable influence on the back of the skull, where the occipital squama is displaced posteriorly. Hence, the value of this measurement lies primarily in a comparison of hominoids with a similar brain volume.

9. Consider, for example, the orientation of the nasal bones in Neandertals and modern humans.

10. Because this measurement is not applicable to most gorillas, our value is based on a small sample of females that do permit the measurement.

11. The deepest point of the porial saddle is more easily defined than Tobias's (1967) "base of the temporal squama."

12. See Clarke's discussion (1977). We agree with his interpretation that the notchlike structure in OH 5 is not a true submalar incisure but the outcome of a spur in the masseteric scar.

13. This strategy is based on our observations of the high correlation between the biorbital distance and many other measurements that reflect skull size.

14. As in the section on the ratio of palatal width to total facial height, we find that although we have previously discussed each of the components of the shape index independently, the index is worth computing because of the valuable insight it provides as a result of the combination of these components.

15. Specimen A.L. 427-1 is from an old individual. The sunken surface of its clivus and the resulting protrusion of the canine juga are probably the outcome of bone resorption due to advanced age.

16. We have used the orbital breadth instead of the biorbital breadth because the interorbital region, which is currently under discussion, is part of the biorbital breadth.

17. The effect of variation in the interorbital distance is somewhat mitigated by the greater contribution of the two orbital breadths to the biorbital distance.

18. We have not used the distance between the supramastoid crests (the maximum breadth seen in posterior view) because they lie on a more posterior coronal plane than the maximum height of the calvaria.

19. The fact of a horizontal distance between the vertex (which, as seen in Figure 3.47, is also the highest point on the calvaria in posterior view) and the slightly lower, more posteriorly located height of the calvaria at the coronal plane of porion (apex)—the point on which we have based our measurements and calculations—is not significant, as the vertical difference between these two points is minimal and remains constant in all the taxa used for comparison.

20. The angle in *Cercopithecus* measures 16°.

21. Since the mean value of a small sample ($n = 4$) of *Cercopithecus* is also 36%, this configuration appears to be the plesiomorphic one.

22. "Elevated" refers to the basal view when the cranium is held upside down.

23. Since the juxtaposition of foramen ovale and the spheno-occipital synchondrosis seems to be fixed in all taxa, the distance between basion and the foramen ovale is correlated with the length of the basioccipital element.

24. Table 3.27 shows that even when we use the biorbital breadth as the yardstick, the values do not change significantly.

25. Note that some measurements presented in this paragraph are based on those made by Dean and Wood (1982), who express certain angular relationships in reference to the coronal plane. However, we use the sagittal plane as the reference for all the angles, including Dean and Wood's. Thus we have recalculated Dean and Wood's angles to adjust them accordingly. See Figure 3.58 and Table 3.28.

26. The forward shift of the articular eminence contributes to an additional small reduction in the distance between the coronal plane of M^3 and that of the tubercle in *A. boisei* and thus slightly influences the indexes that express the relationships within the masticatory system (see "Relationship between Elements of the Masticatory System").

27. The orientation of the pterygoid plates is discernible in a side view and could thus be discussed in the section on the lateral view. However, we prefer to discuss it here because the magnitude of inclination is a direct outcome of structures in the base of the skull and the relationship between them, as described.

28. The length of the dental arcade obviously contributes to prognathism, too, but does not constitute part of the cranial base.

CHAPTER 5

1. Most casts we have seen of the Omo 323-1976-896 temporal bones make it appear as though the mandibular fossae are intact. In fact, however, breakage on both sides precludes actual articulation of the pieces that make up the fossa (see also Alemseged et al., 2001).

2. The maximum width of the palate in gorillas, primarily in males, is usually located in the front, in the vicinity of the canines. Posteriorly, the palate narrows continuously to the point of minimum width, which falls between the lateral aspects of the palate at M^3. In calculating the shape index in the gorilla specimens, however, we chose to employ the width measurement at the site that is homologous to the site of maximum width in *A. afarensis*—the lateral aspect of the palate at M^2.

CHAPTER 6

1. We are not suggesting that phylogenetic polymorphisms are novel in terms of the mechanisms that produce them, only that they are special in terms of *information content* in a macroevolutionary context.

2. We are aware that gorillas and chimpanzees may not be isomorphic for some of the character states identified as plesiomorphic for hominins in this analysis, complicating polarity determination (see, for example, Lockwood et al., 2002, on the temporal bone). Our analysis may require refinement along these lines.

3. Asfaw et al. (1999) contradict themselves on the state of the postorbital constriction in *A. garhi*. In their Table 1 it is listed as "moderate," whereas it is described as "marked" in their text (p. 633). The latter state is assigned to the three robust species in their Table 1.

References

Ahern JCM (1998) Understanding intraspecific variation: the problem with excluding Sts. 19 from *Australopithecus africanus*. *Am. J. Phys. Anthropol.* 105: 461–480.

Alemseged Z, Coppens Y, and Geraads D (2002) Hominid cranium from Omo: description and taxonomy of Omo 323-1976-896. *Am. J. Phys. Anthropol.* 117: 103–112.

Asfaw B (1987) The Belohdelie frontal: new evidence from the Afar of Ethiopia. *J. Hum. Evol.* 16: 611–624.

Asfaw B, White TD, Lovejoy CO, Latimer B, Simpson S, and Suwa G (1999) *Australopithecus garhi*: a new species of early hominid from Ethiopia. *Science* 284: 629–635.

Ashton EH, and Zuckerman S (1956) Cranial crests in the Anthropoidea. *Proc. Zool. Soc. Lond.* 126: 581–635.

Barton RA, and Harvey PH (2000) Mosaic evolution of brain structure in mammals. *Nature* 405: 1055–1058.

Begun DR (1994) Relations among the great apes and humans: new interpretations based on the fossil great ape *Dryopithecus*. *Yrbk. Phys. Anthropol.* 37: 11–64.

Biegert J (1957) Der Formwandel des Primaten-Schädels und seine Beziehungen zur ontogenetischen Entwicklung und den phylogenetischen Specialization in der Kopforgane. *Morph. Jb.* 98: 77–199.

Blumenburg B (1985) Biometrical studies upon hominoid teeth: the coefficient of variation, sexual dimorphism and questions of phylogenetic relationship. *Biosystems* 28: 149–184.

Blumenberg B, and Lloyd BT (1983) *Australopithecus* and the origin of the genus *Homo*: aspects of biometry and systematics with accompanying catalog of tooth metric data. *Biosystems* 16: 127–167.

Boaz, NT (1979) Hominid evolution in eastern Africa during the Pliocene and early Pleistocene. *Ann. Rev. Anthropol.* 8: 71–85.

Broom R, Robinson JT, and Schepers GWH (1950) *Sterkfontein Ape-Man Plesianthropus*. Pretoria: Transvaal Museum.

Brothwell DR (1981) *Digging up Bones: The Excavation, Treatment and Study of Human Skeletal Remains*. Ithaca, N.Y.: Cornell University Press.

Brown B, Walker A, Ward C, and Leakey R (1993) New *Australopithecus boisei* calvaria from East Lake Turkana, Kenya. *Am. J. Phys. Anthropol.* 91: 137–159.

Brunet M, Beauvilain A, Coppens Y, Heintz E, Moutaye AHE, and Pilbeam D (1996) *Australopithecus bahrelghazali*, une nouvelle espèce d'hominidé ancien de la région de Koro Toro (Tchad). *C. R. Acad. Sci.* 322: 907–913.

Cande SC, and Kent DV (1995) Revised calibration of the geomagnetic polarity time scale for the Late Cretaceous and Cenozoic. *J. Geophys. Res.* 100: 6093–6095.

Chamberlain AT, and Wood BA (1985) A re-appraisal of the variation in hominid mandibular corpus dimensions. *Am. J. Phys. Anthropol.* 66: 399–405.

Chamberlain AT, and Wood BA (1987) Early hominid phylogeny. *J. Hum. Evol.* 16: 119–133.

Clark JD, Asfaw B, Assefa G, Harris JWK, Kurashina H, Walter RC, White TD, and Williams MAJ (1984) Paleoanthropological discoveries in the Middle Awash Valley, Ethiopia. *Nature* 307: 423–428.

Clarke RJ (1977) The cranium of the Swartkrans hominid, SK 847 and its relevance to human origins. Ph.D. diss., University of the Witwatersrand, Johannesburg.

Clarke RJ (1988) Habiline handaxes and paranthropine pedigrees at Sterkfontein. *World Archaeol.* 20: 1–12.

Clarke RJ (1994) Advances in understanding the craniofacial anatomy of South African early hominids. In RS Corruccini and RL Ciochon (eds), *Integrative Paths to the Past*. Englewood Cliffs, N.J.: Prentice Hall, pp. 205–222.

Cole TM, and Smith FH (1987) An odontometric assessment of variability in *Australopithecus afarensis*. *Hum. Evol.* 2: 221–234.

Collard M, and Wood BA (2000) How reliable are human phylogenetic hypotheses? *PNAS* 97: 5003–5006.

Collard M, and Wood BA (2001) Homoplasy and the early hominid masticatory system: inferences from analyses of extant hominoids and papionins. *J. Hum. Evol.* 41: 167–194.

Conroy GC, Weber GW, Seidler H, Tobias PV, Kane A, and Brunsden B (1998) Endocranial capacity in an early hominid cranium from Sterkfontein, South Africa. *Science* 280: 1730–1731.

Coppens Y (1983) Systématique, phylogénie, environnement et culture des australopithèques, hypotheses et synthèse. *Bull. et Mem. de la Soc. D'Anthrop. de Paris* 10, série 13: 273–284.

Daegling DJ, and Hylander WL (1998) Biomechanics of torsion in the human mandible. *Am. J. Phys. Anthropol.* 105: 73–87.

Dart RA (1948) The Makapansgat proto-human *Australopithecus prometheus. Am. J. Phys. Anthropol.* 6: 259–281.

Dean, MC (1985) Comparative myology of the hominoid cranial base. II. The muscles of the prevertebral and upper pharyngeal region. *Folia Primatol.* 44: 40–51.

Dean MC (1986) *Homo* and *Paranthropus*: similarities in the cranial base and developing dentition. In B Wood, L Martin, and P Andrews (eds.), *Major Trends in Primate and Human Evolution*. Cambridge: Cambridge University Press, pp. 249–265.

Dean MC, and Wood BA (1981) Metrical analysis of the basicranium of extinct hominoids and *Australopithecus. Am. J. Phys. Anthropol.* 54: 63–71.

Dean MC, and Wood BA (1982) Basicranial anatomy of Plio-Pleistocene hominids from East and South Africa. *Am. J. Phys. Anthropol.* 59: 157–174.

Deane A, and Begun D (2002) Please don't throw the baby out with the bath water: skeletal characters in cladistic analyses of hominoid evolution. *Am. J. Phys. Anthropol.*, Suppl. 34: 61.

de Bonis L, and Koufos G (1993) The face and mandible of *Ouranopithecus macedoniensis*: description of new specimens and comparisons. *J. Hum. Evol.* 24: 469–491.

de Villiers H (1968) *The Skull of the South African Negro: A Biometrical and Morphological Study.* Johannesburg: Witwatersrand University Press.

Drapeau M (2001) Functional analysis of the associated partial forelimb skeleton from Hadar, Ethiopia (A.L. 438-1): implications for understanding patterns of variation and evolution in early hominin forearm and hand anatomy. Ph.D. diss., University of Missouri, Columbia.

Du Brul EL (1977) Early hominid feeding mechanisms. *Am. J. Phys. Anthropol.* 47: 305–320.

Falk D (1985) Hadar AL 162-28 endocast as evidence that brain enlargement preceded cortical reorganization in hominid evolution. *Nature* 313: 45–47.

Falk D (1988) Enlarged occipital/marginal sinuses and emissary foramina: their significance in hominid evolution. In FE Grine (ed.), *Evolutionary History of the "Robust" Australopithecines.* New York: Aldine de Gruyter, pp. 85–96.

Falk D (1990) Brain evolution in *Homo*: the "radiator theory." *Behav. Brain Sci.* 13: 333–381.

Falk D, and Conroy GC (1983) The cranial venous sinus system in *Australopithecus afarensis. Nature* 306: 779–781.

Falk D, Gage TB, Dudek B, and Olson TR (1995) Did more than one species of hominid coexist before 3 Ma? Evidence from blood and teeth. *J. Hum. Evol.* 29: 591–600.

Falk D, Redmond JJC, Guyer J, Conroy GC, Recheis W, Weber GW, and Seidler H (2000) Early hominid brain evolution: a new look at old endocasts. *J. Hum. Evol.* 38: 695–717.

Gower CG (1923) A contribution to the morphology of the apertura piriformis. *Am. J. Phys. Anthropol.* 6: 27–36.

Gowlett JAJ, Harris JWK, Walton D, and Wood BA (1981) Early archaeological traces of fire from Chesowanja, Kenya. *Nature* 294: 125–129.

Greenfield, LO (1990) Canine "honing" in *Australopithecus afarensis. Am. J. Phys. Anthropol.* 82: 135–143.

Grine FE (ed.) (1988) *Evolutionary History of the "Robust" Australopithecines*. New York: Aldine de Gruyter.

Grine FE, and Strait DS (1994) New hominid fossils from Member 1 "Hanging Remnant" Swartkrans Formation, South Africa. *J. Hum. Evol.* 26: 57–75.

Groves CP (1989) *A Theory of Human and Primate Evolution.* Oxford: Oxford University Press.

Groves CP (1996) *Australopithecus afarensis* Johanson, 1978 (Mammalia, Primates); proposed conservation of the specific name. *Bull. Zool. Nomencl.* 53: 24–27.

Groves CP (1999) Nomenclature of African Plio-Pleistocene hominins. *J. Hum. Evol.* 37: 869–872.

Hawks J, and Wolpoff MH (1999) Endocranial capacity of early hominids. *Science* 283: 9b.

Holloway RL (1970) New endocranial values for the australopithecines. *Nature* 277: 199–200.

Holloway RL (1972) New australopithecine endocast, SK 1585, from Swartkrans, South Africa. *Am. J. Phys. Anthropol.* 37: 173–186.

Holloway RL (1973) Endocranial volumes of early African hominids and the role of the brain in human mosaic evolution. *J. Hum. Evol.* 2: 449–459.

Holloway RL (1983) Cerebral brain endocast pattern of *Australopithecus afarensis* hominid. *Nature* 303: 420–422.

Holloway RL (1988) "Robust" australopithecine brain endocasts: some preliminary observations. In FE Grine (ed.), *Evolutionary History of the "Robust" Australopithecines.* New York: Aldine de Gruyter, pp. 97–105.

Holloway RL (1996) Evolution of the human brain. In LA and RC Peters (eds.), *Handbook of Human Symbolic Evolution*. New York: Oxford University Press, pp. 74–108.

Holloway RL (2000) Brain. In E Delson, I Tattersall, J Van Couvering, and AS Brooks (eds.), *Encyclopedia of Human Evolution*. New York: Garland, pp. 141–149.

Holloway RL, and Kimbel WH (1986) Endocast morphology of Hadar hominid AL 162-28. *Nature* 321: 536–537.

Holloway RL, and Post DG (1982) The relativity of relative brain measures and hominid mosaic evolution. In E Armstrong and D Falk (eds.), *Primate Brain Evolution: Methods and Concepts.* New York: Plenum, pp. 57–76.

Hylander WL (1979) The functional significance of primate mandibular form. *J. Morphol.* 160: 223–240.

Hylander WL (1988) Implications of in vivo experiments for interpreting the functional significance of "robust" australopithecine jaws. In FL Grine (ed.), *Evolutionary History of the "Robust" Australopithecines.* New York: Aldine de Gruyter, pp. 55–80.

Johanson DC, and White TD (1979) A systematic assessment of early African hominids. *Science* 203: 321–329.

Johanson DC, White TD, and Coppens Y (1978) A new species of the genus *Australopithecus* (Primates: Hominidae) from the Pliocene of eastern Africa. *Kirtlandia* 28: 1–11.

Johanson DC, Taieb M, and Coppens Y (1982a) Pliocene hominids from the Hadar Formation, Ethiopia (1973–1977): stratigraphic, chronologic, and paleoenvironmental contexts, with notes on hominid morphology and systematics. *Am. J. Phys. Anthropol.* 57: 373–402.

Johanson DC, Lovejoy CO, Kimbel WH, White TD, Ward SC, Bush ME, Latimer BM, and Coppens Y (1982b) Morphology of the Pliocene partial hominid skeleton (A.L. 288-1) from the Hadar Formation, Ethiopia. *Am. J. Phys. Anthropol.* 57: 403–451.

Johanson DC, White TD, and Coppens Y (1982c) Dental remains from the Hadar Formation, Ethiopia: 1974–1977 collections. *Am. J. Phys. Anthropol.* 57: 545–604.

Jolly CJ (1970) The seed eaters: a new model of hominid differentiation based on a baboon analogy. *Man* 5: 5–27.

Kimbel WH (1984) Variation in the pattern of cranial venous sinuses and hominid phylogeny. *Am. J. Phys. Anthropol.* 63: 243–263.

Kimbel WH (1986) Calvarial morphology of *Australopithecus afarensis*: a comparative phylogenetic study. Ph.D. diss., Kent State University, Kent, Ohio.

Kimbel WH (1988) Identification of a partial cranium of *Australopithecus afarensis* from the Koobi Fora Formation, Kenya. *J. Hum. Evol.* 17: 647–656.

Kimbel WH, and Rak Y (1985) Functional morphology of the asterionic region in extant hominoids and fossil hominids. *Am. J. Phys. Anthropol.* 66: 31–54.

Kimbel WH, and Rak Y (1993) The importance of species taxa in paleoanthropology and an argument for the phylogenetic concept of the species category. In WH Kimbel and LB Martin (eds.), *Species, Species Concepts and Primate Evolution.* New York: Plenum, pp. 461–484.

Kimbel WH, and White TD (1988a) A revised reconstruction of the adult skull of *Australopithecis afarensis. J. Hum. Evol.* 17: 545–550.

Kimbel WH, and White TD (1988b) Variation, sexual dimorphism and taxonomy of *Australopithecus.* In FE Grine (ed.), *Evolutionary History of the "Robust" Australopithecines.* New York: Aldine de Gruyter, pp. 175–192.

Kimbel WH, Johanson DC, and Coppens Y (1982) Pliocene hominid cranial remains from the Hadar Formation, Ethiopia. *Am. J. Phys. Anthropol.* 57: 453–499.

Kimbel WH, White TD, and Johanson DC (1984) Cranial morphology of *Australopithecus afarensis*: a comparative study based on a composite reconstruction of the adult skull. *Am. J. Phys. Anthropol.* 64: 337–388.

Kimbel WH, White TD, and Johanson DC (1985) Craniodental morphology of the hominids from Hadar and Laetoli: evidence of *"Paranthropus"* and *Homo* in the mid-Pliocene of Eastern Africa? In E Delson (ed.), *Ancestors: The Hard Evidence.* New York: Alan R. Liss, pp. 120–137.

Kimbel WH, White TD, and Johanson DC (1988) Implications of KNM-WT 17000 for the evolution of "robust" *Australopithecus.* In FE Grine (ed.), *Evolutionary History of the "Robust" Australopithecines.* New York: Aldine de Gruyter, pp. 259–268.

Kimbel WH, Johanson DC, and Rak Y (1994) The first skull and other new discoveries of *Australopithecus afarensis* at Hadar, Ethiopia. *Nature* 368: 449–451.

Kimbel WH, Johanson DC, and Rak Y (1997) Systematic assessment of a maxilla of *Homo* from Hadar, Ethiopia. *Am. J. Phys. Anthropol.* 103: 235–262.

Kimbel WH, Walter RC, Johanson DC, Reed, KE, Aronson, JL, Assefa Z, Marean CW, Eck GG, Bobe R, Hovers E, Rak Y, Vondra C, Yemane T, York D, Chen Y, Evensen NM, and Smith PE (1996) Late Pliocene *Homo* and Oldowan tools from the Hadar Formation (Kada Hadar Member), Ethiopia. *J. Hum. Evol.* 31: 549–561.

Leakey MG, Feibel CS, McDougall I, and Walker AC (1995) New four-million-year-old hominid species from Kanapoi and Allia Bay, Kenya. *Nature* 376: 565–571.

Leakey MG, Feibel CS, McDougall I, Ward CV, and Walker A (1998) New specimens and confirmation of an early age for *Australopithecus anamensis. Nature* 393: 62–66.

Leakey MG, Spoor F, Brown FH, Gathogo PN, Kiarie C, Leakey LN, and McDougall I (2001) A new hominin genus from eastern Africa shows diverse middle Pliocene lineages. *Nature* 410: 433–440.

Leakey REF, and Walker AC (1980) On the status of *Australopithecus afarensis. Science* 207: 1103.

Leakey REF, and Walker AC (1988) New *Australopithecus boisei* specimens from East and West Lake Turkana, Kenya. *Am. J. Phys. Anthropol.* 76: 1–24.

Le Gros Clark WE (1950) New palaeontological evidence bearing on the evolution of Hominoidea. *Quart. J. Geol. Soc. Lond.* 105: 225–264.

Lieberman DE (1995) Testing hypotheses about recent human evolution from skulls. *Curr. Anthropol.* 36: 159–178.

Lieberman DE (1997) Making behavioral and phylogenetic inferences from hominid fossils: considering the developmental influences of mechanical forces. *Annu. Rev. Anthropol.* 26: 185–210.

Lieberman DE (1999) Homology and hominid phylogeny: problems and potential solutions. *Evol. Anthropol.* 7: 142–151.

Lieberman DE (2000) Ontogeny, homology, and phylogeny in the hominid craniofacial skeleton: the problem of the browridge. In P. O'Higgins and M. Cohn (eds), *Development, Growth and Evolution: Implications for the Study of the Hominid Skeleton.* London: Academic Press, pp. 86–122.

Lieberman DE, Wood BA, and Pilbeam DR (1996) Homoplasy and early *Homo*: an analysis of the evolutionary relationships of *H. habilis* (sensu stricto) and *H. rudolfensis*. *J. Hum. Evol.* 30: 97–120.

Lockwood CA, and Kimbel WH (1999) Endocranial capacity of early hominids. *Science* 283: 9b.

Lockwood CA, and Tobias PV (1999) A large male hominin cranium from Sterkfontein, South Africa, and the status of *Australopithecus*. *J. Hum. Evol.* 36: 637–685.

Lockwood CA, and Tobias PV (2002) Morphology and affinities of new hominin cranial remains from Member 4 of the Sterkfontein Formation, Gauteng Province, South Africa. *J. Hum. Evol.* 42: 389–450.

Lockwood CA, Richmond BG, Jungers WL, and Kimbel WH (1996) Randomization procedures and sexual dimorphism in *Australopithecus afarensis*. *J. Hum. Evol.* 31: 537–548.

Lockwood CA, Kimbel WH, and Johanson DC (2000) Temporal trends and metric variation in the mandibles and dentition of *Australopithecus afarensis*. *J. Hum. Evol.* 39: 23–55.

Lockwood CA, Lynch J, and Kimbel WH (2002) Quantifying temporal bone morphology of great apes and humans: an approach using geometric morphometrics. *J. Anat.* 201: 447–464.

Lovejoy CO (1981) The origin of man. *Science* 211: 341–350.

Lovejoy CO, Kern KF, Simpson SW, and Meindl RS (1989) A new method for estimation of skeletal dimorphism in fossil samples with an application to *Australopithecus afarensis*. In G Giacobini (ed.), *Hominidae: Proceedings of the 2nd International Congress of Human Paleontology*. Milan: Jaca Books, pp. 103–108.

Lovejoy CO, Cohn MJ, and White TD (1999) Morphological analysis of the mammalian postcranium: a developmental perspective. *PNAS* 96: 13247–13252.

Lovejoy CO, Cohn MJ, and White TD (2000) The evolution of mammalian morphology: a developmental perspective. In P. O'Higgins and M. Cohn (eds), *Development, Growth and Evolution: Implications for the Study of the Hominid Skeleton*. London: Academic Press, pp. 41–55.

Maddison DR, and Maddison WP (2000) *MacClade 4: Analysis of Phylogeny and Character Evolution*. Sunderland, Mass.: Sinauer Associates.

McCollum MA (2000) Subnasal morphological variation in fossil hominids: a reassessment based on new observations and recent developmental findings. *Am. J. Phys. Anthropol.* 112: 275–283.

McCollum MA, and Sharpe PT (2001) Developmental genetics and early hominid craniodental evolution. *BioEssays* 23: 481–493.

McCollum MA, Grine FE, Ward CS, and Kimbel WH (1993) Subnasal morphological variation in extant hominoids and fossil hominids. *J. Hum. Evol.* 24: 87–111.

McHenry HM (1975) Fossils and the mosaic nature of human evolution. *Science* 190 (4213): 425–431.

McHenry HM (1986) The first bipeds: a comparison of the *A. afarensis* and *A. africanus* postcranium and implications for the evolution of bipedalism. *J. Hum. Evol.* 15: 177–191.

McHenry HM (1991) Sexual dimorphism in *Australopithecus afarensis*. *J. Hum. Evol.* 20: 21–32.

McKee JK (1989) Australopithecine anterior pillars: reassessment of the functional morphology and phylogenetic relevance. *Am. J. Phys. Anthropol.* 80: 1–9.

Moggi-Cecchi J, Tobias PV, and Beynon AD (1998) The mixed dentition and associated skull fragments of a juvenile fossil hominid from Sterkfontein, South Africa. *Am. J. Phys. Anthropol.* 106: 425–465.

Moyà Solà S, and Köhler M (1995) New partial cranium of *Dryopithecus* Lartet, 1863 (Hominoidea, Primates) from the Upper Miocene of Can Llobateres, Barcelona, Spain. *J. Hum. Evol.* 29: 101–139.

Olson TR (1981) Basicranial morphology of the extant hominoids and Pliocene hominids: the new material from the Hadar Formation, Ethiopia, and its significance in early human evolution and taxonomy. In CB Stringer (ed.), *Aspects of Human Evolution*. London: Taylor & Francis, pp. 99–128.

Olson TR (1985) Cranial morphology and systematics of the Hadar hominids and "*Australopithecus*" *africanus*. In E Delson (ed.), *Ancestors: The Hard Evidence*. New York: Alan R. Liss, pp. 102–119.

Picq P (1985) L'articulation temporo-mandibulaire d'*Australopithecus afarensis*. *C. R. Acad. Sci. Paris* 300, Série 2: 469–474.

Puech P-F, Cianfarani F, and Roth H (1986) Reconstruction of the maxillary dental arcade of Garusi hominid I. *J. Hum. Evol.* 15: 325–332.

Rae TC (1997) Mosaic evolution in the origin of the Hominoidea. *Folia Primatol.* 70: 125–135.

Rak Y (1978) The functional significance of the squamosal suture in *Australopithecus boisei*. *Am. J. Phys. Anthropol.* 49: 71–78.

Rak Y (1983) *The Australopithecine Face*. New York: Academic Press.

Rak Y (1988) On variation in the masticatory system of *Australopithecus boisei*. In F Grine (ed.), *Evolutionary History of the "Robust" Australopithecines*. New York: Aldine de Gruyter, pp. 193–198.

Rak Y, and Clarke R (1979) Ear ossicle of *Australopithecus robustus*. *Nature* 279: 62–63.

Rak Y, and Howell FC (1978) Cranium of a juvenile *Australopithecus boisei* from the Lower Omo Basin, Ethiopia. *Am. J. Phys. Anthropol.* 48: 345–366.

Rak Y, Kimbel WH, and Johanson DC (1996) The crescent of foramina in *Australopithecus afarensis* and other early hominids. *Am. J. Phys. Anthropol.* 10: 93–99.

Renne PR, Walter RC, Verosub K, Sweitzer M, and Aronson J (1993) New data from Hadar (Ethiopia) support orbitally tuned timescale to 3.3 Ma. *Geophys. Res. Lett.* 20: 1067–1070.

Robinson JT (1953) *Telanthropus* and its phylogenetic signficance. *Am. J. Phys. Anthropol.* 11: 445–502.

Robinson JT (1956) *The Dentition of the Australopithecinae*. Pretoria: Transvaal Museum.

Robinson JT (1958) Cranial cresting patterns and their significance in the Hominoidea. *Am. J. Phys. Anthropol.* 16: 397–428.

Schmid P (1983) Eine Rekonstruktion des Skelettes von A.L. 288-1 (Hadar) und deren Konsequenzen. *Folia Primatol.* 40: 283–306.

Semaw S, Renne P, Harris JWK, Feibel CS, Bernor RL, Fesseha N, and Mowbray K (1997) 2.5-million-year old stone tools from Gona, Ethiopia. *Nature* 385: 333–336.

Senut B (1983) Les hominidés plio-pléistocènes: essai taxonomique et phylogénétique a partir de certains os longs. *Bull. et Mem. de la Soc. d'Anthrop. de Paris* 10, série 13: 325–344.

Shipman P (1986) Baffling limb on the family tree. *Discover* 7: 87–93.

Skelton RR, and McHenry HM (1992) Evolutionary relationships among early hominids. *J. Hum. Evol.* 23: 309–349.

Skelton RR, McHenry H, and Drawhorn G (1986) Phylogenetic analysis of early hominids. *Curr. Anthropol.* 27: 21–43.

Stephan E, Frahm H, and Baron G (1981) New and revised data on volumes of brain structures in insectivores and primates. *Folia primatol.* 35: 1–29.

Strait DS (2001) Integration, phylogeny, and the hominid cranial base. *Am. J. Phys. Anthropol.* 114: 273–297.

Strait DS, Grine FE, and Moniz MA (1997) A reappraisal of early hominid phylogeny. *J. Hum. Evol.* 32: 17–82.

Suwa G (1989) The premolar of KNM-WT 17000 and relative anterior to posterior dental size. *J. Hum. Evol.* 18: 795–799.

Suwa G (1990) A comparative analysis of hominid dental remains from the Shungura and Usno Formations, Omo Valley, Ethiopia. Ph.D. diss., University of California, Berkeley.

Suwa G, White TD, and Howell FC (1996) Mandibular postcanine dentition from the Shungura Formation, Ethiopia: crown morphology, taxonomic allocations and Plio-Pleistocene hominid evolution. *Am. J. Phys. Anthropol.* 101: 247–282.

Suwa G, Asfaw B, Beyene Y, White TD, Katoh S, Nagaoka S, Nakaya H, Uzawa K, Renne P, and WoldeGabriel G (1997) The first skull of *Australopithecus boisei*. *Nature* 389: 489–492.

Swofford D (1991) *Phylogenetic Analaysis Using Parsimony (PAUP), Version 3.0.* Champaign: Illinois Natural History Survey.

Tardieu C (1983) L'articulation du genou des primates catarhiniens et hominidés fossiles: implications phylogénétique et taxonomique. *Bull. et Mem. de la Soc. d'Anthrop. de Paris* 10, série 13: 355–372.

Tobias PV (1967) *Olduvai Gorge*: Vol. 2. *The Cranium and Maxillary Dentition of* Australopithecus (Zinjanthropus) boisei. Cambridge: Cambridge University Press.

Tobias PV (1980) "*Australopithecus afarensis*" and *A. africanus*: critique and an alternative hypothesis. *Palaeontol Afr.* 23: 1–17.

Tobias PV (1991) *Olduvai Gorge*: Vol. 4. *The Skulls, Endocasts and Teeth of* Homo habilis. Cambridge: Cambridge University Press.

Tobias PV, and Falk D (1988) Evidence for a dual pattern of cranial venous sinuses on the endocranial cast of Taung (*Australopithecus africanus*). *Am J. Phys. Anthropol.* 76: 309–312.

Tobias PV, and Symons JA (1992) Functional, morphogenetic and phylogenetic significance of conjunction between cardioid foramen magnum and enlarged occipital and marginal venous sinuses. *Perspect. Human Biol. 2 Archaeol. Oceania* 27: 120–127.

Walker AC, and Leakey RE (1988) The evolution of *Australopithecus boisei*. In F Grine (ed.), *Evolutionary History of the "Robust" Australopithecines.* New York: Aldine de Gruyter, pp. 247–258.

Walker AC, Leakey RE, Harris JM, and Brown FH (1986) 2.5-Myr *Australopithecus boisei* from Lake Turkana, Kenya. *Nature* 322: 517–522.

Walter RC (1994) Age of Lucy and the First Family: laser $^{40}Ar/^{39}Ar$ dating of the Denen Dora member of the Hadar Formation. *Geology* 22: 6–10.

Walter RC, and Aronson JL (1993) Age and source of the Sidi Hakoma Tuff, Hadar Formation, Ethiopia. *J. Hum. Evol.* 25: 229–240.

Ward CV, Leakey MG, and Walker A (1999) The new hominid species *Australopithecus anamensis. Evol. Anthropol.* 7: 197–205.

Ward CV, Leakey MG, and Walker A (2001) Morphology of *Australopithecus anamensis* from Kanapoi and Allia Bay, Kenya. *J. Hum. Evol.* 41: 255–368.

Ward SC, and Kimbel WH (1983) Subnasal alveolar morphology and the systematic position of *Sivapithecus. Am. J. Phys. Anthropol.* 62: 157–171.

Weidenreich F (1943) The skull of *Sinanthropus pekinensis. Palaeont. Sin.* 127: 1–486.

White TD (1977) New fossil hominids from Laetoli, Tanzania. *Am. J. Phys. Anthropol.* 46: 197–230.

White TD (1979) Evolutionary implication of Pliocene hominid footprints. *Science* 208: 175–176.

White TD (1980) Additional fossil hominids from Laetoli, Tanzania. *Am. J. Phys. Anthropol.* 53: 487–504.

White TD (1985) The hominids of Hadar and Laetoli: an element-by-element comparison of the dental samples. In E Delson (ed.), *Ancestors: The Hard Evidence.* New York: Alan R. Liss, pp. 138–152.

White TD, and Johanson DC (1982) Pliocene hominid mandibles from the Hadar Formation, Ethiopia: 1974–1977 collections. *Am. J. Phys. Anthropol.* 57: 501–544.

White TD, Johanson DC, and Kimbel WH (1981) *Australopithecus africanus*: its phyletic position reconsidered. *S. Afr. J. Sci.* 77: 445–470.

White TD, Suwa G, Hart WK, Walter RC, WoldeGabriel G, de Heinzelin J, Clark JD, Asfaw B, and Vrba E (1993) New discoveries of *Australopithecus* at Maka in Ethiopia. *Nature* 366: 261–265.

White TD, Suwa G, and Asfaw B (1994) *Australopithecus ramidus*, a new species of early hominid from Aramis, Ethiopia. *Nature* 371: 306–312.

White TD, Suwa G, and Asfaw B (1995) Corrigendum. *Nature* 375: 88.

White TD, Suwa G, Simpson S, and Asfaw B (2000) Jaws and teeth of *Australopithecus afarensis* from Maka, Middle Awash, Ethiopia. *Am. J. Phys. Anthropol.* 111: 45–68.

Wolpoff MH (1974) Sagittal cresting in the South African australopithecines. *Am. J. Phys. Anthropol.* 40: 397–408.

Wood BA (1991a) *Koobi Fora Research Project IV: Hominid Cranial Remains from Koobi Fora.* Oxford: Clarendon Press.

Wood BA (1991b) Paleontological model for determining the limits of early hominid taxonomic variability. *Palaeont. Afr.* 28: 71–77.

Wood BA, and Chamberlain, AT (1986) *Australopithecus*: Grade or clade? In BA Wood, L Martin, and P Andrews (eds), *Major Topics in Primate and Human Evolution.* Cambridge: Cambridge University Press, pp. 220–248.

Wood BA, and Chamberlain AT (1987) The nature and affinities of the "robust" australopithecines: a review. *J. Hum. Evol.* 16: 624–641.

Wood BA, and Collard M (1999) The human genus. *Science* 284: 65–71.

Wood BA, and Richmond BG (2000) Human evolution: taxonomy and paleobiology. *J. Anat.* 196: 19–60.

Wood BA, Wood C, and Konigsberg L (1994) *Paranthropus boisei*: an example of evolutionary stasis? *Am. J. Phys. Anthropol.* 95: 117–136.

Yuan MS, and Holloway RL (2000) New endocast reconstructions of *Australopithecus africanus* (Type II and Type III) from Sterkfontein, South Africa. *Am. J. Phys. Anthropol.* Suppl. 30: 330 (Abstract).

Zihlman AL (1985) *Australopithecus afarensis*: two sexes or two species? In PV Tobias (ed), *Hominid Evolution: Past, Present and Future.* New York: Alan R. Liss, pp. 213–220.

Zollikofer CPE, Ponce de Leon MS, Martin RD, and Stucki P (1995) Neanderthal computer skulls. *Nature* 375: 283–285.

Index